IMPROVING
the
QUALITY
of
URBAN
MANAGEMENT

Volume 8, URBAN AFFAIRS ANNUAL REVIEWS

IMPROVING
the
QUALITY
of
URBAN
MANAGEMENT

Edited by

WILLIS D. HAWLEY

and

DAVID ROGERS

Volume 8, URBAN AFFAIRS ANNUAL REVIEWS

 SAGE PUBLICATIONS / BEVERLY HILLS / LONDON

For information address:

SAGE PUBLICATIONS, INC.
275 South Beverly Drive
Beverly Hills, California 90212

SAGE PUBLICATIONS LTD
St George's House / 44 Hatton Garden
London EC1N 8ER

Printed in the United States of America

International Standard Book Number 0-8039-0292-1

Library of Congress Catalog Card No. 72-98108

FIRST PRINTING

CONTENTS

IMPROVING
the
QUALITY
of
URBAN
MANAGEMENT

Volume 8, URBAN AFFAIRS ANNUAL REVIEWS

1

The Mismanagement of the Cities and Defective Delivery Systems

DAVID ROGERS
WILLIS D. HAWLEY

INTRODUCTION

☐ BY THE END OF THE 1960s, one could finally point to a national awareness that America faces an urban crisis of quite substantial proportions. Indeed, most domestic social problems, whether rural, suburban, metropolitan, or inner city, are seen by most sophisticated observers as part of it. Its dimensions include a concentration of poverty in inner cities; a severe gap between municipal revenues and costs; runaway cost inflation of public services with salary increases generally not tied to productivity; public service union strikes and breakdowns; decaying housing, schools, and other physical plant and equipment; increasing crime, delinquency, and drug addiction; intensified conflicts among citizen groups for scarce services, housing, and neighborhood control, often along racial, ethnic, and religious lines, and extending to those between minority communities and city agencies; and an accompanying lack of hope or expectation by many citizens and city officials that they can do anything about it. The political and management process in big cities is much like a "nobody wins" or "zero-sum" game, with heightened distrust and conflict among the main participants.

As has by now been documented so well, this crisis is a result of numerous forces, many well outside the control of cities, such as rapid and unplanned urbanization, industrialization, technological change, and outdated political boundaries. The following specific changes are of particular importance: (1) the automation of agriculture in the South, starting a massive migration from rural areas to inner cities in the North. The migration was, in turn, supported by the myth of the North as the "promised land," with an abundance of jobs; (2) advances in transportation technology, leading to the development of a national highway system and speeding the exodus of industry and the white middle class from the increasingly poor central cities to suburbs; (3) federal legislation supporting suburban housing (and, perhaps, revenue-sharing) that furthered suburbanization while contributing to increased segregation as well; (4) the automation of production, reducing certain types of employment opportunities for less-well-educated city dwellers; and (5) the development of television and other mass media that spread information nationally about affluence, ghetto riots, and inequality, thereby heightening frustrations and raising aspirations among the poor minorities, while increasing the anxiety of all city residents.

One political change exacerbating the crisis was the decline of the urban machine that had contributed in the past to the sharing of political power in big cities on the part of white ethnic groups. Francis Piven's analysis of the "ethnic stakeout," pointing out how these groups were given jobs and control over entry into municipal bureaucracies in return for votes, makes the important point that, when blacks, Puerto Ricans, and Chicanos arrived in numbers, the machine and ethnic patronage were out of style. With the exception of the poverty program now being phased out (where charges of black and Puerto Rican patronage were made so often), there were few jobs available in local government for the new arrivals. The emergence of community control as an issue in the 1960s must be interpreted in this context, as must the many conflicts between white civil servants (teachers, police, firemen, sanitation men, and so on) and minority communities.

All these new demands for jobs and city services were being made at the same time in history that many big cities were experiencing a financial squeeze and an exodus of many jobs. To be sure, not all cities—not even all big cities—have been experiencing each of the maladies just outlined. But throughout the nation there is a sense of unease about the capacities of local governments to meet the challenges of old demands and rising expectations. Perhaps we have begun—for the first time since the Depression—to believe that the problems our governments have been or should have been grappling with might not be solvable.

It would be quite wrong to trace the burden for the social problems we have been discussing solely—or even primarily—to failures in urban management. And it would be foolish to imply that these difficulties can be eliminated through more effective strategies and techniques for delivering public services.

Nevertheless, it is generally true that the machinery of city governments has been inadequate and inflexible in the face of these new and continuing challenges. Too many city agencies at best functioned in a caretaker capacity, unable and unwilling to deal with change and trying instead to keep going and to conserve what was there. Too few city governments have had the managerial capability to plan, set goals, develop cost-effective programs, monitor and evaluate them, develop control systems that permitted a constant appraisal of employee performance and accountability, and become more responsive to changing client needs. In many cases, city, county, and state agencies have become extreme examples of static, traditional bureaucracies. Moreover, agencies responsible for urban problem-solving are increasingly fragmented both from one another and internally, further hampering their capacity to adapt and to deliver services.

Unfortunately, much public discussion of the urban crisis has been in the nature of political rhetoric, apocalyptic forebodings of imminent doom, and headline mongering. What is needed now is more considered analysis. Hopefully, we may soon begin to ascertain the extent to which decay or stagnation actually exist and in what aspects of urban life. Conceivably, such decay may be much more prevalant in older, more industrialized cities in the Northeast and Midwest (Baltimore, Washington, Newark, Philadelphia, New York, Boston, Buffalo, Cleveland, Detroit, and so on) than in many cities in the South and Southwest, at a much earlier stage of industrialization and urbanization; and where there is less population density. On the other hand, the older cities may be the nightmare toward which others will move, unless there are major changes in public policy and private sector practices to reverse the trend.

But things are changing. Social scientists are becoming more concerned with doing policy-oriented studies (in part because that is where research funds are). Some urban managers have developed substantial analytical skills that not only allow them to translate but to be more comfortable with the work of academicians. In recent years, the search for better ways of analyzing and describing problems and for facilitating the development and application of more sophisticated ways of problem-solving has accelerated. It is not as though we need start from scratch. For example, a good deal of intelligent and useful research has been done in response to the urban riots of the mid-sixties and the massive growth of federal programs focused on the cities. While these studies—some done by or for city governments themselves—and other analyses of urban problems and efforts to solve them have resulted in a virtual avalanche of publications, too little of the knowledge gained has worked its way to the operational levels of policy-making and management. No small part of the reason for this is that social scientists and urban managers are seldom comfortable with each others' priorities and perspectives. The former lack the capacity or interest to view problems from the perspectives of those experiencing them and responsible for their solution. Too often, researchers—academicians, in particular—shun the context of public policy and application as if social theories, models or behavior, or methodologies can be justified or validated without testing them on "the real world" and examining their consequences. Public managers, for their part, too often have inadequate training or time to translate research findings into viable programs and plans for implementation and harbor the conviction that only those on the firing line can really understand the problems, much less provide "realistic" solutions.

While there is substantial agreement that the quality of urban life is less than it should or could be, there are markedly different diagnoses of the problems and consequent prescriptions of social change strategies. Indeed, the issues have become much more complex than they were in the past. In the 1930s, for example, liberals and conservatives battled over whether the federal government should spend much more or much less for domestic programs. As we know, the liberal position, that Washington should take over leadership where state and local governments could not or would not provide it, gained increasing acceptance. But if it was not clear before, it became even less so after the decade of the sixties. Pouring resources into urban development programs may not help that much if the cities, and the public sector in general, do not have the management capability and delivery systems to use them effectively. Too many domestic programs in the sixties were poorly managed and implemented and in no way cost-effective. Title I programs under the Elementary and Secondary Education Act, for example, are a good case in point. In many states, some of the funds did not even go into education. In others, even where they did, they were poorly used. The New York City schools, for example, were receiving between $75 and $100 million in Title I funds in the late sixties and much of that money went to creating more jobs for coordinators and other bureaucrats, many of them middle-class whites. Swelling the already dysfunctional bureaucracies of school systems and providing jobs and promotions for educators whose limited competence was sometimes part of the problem was not the main intent of the law.

More generally, providing additional resources is much too simplistic a strategy for improving the delivery of needed services in inner cities. In many instances, its main effect is to spread inefficiency. And yet this remains at the core of many proposed solutions. Changed national priorities, more money for the cities, more and better programs, technology, and more managers might be important, but they will be hopelessly inadequate unless accompanied by major changes in the management and delivery systems of cities and in the politics that affect them. In a word, it is not enough for cities just to have new resources. They must be organized and disposed to use them well.

Indeed, most cities require not just organization, but a fundamental reorganization if they are to develop and implement cost-effective programs. Many social scientists and practitioners of city government speak increasingly of new approaches as essential for improving the delivery of services. In large part, this book constitutes an assessment of that thinking. This chapter provides an overview of a number of the new organizational change strategies that have been recommended, of the diagnoses on which they are based, and of the assumed benefits that will result from their implementation. Thus, we seek here to put the articles between these covers in context. Of course, no single book can survey all aspects of urban management, and we make no pretense that all the important questions are assessed here. Nonetheless, the book covers substantial and important ground. Part I examines the role of information as a source of increased effectiveness. Part II deals with the related question of the strengths and limits of new planning and decision-making strategies.

The notion that decentralization is the answer to mismanagement and unresponsiveness is addressed in Part III. Part IV considers the uses and direction of efforts to restructure city governments in response to claims that at the root of the service delivery problem is the rigidity and overly bureaucratic character of public organizations. Part V examines various ways that we might privatize or otherwise involve the private sector in delivering the goods to city residents. Finally, some barriers to change are examined in Part VI, with special attention to the role of political power in achieving effective urban management.

INFORMATION AS A SOURCE OF CHANGE

One way to facilitate change is to improve the quality of information we have about how well or poorly services are being delivered. If this seems obvious, the plain fact is that most cities collect little systematic performance data. If such data were available, they might serve several functions. First, they could provide managers with ways of controlling the organizations they presumably lead. Performance information, especially if linked to subunits, could be used to spot both strengths and weaknesses. Second, performance data could allow individuals or work groups to adjust their own behavior to better achieve the goals with which they identify. Third, such information might provide some basis for legislative review. Fourth, performance data may provide interest groups and other interested citizens with a way to hold government officials accountable. And fifth, information on the amount, quality, and impact of services being delivered allows agencies to be more efficient and effective and respond to attacks by those unhappy with the services being received.

In short, better information about performance of urban delivery systems is a potential source of change and improved performance in urban service delivery.

In Chapter 2, Harry Hatry outlines some of the problems and possibilities of assessing the performance of city governments. Lewin and Blanning suggest the kinds of information that city governments might be required to report and some ways those data might be organized to facilitate its use by policy makers and the interested public. In assessing the performance of public agencies, one is heavily interested in knowing how differences in organizational arrangements account for variations in agency efficiency and effectiveness. In Chapter 4, Dennis Young suggests how the evaluation of organizations might most usefully be undertaken.

TOP-DOWN STRATEGIES FOR IMPROVING URBAN MANAGEMENT

Another set of general strategies for more effective service delivery might be characterized as *top down* in nature. There are two complementary versions of this approach, one concentrating heavily on structural solutions and the other on procedural ones. Proponents of each version, however, are equally convinced that the way to improve the outputs of city government is to create a strong center or top management with consolidated authority and power, and considerable professionalism in administration, program analysis, and political bargaining. The structural version is based on the diagnosis that the public sector is hampered by its extreme fragmentation—as, for example, across federal, state, county, and local governments, among agencies within a single jurisdiction, and even within agencies themselves. Its supporters argue that if there were only more integration or consolidation of these diverse governmental units, the delivery of services would improve. Under a fragmented structure, so the argument goes, it is impossible to deliver services in any efficient, effective, or responsive way. Each agency and jurisdiction tends to guard its own autonomy and views collaboration as a potential threat. This, in turn, prevents comprehensive planning and policy-making. It also leads to much duplication in programs and administrative services; to constant buck-passing and avoidance of accountability; to long delays in implementation of programs, since with so many agencies there are numerous check points and veto groups; and to increasing frustration for citizen groups that see no way to change the process, not knowing where decisions are made and who is accountable. New York City was a dramatic case of many of these problems in the 1960s, as Fred Hayes indicates in Chapter 5.

The forms most often suggested that reflect this integration strategy are superagencies, metropolitan authorities, Councils of Government, and interagency coordinating councils. New York City is the best single example of the superagency strategy, with Mayor Lindsay having consolidated the very fragmented, fifty-odd agency machinery of city government into eleven "superagencies." Councils of Government at the metropolitan level have, of course, become much more prevalent in recent years in response to both federal funding requirements and a recognized need to consolidate. The development of metropolitan government, by contrast, has been very limited, due, among other things, to the intense protectionism among many suburbanites, desirous of not entering into any involved relationships with cities, and, in some cases, to the fears of inner-city blacks that their power will be diluted.

Interagency coordinating councils have been on the rise since the community action programs of the sixties with their emphasis on comprehensiveness. A good example at the agency level is the Cooperative Area Manpower Planning System (CAMPS) established at the initiative of the Department of Labor for coordinated planning of the delivery of manpower training and related services. The Department of Labor, HEW, the Office of Education, OEO, HUD, the Department of Transportation, and the Department of Commerce all participate, and they do so at the federal level as well as in states, counties, cities, and various regional offices throughout the country. Furthermore, in the late sixties, the federal government set up uniform regions and regional offices for all those agencies involved in domestic programs, so they would begin to engage in more coordination and comprehensive planning.

There is every indication, then, that this strategy has increasing support. It faces many political obstacles to effective implementation, however, as indicated in the experience of New York City's superagencies, which can probably be generalized to that of agency coordination elsewhere as well. The idea of grouping various operating agencies involved in similar or related activities was admirable, but city officials did not develop an adequate implementation strategy. Specifically, they did not always decide what tradeoffs and incentives might make previously separate agencies want to collaborate, nor did they consistently develop any coherent program directions around which such collaboration might take place. And as far as can be determined, the unions and civil service groups were not made part of the planning or later implementation, and yet the involvement of these key participants was essential.

The results of this neglect were often quite negative. Many superagencies were merely collections of particular operating agencies that just happened to be under the same administrative umbrella. They did not engage in that much coordinated planning and program development. Indeed, for the most part, they hardly interacted. Furthermore, the superagencies often proliferated more bureaucratic layers between the citizenry and top city officials than had existed before, making for less rather than more political accountability. These layers included many superagency headquarters officials who duplicated, made policy for, and sometimes obstructed officials in the operating agencies, creating considerable confusion. Many of these problems were transitional ones, but they were often quite severe. A potentially good idea, then, became unfortunately discredited and produced fewer benefits than it might have, largely because those who developed it had not thought through how they were going to get it implemented. Though New York is not America, its problems with superagencies are seemingly illustrative of those elsewhere.

A strategy complementary to this structural approach is to improve the capacity of top agency officials for analysis and for program and organizational development by introducing them to new management techniques. Indeed, in New York City, this second strategy was combined with the superagency as part of a phased strategy of management reform. The first was to develop a new, more consolidated structure, to create a planning environment that was indispensible for improved management, forcing the separate agencies and departments out of their secrecy and protectionism so that they might exchange information more readily and collaborate. After that, new "management science" approaches could be used, including PPBS (Program-Planning-Budgeting System), the use of "project management" techniques to speed decision-making and implementation, the development of management information systems, and the use of "systems analysis."

The diagnosis that these strategies was based on suggested that a major source of city governments' poor performance was their lack of management capability, especially in goal-setting, program-planning, auditing, evaluation, and the like. The proposed solution was to improve that capability, often through a newly constituted budget bureau and through the continued use of outside management consultants. The Office of Management and Budget does that at the federal level, and it has numerous functional equivalents in states, counties, and cities. Taking the New York City case again, exemplifying a city that is farthest along in this process, its Bureau of the Budget was transformed during the sixties under Mayor Lindsay from a traditional cost accounting unit into a change agent and management development agency. In addition, it brought in many outside consulting firms to help. They worked intensively on the technologies, programs, management procedures, and structures in particular agencies, as well as in actual management training and development. The Bureau of the Budget often spun off some of its staff to particular agencies to help ensure that they would develop their own internal management capability so that they would not have to rely on continued outside prodding. Fred Hayes's chapter on New York City reviews these many efforts. As Selma Mushkin has reported in an earlier Urban Affairs Annual Review, such efforts have gone on in many states and localities on an increasing scale. None is nearly as far along as New York City, however, and there are many lessons to be learned from an in-depth analysis of its early experiences.

In Chapter 6, Garry Brewer takes a constructively critical look at various aspects of system analysis. His efforts to identify the limits of this general approach to management are aimed not at debunking "new" management techniques but in clarifying their potential.

The basic unit of action for management activity is, of course, the decision of the administrator or policy maker. James Vaupel outlines some ways individual decision makers can apply systematic analysis to the day-to-day choices they have to make. He argues that while comprehensive rationality is an unrealistic objective in decision-making, there are simple quantitative and logical procedures that can be used to reduce the uncertainty which necessarily characterizes much of a city official's efforts at problem-solving.

Top-down strategies sometimes seem, at least implicitly, to reflect a "trickle-down" theory of social change, assuming that policy makers and top government officials, in alliance with management experts, social scientists, and citizen groups who provide them with more information, are the best hope for significant economic and political change. The major emphasis, then, is on improving what Etzioni refers to as "societal guidance and control" from the top, from superordinate agencies and elites. Often, the intervention efforts are justified by criteria of administrative efficiency rather than of equity, political democracy, or other concerns.

DECENTRALIZATION—CHANGING MANAGEMENT
FROM THE BOTTOM UP

Jack Newfield (1971), social critic and commentator on urban problems, has seriously questioned the top-down strategy as anti-people. He suggests that a lot of the management science experts know little about the agencies they work with—about their traditions and culture, their programs, their politics, or the needs of citizens with regard to agency services. Furthermore, he argues, these consultants, as a predominantly WASP elite group of technocrats, are not sensitive enough to the needs of various neighborhood groups to know what kinds of changes in program and structure would significantly improve the delivery of services.

This perspective is one set of justifications for a third general strategy for change that we will characterize as "bottom-up."

Some management science proponents would probably view Newfield's position as anti-intellectual or anti-rational. It seems to us, however, that top-down and bottom-up change strategies are not mutually exclusive unless taken to extreme. At least as the authors of the first half of this volume see it, neither general approach is without serious weaknesses. While their analyses do not explicitly address the problem of finding the appropriate mix of change strategies, many of the articles provide stepping stones to such integration.

The bottom-up strategy, reflected in such slogans as *participatory democracy* and *community control,* aims at improving services by putting both administrative and elected officials in closer contact with, and making them more vulnerable to, citizen demands. It reached its high point with the community action agencies of the sixties. The black power movement, the New Left, the youth counter-culture, and the many related anti-Establishment, anti-bureaucracy movements of the decade gave strong political support to this strategy.

Bottom-up change strategies are based essentially on the diagnosis that professional and civil service power had prevented city agencies from adapting to change by permitting them to become insulated, rigid, and nonresponsive to citizen demands. With civil service reform, agency officials had become protected from outside review and demands for accountability. What began as a strategy to eliminate patronage and corruption by preventing political party interference in agency appointments and operations created a new civil service machine that was equally as insidious in its effects. City agencies became isolated islands of power with civil service leading to in-breeding, outdated and irrelevant examinations, and highly politicized unions and professional associations. Advocates of the bottom-up strategy asserted, then, that the power of city officials to block change and perpetuate inefficient programs was being used against the democratic rights of citizens.

The distinction is often made in discussions of this subject between administrative and political decentralization. The former does not involve shifting power to citizen groups at all, except in the most indirect sense, and is rather a delegation of authority within city agencies from headquarters bureaucrats to those in district offices. It is in many respects an extension of top-down strategies and is meant to provide for greater administrative flexibility. Political decentralization, with which the papers in Part III are primarily concerned, obviously goes much farther than that.

There are many assumed benefits that advocates of this strategy see as resulting from it. They include more relevant programming and accountability, more flexibility, innovation, and efficiency, more legitimacy for the agency, and, hence, more social peace about its operations, and more jobs within the agency for community residents. Annmarie Walsh deals extensively with these points in her paper on decentralization and suggests that many of the asserted benefits may well not accrue. And Douglas Yates reviews the politics and logic of decentralization indicating the costs of fragmentation while arguing that more political and administrative power sharing is necessary.

There are many obvious ways in which the bottom-up strategy differs from the top-down one. There are more and different participants in the former, most of them dispossessed client groups and their leaders who question the legitimacy and effectiveness of existing agencies, regard "lay" judgments as important for the agencies to take into account, and want power to determine agency policy and programs. Their goals are much more diffuse than those of the top-down advocates and include all the assumed benefits mentioned above, especially jobs, accountability, policy and program power, and local control over the agencies.

As implied above, there is disagreement between many advocates of each of these two points of view over the relative importance they respectively attach to alternative diagnoses of the source of the inadequate delivery of services that both acknowledge. The top-down advocates tend to argue that *incomplete bureaucratization* is the key problem, correctly pointing out how many city agencies function without even the most primitive management tools, job descriptions, lines of authority, information systems, and the like. For them, creating a strong center with vastly improved management capabilities is essential. Once that is accomplished, these top-down advocates often argue, then *administrative* decentralization could be encouraged to make the bureaucracy more flexible.

By contrast, most bottom-up advocates argue that *overbureaucratization* is the problem, pointing out that the proliferation of rules, procedures, programs, bureaus, and agencies, has prevented the effective delivery of services. They suggest that many of the benefits assumed to result from centralization and bureaucratization did not result. There was not more professionalism, not economies of scale, not area- and citywide planning in most instances, and the costs, as already discussed, were quite substantial.

Among the questions which have received relatively little discussion in the debate over decentralization are those dealing with its fiscal aspects. For example, what impact will decentralization have on the costs of public services and how will fiscal constraints, in turn, affect the political consequences of such changes? Jonathan Sunshine addresses himself to these matters in Chapter 10.

Many of the issues that surround the decentralization controversy can be seen quite clearly when one considers the debate over the way law enforcement services should be organized for delivery. Elinor Ostrom and Gordon Whitaker's article here reviews this debate and empirically examines some of the arguments made on both sides.

ORGANIZATIONAL CHANGE AND CHANGING ORGANIZATIONS

One perennial strategy for improving the delivery of urban services is government reorganization. Three major questions in this respect are: (1) how organizational change can be achieved; (2) what its general direction should be; and (3) how change can be institutionalized.

The social science literature on organizational intervention is none too conclusive, as two antithetical points of view tend to gain support. One strategy that has the greatest support among many human relations advocates might most appropriately be labelled the participative approach. It views malfunctioning organizations in terms of their pathologies, coping mechanisms, and the like, and places great emphasis on outside change agents playing a supportive role in helping officials in the target agency improve its operations. The establishment of trust by outsiders with target agency officials, leading to the systematic exchange of valid information and effective group problem-solving, are seen as critical for organizational renewal.

This is in sharp contrast to an organizational intervention strategy stressing *power* and *conflict.* Proponents of that view, including many of those who view bottom-up strategies for change as the only significant alternatives, regard public organizations as coalitions of conflicting power groups, having different interests and values. The way to change organizations, they assert, is to change the balance of power among their coalitions, preferably through the intervention of outsiders who at least implicitly support those coalitions that want such change.

Proponents of this view argue that the distribution of power in organizations is a much more systemic variable than interpersonal relations and communication, which are at most only peripheral. Organizational therapy advocates counter with the argument that though power is certainly a fundamental characteristic of organizations, no significant change in its distribution and in the consequent performance of organizations is possible without increasing the extent of trust among the participants. Furthermore, they argue, attempts to change organizations through such a power-conflict strategy may only intensify conflict and freeze the organization in its status quo position. Communication, trust, and group problem-solving will break down further, preventing either the initiation or the implementation of new structures or programs.

It may well be that both advocacy and collaboration, at different times, with different strata in organizations, are essential. Collaboration strategies alone may neglect critical issues of interest-group conflict and coalition politics within organizations and in their relations with dissatisfied clients. Advocacy alone, however, may fail to provide for receptivity within organizations for change.

This book is concerned with intervention strategies, but even more with new forms that they should support. For example, Hawley's chapter on debureaucratization reflects an emerging line of thinking in the literature on organizations whose best-known proponents include Eugene Litwack, Victor Thompson, and Warren Bennis. The key concept Bennis and others espouse is that the traditional bureaucratic model—emphasizing the importance of hierarchy, fixed rules and procedures, and continuity of operations is hopelessly inadequate in a society of accelerating change and increasing complexity. They suggest that organizations designed and managed in general accordance with the traditional model stifle adaptability, do not easily take in new technology, provide little opportunity for personal growth, and foster a narrow kind of conformity. The new form Bennis advocates is for more decentralized and adaptive organizations, sometimes structured as temporary systems around well-defined problems, rather than around longer-term programs and institutional (departmental, agency) survival considerations. Hawley outlines the potential contributions of such organizational systems and tries to bring attention to the difficulties involved in sustaining them and maximizing their effectiveness.

Argyris' chapter speaks to the need for more adaptive organizations in the public sector but reaches more broadly to the rationale for restructuring traditional hierarchical organizations and discusses the problems, as well as the strategies, for achieving change. He focuses in particular on the role urban managers must take in instituting and sustaining the change process.

BUSINESS INTERVENTION, PRIVATIZATION, AND COMPETITION

There are some students of urban management who argue that these various strategies will not significantly improve the delivery of services because they leave untouched the main obstacle to such improvement—namely, the monopolistic position of city agencies. The public is the captive client of such agencies and has few if any options to seek alternative sources for the services. Yet, monopolists have no incentive to innovate or improve. The strategy thus urged by these advocates is to create *alternative, parallel systems* to the existing agencies as a way to revitalize them. The argument is that large, monopolistic organizations—in this instance, municipal agencies—almost never make major changes in established patterns of behavior unless strongly pressured to do so by outside forces. They generally view innovations negatively, as upsetting routines to which they have become accustomed. Even an active consumer movement, such as community control advocates had hoped to develop, is not enough to make a difference, because the monopolistic position of the agencies will still be maintained. The negative power to block change and insulate agencies from responding to citizen demands that such a position permits can only be reduced by creating competitive agencies to provide the service. This, it is argued, will generate sustained, almost automatic pressure on the agencies involved to keep adapting to what consumers want, without constant vigilance by the consumers themselves. Anthony Downs (1970: ch. 11), one of the advocates of this strategy, suggests that even a small dose of competition might conceivably produce important, perhaps even radical, changes in the nature and quality of services. In Chapter 4, Dennis Young reviews the evidence on the effectiveness of this strategy in a few limited instances where it has been tried in education.

The argument continues that competition gives the consumer or client more options than he had before, and that it forces monopolists to change, as the new agencies begin to draw away clients and funds. This can only work effectively, however, if there is an accurate and comprehensive evaluation system whose results are well publicized to the various client groups. Under those conditions, competition might indeed work, as new agencies develop a diversity of approaches to delivering the service that the older mainstream (monopolistic) agencies will feel compelled to incorporate, if they want to maintain their privileged position.

Good examples of this strategy come from the poverty and manpower training programs of the 1960s. The U.S. Department of Labor and the Office of Economic Opportunity both consciously embarked on the strategy of bypassing such mainstream agencies as the Employment Service and vocational education departments (federal, state, and local) in developing new manpower training programs for the disadvantaged. Top federal officials, including Congressmen, felt that the mainstream agencies were too encrusted with tradition and too insulated from minority populations to do an effective job. They provided funds to have the programs developed instead in neighborhood-based community action agencies that might be more responsive to local needs. Traditional social service agencies, like settlement houses, also saw their funds and programs transferred to many Community Action Agencies (CAAs), as the federal government temporarily gave up on them as well.

A whole new set of problems can then arise from that strategy, however. Tremendous duplication, competition, political conflict, and limited coordination soon follow. The traditional agencies do not give up the fight and begin, instead, to issue sharp criticisms of the new agencies as badly managed and inefficient, which they often are. The new agencies fight back and accuse the old ones of being racist, insensitive, and rigidly bureaucratic, which they often are as well. A "no win" political struggle then ensues, which consumes already limited agency resources and may hurt clients and community groups most of all. That is why it is important to keep the agencies in constant contact through a superordinate planning body like Cooperative Area Manpower Planning Systems (CAMPS) and perhaps through some form of superagency as well.

It may well be that these strategies should be pursued in a cyclical way. In some instances, where one or a few agencies have a monopoly and are not responding to new pressures for change, the alternative, parallel systems strategy may be effective. Though no systematic studies seem to have been done on the extent to which that strategy had an effect on traditional manpower and education agencies, many observers of these programs, including the authors, have noted changes. As Ruttenberg with Gutchess (1970a) and others have indicated, the Employment Service and vocational high schools did, indeed, open more outreach centers in the ghetto, provide more remedial education and supportive services, hire more minority staff to help the agency relate better to that clientele, and pursue a more aggressive job development strategy with employers (see also Ruttenberg and Gutchess, 1970b). It is doubtful if such changes would have taken place in the absence of the competition that federal manpower programs engendered.

The same reasoning can be applied to the centralization versus decentralization issue. Taking big-city school systems as an example, it was quite functional for reform and efficiency to move from a highly decentralized, ward-based structure that was very vulnerable to spoils, patronage, and outright corruption to a more centralized, civil-service-oriented system. This trend took place in every big city in the nation, as documented in Joseph Cronin's (1973) recent study. By the fifties and sixties, however, the centralized systems had become increasingly insulated from the diverse and constantly changing subcommunities of big cities. They were not at all responsive to the needs of many of these subcommunities, especially in ghetto areas. And the civil service system that had originally been an instrument for reform became heavily politicized and perverted in the process, promoted its own in-breeding and parochialism, and actively prevented the schools from developing the new programs required in the face of new ghetto populations and the technological and labor force requirements that they faced. A move back to decentralization and a more community-based system seems appropriate now.

There are different versions of this strategy of establishing alternative systems that merit separate treatment. In education, for example, the publicly financed "voucher" system has been widely discussed. It involves providing parents an opportunity to send their children to alternative community schools if they so desire. Most importantly, it provides to ghetto parents the same option as middle- and upper-income groups have of buying their way out of the public schools. Both Young and Fitch discuss the benefits and costs of this system.

The more generalized version of this strategy including hospitals, universities, and newly established community development corporations, involves subcontracting out for the management and delivery of a public service to a private sector organization like a business firm. Peter Drucker (1968) has referred to this as the "reprivatization" strategy, which he strongly urges for many domestic social programs. This is based on the diagnosis that the public sector has a number of characteristics that preclude it from ever being an effective manager. They include the fact that policy-making is fragmented, that it is divorced from execution, which is governed instead by the inertia of large bureaucratic empires—departments, divisions, and special programs—that become autonomous ends in themselves, directed by their own narrow vision and desire for power; that these bureaucracies are often separate from elected public officials and cannot set priorities; that they have no automatic abandonment process for programs that are unproductive, as each beneficiary of a program becomes a highly politicized constituent; and that much of this is perpetuated by civil service for reasons already discussed.

Though many people have discussed this strategy in the abstract, few actual cases of subcontracting or private sector management have been systematically evaluated. The strategy has been tried, however, in education, health, day care, manpower training, transportation, and other public services. E. S. Savas's piece in this volume presents a rationale for the strategy and indicates how it was further developed and implemented in New York City. There is nothing unique about the way it was done there, and much that he has to say can be generalized to other cities, including counties and states. Fitch discusses the strategy in an even broader context, reviewing the wide range of subcontracting strategies developed to improve the delivery of public services. He also effectively disabuses us of the long-held myth that the private sector is invariably more efficient than government.

There are potential costs to this strategy, many of them resulting from the profit-seeking of the private sector firms, and they must be minimized. Findings from the performance-contracting experiments in Texarkana to the effect that the firms "taught to the test" and in some instances fabricated the results, indicate how the strategy can go awry. Yet it does offer enough promise in terms of potential economies and contributions to agency performance to make more experimentation with it worthwhile. At the same time, strong government controls just have to be maintained to ensure that the profit goals of the firms do not result in practices that are contrary to the interests of the public.

There remain other important contributions that the private sector can make to improving city government, as Rogers' chapter indicates. Reviewing a major effort by a private group to provide management assistance to city agencies, he explores the future prospects of this strategy of transferring the management expertise of business to government, concluding that the strategy does have important benefits in some cases, though it needs more development.

BARRIERS TO CHANGE

A theme which underlies almost all analyses of the so-called "urban crisis" is that the power necessary to achieve change is institutionally fragmented and inadequate. Cities, of course, do not exist in a vacuum and are part of a larger system in which county, state, and federal agencies—as well as local ones—play important roles in the delivery of services in cities. There has been increasing concern among political scientists and public officials about some of the dysfunctional aspects of this federal system. The Committee on Intergovernmental Relations headed by Senator Muskie in the late 1960s pointed up many of these aspects. Often the federal government has exacerbated problems of delivering services in cities by spawning numerous agencies and programs that duplicate, overlap, and often come into political conflict with one another. The states have created problems as well. Responsive more to rural and suburban than to inner-city interests, they have not provided the financial assistance to cities that is needed and have blocked efforts by city officials to gain more taxing powers. They have also competed with cities for control over agencies in a way that has hampered the cities' flexibility. Many state agencies—in education, welfare, health, correctional, and other services—are at least as rigidly bureaucratic and unresponsive to local and minority needs as are city agencies, and usually more so.

A major new strategy for the 1960s and 1970s involves decentralizing the entire federal system, through block grants to states and cities. Referred to in such hopeful language as the "new federalism," "creative federalism," and "the New American Revolution," this is a strategy of political and administrative decentralization, decategorization of federal grant-in-aid programs, and revenue-sharing. It is based on the diagnosis that too much centralization of the federal system in Washington, leading to an overly restrictive and inefficient management of federal programs, has hampered states and cities in their attempts to improve the delivery of services. Furthermore, it is increasingly recognized that more federal funds or revenues that funnel from cities and states to Washington should be funneled right back again, given the limited taxing powers of these levels of government and their vastly increased responsibilities and service demands. The Nixon Administration's revenue-sharing push reflects this strategy.

Several benefits are assumed to follow from this reform. Theoretically, it is supposed to be a way of getting more funds to states and cities, though that all depends on how it is done. From an administrative point of view, it supposedly gives states and localities greater flexibility. Under the old system of categorical programs, federal agency officials develop national guidelines that place major constraints on local agency administrators. Localities differ so much in needs, resources, institutional capabilities, and politics that it makes little sense to impose national program guidelines on them in such a top-down manner, without taking such local variations into account. Furthermore, with decentralization, decategorization, and revenue-sharing, local agencies may be able to bypass federal bureaucracies and the delays that they and various congressional appropriating committees impose. Finally, this reform could contribute to more political accountability than the traditional system did and satisfy the assumed needs of many citizens for more power to affect government decisions.

Several problems are raised, however, by such a strategy. It may in fact give less rather than more "power to the people," or at least to minorities and the poor, since they are much less adequately represented in many cities, and certainly in states, then they are through the federal government. Bloc grants to states and cities puts this money up for grabs, and the poor and minorities might well lose out, unless the federal government maintains strong controls. Furthermore, it appears that revenue-sharing has fewer redistribution effects, in the sense that the federal government plays a less important role in the allocation of funds than it did in many of the programs the revenue-sharing displaces.

In addition, there is serious question as to whether states and localities have the capability to manage and deliver services, if given that degree of power and autonomy. Though adequate data do not yet exist, Etzioni (1970) has suggested that, on a whole series of indicators—including the extent of nepotism, corruption, professionalism, accountability, and responsiveness to minorities—state and city agencies are far inferior to those at the federal level. It is easier, both administratively and politically, to purchase machinery than to develop social programs.

Revenue-sharing could, then, conceivably lead to a decline in the quantity and quality of services. In order for it to work, determined efforts will have to be made to substantially improve the staff and workings of local agencies. Rather than simply moving ahead fast and dumping these revenue-sharing funds on states and cities with few constraints, the federal government should pursue a more gradual phasing strategy that gives these local agencies more funds and power only when they demonstrate their capability. This is, of course, much easier said than done, given the strong pressures from localities for immediate change.

This discussion of revenue-sharing is meant to suggest one way that federal and state programs affect the delivery of urban services. In the final part of this book, Jeffrey Pressman focuses attention on additional ways that federal actions have shaped the capacity of cities to adapt to new demands. The irony of his analysis is that the federal programs in question were themselves efforts to encourage change.

As we consider all of the possible things that might be done to improve urban management, it is clear that it is much easier to develop ideas for new structures and procedures than it is to get them adopted and implemented. Indeed, a major reason for the failure of municipal agencies to deliver services more effectively and to change is a political one. The power of civil service groups and municipal employee unions along with the fragmentation among reform groups often has made it difficult to achieve change. A major item on the municipal reform agenda, then, is how to put change coalitions together.

There were some attempts to do so in the 1960s and early 1970s, though none was very successful. One was the alliance of the very rich and the very poor, in collaborative urban development efforts, usually involving large corporations, ghetto blacks, and Puerto Ricans. Manpower training, education, housing, economic development, and community organization were among the arenas where such alliances were tried. Business depended on minorities to meet its manpower needs, and it certainly saw the continuation of ghetto riots, crime, and political unrest as a threat to its economic interests. The National Urban Coalition represents a signal effort to find common interest upon which action could be based. Many local branches went out of existence after a year or two, however, for lack of business support. Jules Cohn (1971) has documented in his study of corporate social responsibility how much big corporations cut back from urban affairs contributions, after their disenchantment with urban coalitions and after the recession of the late sixties. Individual business organizations in particular cities, like Chambers of Commerce, also developed close linkages with community-based agencies, but the extent of improvement in service delivery resulting from these relationships was usually minimal. The business associations committed limited resources to the task, at least relative to need. Many of them did not know what to do and often dissipated the funds they did provide. The Philadelphia Chamber of Commerce, for example, gave one million dollars to a black community group after Martin Luther King's assassination, and little of that money was funneled into constructive programs.

A second kind of coalition involved alliances of community groups against civil service and municipal employee unions, in efforts to make city agencies more responsive and accountable. That kind of coalition was often organized around the demand for community control, but it never did coalesce to any significant extent. Factionalism among racial, ethnic, and neighborhood groups wanting to "do their own thing" and maintain a separate identity hampered these efforts. Often, leadership struggles among these groups contributed to their failure to come together.

Recently, various Establishment coalitions of private and public agencies have been formed on a limited basis, constituting another form of alliance that may improve the delivery of services in cities. Rogers' chapter on the Economic Development Council of New York reports on one such effort, involving task forces of businessmen working with various city agencies. In a way, it is a kind of partial privatization strategy, involving the private sector, not so much in the actual management of programs, but rather in a technical assistance role to improve that management by public sector officials. This may not be a political coalition in the conventional sense of that term, but it functions somewhat in that fashion. It constitutes an alliance or collaborative relation between government and Establishment organizations to effect change, thus linking the public and private sectors in potentially productive ways.

Another kind of political coalition that may be necessary for improved delivery of services in cities is between inner cities and adjacent suburbs. Though this may seem like a utopian dream, given the limited success at metropolitan government, conditions in metropolitan areas may be changing in a way that cities and suburbs increasingly face common problems. Indeed, the urban crisis is being exported in the same unplanned and haphazard way in which it originally developed. Adjacent, inner-ring suburbs not only have many problems in common with the cities, but they have a strong stake in keeping the problems within limits through collaborative planning. Problems of economic development, mass transit, education, housing, jobs, welfare, race relations, and many others may have to be handled on such a metropolitan basis to be at all manageable.

The political context of urban management is the prime concern of the chapters that end this book. Pressman indicates how federal programs have diminished the capacity of cities to develop adequate delivery systems. Rosen deals with civil service unions and their role in blocking or facilitating change. Contrary to most views on this subject, he sees these unions as a positive force for change by suggesting the effects their demands have on forcing agency management to rationalize its operations. Finally, Lupsha presents a compelling synthesis of studies indicating how the limited power of mayors cripples many cities in efforts at initiating and implementing new programs.

Obviously, these three articles do not define the limits of the barriers to changes that would improve the capacity of city governments to "deliver the goods" more effectively. They do, however, have a common theme—seen most clearly in the work of Lupsha and Pressman—that lies at the heart of the difficulties of urban management. This theme is that we cannot seem to understand that political leadership is essential to effective management.

One paradox of democratic systems is that a broad sharing of resources and power probably requires the centralization of political power in the hands of a few elected officials. The dilemma that underlies the paradox is that of keeping the powerful responsive. The difficulty of assuring that power is not abused has led us to avoid giving our leaders sufficient power to bring about change, the consequence of which is a kind of de facto privatization strategy. In short, political decisions will be made, but we seem to prefer that they not be made by elected politicians. And, in democratic systems, the most serious barrier to change is the absence of politics.

CONCLUSION

Though there are many dimensions to the urban crisis, clearly the task of improving the capacity of government to deliver services is critical. Solving this problem alone may not restore the nation's cities and metropolitan areas, but it would at least help a lot in getting on with that task. The purpose of this book is to shed more light on (1) some of the alternative ways that policy makers and administrators might more effectively deliver the goods to those they are charged to serve and (2) a number of the obstacles that stand in the way of more responsive urban government.

Certainly, there are no easy answers, no simple or final solutions, and no prescriptions that apply in all or even most cases. While most of the authors whose views are represented in this book have concerned themselves with particular approaches, we are confident they would agree that no one single strategy, however sweeping, is enough. The problems are complex, and it will be necessary to devise comprehensive formulae for increasing the effectiveness of urban governments. We will need more high-powered management techniques and structural reorganization; metropolitanwide authorities for some problems and decentralization for others; the introduction of more service choices and an increase in the capacity of administrators and—more importantly—elected leaders to act decisively. The appropriate mix of strategies will depend, of course, on the needs of the people in particular locales. Whether there is the wisdom and the imagination around to develop and implement multidimensional approaches to the problems of urban management is uncertain. We hope this volume will be helpful to a broad range of people who are interested in improving the quality of urban life.

REFERENCES

COHN, J. (1971) The Conscience of the Corporations. Baltimore: Johns Hopkins Press.

CRONIN, J. M. (1973) The Control of Urban Schools. New York: Harper & Row.

DOWNS, A. (1970) Urban Problems and Prospects. Chicago: Markham.

DRUCKER, P. F. (1968) The Age of Discontinuity. New York: Harper & Row.

ETZIONI, A. (1970) Article in the New York Times (January 31).

NEWFIELD, J. (1971) "The consultant hustle: $10,000,000 caper." Village Voice (April 1).

PIVEN, F. S. (1969) "Militant civil servants in NYC." Trans-action (November): 24-28, 55.

RUTTENBERG, S. H. with J. GUTCHESS (1970a) Manpower Challenge of the 1970s: Institutions and Social Change. Baltimore: Johns Hopkins Press.

――― (1970b) The Federal-State Employment Service: A Critique. Baltimore: Johns Hopkins Press.

Part I

ON KNOWING WHEN THINGS
ARE GETTING BETTER

2

Measuring the Quality of
Public Services

HARRY P. HATRY

☐ CITIZEN CONCERN ABOUT THE QUALITY of the public services they receive—from their municipal, county, state and federal governments and from their schools—is nothing new. Surprisingly, however, until very recently, governments themselves and others have done little to measure the quality of these services. Even readily available information, such as citizen complaints, has seldom been recorded and tabulated in useful ways.

Massive criticisms of the quality of public services gained momentum in the post-World-War-II years as the civil rights and anti-poverty movements grew. Middle- and upper-class citizens joined in the questioning of service quality in recent decades when they sensed that their interests were threatened by exploding population, pollution, blight, crime, and uncontrolled urban growth. The urgency of the issues, combined with the escalation of government costs and the resulting tax increases, have tended to force government officials to justify their activities in terms of the quality of services they are providing.

Support for measuring service quality, sporadic in the past, has thus grown significantly in recent years. (One of the earliest published examples of this interest is Ridley and Simon, 1938; a

AUTHOR'S NOTE: *This chapter is based to a large extent on the findings of projects sponsored by the National Science Foundation and the U.S. Department of Housing and Urban Development.*

more recent example is International City Management Association (ICMA), 1970.) Many state and local officials are beginning to seek measurements of progress toward their goals (for example, see the findings in Council of State Governments, 1973; International City Management Association, 1972). To some extent, this reflects the past efforts to implement PPB (Planning-Programming-Budgeting) systems and the continuing emphasis on program budgeting, both of which stress explicit identification of objectives. The 1973 report by the International City Management Association's Committee on the quality of municipal services put considerable emphasis on measuring the *effectiveness and productivity* of local government services (International City Management Association, 1973). The General Accounting Office's 1972 report on audit standards for governmental organizations (U.S. Office of the Comptroller General, 1972) calls for audits of program *results* as well as the more traditional checks on legal compliance and fiscal regularity. The GAO report has already encouraged state legislative units to focus more directly on program performance, thus providing the executive branches of both state and local governments with an added incentive to follow suit.

This chapter focuses primarily on measuring the quality of the activities of state and local general units of government; nevertheless, much of what follows is also applicable to federal programs and school districts. First, by way of background, the various reasons for measuring service quality will be discussed briefly. Following sections discuss the major quality aspects that might be measured regularly, some of the existing and emerging ways to measure the quality of services, how realistic targets or standards of quality might be constructed, and, finally, who should be responsible for measuring the quality of public services. Note that it is not the role of this chapter to discuss the methodology of designing program evaluations (including social experimentation).

BENEFITS FROM MEASURING
PUBLIC SERVICE QUALITY

The information derived from measuring public service quality can serve a number of potential uses. Among the important ones are the following:

- Indicate where problems exist. Measurement results will often suggest where government attention should be directed.

- Provide feedback to government officials on the performance of programs and policies. This can be particularly instructive after new services or broad programs have been initiated. Insofar as the measurements focus on specific programs, the examination is commonly labeled "program evaluation."

- Assist in determining priorities for allocating government funds and manpower. This, of course, follows from the two previous points; indications of problem areas, or of the success or failure of specific programs, will be among the key elements that guide those responsible for setting new budgets and plans.

- Help evaluate government management and establish employee incentives. Quality-of-service measurements can serve this function if they are undertaken regularly and systematically. New York City in 1972-1973 set productivity targets and used information on the progress toward meeting those targets to encourage agencies to perform better (New York, City of, 1973). In 1973, two localities established incentive formulas for union employees based in part on quality-of-service indicators: in Detroit, the earnings of sanitation workers were linked to measures of street cleanliness and cost savings in their operations; in the city of Orange, California, policemen's earnings were linked to measurements of selected crime rates. Elementary schools were involved in some of the earliest incentive experiments, with performance bonuses to teachers based partially on their success in boosting pupil achievement, according to standard test scores. As reliable and comprehensive measurements become available, their use in incentive programs for management and employees are likely to mark an important new trend in local government.

- Permit greater community involvement in determining the priorities of government activities. As quality-of-service measurements come into wider use, they inevitably will become public. Citizens could then respond and make requests that take such measurements into account. The quality of citizen interaction and interest in government activities would be likely to increase. Not everyone believes this is necessarily a good thing, but at least up to a point, it is essential to the democratic system.

Another way to view the usefulness of quality-of-service measures is to consider the current data about services that are available in most city, county, and state governments. Such information tends to be mostly technical and financial—and often obscure. It offers little insight into whether the specific services have functioned appropriately and served the public well. For such evaluations, most of us must turn to our own personal experiences with those services with which we come in contact and to reports by the communications media. Yet personal experiences and those reported by the press may be unusual, providing a distorted picture of the general situation. Quality-of-service measurements, in contrast, are a potential means of providing interested citizens and officials alike with representative, reliable information.

COMPONENTS OF QUALITY IN
PUBLIC SERVICES

"Quality of service" may have many different meanings. As used here, the term is interpreted broadly to encompass a variety of qualitative and quantitative elements. The dictionary definition of quality—"degree of excellence"—is a useful starting point. "Effectiveness" is a frequently used term to describe the quality of a public service or program. For the purposes of this chapter, no sharp distinction between effectiveness and quality will be drawn, although the terms do have various nuances of difference.

The meaning becomes clearer as one thinks of specifics. Let us identify a number of elements or attributes of public services that describe quality. These are summarized in Table 1. These categories of quality aspects perhaps can serve as a checklist for those wanting to establish a measurement system. Before describing them, it should be noted that even in cases where an aspect of quality does apply, the state of the art will not *always* offer a fully satisfying means of measurement that is within the technical or financial capacity of most governments. Possible data collection approaches to help overcome certain practical measurement problems are discussed in the next section. Nine major components of quality are described below, by attributes relating to these components.

One. Intended purposes of the service. These purposes may be explicitly or implicitly stated. Following are some examples of the purposes of a few common public services for which measurements are likely to be needed. Some specific measurements appropriate for each attribute cited are indicated in parentheses (for more extensive

TABLE 1
ASPECTS OF QUALITY IN GOVERNMENT SERVICES

1. **Intended purposes** of the service activity.
2. **Negative effects** that may be involved in the provision of the service.
3. **Adequate quantities** of the service.
4. **Equitable distribution** of the service.
5. **Courtesy and respect** to citizens receiving the service.
6. **Response time** in providing the service.
7. **Amounts of citizen use** of a service.
8. **Perceived satisfaction** among citizens receiving the service.
9. **Efficiency (productivity, economy, or cost).**

listings and discussion of service quality attributes and associated measurements, see Blair and Schwartz, 1972; Hatry and Dunn, 1971; Winnie and Hatry, 1972).

- Solid waste collection—Removal of health hazards (incidence of certain types of related diseases such as salmonella, or the rate of health hazards detected during sanitation inspections); providing streets that are reasonably free of garbage and litter (street cleanliness index).

- Crime control—Reduction of crime (crime rates); apprehension of criminals (clearance rates for particular categories of crimes or, as an improved measurement of quality, clearance rates that lead to prosecution and conviction).

- Fire control—Prevention of fire (fire rates); rapid control of fires to reduce loss of life, injury, and property damage (numbers of fire-related deaths and injuries, and amount of property damage).

- Transportation—Moving traffic rapidly and safely (travel times between selective origin and destination links); moving traffic safely (accident rates); providing accessibility to major work, recreation, and shopping areas (percentage of families without automobiles who are within, say, 45-minute public transit time of key job centers in the area); providing smooth-riding streets (street "bumpiness" index).

- Welfare assistance—Meeting the basic economic needs of families unable to provide for themselves (percentage of poverty families raised to the minimum subsistence level through welfare aid).

Two. Negative effects in the provision of the service. Governments and analysts have been particularly deficient in dealing with these negative side effects in their evaluations of services. Following are examples of such quality attributes (with appropriate measurement again cited in parentheses):

- Solid waste collection—Noise or property damage resulting from collection activities (number of families reporting valid complaints of these nuisances).

- Crime control—Harassment of citizens, as through "overdiligent" law enforcement (number of households reporting such instances as determined through citizen surveys).

- Fire control—Excessive property damage due to water or breakage occurring in the act of putting out fires (estimates of dollar value of property damage from such causes).

- Transportation—Air and noise pollution from public transit or automobile vehicles (levels and duration of pollution in particular transportation corridors); public inconvenience due to road maintenance (number of families reporting excessive disruptions and dust).

- Welfare assistance—Excessive payments from cheating or mistakes (percentages and total amounts of such instances from a sampling of cases); discouraging job-seeking by those able to work (estimates of the extent of such instances from work histories of a sample of welfare families).

Three. Adequate quantities of the service. Intentionally or unintentionally, sometimes only a limited portion of the population may be served. Welfare, day care, recreation, health care, manpower training, and so forth may reach only a fraction of the potential clients. This may be due to limited resources. It also may occur because clients are not made aware of the available service. The quantity limitation may also be in the form of not providing the full range of service ideally desirable—e.g., providing only limited financial assistance for a particular emergency case, and so on. In some instances, government officials may feel it necessary not to make fully clear the availability of certain services if their resources are so strained that such publicity would likely lead to exceeding the capacity of the system. Regardless of the rationale for limitation, it seems desirable to address such deficiencies explicitly in the consideration of policies and programs. Measurements of inadequate amounts of service may be expressed in many instances as estimates of the percentage of the total potential clients who are actively being served (e.g., the percentage of the total number of poverty-level families that are receiving welfare payments).

Four. Equitable distribution of the service. This issue has become of increasing importance in the United States, with the growing recognition that various minorities—low-income groups, racial groups, the aged, children, and so on—were not receiving appropriate shares of certain public services. The concern with environmental and growth issues have substantially increased attention to the distribution of services to particular *geographical* areas within a city, county, or state. Because citizens residing in different areas may have dissimilar needs or problems, identical government services to everybody may not be suitable or fair.

It is beyond the scope of this chapter to handle the complexities of equity—what it is and how to deal with it. The concern here is to suggest the need for measuring the quality of services, not only in the aggregate (that is, for all citizens taken together in the community), but also for various clientele groups within the jurisdiction. Such information would seem to be indispensable to the debate over the proper resource allocations that provide equitable or fair shares.

This data disaggregation may add considerably to the cost of the measurements. Some government officials, moreover, will resist obtaining or publicizing such information for fear it may raise issues that will add to their problems. If data, for example, show the measurements of services by neighborhoods, citizens in neighborhoods that appear to be getting lower-quality services may be expected to protest. Nevertheless, thus far, there appears to be surprisingly high acceptance by public officials of the desirability of such information, indicating a growing willingness on their part to "face the music" and try to anticipate problems that, if untreated, will probably erupt later in even more bothersome form.

Five. Courtesy and respect with which the service is provided. Many public services involve direct contacts between government employees and citizens, as well as contacts by telephone or by written communications. The manner in which these contacts are handled affects the satisfaction of the clients and therefore is an important aspect for quality measurement. Citizens may encounter welfare employees, medical employees, policemen, firemen, librarians, and solid waste collectors, to mention only a few of the service personnel who meet the public. Concern about the quality of these encounters is illustrated by the recent criticisms of the handling of rape victims by police and medical personnel. City and county governments provide the widest range of services that require direct and regular contacts with the bulk of the population. However, state employees also deal directly with *special* populations, such as those in various state health and penal institutions and those who make use of health and employment services and motor vehicle licensing agencies.

Six. Response time in providing a service. Many, if not most, government services have some aspect for which it is better to provide a service more quickly. There are the obvious situations, such as police and fire response times to calls, where speed of response maximizes the likelihood that a crime will be stopped in the process, that the criminal will be apprehended, or that reduced fire damage will occur. In addition to acting as a "proxy" for quality service, speedy arrivals also can do much to alleviate citizen anxiety.

Speed of response is also an important factor in other services—in the repair of water and sewer breaks, providing a book requested at the library, delivering welfare or unemployment checks, treating health problems (whether physical or mental), providing emergency ambulance and rescue services, adjudicating expeditiously both civil and criminal cases, and the handling of citizen complaints for any public service. Response-time considerations include both the initial waiting period needed to obtain government attention (such as the queues for drivers' licenses or to obtain public medical attention) as well as the time it takes to obtain satisfactory service after the initial contact has been made.

Seven. Amount of citizen use of a service. One way in which citizens express their desires and satisfaction with a service, in those instances when they have a choice, is whether or not they utilize the specific service. Thus, for certain government services, measurements of attendance—or participation or usage—become important indicators of the quality of those services. This kind of service includes recreation programs, libraries, public transit, and public health facilities. If a community swimming pool is not widely used, presumably the citizens do not like something about the swimming pool. It may not be easily accessible, or those who do have access to it may be unaware of its existence or may dislike some aspect of the pool or its surroundings.

Attendance figures are often currently used as measurements of the effectiveness of recreation and cultural activities. A special problem exists with attendance counts that represent accumulations over time, rather than participation in single events; normally, these counts do not distinguish how many *different* persons or families were users. It remains unclear whether a large portion of the population attended once in a while or whether a small percentage of persons used the facilities frequently. Participation rates which reflect the number and percentage of different persons utilizing the activities are also desirable. A still fuller perspective would be provided by relating actual usage to latent demand. Such data are usually extremely difficult to obtain; a practical compromise is to relate actual usage to the total population which the program or facility might serve.

Eight. Citizen perceptions of their satisfaction with a service. The quality-of-service attributes discussed to this point have emphasized aspects that could be measured by official counts of events or observations of actual situations. Citizen satisfaction (partially and indirectly covered above in the amount of usage of services) is a subjective matter, but to ignore it is to avoid an important aspect of the adequacy of the service. For example, in the matter of crime protection, citizen feelings of security are also quite important (to the citizen) as well as actual crime rates in a community (even if the rates include estimates of unreported as well as reported crimes). One would expect these citizen perceptions of security and crime rates to be roughly correlated, at least over long periods of time. However, it seems such measurements may differ considerably in the short run, especially in terms of the rate at which each is rising or falling. Officials charged with providing services to maintain community security should be interested in measuring how citizens feel about the safety of their persons and property, even though such data may not agree with "objective" data on crime rates.

To a greater or lesser degree in most government services, this dimension—citizen perception—is an important aspect of the quality of the services and should be measured. Some have argued that consideration of citizen perceptions would lead governments to emphasize publicity campaigns to increase citizen satisfaction with a particular service rather than to put resources into truly improving them. This view is not without merit, but good government seems to require a balance of both functions—offering quality services and encouraging citizens to use and appreciate them.

One form of information about citizen perception is already commonly available to local governments: counts of reported citizen complaints. These must be used with caution because reported complaints are not likely to be fully representative of the entire population. Some citizens do not know how to lodge a complaint or to whom. Others are silent because they feel complaining will do no good or they do not like to complain. On the other hand, chronic complainers may give an exaggerated picture of the degree of dissatisfaction. In addition, an increase in complaints from one time to the next could result from a greater willingness to complain rather than from an increase in annoying incidents. For these reasons, a more reliable approach to measuring citizen perception is likely to be a survey of a cross-section of citizens in the jurisdiction.

It is not currently known how closely correlated are citizen satisfaction measurements (such as "feeling of security" and "perception of cleanliness of neighborhood") to more "objective" measurements (such as crime rates or systematic cleanliness ratings by inspectors). Research is only now beginning on such questions (Ostrom, 1973). However, it seems unlikely that such research will turn up sufficient evidence for excluding direct citizen perception measurements as a desirable form of measurement.

Nine. The efficiency (productivity, economy, or cost) of providing the service. Although these aspects often are not considered part of quality, it can be viewed that an efficient, highly productive, economical operation is an aspect of the quality of a governmental service. Certainly this conforms with the general concern of citizens that services be provided at the lowest reasonable cost in dollars and manpower.

The term "productivity" has begun to receive considerable emphasis, spurred in part by the establishment of the National Commission on Productivity, which itself came into being in 1970 because of federal concern that productivity in the private and public sectors has not been adequate. In the National Income Accounts, government productivity has never been directly measured; rather, it has been assumed to be constant. Surprisingly few states or local governments have kept systematic data on the productivity of many of their activities in even the more common and simplest form—cost per unit of output—such as "man-hours per ton of garbage collected" or "man-hours per mile of streets repaired" (National Commission on Productivity, 1971).

As simple as these productivity measurements may sound, there are many difficulties in making comparisons even within the same government from one time period to the next. Conditions may change, such as the composition of the garbage. Special problems may arise—as in street maintenance—that did not occur before. Sometimes what seems to be an efficiency change may, on further examination, prove to be a change in service quality. For example, when a community sanitation department switches from back-door to curb collection, this almost invariably brings a significant drop in the cost or man-hours per ton of garbage collected. But this is not, in the broadest sense, an efficiency improvement. It reflects a service reduction, as citizen instead of government employees haul solid waste from their back door to the curb pickup points. Despite these difficulties and the care that must be taken with analysis and interpretation, measurements of productivity (e.g., output per unit of input) or efficiency (e.g., amount of inputs per unit of output) seem highly desirable for evaluating public service quality in its broadest sense (for further discussion of productivity measurements, see National Commission on Productivity, 1971, 1972).

These nine categories of quality (see Table 1 for summary) indicate that to fairly describe any public service, multiple measurements will be needed. It will be tempting to many analysts and others to attempt to develop a single quality *index* that encompasses the various quality attributes of a particular service or program. Such single indexes seem to be much easier to handle. However, the warning is necessary that such indexes are synthetic, they necessarily include value judgments as to the weightings of each separate measurement comprised by the index, and they have the danger of burying needed information and oversimplifying issues that are inherently complex.

UNDERTAKING THE MEASUREMENTS

One of the major obstacles to undertaking quality-of-service measurements has been the belief that such measurements are extremely difficult, if not impossible. Another important obstacle is that their costs may be substantial. It is clearly impossible to measure all conceivable aspects of quality for any given service. For formal program evaluations, it is not feasible to assess every program every year. Considerable selectivity is needed in measurement.

However, the current state of the art seems to permit much more expanded measurement than is generally realized. The increased ability to carry out quality-of-service measurement reflects the following circumstances: (1) the tremendous development in data processing capabilities over the past years, (2) increased development of the applications of statistical sampling, especially as applied to surveys of samples of citizens of a jurisdiction, and (3) development of practical program evaluation techniques. These, combined with ingenuity, have led to substantial inroads into measuring aspects of services that previously have been considered unmeasurable. Items presumed to be completely subjective and qualitative often can be at least partly, but significantly, translated into systematic quantifications of a meaningful nature.

The following paragraphs describe four approaches to measuring the quality of service. Only the main outlines are noted, since it is beyond the scope of this report to describe measurement procedures in detail.

Systematic inspections of quality attributes with physical features. For example, instead of the traditional measurement for solid waste collection—tons of garbage collected—street cleanliness itself can be roughly measured. The District of Columbia and New York City have been testing procedures for this approach. They use inspectors trained to rate city streets against a preselected set of photographs representing different degrees of cleanliness (for example, see Blair and Schwartz, 1972).

For street maintenance activities, local government measurement has typically focused on such workload measurements as the "number of potholes fixed" or "miles of roads maintained." Complaint data have also been used, with some of the disadvantages of misrepresentation noted earlier. One physical measurement alternative, for road maintenance, makes use of a device called a "roughometer" which indicates the roughness of the street for passengers (Boots et al., 1972). The roughometer readings were previously calibrated against degree of comfort as rated by a test group of citizens.

These physical counts—from the physical measuring device for street roughness or the photograph rating guide for street cleanliness —because of the precalibration, permit public officials and citizens to readily interpret the figures in terms of familiar reference points. In certain technical areas such as air and water quality, technology has increased to permit meaningful measurement of various pollutants. Procedures are being improved steadily to increase this capability. As with the case of the roughometer readings, these direct measurements should be translated into terms that are meaningful to public officials and citizens. Parts per million of various polluting substances is not understandable except to technicians and needs to be expressed as some degree of health hazard or as conditions requiring various degrees of restriction on human activity. To do this is complicated by the fact that simultaneous interactions of pollutants, weather, and other conditions give rise to a great number of combinations. Until full satisfactory measurement becomes possible, however, best approximations will prove far more helpful to society than no measurement at all.

Sample surveys of citizens. Among the quality-of-service attributes listed in the previous section, a number of them—perceived satisfaction by citizens, the courtesy and respect with which government employees provide services, and citizen participation rates—can probably best be measured through the use of citizen surveys. Properly conducted surveys overcome the potentially distorted views that come to officials who rely primarily on complaint records or their personal contacts with citizens and vested interest groups.

The use of professional sampling techniques reduces costs while maintaining sufficient precision for most purposes. The Gallup and Harris Polls typically use samples of about 1,500 adults to represent the complete adult population of the United States. Their success in predicting elections has indicated that considerable precision is possible. For most purposes, state and local governments do not need such precision as is desirable for predicting close elections. Sample sizes of perhaps 500 to 600 families may well be adequate for most purposes in the smaller cities.

The great majority of sample surveys used by governments to date have been undertaken for special, ad hoc purposes, such as for transportation or housing planning. However, for monitoring service quality, surveys should be conducted regularly, at least annually, utilizing similar questions each period. The surveys, if properly constructed and worded, can yield valuable information. For example, users of libraries, recreational facilities, and transit facilities can be asked to rate various characteristics such as safety, hours of operation, courtesy of employees, adequacy of the activities contained within the facilities, and so on. The survey can identify the percentage of users and nonusers. Reasons for nonuse can also be determined, at least to some degree, and further analysis can distinguish those reasons that are entirely or partly controllable by government actions (and which officials may then remedy; see U.S. Department of Interior, Bureau of Outdoor Recreation, 1973; Winnie and Hatry, 1972, for examples of these uses for recreation and transportation services).

Use of regular citizen surveys related to governmental services has been or is being undertaken in such cities as Dayton, Ohio (by the Dayton Public Opinion Center), and St. Petersburg, Florida. Whether these will be continued at regular intervals, permitting time trend identification and comparisons, is yet to be seen. Gradually, however, the citizen survey seems destined to become a major management tool over the next decade. Citizen surveys are not without their pitfalls and dangers. This is not the place to discuss survey problems in detail, except to say that the difficulties can probably be alleviated sufficiently to permit the effective use of the citizen survey for measuring selected aspects of the quality of government services (Webb and Hatry, 1973).

The sample survey can also be used for feedback from particular clientele groups relevant to individual services. This can provide not only information on clientele perceptions of the quality of service they have received, but also help in measuring the long-term impact of programs. For example, feedback on ex-clients can be obtained for such programs as drug and alcohol treatment, correctional treatment, manpower training, vocational rehabilitation, and a variety of health programs. The state, county, and city governments that manage these treatment programs have often, at best, tracked their clients only while they are in their systems. Statistics on numbers of persons treated, number of persons graduated, and percentage of dropouts are generally available. However, governments have seldom measured the longer-term impacts of these programs—which, after all, is the real key to success of such programs—such as by survey samples of clients who have left the government's system. Interviewing at least a sample of graduates and dropouts at intervals to determine their current condition seems essential to adequately estimate the quality of such programs.

Linked records. Procedures can be developed that enable a government to trace clients through one or more agency and thus to assess a variety of quality aspects of related services. Police agencies, for example, seldom obtain information on the disposition of persons they arrest in the adjudication process, yet such information is relevant to the quality of the police apprehension-of-criminals function. More data linkages among the police, the courts, and the jails and prisons would be mutually beneficial to them. More insights would become available about recidivism, an important measurement of success in any type of correctional program for adults or juveniles. When a person returns to the same agency that provided treatment in the first place (whether a prison, mental or health institution, or any of the "treatment" programs), measurements of this recidivism become readily available. However, often this is not the case, and estimates of recidivism may depend on linked records. Another example: programs aimed at improving the employability of adults might need information from unemployment and welfare offices to assess success.

Linked records were used in Nashville in a 1971 study of neglected and dependent children in Nashville. The flow of children was traced from intake units through various short-term holding facilities and the courts to determine the delay times as well as the apparent quality of the short-term care (Burt and Blair, 1971).

An important consideration is the need to preserve confidentiality of information on individuals. If adequate control cannot be found, the measurements affected will have to be foregone.

Better use of existing data. For example, with appropriate data collection and processing techniques, existing data can often be identified by neighborhood or other type of clientele group (such as sex, age, or race) to provide the disaggregated analysis discussed earlier. Important aspects of service performance become apparent once statistics are classified and disaggregated that do not show up otherwise. For example, the number of fires in buildings that were recently inspected and passed by fire inspectors could be distinguished from other fires to measure fire rates for inspected buildings. This applies to any type of inspection, including building and housing code inspections. Another example: consumer affairs agencies, rather than concentrating solely on complaints and on "haphazard" inspection processes could undertake systematic inspections of commercial establishments. These could provide estimates of the percentage of overpricing or false advertising. If monitored over time, such data would indicate problem areas and suggest in part the effectiveness of the agency's service—if an appropriate systematic sampling approach was utilized.

The above discussion has touched on only some of the tools and techniques for measurement available from the social sciences. The principal point of this section has been to suggest that many of the often-perceived limitations of quality-of-service measurements may in practice be overcome. There appears to be considerable potential for such measurement.

CAN QUALITY TARGETS OR STANDARDS BE SET AND, IF SO, HOW?

The measurement of a service, or of some aspect of it, does not stand alone. We immediately would like to know more: Is it good, bad, or indifferent? What level of quality should be sought for each quality measurement?

To answer these questions, standards or targets are desirable. In this section, we examine six approaches to standards or targets—the use of absolutes, engineering standards, comparisons with other times and places, comparisons with the private sector, analytic techniques for developing standards, and the use of judgment.

The use of absolutes—striving for perfection. For most measurements, a figure representing perfection can be specified. For example, targets for fire and crime rates can be set at zero. Quantities of service could be specified that would be ample for all citizens (or for all who are classified as in need of the service). Similarly, targets for participation could be set that would involve all citizens during a particular time period. Such absolute standards, however, are not generally useful. They tend to be impossible of practical achievement because of limited resources and other difficulties. There are some exceptions. For a very few measurements, it may be appropriate to set the target of "zero defects" (such as "zero" incidents of smallpox in the United States). But generally the goal of perfection is so unrealistic that its use as a standard would provide a misleading set of expectations.

Development of engineering standards. The process of engineering certain targets or standards is familiar in industry, where work standards have been established. Local governments are increasingly using similar techniques, usually for routine tasks. Time requirements, for instance, can be set for repairing potholes, for processing standard forms, and for completing certain data processing activities. Target response times also could be established for particular conditions, based perhaps on the averages of test runs. For the most part, such engineered targets seem most appropriate for measuring workload outputs for repetitive, routine activities. More complex matters such as crime or usage rates for government facilities usually do not lend themselves to this engineering approach.

Comparisons with other times and other places. This approach, the most widely used at present, probably has the most general applicability. At least three important types of comparison need discussion—comparisons of current versus prior years; comparison with the performance of other similar jurisdictions; and comparisons among clientele groups within the jurisdiction.

(a) *Comparisons with prior year measurements.* Any governmental jurisdiction should find it useful to see how its current performance measures up to that of previous years. In a sense, the preceding year's value becomes the standard for comparison. In interpreting the comparisons, however, it is necessary to identify events that have transpired which change important characteristics from one year to the next. For example, if measurement procedures have been improved, so that more crime is reported or if more extensive measurements lead to the detection of greater water and air pollution, the apparent worsening in the data for the current year may be explained in part by these developments. Nevertheless, tabulation of the changes are important to the government, tempered, of course, by other pertinent information.

(b) *Comparisons with other, similar jurisdictions.* For understandable reasons, governmental officials are ambivalent toward such comparisons. Those who feel their governments will perform well look forward to them; those who feel that their own performance will look bad compared to those of other governments are naturally skeptical. The central problem is whether or not such comparisons are meaningful and fair. Inevitably, scores of factors affect a measurement. Different conditions exist in different jurisdictions. Further complications arise frequently because of the existence of partially or significantly different measurement procedures. Sporadic attempts have been made to obtain compatability of definitions and procedures from jurisdiction to jurisdiction for certain services. Yet subtle differences at the outset often result in important differences in the conclusions. The most familiar example of intergovernmental comparisons are the FBI Uniform Crime Reports. Despite major attempts to achieve commonality of data collection procedures for these reported crime rates, many criticisms continue to be leveled against the measurements on a variety of grounds. However, few if any public officials seriously propose that comparative crime data be dropped; suggestions usually call for their reform, not their elimination. Other than data on crime and fiscal affairs (revenues and expenditures), there are surprisingly few comparative data on local or state government performance. This is not to imply that making such comparisons is easy. Attempts to derive, for example, the "cost per capita for sanitation services" are fraught with numerous problems: differences in the way jurisdictions allocate funds for the direct support of services and for overhead; the degree to which the government itself performs waste collection or relies on private organizations; the mix of residential versus commercial collection; whether the collection occurs at the back door or at the curb; whether the collection frequency is once, twice, or more times per week; weather conditions within the cities; city household density characteristics, and so on. There has recently been a resurgence of interest in intergovernmental—particularly intercity—comparisons. The International City Management Association's Committee on quality of municipal services in 1973 recommended that attention be given to such comparisons, suggesting that the initial emphasis be regional, evolving subsequently into nationwide comparisons (International City Management Association, 1973). This is reasonable because governments within the same regions presumably are more likely to have similar characteristics. The United States as a whole is currently far from having dependable comparative data available on a wide range of local services, but the interest now being expressed by government officials makes it appear likely that this will receive substantial attention in the coming years.

(c) Intrajurisdictional comparisons. One of the quality-of-service attributes identified earlier was the equity of distribution among clientele groups within a jurisdiction. As governments undertake more comparisons of the level of services received by various groups of citizens, target setting to assure fair distribution becomes possible. One important basis for comparison is geographical, looking at various neighborhoods within a city, county, or state jurisdiction. When measurements on all areas reveal the average, median, and highest performance level, these readings can be used as the targets or standards for other parts of the jurisdiction. For example, the District of Columbia system of rating street cleanliness provides ratings for each part of the city. City officials have identified the dirtier areas and set targets for improvement based in part on the higher quality

achieved in other parts of the city (limited, of course, by judgments as to resource availability). Agency service districts also provide natural geographical comparisons for quality-of-service ratings; city agencies often utilize such geographical arrangements as police, sanitation, or recreation districts.

Comparisons with the private sector. In some limited instances, data are available for similar activities in the private sector. This applies primarily to measurements of productivity or efficiency. Thus, comparisons have been used by New York City of its own performance on unit costs (and vehicle downtime) versus those of the private sector for solid waste collection and vehicle maintenance activities. Other aspects of service quality may be compared in special cases such as in comparisons of private fire departments (such as in Scottsdale, Arizona) or, with more contrived examples, such as with certain private police services. These comparisons may suffer from even more difficulties than those in making intergovernmental comparisons. However, if government interest in purchase-of-service arrangements increases, such comparisons will become more important for target setting.

Use of analysis to estimate what targets should be. The special characteristic of this approach is the explicit analysis in some detail of the particular service—its characteristics, what has happened in the past, and what is likely to occur if particular programs and policies are implemented. A variety of techniques such as statistical extrapolations, systems analysis, and cost-benefit analysis can be utilized to make these estimates. Such estimates theoretically should be better than those based solely on any of the other approaches. This is because good analysis would take into account the data from the previously described approaches, such as comparisons with other jurisdictions and past performances, and make realistic estimates of what can be accomplished and for what costs. The major drawback of detailed analysis is that it can be expensive and requires especially trained and experienced personnel, currently in scarce supply.

Targets established by judgment. Individual government agency heads, or others, may fall back on their own judgments as to what specific targets they should seek—perhaps using input information from one or more of the previous approaches to target setting. There, of course, almost always will be some element of judgment in any setting of targets or standards.

Thus far, there has been only scattered use of targets or standard setting by governments, but reliance on them is likely to emerge as an important feature of public management. Probably most governments will use a combination of approaches in developing their standards or targets. The management by objectives (MBO) approach and the evolution of PPBS and program budgeting have focused on performance measurement; the logical next step is the setting of targets and provision of follow-up to measure accomplishments relative to the targets. The most ambitious reflection of this as of this writing has been the New York City government's productivity measurement effort, cited earlier (New York, City of, 1973).

WHO SHOULD BE RESPONSIBLE FOR MEASURING PUBLIC SERVICE QUALITY?

Some possibilities are discussed in the following paragraphs.

The government role in assessing its own public services. It seems generally accepted that a government providing a service should attempt to measure the quality of that service as a regular management function. Some reservations about this assumption merit attention:

- It is difficult for someone to evaluate themselves objectively, and measurement is a major part of evaluation. This can pose a serious problem, particularly if the group doing the measuring has an important vested interest in the subject area being measured (for an extensive discussion of governmental self-evaluation, see Wildavsky, 1972). However, in any government, there are various levels of management. The actual vested interest will normally lie at a lower level in the program management. In upper or parallel levels of city, county, state, or federal government, there is likely to be an agency that in a majority of cases will be able to undertake evaluation without jeopardizing its own status. There are, of course, exceptions—as when upper-level management has publicly boxed itself into a corner by basing its reputation or political future on a particular program. Any type of self-assessment from that source is likely to be suspect, regardless of its actual objectivity. One way to alleviate this, as well as to maintain the credibility of measurements, is to subject the measurents to periodic auditing, much in the same manner that financial data is audited. This would be particularly appropriate for measurements used in reaching major public decisions.

- Some quality measurement requires special skills not currently available in most governments (other than perhaps at the federal level) and may not be available to them, even over the long run. For example, some skills needed for undertaking sample surveys are too specialized, and too infrequently required for other purposes, for the smaller state and local government to obtain as permanent staff. Governments can still obtain such skills through the use of outside contractors.

- Suggesting that measurements be undertaken implies that something can be done about the findings once they are achieved. If this is not the case, there is little sense in spending the considerable effort and resources to provide data that will not be used. This is not solely a question of government funds or will. A major issue in measuring quality of services is that there are few quality-of-service aspects that are completely under the control or responsibility of a particular governmental body. Many important measurements of urban quality effectiveness will be affected by external factors that officials cannot fully change or influence. For example, crime rates and the incidences of various diseases reflect circumstances that are not fully controlled by governmental bodies. To the extent that nongovernmental factors are contributing to the crime rate or health incidences, the government is not fully to blame if these rates worsen, or fully deserving of praise if the rates improve. The likelihood of being blamed for conditions somewhat beyond their control understandably tends to discourage government bodies from undertaking measurements of service quality. Yet, whether a government does or does not have *complete* control over an aspect of service quality, information about the quality level should be useful—even essential—to government decisions. In general, governments have most control over aspects that do not directly measure service quality or effectiveness. Examples of these are workload and certain intermediate outputs such as the "response time from a call to arrival at the scene" by police, fire, or emergency rescue service.

- In any case, individual governments, particularly the smaller or poorer ones, will be financially able to undertake only limited amounts of measurement of their own acitivities.

The role of higher levels of government than the one providing the service. The federal government and, for local governments, state governments might undertake, sponsor, or at least monitor measurement efforts. For example, the U.S. Bureau of the Census has considerable capability for obtaining data through sample surveys of citizens. Perhaps the federal government should undertake certain measurements regularly for state and local governments. Some of these, such as crime victimization and health surveys, are already being undertaken, but unfortunately the federal government's efforts have been ad hoc or have concentrated on aggregative data rather than on data that can be used to provide measurements for individual localities. Also, the federal efforts have utilized the most precise and costly techniques. The less precise needs of many local governments could be met with smaller samples and possibly less elaborate procedures. *State* governments generally have a long way to go before they will have the capability to help local governments significantly with regular measurements, but this capability could be developed.

Citizen groups. There are many citizen groups in any community that have interests in measuring quality of public services. Organizations such as the League of Women Voters, Chambers of Commerce, and neighborhood citizens groups are often concerned about measurement. Unfortunately, citizen groups are not always stable as to their composition or their interests. Nor do they normally have major resources to do measurement on a systematic and regular basis. However, they can play an important role, both in spurring interest in measurement work and reviewing work done by others for comprehensiveness and objectivity. They also can, on occasion, provide resources where public resources are scarce. For example, the Arlington County (Virginia) League of Women Voters in 1972 provided the interviewers for the county to undertake a survey of residents on various aspects of law enforcement quality.

Universities. Universities often have considerable resources for making quality-of-service measurements. However, again, there are major limitations as to their interest in undertaking regular, annual assessments. Of course, universities in the role of consulting firms may often be available on a contractual basis to help governments do the measurements if governmental funds are available. University programs—such as in public administration, business, industrial engineering, economics, and the other social sciences—may at least periodically be interested in assessing the quality of services. On occasion, university resources can be used as an added resource at either no or low cost to a government. Help to state and local governments is already occurring with greater frequency. However, this tends to be irregular and cannot always be depended upon.

Legislatures. The interest stirred by the 1972 General Accounting Office's report (U.S. Office of the Comptroller General, 1972) is beginning to widen the interest of legislative bodies at all levels of government in providing periodic audits of "program results" as well as "economy and efficiency." State legislative audit agencies are already expressing considerable interest in performance audits of executive branch programs as well as the more traditional fiscal and compliance audits. The amount of actual evaluation measurement effort done by them is likely to be severely limited by the staffing of legislative auditing bodies and their level of funding. However, these units seem likely to play a major role in spurring the executive branch to carry out assessments. In addition, they can assume a major role in reviewing the measurement procedures for objectivity and comprehensiveness.

Considering all these facets of responsibility for the measurement of service quality, it appears that, if the job is to be done on a regular and systematic basis, governmental bodies will have to take much of the lead. However, the other resources discussed will probably be needed periodically. Auditing by some external organizations also seems advisable.

SUMMARY

Essentially, the point of view of this chapter can be summarized as follows:

(1) Interest in obtaining information on the quality of individual public services has substantially expanded recently among governments and their citizens.

(2) For any given service, a number of attributes need measurement for a fair assessment of the quality of that service. Yet, traditionally, few of these attributes have been subjected to systematic measurement.

(3) Though far from perfect, the current state of the art does permit significant measurement of many quality characteristics. The cost of such measurement need not be prohibitive.

(4) Measurement of quality characteristics will encourage jurisdictions to set performance targets or standards that generally do not

exist at this time. Various reasonable approaches can be used to formulate standards. When these standards are used to evaluate agency and program performance, focusing more attention on quality-of-service measurements, there may be temptations to play with the data; regular auditing of measurement procedures may be advisable to guard against this.

(5) It appears inevitable that individual governments will have to be responsible for the major amount of their own quality-of-service measurement. However, the occasional participation of higher levels of government, citizen groups, universities, and legislatures—to provide both encouragement and controls for comprehensiveness and objectivity—would appear to strike a proper balance.

Despite certain pitfalls and limitations, the need and potential utility for quality-of-service measurement is clear. With the emergence of practical, though not perfect, tools, greatly expanded measurement of quality-of-public-service appears to be beginning.

REFERENCES

BLAIR, L. H. and A. I. SCHWARTZ (1972) How Clean Is Our City? A Guide for Measuring the Effectiveness of Solid Waste Collection Activities. Washington, D.C.: Urban Institute.

BOOTS, A., III, G. DAWSON, W. SILVERMAN, and H. P. HATRY (1972) Inequality in Local Government Services: A Case Study of Neighborhood Roads. Washington, D.C.: Urban Institute.

BURT, M. R. and L. H. BLAIR (1971) Options for Improving the Care of Neglected and Dependent Children/Nashville-Davidson, Tennessee (Program Analysis Applied to Local Government). Washington, D.C.: Urban Institute.

Council of State Governments (1973) State Government Program Evaluation Activities. Lexington, Kentucky.

HATRY, H. P. and D. R. DUNN (1971) Measuring the Effectiveness of Local Government Services: Recreation. Washington, D.C.: Urban Institute.

International City Management Association (1973) "Achieving quality local government." Public Management (September): 19-23.

——— (1972) "Local government budgeting, program planning, and evaluation." Urban Data Service Report, May.

——— (1970) "Measuring effectiveness of municipal services." Management Information Service, August.

National Commission on Productivity (1972) The Challenge of Productivity Diversity: Improving Local Government Productivity Measurement and Evaluation. Washington, D.C.

——— (1971) Improving Productivity and Productivity Measurement in Local Governments. Washington, D.C.

New York, City of (1973) Productivity Program: Second Quarter Progress Reports. New York.

OSTROM, E. (1973) "The need for multiple indicators in measuring the output of public agencies." Policy Studies J. (December).

RIDLEY, C. E. and H. A. SIMON (1938) Measuring Municipal Activities: A Survey of Suggested Criteria for Appraising Administration. Chicago: International City Managers' Assn.

U.S. Department of Interior, Bureau of Outdoor Recreation (1973) How Effective Are Your Community Recreation Services? Washington, D.C.: Government Printing Office.

U.S. Office of the Comptroller General (1972) Standards for Audit of Governmental Organizations, Programs, Activities and Functions. Washington, D.C.: U.S. General Accounting Office.

WEBB, K. and H. P. HATRY (1973) Obtaining Citizen Feedback: The Application of Citizen Surveys to Local Governments. Washington, D.C.: Urban Institute.

WILDAVSKY, A. (1972) "The self-evaluating organization." Public Admin. Rev. (September/October): 509-520.

WINNIE, R. E. and H. P. HATRY (1972) Measuring the Effectiveness of Local Government Services: Recreation. Washington, D.C.: Urban Institute.

3

The Urban Government
Annual Report

ARIE Y. LEWIN
ROBERT W. BLANNING

> *Governmental programs rarely have an automatic regulator that tells us when*
> *an activity has ceased to be productive or could be made more efficient, or*
> *should be displaced by another activity. In private business, society relies*
> *upon profits and competition to furnish the needed incentives and discipline*
> *and to provide a feedback on the quality of decisions. The system is imperfect*
> *but basically sound in the private sector—it is virtually non-existent in the*
> *government sector. In government we must find another tool for making the*
> *choices which resource scarcity forces upon us [Schultze, 1967: 21].*

OVERVIEW

□ THE OBJECTIVE OF THIS CHAPTER is to demonstrate the need
for and the feasibility of requiring municipal governments to publish
an annual report to their residents similar to the annual report
published by corporations for their stockholders.

The corporate annual report to stockholders and the rules setting
disclosure requirements resulted from the financial collapse of 1929
and the legislation which followed—the Securities Act of 1933 and
the Securities and Exchange Act of 1934. The objective of that

landmark legislation and the legislation which has subsequently been enacted was to make the management of corporations accountable to their stockholders. Accountability was achieved by requiring top management to disclose periodically to stockholders the results of operations and to have the reports certified to by independent auditors. The task of the auditors is to certify that the management report fairly reflects the results of operations and that the accounting conforms to generally accepted accounting principles. The auditors are not expected to evaluate management's stewardship. It is recognized that there is no single way to evaluate management decisions. Therefore, the judgment of the management stewardship function is left with the investing public, which signifies its satisfaction or dissatisfaction by bidding for a firm's shares in the market. Thus, the intent of the legislation was to make explicit the financial consequences of operations so that investors can evaluate management performance for themselves.

A similar situation exists in a city. The public, through its exercise of the ballot, is called upon to evaluate the performance of the mayor and the city council and to decide whether to keep them in office or to replace them. Unlike the investing public, citizens of a city do not presently receive from the "city executives" a report similar to the corporate annual report to aid them in evaluating the performance of the city administration. Because of the size of many city governments, the tendency of the bureaucracy to insulate itself from the public, the complexity of functions and services performed for city residents, and the relative importance of municipal demand for resources in a city economy, it is almost impossible for citizens to evaluate the performance of their government. In a sense, citizens face a dilemma similar to that which confronted the investor in 1929. Their inability to perform an objective evaluation of municipal performance or to influence the city administration's goals and programs has contributed to apathy and a lack of confidence in municipal government. It is with the objective of making local government more responsible that we examine the feasibility of requiring municipalities to provide annual reports to their publics.

The remainder of this section reviews current municipal reporting practices and discusses the objectives that should underlie the proposed annual report. The rest of this chapter discusses the information needs of citizens and their representatives; the content of the annual report; its independent certification; an example of a departmental report (for a police department); and the implementation and political feasibility of a municipal reporting system.

MUNICIPAL REPORTING PRACTICES

Almost all municipalities in the United States are mandated either by state law or by custom to publish an annual financial report certified by an independent auditor or an independent elected official such as a city comptroller. The purpose of current municipal financial reports, however, differs from that of corporate annual reports. A corporate financial report is primarily a management report to stockholders presenting audited financial results of operations and other nonfinancial information, but a municipal financial annual report is primarily a report of financial operations carried out by all agencies showing compliance with applicable laws. As a result, municipal financial annual reports make use of fund accounting, because traditional municipal accounting systems operate on a fund basis.[1]

The guidelines for municipal financial annual reports prepared by the National Committee on Governmental Accounting (1951) are indicative of the intended audience for these reports; the guidelines suggest that the report include a financial section and a statistical section. The financial section contains statements of revenues and expenses by funds. The statistical section includes various schedules on revenue sources, tax collections, debt limits, and the like.

An analysis of fifty municipal annual reports by Cahill (1973) suggests that most such reports follow the guidelines of the National Committee on Governmental Accounting. These guidelines, however, emphasize the needs of credit rating organizations and financial institutions rather than those of the electorate. The reports are highly technical and complex. They are intended to present the financial condition and operation of funds and to demonstrate compliance with legal regulations. Thus, traditional municipal accounting systems have become mechanisms for validating expenses, and the resulting financial reports are not oriented to the needs of the electorate. That is, the reports do not contain information which permits citizens, neighborhood groups, and civic associations to assess citizen benefits of government policies.

THE INFORMATION NEEDS OF
CITIZENS AND THEIR REPRESENTATIVES

An urban annual report, like a corporate annual report, must be designed to help certain persons make decisions. A corporate annual report is intended to help shareholders evaluate the performance of top management and to help investors decide whether and how much to invest in the corporation. Similarly, an urban annual report should be useful to residents of the city and their representatives in their decision-making. There are several distinct types of people and organizations that have a legitimate need for information about the activities of a city government (for a more complete discussion of these participants in the urban political process, see Banfield and Wilson, 1966; Baker, 1971). An annual report cannot supply all of these information needs, but it can offer a "data base" that will be of use to the variety of people interested in the city government.

There are five principal types of decision makers (other than officials of the city government) who may use a municipal annual report to make decisions or to attempt to influence the decisions of others. The *first* is the collection of persons who live, work, vote, or pay taxes in the city. Although these persons are generally diffuse in their opinions and actions, they occasionally make decisions —individually and collectively—that have a substantial impact on the city and its government. The most important of these decisions is to vote for the mayor, the city council (or other legislative body), and any other elected officials (such as the city comptroller). In addition, these persons may also vote in referenda and may initiate taxpayer's suits.

The *second* type of decision maker is the special interest group. These groups attempt to influence the mayor, the legislature, and the executive bureaucracy to take actions that will accommodate the economic, social, and political objectives of interest to them or to their constituencies. The economic interest groups include financial, industrial, and commercial groups (especially banks, construction companies, and merchants), and labor unions (including unions of city employees). The social interest groups include neighborhood groups, racial groups, and groups interested in specific social services (such as public education). Political interest groups are the established and independent political parties and any ad hoc political groups formed to support a specific candidate.

A *third* type of decision maker is the city legislator, who votes in the city council for appropriations and for other city laws and who often attempts to influence the executive on behalf of his constituents. Although legislators are a part of the city government, they typically do not have access to the detailed data, information, analyses, and projections available to the executive. Even when they do have access to such information, they do not have the staff resources to summarize and analyze it. Thus, the legislature can exert only a general or aggregate control over the actions of the executive.

A *fourth* type of decision maker is the officials in county, state, and federal governments (especially the latter two). These higher-level governmental bodies (both executive and legislative) may take three types of actions that affect the city (for an examination of these actions in more detail, see Mitau, 1966; Break, 1967; Derthick, 1970). First, they may enact legislation restricting the activities of the city government. Second, they may appropriate certain restricted or unrestricted funds for the city. Third, they may attempt to bring direct influence on the city bureaucracy, using the withholding of funds or the enaction of special legislation as a veiled (and sometimes open) threat.

The *fifth* type of urban decision maker considered here is the journalist, who investigates certain activities of the city government and reports them to the people. His diligence in investigating and reporting government activities may have a significant influence on the practices of the executive and the legislature, and he may interpret the annual report for the groups described above. However, for the purposes of this analysis, the journalist will be considered a medium of communication, rather than an active participant in the political process.

In order to perform their analyses, the political actors described above require uniformly prepared periodic information about the performance of the city government during a previous period and the status of the city at the end of the period. In this chapter, we will assume that the reporting period is one year. However, it can also be one quarter, two years, and the like.

In designing an annual (or other periodic report of the performance of the city government and the status of the city, two principal problems must be considered. Both concern the difficulty of establishing proper measures of performance and status. The first problem is that the annual report must satisfy the wide range of objectives of the various decision makers outlined above. The second problem is that most of the decision makers do not know what their objectives are, or at least they are unable to articulate them. In this respect, urban annual reports differ significantly from corporate annual reports. It can be safely assumed that there is one principal type of user for a corporate annual report (the stockholder or potential stockholder), and that his objectives are simply defined (dividends and the likelihood of capital appreciation).

Because of the variety of conflicting objectives and the inability to articulate objectives, it is not possible to define appropriate measures of performance and status in the abstract. However, we can outline major categories of performance and status information and suggest in some detail the types of information that might be subsumed under these categories. We feel that the preparation of the report by a municipality will be an evolutionary process in which government officials prepare a report, the users outlined described above assess the usefulness of the report and propose modifications, government officials revise the report (either in the same year or during the following year), the users evaluate the new report, and the cycle begins anew.

In the remainder of this section, we examine the principal types of information that might be included in the report. In the following section, we examine in some detail the major components of the report. There are three principal types of information that might be useful to the decision makers outlined above. The *first* is the present financial position and previous year's financial operations of the city government, the departments and agencies that make up the government, and any funds maintained separately from the general fund or city budget (see, e.g., Tenner, 1955).

The *second* type of information is ex post cost and performance analyses of the programs and other activities of each agency or department in the government. These would be similar to the cost/benefit analyses performed by government analysts prior to the initiation of a program (see, e.g., Dorfman, 1965; Rossi and Williams, 1972).

The *third* type of information consists of a set of indicators that measure the aggregate consequences of agency and government performance—for example, crime rates or the distribution of economic wealth. These aggregate indicators will differ from the agency performance measures in that (1) a single indicator (such as income distribution) may depend on the actions of several agencies (such as public assistance and tax policies), and (2) the indicators will often depend partly on actions that are beyond the control of the city government. The annual report should contain two types of aggregate indicators. The first is a set of economic indicators that describe the economic state of the city (see, e.g., Perloff and Wingo, 1968). The second is a set of social indicators that describe the social state (or quality of life) in the city (see, for example, Bauer, 1966; U.S. Department of Health, Education and Welfare, 1970; U.S. Environmental Protection Agency, 1973; Sheldon and Land, 1972).

It has been suggested that citizens should also be made aware of the impact of various agency programs on social goals (see Terleckyj, 1970; Charnes et al., 1973). This might be accomplished by constructing a matrix whose coefficients measure the ratios of goal indicators (such as increase in leisure hours per person per year) to levels of program effort (such as dollars spent in improving intercity transport). This would allow citizens to form their own judgments about the proper allocation of financial resources to achieve social goals.

We see two possible disadvantages of including in an urban annual report a matrix that relates levels of program effort to social goals. First, the relationship between program effort and social goals may be too complex to be described by a linear relationship in a matrix. That is, there may be interactions and nonlinearities in the relationship that require a more sophisticated representation (for example, a time-staged simulation). Second, even if the relationship is amenable to matrix representation, the values of the coefficients will certainly be subject to question. That is, different analysts may reasonably propose quite different numerical values for the coefficients. Furthermore, various political interest groups may attempt to influence those who prepare the annual report to use coefficients that would make the programs they favor appear more attractive.

In our view, this is inappropriate for an urban annual report. Such a report, like a corporate annual report, should be subject to attestation by independent evaluators (or auditors). It is difficult enough to establish workable standards for an independent evaluation of measures of social performance (such as crime rates and environmental pollution), but it will be almost impossible for some time to develop such procedures for evaluating social models.[2] An urban annual report that contains such questionable data can easily become a political football, prepared in part to justify past or proposed policies. Therefore, the urban annual report, like the corporate annual report, should report program-oriented inputs and goal-oriented outputs, but should not attempt to report relationships between them.

We are not suggesting that such matrices or social models should not be constructed or that citizens should not have access to them. However, in preparing reports for citizens, municipal officials should separate that information with which people of varied analytical and political persuasion are in substantial agreement from that information with which they are not. For this reason, we feel that the information contained in the urban annual report should be confined to the three types of information described above.

THE CONTENT OF THE REPORT

In the preceding section, we suggested that the urban annual report contain a set of comparative reports to include consolidated financial statements, departmental costs and benefit reports, and a report of economic and social indicators. The resulting five components of the report are outlined in Table 1.

The annual report outlined below omits two items that might be of use to the urban decision makers described above. The first is a set of projections of such quantities as population, tax revenues, as so on. The second is an analysis (or simulation) of the consequences of alternative city policies, such as housing policies, transportation policies, and the like. However, as we argued in the previous section, the annual report should contain only historical information, because this information can be examined with reasonable objectivity by a group of independent evaluators (auditors), who can attest to the validity of the data.

Unfortunately, projections and simulations must contain assumptions and estimates that are often difficult to verify and that are sometimes not made explicit.[3] Since the officials preparing the annual report may deliberately or inadvertently include questionable assumptions in their analyses, the urban annual report, like corporate annual reports, should contain only historical data. However, an inspection of the accumulated information in a sequence of annual reports covering several years may provide some of the data necessary to perform analyses.

The five components of the annual report are as follows:

(1) Statements of revenue and expense: The purpose of these statements is to report the revenues and expenditures of each department or agency at the lowest level of the government hierarchy for the year then ended and for several prior years. Each such statement should use the format of the budget for the department or agency, reporting the amount budgeted, the amount spent, and the reason for any significant difference.

Consolidated statements of revenues and expenses should be prepared at all levels of government, such as the department level, the agency level, and the city level.[4] The consolidated reports will not be presented in the same level of detail as the departmental statements (that is, by line item). The level of detail will depend on the particular needs of the report users and on the reporting agency.

A sample consolidated statement for an entire city government appears in Figure 1. A more detailed statement of revenue and expense should be prepared at the lower levels of the city government.

The preparation of a consolidated statement should be straight-forward, because most city governments prepare similar statements at present. However, most governments do not identify and report revenues and expenditures at lower levels of the government. Thus, the principal problem in preparing this component of the annual report is to develop a standard reporting system for all departments so that all such reports can be prepared on a consistent basis. That is, it is essential that the figures in the departmental reports "add up" to those on the consolidated report. This in turn requires that a

TABLE 1
COMPONENTS OF AN URBAN ANNUAL REPORT

	Financial Reports	Cost Performance Reports	Report of Indicators	
			Economic	Social
Comparative Past Performance (during past and prior years)	(1) Statements of revenue and expense (by department and consolidated)	(3) Statements of cost and perform-ance (by program and by department)	a	a
Present State (at end of previous year)	(2) Statements of financial position (by fund and consolidated)	a	(4) Report of the economic state of the city (consoli-dated)	(5) Report of the social state of the city (consoli-dated)

a. No report.

Revenues	Current Year	Previous Year
Taxes		
Fines & Forfeits		
Licenses & Permits		
Interdepartmental Charges for Services		
Miscellaneous Receipts		
Total Revenues -		
Expenses		
General		
Debt Service		
Depreciation		
Accrued Pension Liabilities		
Miscellaneous		
Total Expenses -		
Total Surplus (Deficit) -		

Figure 1: CONSOLIDATED STATEMENT OF REVENUE AND EXPENSES

standard cost accounting and reporting system be established at all levels of the city government including a common chart of accounts.

(2) Statements of financial position: The objective of these statements is to present the assets and liabilities (that is, a collection of balance sheets) of each of the funds established by the legislatures, including the general fund, the working capital fund, and any trust, agency, enterprise (such as utility), and special assessment funds. A statement of financial position will also be prepared for each fund, and a single consolidated statement will be prepared for the city. A sample consolidated statement for the entire city government is presented in Figure 2. The statements prepared for each fund might be further broken down into more detailed categories.

These statements, like the statements of revenue and expense, should not be difficult to prepare, because most city governments prepare balance sheets for some or all of their funds. However, the separate reporting systems should be standardized so that a consolidated statement can be prepared.

(3) Statements of cost and performance: Each department would prepare a statement of the cost and performance of each major project in the department. These reports can also describe resource usage by program and program results or performance. Where the

Assets	Current Year	Previous Year
Fixed:		
Land		
Plant & Equipment		
Parks & Public Works		
Total fixed assets-		
Current:		
Cash		
Investments		
Accounts Receivable		
Taxes Receivable		
Loans & Grants Receivable		
Materials, Commodities & Supplies		
Prepaid Expenses		
Total current assets -		
TOTAL ASSETS:		
Liabilities, Net Worth		
Current:		
Accounts Payable		
Interest Payable		
Taxes Payable		
Current Debt		
Other current liabilities		
Total current liabilities -		
Long Term:		
Long term debt		
Accured Pension Liabilities		
Total long-term liabilities -		
TOTAL LIABILITIES & NET WORTH:		

Figure 2: CONSOLIDATED STATEMENT OF FINANCIAL POSITION

municipal government has been required to submit cost-benefit reports to the city council prior to funding, the anticipated and actual costs and performance should be reported and any significant differences explained. Like the statements of revenue and expenses, these statements could also be consolidated at the department and agency level.

As suggested previously, these reports should describe only those measures of performance that are directly controllable by the appropriate department. They should not contain measures of social performance. Such measures will appear in the report of the social state of the city. For example, the statement of cost and performance prepared by the police department might contain the number of police assigned to various duties and the response time of police to various types of complaints, but crime rates and arrest rates will appear with the other social indicators. An example of a cost performance statement for a police department is discussed later in this chapter.

(4) Report of the economic state of the city: A consolidated set of economic indicators will be presented in the report on the economic state of the city. This report will describe distribution of material wealth among city residents and the availability and price of capital, industrial goods and services, and consumer goods and services in the city. A single report will be prepared for the entire city. That is, the report will not be partitioned into measures appropriate to each department and agency. The report will also include comparative statistics for prior years. The number of prior years depends on the availability of data, but should be sufficient to highlight trends.

The report of the economic state of the city will certainly change in content and in format in response to economic issues of concern to the citizens. In this respect, the format of the economic report, like the statements of cost and performance and the report of the social state of the city, will be far less stable than the formats of the financial statements described above.

It is essential that the economic report be modified in response to citizen information needs.[5] Also, it should be noted that much of the information in this report is compiled by agencies (federal, state, and county) outside the city, such as the Bureau of Labor Statistics, the FBI, and so on. An example of such a report is given in Figure 3. It is assumed in this example that the citizens are interested in the total level of economic activity in the city, the price levels in the city relative to those in the nation, the rate of change of prices, and the job opportunities in the city by age, race, and sex. If the decision makers described earlier find this information of little use or if they find that this information gives rise to a need for more information, the report should then be modified.

(5) Report of the social state of the city: This report will present a set of social indicators for the city. A single report will be prepared for the entire city, and it should contain comparative statistics for prior years. The report should contain social indicators for each

Net Metropolitan Product _____

City Prices / National Prices _____

Rate of Increase of Prices _____

Number of jobs	White		Non-White	
	Male	Female	Male	Female
Juvenile Young Adult Middle Aged Senior				
Unemployment				
Juvenile Young Adult Middle Aged Senior				

Figure 3: REPORT OF THE ECONOMIC STATE OF THE CITY

major category of social interest. These categories might include human resources, transportation, housing and property, public safety, health services, sanitation and environmental protection, and community development.

VALIDATION OF THE ANNUAL REPORT

An independent audit of the municipal annual report would be limited to certification of the financial statements and the performance evaluation statistics it contained. The audit function should be performed by an independent certified public accounting firm, independent elected official, state audit agency, or some other legally mandated group.

The American Institute of Certified Public Accountants (1963) has developed and continues to revise a set of audit standards and procedures. In addition, further audit guidelines for government activities have been published by the U.S. Comptroller General (1972). These standards, when used in conjunction with generally accepted accounting principles, provide the framework and basis for the certification of financial statements.

An examination of the proposed municipal annual report indicates that it consists of three classes of information: pro forma financial statements; definitions of municipal functions and programs; and performance statistics.

The auditing of financial statements should present no special difficulties. Since most of the information presented is taken directly from the books and accounts of the municipality, the auditing procedures should be familiar to any auditor. The nature of the adjustments to the financial statements may be different from those found in most audits, but the procedures for handling these are similar to those found in corporate registration statements. The certification of the financial statements would therefore appear to follow routine auditing procedures.

The definitions of municipal functions contain nonquantitative information and require that the auditor determine their reasonableness in terms of the consistency of definition from period to period and in terms of the completeness of the definitions. Since the scope of the audit does not require the auditor to determine whether the definitions represent the most logical set, it does not appear that the area of definitions should present any undue problems.

With regard to certification of the performance statistics contained in the municipal annual report, the auditor must ascertain the accuracy of the information in the various statements and the relevance of the information to the function being reported on.

Ascertaining accuracy of information is a familiar and routine task. The auditor investigates the source of the statistics being presented, such as departmental reports and other documents, and, through interviews, mailing of confirmations, and so on. Although the subject matter under investigation may be unfamiliar to the auditor, it is in the form of quantified data, and the techniques for determining accuracy follow accepted standard operating procedures.

Evaluating the relevance of the performance statistics as they relate to the function being described appears to present more of a problem. The procedure for handling this requires the auditor to express an opinion about whether the statistics being presented are related or unrelated to the funciton being described. If the auditor concludes that the performance statistics are totally unrelated, then his report should contain a disclaimer to that effect. However, in reality, such a situation would be rare because, normally, the reporting department would be willing to make necessary adjustments.

In summary, therefore, it appears that conventional auditing standards, somewhat modified to the needs of public agencies, would provide the necessary framework for auditing the proposed municipal annual report. Undoubtedly, more precise standards would evolve over time. The requirement for an indpendent audit would minimize political manipulation of the report and place greater confidence in the information which it contained. Finally, it should be pointed out that some cities are already quite advanced in their internal auditing methods, through the office of an independent comptroller, and these cities would have few, if any, problems performing the audit as visualized here (see, e.g., Beame, 1973).

AN EXAMPLE OF A DEPARTMENTAL REPORT

Ideally, an example of a departmental report should be based on real data from some municipality. Such data, however, are not readily available because of the restrictions that municipalities place on the release of the information. The example which follows for a city police department is largely illustrative, but it is based in part on some actual data available for the period 1970-1971 from the cities of Ann Arbor, Michigan, and Milwaukee, Wisconsin.

The departmental report consists of three statements: (1) statement of revenues and expenses; (2) statement of resource allocation by program; and (3) statement of performance by program.[6]

The statement of revenues, budget, and expenses should follow the same format for all departments. The major categories in the statement consist of (1) salaries, overtime pay, and allowances; (2) general and administrative; (3) equipment and supplies; and (4) miscellaneous. Table 2 is an illustrative example of a police department expense report and Table 3 illustrates the comparative ten-year experience (for another approach to departmental financial reporting, resource allocation by programs and statement of program results, see Lawrence, 1972). Figure 4 presents the police department

TABLE 2
ILLUSTRATIVE STATEMENT OF BUDGET AND EXPENSES
FOR A POLICE DEPARTMENT—FISCAL YEAR ENDING
DECEMBER 31, 1972

	Budget	Actual Expenses	Variance
Salaries			
Uniformed	15,600,000	15,300,000	−300,000
Civilian	3,200,000	3,650,000	− 50,000
Overtime	4,600,000	4,650,000	+ 50,000
Allowances	780,000	813,000	+ 33,000
General and Administrative			
General supplies	15,000	13,640	− 1,360
Office supplies	18,000	15,700	− 300
Office equipment	14,000	14,120	+ 120
Commissioner's Office	150,000	132,000	− 18,000
Public affairs	120,000	99,600	− 20,400
Equipment and Supplies			
Ordnance	65,000	59,400	− 6,600
Radio supplies	42,000	39,000	− 3,000
Laboratory supplies	38,000	38,540	+ 540
Gasoline and oil	636,360	718,210	+ 81,850
Tires	320,000	278,000	− 42,000
Vehicle repairs	148,000	137,000	− 11,000
Equipment maintenance	25,000	27,000	+ 2,000
Lab equipment	28,000	27,400	− 600
Radio equipment	31,000	30,875	− 125
Vehicles	260,000	279,500	+ 19,500
Miscellaneous			
Travel	11,000	10,850	− 150
Insurance	18,000	17,670	− 330
Subscription and dues	1,700	1,635	− 65
Other	14,000	8,150	− 5,850
total	$26,635,060	26,362,290	−272,770

allocation of resources by major departmental programs. In reporting
the allocation of such resources, it is important to consider what the
relevant resources are. The report may be in financial (cost) terms, or
it may be organized around other key parameters of resource
utilization. The illustrative example in Figure 4 presents police
department resource allocation as a function of direct labor costs
(salaries) and as a function of personnel utilization.

The reporting of performance results must reflect the operations
for which a department has direct responsibility. Therefore, the
departmental performance report must be specially designed. In the
case of a police department, the illustrative examples in Table 5A,

TABLE 3
ILLUSTRATIVE TEN YEAR COMPARATIVE
EXPENSE REPORT FOR POLICE DEPARTMENT

	Budget[a]	Estimated Actual Expenses 1973	Actual Expenses	
			1972	1971-1963
Salaries				
Uniformed	$17,100,000		15,300,000	
Civilian	4,200,000		3,650,000	
Overtime	4,700,000		4,650,000	
Allowances	820,000		813,000	
General and Administrative				
General supplies	15,000		13,640	
Office supplies	17,000		15,700	
Office equipment	16,000		14,120	
Commissioner's office	140,000		132,000	
Public affairs	110,000		99,600	
Equipment and Supplies				
Ordnance	62,000		59,400	
Radio supplies	40,000		39,000	
Laboratory supplies	39,000		38,540	
Gasoline and oil	725,000		718,210	
Tires	300,000		278,000	
Vehicle repairs	140,000		137,000	
Equipment maintenance	29,000		27,000	
Lab equipment	28,000		27,400	
Radio equipment	32,000		30,875	
Vehicles	295,000		279,500	
Miscellaneous				
Travel	11,000		10,850	
Insurance	18,000		17,670	
Subscriptions and dues	2,000		1,635	
Other	11,000		8,150	

a. Fiscal year beginning January 1, 1973.

5B, and 5C summarize frequencies of offenses and arrest records; police response time by precinct; stolen property and recovery; and summary of offenses investigated and arrest records by precinct. The departmental reports do not include those social indicators that reflect the mission of a department, because they are not under the direct control of a department. For example, FBI crime statistics which reflect on the quality of law enforcement in the city are not solely a by-product of the actions of the police department. Other factors—such as economic, social, and environmental—may be involved, as well as the operations of other agencies—such as the courts. Therefore, social and economic indicators should be reported for the whole city in separate reports.

TABLE 4
ILLUSTRATIVE EXAMPLE OF A RESOURCE ALLOCATION REPORT FOR A POLICE DEPARTMENT FOR THE YEAR 1972

	Personnel:	Civilians	Radio Patrol Cars	Foot Patrol	Detectives	Tactical Patrols	Parole Officers	Juvenile	Administrative	Public Affairs	Direct Labor Cost (Salaries and Overtime)
Precincts											
1											
2											
3											
4											
5											
Investigation of Offenses											
Criminal homicide											
Rape											
Robbery											
Assault I											
Burglary											
Larceny											
Auto theft											
Assault II											
Arson											
Forgery and counterfeiting											
Embezzlement											
Stolen property											
Malicious destruction											
Weapons violation											
Prostitution and commercialized vice											
Sex offenses											
Narcotics laws											
Gambling											
Against family and children											
Driving offenses											
Liquor laws											
Drunkenness											
Disorderly conduct											
Vagrancy											
All other offenses											
Deceased person											
Traffic Control											
Administration											
Communication center											
Identification											
Public affairs											

TABLE 5

**ILLUSTRATIVE EXAMPLE OF PERFORMANCE REPORTING
FOR A POLICE DEPARTMENT FOR THE YEAR 1973[a]**

A. Classification of Offenses	Offenses Reported	Unfounded	Actual	Cleared by Arrest	Not Cleared
Criminal homicide	1			1	
Rape	26		26	11	15
Robbery	213	2	212	63	148
Assault I	170		164	87(1)	77
Burglary	3033	7	3026	248	2778
Larceny	6040	32	6008	1096	5012
Auto theft	342	22	320	51	269
Assault II	642		642	466(3)	179
Arson	42		42	9	33
Forgery and Counterfeiting	1117	7	1110	505(3)	608
Embezzlement	9	1	8	5	3
Stolen property	69	1	68	48	20
Malicious destruction	871		871	101	770
Weapons violation	67		67	57(2)	12
Prostitution and commercialized vice	2		2	3(1)	
Sex offenses	83		83	31(3)	55
Narcotics laws	309	6	303	24(13)	69
Gambling	1		1	2(1)	
Against family and children	22		22	23(3)	2
Driving offenses	399		399	343(3)	59
Liquor laws	68	1	67	56(6)	17
Drunkenness	318		318	312	6
Disorderly conduct	107	1	106	75(2)	33
Vagrancy	2		2	2	
All other offenses	1895	20	1875	1490(8)	393

a. Adapted from Ann Arbor Police Department Report.

B. Precincts	Average Response Times (Minutes)	Property Value: Stolen $	Recovered $
1			
2			
3			
4			
5			

TABLE 5 (Continued)

Precincts:	1		2		3		4		5	
C. Classification of Offenses	Actual Offense	% Cleared by Arrest	Actual Offense	% Cleared by Arrest	Actual Offense	% Cleared by Arrest	Actual Offense	% Cleared by Arrest	Actual Offense	% Cleared by Arrest
Criminal homicide										
Rape										
Robbery										
Assault I										
Burglary										
Larceny										
Auto theft										
Assault II										
Arson										
Forgery and counterfeiting										
Embezzlement										
Stolen property										
Malicious destruction										
Weapons violation										
Prostitution and commercialized vice										
Sex offenses										
Narcotics laws										
Gambling										
Against family and children										
Driving offenses										
Liquor laws										
Drunkenness										
Disorderly conduct										
Vagrancy										
All other offenses										
Deceased person										

IMPLEMENTATION AND POLITICAL FEASIBILITY

The discussion of the feasibility of annual reporting by municipalities would be incomplete without an analysis of the implementation process and of factors determining political feasibility. Often the failure of efforts to implement public and business policies proposals can be traced to the failure of their proponents to consider the political and organizational feasibility of their proposals or to an inability to create the political coalitions necessary for the adoption and implementation of these policies (Rogers, 1971). In this section, we will examine the feasibility of municipal annual reporting by examining the organizational and political factors which affect eventual implementation.

The initial acceptance of the concept of municipal annual reporting will be facilitated by avoiding major departures from existing municipal record-keeping procedures and organizational practices. It should be recognized that the accounting principles, disclosure requirements, and current accounting and auditing practices underlying the present corporate annual report have been evolving continuously since 1933, when reporting and disclosure requirements were mandated by the Securities and Exchange Act. We expect that a similar evolution of appropriate public accounting principles, disclosure requirements, reporting objectives, auditing procedures, and so on, will occur in the field of municipal reporting.

The initial municipal report may not correspond to the ultimate desired municipal functions to be presented in a report. For example, in most municipal systems, the safety and law enforcement function may involve a number of departments (such as the police department, transit police, sanitation police, fire department arson investigator, and so on). However, in the initial phase, it may be impossible to aggregate expenditures within and across departments and to associate them with a single function. Furthermore, an acceptable definition of municipal functions can only come about from an ongoing exchange between the various report users and the municipal executive branch.

In the initial phase, the content of the municipal annual report will depend on the information available in the city's ongoing transaction-processing system. The recent experience of a number of local governments in redesigning and computerizing their accounting, record-keeping, and reporting functions suggests that it makes no sense to design a report which requires a municipality to design an information-processing system solely for the purpose of preparing this report. Therefore, during initial implementation, the report should use data already in the municipality's information system. It is expected that communication between report users and the municipal executive will give rise to a reporting system that incorporates new categories of information and procedures for the creation and management of municipal data bases. The important thing, however, is to establish an initial reporting system and to avoid placing on it such burdens as requiring changes to existing municipal information systems which would render the publication of an annual report infeasible.

Preparation of the annual report makes it necessary for the municipality to define its reportable service units. For example, a municipality may structure its report by departments or super-agencies (which contain several departments), or according to some other organizational or program structure. It can be expected that program performance criteria will develop over time in accordance with the needs of report users subject to the capacity of the municipal information or the ability to redesign it.

We started this section by noting that policy alternatives must be organizationally and politically feasible. Clearly, implementing the concept of a municipal report will encounter political difficulties.

A review of research on governmental bureaucracies suggests that the unresponsiveness of bureaucracies to their publics is largely an unanticipated outcome of the democratic process. In this process, it is assumed that direct election of a mayor (or the city's Chief Executive Officer), the City Council, and sometimes an independent comptroller, will result in bureaucratic responsiveness. It appears, however, that the stability of municipal decision processes and the desire of the bureaucracy for independence from external control often results in public bureaucracies being largely unresponsive to their publics.

Bureaucracies have been described as being motivated to secure independence from outside control (Leiserson, 1947; Freeman, 1955). Thus, administrative agencies seek and often obtain statutory, organizational, and fiscal independence. For example, educational agencies on both the local and state level have been successful in securing such autonomy (Eliot, 1959; Martin, 1963; Moos and Rourke, 1959; Rogers, 1968).

Governments have been described as consisting of a collection of quasi-independent agencies and departments depending on legislative bodies for resources (Thompson and McEwen, 1958). The city of New York, for example, has eleven or twelve superagencies, encompassing over one hundred departments. To maintain its independence, every agency or department must secure support for its programs and activities from the legislative bodies which control the allocation of fiscal resources and policy decision-making. In the case of a city agency, the commissioner looks to the mayor for support of his programs; however, he must also cultivate the support of lower-level bureaucrats. In some instances, however, an agency may strive for and achieve independent political power by obtaining strong legislative support. Maass (1951) and Stein (1952) have described the virtual independence from hierarchical control of the Corps of Engineers and the FBI, respectively. These, however, are rare exceptions.

Hierarchical control can depend on the ability of the chief executive (president, governor, mayor) to exercise control and on the administrative structure that he creates for policing the activities of agencies under his jurisdiction. However, research on administrative decision-making suggests some of the reasons why hierarchical control can at best achieve limited effectiveness. Lindblom (1959) has argued that policy decisions are made when various decision makers reach concensus among themselves and do not come about from an exhaustive examination of alternative policies and their potential outcomes. Furthermore, the capacity of the organization to analyze problems is severely limited, and decisions are therefore incremental. The incrementalism of decision-making, the resulting inertia of the internal decision-making, and the resistance of the bureaucracy to change have been further described by Gore (1956), Wildavsky (1961), Crecine (1967), and Gerwin (1969). Crecine (1967) has demonstrated the stability of municipal resource allocation and the incrementalism and stability of the budget process. He has also noted the limited control the mayor has over the requests for appropriations by city departments and the limited power that the city council has over approval of the budget.

It would seem that, because of the incremental nature of government decision-making, the politics of the process, and the resulting bureaucratic inertia, public officials resist closer scrutiny by their publics. Indeed, in a series of interviews with senior officials in local governments, Cahill (1973) has noted strong support for the principle of requiring municipal annual reports, and major concern that it would be politically difficult to implement.[7]

In considering the political contingencies which could determine the adoption of the municipal annual report concept, we believe that this is an idea whose time has come. The crisis of the 1970s is one of a general lack of confidence in government. In large urban areas, this crisis is manifest in the unresponsiveness of the bureaucracy and the breakdown in the delivery of municipal services. The pressure is on to make government more responsive and accountable to its public.

Mandating municipal annual reporting could, just as in the case of the corporate report, provide citizens with the information necessary to evaluate the effectiveness with which municipal management administers its stewardship.

There are many political factors and situations which could make the concept of a municipal annual report politically feasible. Ideally, it should be first demonstrated as feasible in an experimental situation. Such an experiment could be funded or stimulated by one of the large foundations or by an agency of the federal government. The concept of accountability is of concern to the federal government in view of its many and large grant and incentive programs and its recently inaugurated revenue-sharing program. It would seem that the federal government and Congress can exercise pressure on local governments to begin to issue municipal reports. It could also underwrite, through the General Accounting Office (GAO), experiments on annual reporting by municipalities. Finally, Congress could require annual reporting by municipalities as a condition for receiving noncategorial funds. A similar situation can occur at the state level between the state government and municipalities in the state. In New York State, for example, the state government has repeatedly intervened in the affairs of New York City by auditing the performance of a number of city agencies.

It is also possible that an annual report could be implemented or experimented with by a municipality on its own, or as a result of local pressure. However, this would not lead to wide adoption of the concept.

In all these situations, it must be noted that the success of this concept also depends on acceptance by the potential users of the report. In the case of the corporate report, it is the Securities and Exchange Commission (SEC) which regulates and monitors the corporate disclosure process. It may be necessary to establish a similar agency at the federal or the state level. Such an agency would be mainly concerned with regulating disclosure practices and stimulating the development of accounting principles for municipal reporting. At the federal level, the GAO could provide the necessary staff support to such a legislative agency.

As in the case of the corporate report, issuing the municipal report will be a futile exercise if its potential users fail to make use of it. The investing public relies on a variety of financial sources and research services for the in-depth evaluation of corporate results. The use of the municipal annual report would be similarly enhanced if it were used by citizen action groups, university research centers, and possibly by an urban annual report ombudsman. Without such active political use of the annual report, it is doubtful whether it can become the means for greater responsiveness and accountability by municipal governments.

NOTES

1. The types of funds usually recognized in municipal accounting include the General Fund, Special Revenue Funds, Debt Service Funds, Capital Project Funds, Enterprise Funds, Trust and Agency Funds, Intragovernmental Service Funds, and Special Assessment Funds.

2. It is worth noting that the proponents of the social matrix concept (Terleckyj, 1970; Charnes et al., 1973) have not yet reported procedures for developing the matrix coefficients, much less for providing an independent attestation to their numerical values.

3. An example of an urban simulation whose assumptions and conclusions have been questioned is contained in the work on urban dynamics by Forrester (1969).

4. Generally, however, revenues may not be a meaningful statistic for most departments.

5. A useful starting point is the Economic Report of the President prepared annually under the direction of the Council of Economic Advisors for the U.S. Congress. However, a national economic report will certainly contain categories of information that do not apply to city governments (e.g., defense information) or that are of little interest to the residents of some cities. The national report would have similar disadvantages.

6. The city of Pittsburgh as part of the mayor's budget message for 1972 published a municipal financial report. The report, however, does not present statements of resource allocations by departmental programs or statements of performance by program.

7. Among others, see Cahill's reports on interviews with the Honorable Alfred Del Bello, mayor of Yonkers, New York, the Honorable Robert Wagner, ex-mayor of New York City, and Mr. Robert Postel, City Council, New York City.

REFERENCES

American Institute of Certified Public Accountants (1963) Statement on Auditing Procedure No. 33. New York.

BAKER, J. H. (1971) Urban Politics in America. New York: Charles Scribner's.

BANFIELD, E. C. and J. Q. WILSON (1966) City Politics. New York: Vintage.

BAUER, R. A. (1966) Social Indicators. Cambridge, Mass.: MIT Press.

BEAME, A. (1973) "Comptrollership in a major U.S. city." Armed Forces Comptroller 18 (Spring): 13-16.

BREAK, G. F. (1967) Intergovernmental Fiscal Relations in the United States. Washington, D.C.: Brookings Institution.

CAHILL, G. A. (1973) "An investigation into the feasibility of annual reporting by municipalities." Ph.D. dissertation. New York University.

CHARNES, A., W. W. COOPER, and G. KOZMETSKY (1973) "Measuring, monitoring and modeling the quality of life." Management Sci. 19 (June): 1172-1188.

CRECINE, J. P. (1967) "A computer simulation model of municipal budgeting." Management Sci. 13 (July).

DERTHICK, M. (1970) The Influence of Federal Grants. Cambridge, Mass.: Harvard Univ. Press.

DOFFMAN, R. [ed.] (1965) Measuring Benefits of Government Investments. Washington, D.C.: Brookings Institution.

ELIOT, T. H. (1959) "Toward an understanding of public school politics." Amer. Pol. Sci. Rev. 53: 1032-1051.

FORRESTER, J. W. (1969) Urban Dynamics. Cambridge, Mass.: MIT Press.

FREEMAN, J. L. (1955) The Political Process, Executive Bureau-Legislative Committee Relations. Garden City, N.Y.: Doubleday.

GERWIN, D. (1969) "A process model of budgeting in a public school system."

GORE, W. J. (1956) "Decision making in a federal field office." Public Admin. Rev. 16: 281-291.

LAWRENCE, C. (1972) "The study of a program budget for a small city." J. of Accountancy 68 (November): 52-57.

LEISERSON, A. (1947) "Political limitations on executive reorganization." Amer. Pol. Sci. Rev. 41: 68-84.

LINDBLOM, C. E. (1959) "The Science of 'muddling through.' " Public Admin. Rev. 19: 79-88.

LOOMIS, C. J. (1973) "An annual report for the federal government." Fortune 43 (May): 193-199, 322, 324.

MAASS, A. (1951) Muddy Waters. Cambridge, Mass.: Harvard Univ. Press.

MARTIN, R. C. (1963) Government and the Suburban Schools. Syracuse: Syracuse Univ. Press.

MITAU, G. T. (1966) State and Local Government: Policies and Processes. New York: Charles Scribner's.

MOOS, M. and F. E. ROURKE (1959) The Campus and the State. Baltimore: Johns Hopkins Press.

National Committee on Governmental Accounting (1951) Governmental Accounting Auditing and Financial Reporting. Chicago.

PERLOFF, H. S. and L. WINGO, Jr. [eds.] (1968) Issues in Urban Economics. Baltimore: Johns Hopkins Press.

ROGERS, D. (1971) The Management of Big Cities. Beverly Hills: Sage Pubns.

――― (1968) 110 Livingston Street. New York: Random House.

ROSSI, P. H. and W. WILLIAMS (1972) Evaluating Social Programs. New York: Seminar.

SCHULTZE, C. (1967) Statement in Hearings on Planning-Programming-Budgeting. U.S. Congress, Senate Committee on Government Operations, Ninetieth Congress, First Session.

SHELDON, E. B. and K. C. LAND (1972) "Social reporting for the 1970s." Policy Sciences 3 (July): 137-151.

STEIN, H. [ed.] (1952) Public Administration and Policy Development. New York: Harcourt, Brace.

TENNER, I. (1955) Municipal and Governmental Accounting. Englewood Cliffs, N.J.: Prentice-Hall.

TERLECKYJ, N. E. (1970) "Measuring progress towards social goals: some possibilities at national and local levels." Management Sci. 12 (August): B765-B783.

THOMPSON, J. D. and W. J. McEWEN (1958) "Organizational goals and environment: goal-setting as an interaction process." Amer. Soc. Rev. 23: 23-31.

U.S. Comptroller General (1972) Standards for Audit of Governmental Organizations, Programs, Activities, and Functions. Washington, D.C.

U.S. Department of Health, Education and Welfare (1970) Toward a Social Report. Ann Arbor: Univ. of Michigan Press.

U.S. Environmental Protection Agency (1973) "The quality of life concept—a potential new tool for decision-makers." Office of Research and Monitoring Environmental Studies Division, March.

WILDAVSKY, A. (1961) "Political implications of budgetary reform." Public Admin. Rev. 21: 183-190.

4

Evaluating Organizational Change in Public Services

DENNIS R. YOUNG

INTRODUCTION

> *Economists, of course, have always had theoretical reasons for questioning the organization of public services. In general, they are produced under monopolistic conditions. The consumer can rarely shop around among public schools or hospitals or welfare departments. Even if he could, he would have scant information by which to measure and judge the quality of health or education or other social services. Conservatives have always made these points in arguing for less public activity and more reliance on the private market. But more recently, the liberals, who generally favor more public activity, have been among the sharpest critics of public services.*
>
> Alice M. Rivlin [1971]

☐ AS DR. RIVLIN SUGGESTS, public policy makers and analysts have become freshly aware, in recent years, that more effective delivery of public services often requires more than just the discovery

AUTHOR'S NOTE: *I would like to acknowledge the Urban Public Finance Group of the Urban Institute, under the direction of Harold Hochman, and the Program for Urban and Policy Sciences, State University of New York at Stony Brook, under the direction of Robert Nathans, for support of this work. I would* and implementation of better programs or production methods. For what increasingly seems necessary are improvements in the organizational arrangements under which production and consumption decisions are made. Thus, consideration is now given to alternatives such as "privatization" of municipal garbage collection, as well as better route design or manpower scheduling. In the medical field, health maintenance organizations and prepayment for care have been the subject of policy discussions as well as improvements in medical technology. And, in education, the emphasis has shifted from a search for better teaching techniques to organizational innovations such as voucher systems, community control, and performance-contracting.

The underlying reasons for the shift in emphasis from new methods to changes in organization are fairly clear. We are learning that most important decisions in service delivery, and even the impetus for change, must come from within the service sector, not from research agencies or technical consultants. Even for services associated with auxiliary research and development (R & D) sectors, such as health care or transportation, the day-to-day decisions on what methods to use, which customers to serve, and what quality of services to provide, are largely made by practitioners, regulators, customers, and other key decision makers from an economic

like to thank T. Owen Carroll of the Urban Science Program at Stony Brook, and Alan Weisberg, Joe Nay, John Scanlon, and Peter Bloch of the Urban Institute, for their helpful comments on this paper. I also want to thank George Peterson and Harvey Galper of the Urban Institute, Ned Gramlich of the Brookings Institution, and Richard Nelson of the Department of Economics at Yale University, for their constructive suggestions on an earlier paper ("Evaluation of Organizational Change: The Case of Performance Contracting In Education," International Institute of Public Finance, Columbia University, September 1972) in which some of the ideas developed here were first presented. In addition, I want to express my appreciation to Richard Holt of the Office of Economic Opportunity, Frank Overlan of the Center for the Study of Public Policy at Harvard University, Judith Areen of Georgetown Law School, Stephen Wiener of the Rand Corporation, and Paul Hutchinson of the Alum Rock School District Voucher Project, for information and suggestions they supplied me on the OEO experiments in education.

Finally, I would like to note that although this paper contains some critical remarks with respect to certain analytical procedures carried out by OEO and the Rand Corporation in evaluating experiments in education, I would like also to note my respect for the competence of the analytical staffs of these organizations and the importance of their good work in public services. In view of the current political jeopardy of OEO, it seems particularly important to urge that such efforts be allowed to continue.

standpoint—e.g., if the incentives they face encourage inefficiency or maldistribution of services—then the resulting pattern of service delivery will tend to become, or remain, unsatisfactory—especially where the supply system is subject to increasingly severe demands over time. In particular, consumers will tend to use services inefficiently, and, of possibly greater concern, suppliers will tend to do a poor job of selecting and administering the programs through which they provide services.

While there is a dearth of conclusive research on the question, there is some empirical evidence that the effectiveness of programs is highly dependent on nonprogram variables (with organizational variables an important subset of these). Researchers find that the "same programs" administered in different places, by different people, in different agencies, in different organizational contexts, produce widely varying results. Sometimes the variations are so dramatic as to dwarf any "average" differences in across-the-board performance of alternative programs. For example, this phenomenon is clearly demonstrated in a recent Urban Institute study of cost/benefit analyses of federal manpower programs (Nay et al., 1971), which found that differences in average effectiveness (benefit/cost ratios) of alternative manpower programs—Neighborhood Youth Corps, Concentrated Employment Program, Job Corps, and others—were small compared to the within-program variations stemming from differences in effectiveness of individual "projects" administering nominally the same programs. Part of the in-program variations are undoubtedly explainable by differences in project clientele and local economic conditions. But it seems likely that other variables, factors which we shall argue are at least partially dependent on organizational arrangements, such as the competence and motivation of personnel, and the sensitivity and responsiveness of decision makers to local circumstances, were also responsible.

Despite a new awareness that the effectiveness of service delivery is sensitive to organizational factors, the question of how one should *evaluate* (past and proposed) organizational changes in the delivery systems of public services seems not to be well understood. It is our purpose in this paper to consider some crucial issues pertinent to structuring evaluations of organizational changes. This purpose is motivated both by the realization that organizational change is increasingly being sought as an avenue of reform in public service delivery, and the belief that there are crucial differences between evaluating programs and evaluating organizational arrangements, ignorance of which risks improper design of evaluation efforts and, hence, ill-conceived inferences and policy decisions regarding proposals for organizational change. In order to pursue our objective, it is first necessary to develop more precise notions of the key concepts—program and organization—and an understanding of how these concepts are related to one another.

PROGRAMS AND ORGANIZATION

By our definition, a program is a *means*—i.e., a production method or technology—for delivering a service; it is a part of a total production function for transforming resource inputs into service outputs. A complete production function incorporates many programs. In a school, for example, programs include the (combinations of) methods used for teaching, such as computer-assisted instruction, open classrooms, or conventional classroom instruction. Examples of programs in other areas include the methods or technologies used for timetable and route development in bus transit, patient diagnosis, monitoring, and care in hospitals, route and manpower scheduling in waste collection, or offender counseling and rehabilitation in criminal corrections. In short, programs are the immediate instruments which are manipulated in the process of carrying out the activities required in delivery of a service. As such, programs do *not* solely determine the effectiveness of service delivery, since much depends on how programs are selected and administered. This, in turn, is influenced by organization, as we shall soon discuss.

The concept of a program can be illustrated with the use of the conventional "production possibility" curve or "frontier" of microeconomic theory. In Figure 1, the solid convex curve AA' represents such a frontier; the frontier itself can be interpreted as the set of possible outputs (O_1, O_2) of a particular program or production method if it operated at maximum internal efficiency within a fixed resource budget. Points within the frontier are inefficient from a production standpoint. The output dimensions (O_1, O_2) represent a product "mix" of a given service—e.g., achievement in reading and math in education; or these dimensions may represent the level of service at the same or different quality levels delivered to different groups of recipients. (In reality, there would be many such dimensions of output, not just two.)

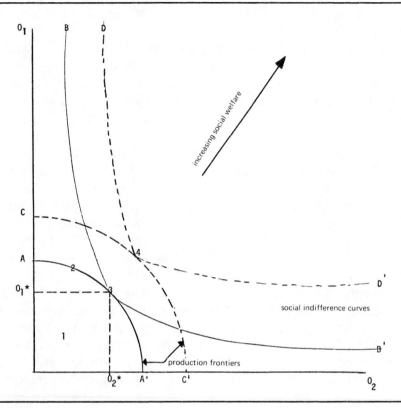

Figure 1.

In considering the production frontier representation, the key concepts are that: (a) there is a *range* of possible service outputs produceable under a given program all contained *within* the production frontier; (b) operation anywhere on the frontier (AA′) itself is maximally efficient from a production point of view—i.e., a selected output mix on this curve could not be produced with fewer resources under the stipulated technology (program) represented by the frontier; and (c) that the output mix can be varied (along the frontier) to conform with consumption preferences of consumers or society collectively. We can represent social preferences by a series of concave indifference curves each representing a particular (constant) level of social welfare. The solid concave curve BB′ tangent to the production frontier at point 3 in Figure 1 represents the highest social indifference curve achievable with the stipulated program. Thus, if production were executed at point 3 (with an output mix $O_1 * O_2 *$) services would be both maximally efficient in production, since output falls on the production frontier, and also maximally responsive to social preferences, because it achieves the highest possible value of social welfare.

The production curve construction illustrates why the program itself, as represented by the production frontier, is not solely responsible for ultimate results. The program merely defines a range of possibilities. Within that range, service delivery may be internally inefficient, as at point 1 in the diagram, or efficient in production but unresponsive to demand, as at point 2, or both efficient and responsive, as at point 3. Furthermore, while the selected program may be *presumed* to be the most effective one available, a better program or technology, existing or yet to be invented, may be found. That is, some better program, or program modification as represented by a new production frontier outside the original one may eventually be conceived (see the dashed convex curve CC' in Figure 1). Hence, operation at a point like 4 in the diagram may become possible, yielding greater welfare than was feasible earlier.

Thus the key question is, given a particular set of social preferences or "demands" for service, what influences selection of the operation point (1, 2, 3, or possibly 4) at which a service is provided?[1] The answer has two parts relevant to this discussion. First, the quality of input resources used in production, especially the quality of management but also other personnel and material resources, may differ among agencies providing the same service even under the same fiscal resource constraints. This will affect short-run decisions. For example, "better" principals and teachers will more competently select and execute education programs than will their less-gifted colleagues. And, in the long run, these superior personnel may be more successful in developing technological innovations (or in alleviating budget constraints by attracting more resources to the system), thus expanding the set of production possibilities. But, one must then ask, what is responsible for certain agencies attracting superior personnel? One important factor is the "reward structure" created by economic arrangements under which the service is organized; that reward structure determines whether (and how) good performance is encouraged and poor performance penalized. A market structure, for example, has vastly different implications for personal rewards than governmental supply.

Certainly, there are stochastic elements in any presumed sequence that dictates that effective program selection and administration is brought about by good personnel, and better personnel are attracted to rewarding organizational arrangements. Poor managers can hit upon good programs, and good managers can be found in unrewarding places. But it seems reasonable to expect that, in general, the efficacy of organizational arrangements will be reflected to some extent in the quality of inputs attracted.

Perhaps more importantly, however, the influence of organizational arrangements is not felt solely through the quality of resources, but also through the incentives that organization creates for how these resouces are used. It seems reasonable to assert that agencies having equal quality of personnel and other resources, but operating in different organizational environments, would still exhibit systematic differences in performance. To understand this, let us consider the nature of organizational arrangements more closely.

For purposes of discussion, organizational arrangements can be described in two parts. On the one hand, organization of "demand" gives expression to individual and collective preferences of citizens and consumers, and their satisfaction with present services. Demand is organized through financial arrangements and decision-making mechanisms ranging from individual consumer choice in the marketplace to collective processes of voting and governmental representation. Demand organization is the vehicle for notifying suppliers about the level and quality of services desired for consumption. Certainly, the demand "signals" differ under different organizational regimes. To provide a highly simplified illustration: where decision-making is vested in the individual consumer, through a system of purchasing in the market, demand will be expressed in economic terms and will depend upon consumers' aggregate willingness to pay as a function of price and quality; where consumption decisions are made collectively, on the other hand, effective demand will be a complex function of political decision-making mechanisms, including voting, lobbying, and log-rolling.

But articulation of consumption preferences is only one function of demand organization; another is the *leverage* provided to control the activities of suppliers. Where consumers have a choice of suppliers in a marketplace, changes in consumer patronage provide financial incentives for suppliers to respond to consumer preferences. Demand expressed collectively through political channels, on the other hand, may sensitize suppliers to be responsive to governmental or political leadership, and unresponsive to citizens as individuals.

Just as demand organization provides for the articulation and pursuit of public and private preferences for services, organization of *supply* determines how well suppliers respond to these preferences. Supply may be organized into various kinds of competitive and noncompetitive arrangements among profit-making, nonprofit, or governmental suppliers. The economic nature of a supplying agent —i.e., whether it is profit-making or not, delineates the objectives and legal constraints of that agency. The supply *environment,* on the other hand, whether competitive, regulated, contractual, or internal to government, will in conjunction with demand organization, provide the external incentives that influence the behavior of the supply agent within the confines of its charter. A private profit-making firm, for example, can be expected to behave differently under competition than under regulation; the same holds for other types of supply agencies.

It should be clear, then, that the effectiveness of service delivery will not solely depend on technology or program selection, nor even on the quality of resources devoted to provision. The same personnel, for example, can be expected to perform differently or face different constraints and opportunities, in different organizational environments. Thus, in the short run, organizational arrangements can be expected to affect how efficiently and responsively services are provided, within the current set of program alternatives and resources. In the longer run, of course, organization will affect how well service performance is maintained and improved, in particular whether suppliers learn to evaluate and adjust or replace their programs, or whether they allow operations to deteriorate. When conditions change—i.e., when there is a technological break-through or a shift in the values or tastes of consumers, the motivational implications of organization may become especially important; for organization will influence whether programs are effectively modified or replaced in order to adapt to new circumstances. Further than this, organization will affect whether suppliers are motivated to *create* as well as simply react to change, by engaging in research and development. To recapitualte, these various longer-run effects of organization can work through different channels—by affecting the quality of (personnel) resources drawn to the service, as well as by providing the operating incentives within which these resources work.

For analytical purposes, we have attempted to draw a sharp distinction here between organization and the resources and programs that depend in part on organization. It must be admitted, however, that such distinctions are sometimes blurred in real circumstances, especially in cases of public legislation that set up new machinery for services administration and financing, as well as imposing a particular program methodology. The Headstart Program, which introduced a particular approach to preschool education and child care services, as well as a new set of organizational units to deliver these services, is an example of this. Such cases make the job of evaluating new enterprises more difficult. For one must decide whether it is the new organizational arrangement one is analyzing or the new program or perhaps both. Even aside from such cases, it is the inherently close and interwoven relationships of program, resources, and organization that make it so difficult to evaluate organizational changes per se. But, however complex, this question is singularly important if evaluation efforts are to serve as useful guides for future decisions regarding organizational reform.

IMPLICATIONS FOR EVALUATING ORGANIZATIONAL CHANGE

The primary goal of a scientific evaluation is to determine the effects (on output or performance) of changes in particular policy variables of interest. One mark of a competent evaluation is how convincingly it eliminates alternative explanations of results it attributes to the policy variables under test. Thus, in our case, evaluations of organizational change should convincingly eliminate alternative explanations of results attributed to organizational factors.

In addition to ensuring valid inference of results, evaluation designs may also attempt to accommodate the goals of providing as much information as possible with respect to the effectiveness of *varieties* of organizational designs, and to maximize the *generalizability* of results to circumstances outside the domain of the experiment or evaluation itself. In the following discussion, we will focus on a number of elements of experimental design which seem particularly crucial to achieving these goals.

WHAT IS TO BE VARIED AND WHAT IS TO BE CONTROLLED?

In order for any evaluation to be successful in determining the impacts of particular policy variables on dependent output variables, and to determine what settings of the policy variables produce the most effective results, it is necessary that (a) extraneous variables be "controlled" or held constant while (b) policy variables are systematically varied. To evaluate organizational changes, it can be generally agreed that certain kinds of variables ought, if possible, to be controlled—for example, the social environment in which services are delivered, the socioeconomic character of the target clientele, and the particular time in history over which delivery takes place. These are all extraneous factors which must be controlled if their effects on service performance are to be separated from the effects of policy variables.

On the other hand, it can also be generally agreed that it is necessary to fix or systematically vary certain policy variables —namely, organizational parameters, among sample points or evaluation sites. Systematic variation allows the determination of differences in performance of services produced under radically different organizational arrangements or can help derive information on how particular arrangements may be "fine tuned" or "optimized." For example, in evaluating the concept of performance-contracting in education, one can compare schools using performance contracts with otherwise similar schools using conventional arrangements. In addition, one can also compare results among similar schools using different *variants* of performance contracts to see if some forms of contracts are more effective than others.

Thus far, our discussion is a straightforward application of experimental design theory (see Campbell and Stanley, 1972). The problem comes in attempting to develop an evaluation design that will *separate* the effects of two different kinds of policy variables we have discussed—*program* variables and *organization* variables. In particular, it must be resolved whether, in an evaluation of organization, program variables are to be held fixed or systematically varied, or left uncontrolled. Here we face an apparent *dilemma.* If program variables are left uncontrolled, and suppliers operating under different organizational arrangements are allowed to choose whatever methods they want to achieve results, then how is one to separate the effects of organizational arrangements from the effects of the selected programs? On the other hand, if we hold program choices fixed—by imposing a particular delivery method or a systematic set of variations of delivery methods upon particular suppliers operating under different organizational arrangements, we do not allow organizational arrangements to work as they are supposed to. In particular, suppliers will not be permitted, within the context of the incentives they face, to reevaluate their methods and change course in response to perceived problems in performance; yet the inclination and capability for such evaluation and adjustment behavior will depend on incentives implicit in the organizational arrangements. Thus, if program choices are controlled, only a *very limited* evaluation of the short-run effectiveness of an organizational regime can be obtained. The crucial long-run effects of organization—on program choice, innovation, and change—would be blocked. Thus, it seems necessary to opt for an evaluation strategy which does *not* tightly control program choices. *Other* means, especially the selection of an appropriate time period for conducting the evaluation, must be exploited to help separate the effects of program and organization.

THE TIME FRAME FOR EVALUATION

If program choices are to be left relatively unconstrained, the choice of an appropriate time period over which an organizational arrangement is to be evaluated becomes especially important. In this regard, it is necessary to consider what the critical "time constants" are, to borrow an engineering term, for various adjustments a delivery system is required to make over a period of operation. (These adjustments are associated with normal operation and are separate from other effects related to experimentation per se, such as the Hawthorne phenomenon.) The rationale here is that, in order to evaluate an organizational arrangement relative to particular performance capabilities, enough time must be allowed to accommodate these time constants.

Start-up. If a new organizational arrangement is implemented or if a major innovation in programs is undertaken, time is required for the delivery system to master its methods. Only after some minimum adjustment time will it be appropriate to evaluate how efficiently and responsively the organization administers a given program, even over the short run. Conceivably, subsequent to the start-up adjustment period, a short-term evaluation could be made to assess organizational effectiveness in administering a particular program. However, such short-run evaluation would yield doubtful promise of separating the effects of organization from those of particular programs or personnel, unless the latter are controlled, a prospect which would severely limit the potential for evaluating the more important, longer-run effects of organization.

Learning and correction. A certain minimum time period, dependent on the nature of the service, is required for a delivery system to evaluate its own performance, through internal measurements or external signals, and then to adjust its operations accordingly. A service like garbage collection or police protection may require only a matter of months to evaluate complaints or losses in revenues, while services like education or criminal corrections may require several years to sense important effects and make adjustments. The objective for evaluation here is to designate an analytical time frame, long enough to permit systematic learning and adjustment to take place, over which data are collected.

Personnel turnover. As we discussed earlier, one possible explanation for a service's performance quality, at least over the short run, may be the quality of input resources attracted to production, especially that of key personnel. We also noted that, over the longer run, the quality of personnel will be dependent upon the reward structure implicit in the organizational regime. Thus, a critical parameter is the time expected for substantial turnover of personnel. This seems particularly important for new services which have been set up in an innovative or experimental mode. In such cases, the initial group of personnel is liable to constitute a somewhat elite, progressive, highly intelligent, or highly motivated cadre of "pioneers." A key question, then, for evaluating the effectiveness of an organizational regime is what happens when these pioneers leave and the system settles down to operations with a more typical cast of characters.

Innovation cycle. Finally, a key characteristic of service delivery by different organizational regimes is the ability and motivation to innovate, to find new and better programs, to evaluate these, to implement promising innovations, and to discard old programs which no longer seem effective. This is perhaps the longest and most uncertain time constant of interest here, and it is highly dependent on the state of knowledge and technology in the given service area.

In summary, it seems desirable to attempt to accommodate all the crucial time constants in the design of evaluations of organizational arrangements. This means, essentially, that the longest of these, probably the innovation cycle, is the determinant. Following such a strategy is more tricky than it may seem, however, since the magnitude of the time constants will depend not only on the inherent nature of the service, but also on the organizational arrangements themselves. Services organized under effective arrangements may start up quickly, exhibit rapid learning and corrective behavior, maintain a particularly stable pattern of personnel turnovers, and regularly produce innovations. Poorly organized services may behave quite differently. The key for evaluation is to allow at least enough time for the particular forms of adjustment to take place, under the assumption that the arrangements under evaluation are reasonably effective. Then, if the particular adjustments fail to occur, one may reasonably attribute such failure to the organizational arrangements themselves, rather than to the restricted time frame of the evaluation design.

Unfortunately, this ideal strategy must be modified by other important considerations, principally: (a) limitations of resources available for the evaluative effort; and (b) the need to produce timely evaluation information for use in imminent policy decisions.

In the (likely) event that only limited resources are available for evaluation, or that evaluation information must be produced quickly for decision-making purposes, one must note both the threats to valid inference caused by restricting the time frame and the positive advantages of extending it. These must be weighed against the research and opportunity costs (of failing to affect short-run decisions) involved in evaluation over a lengthy period.

The major danger in restricting the evaluation time frame is to confound the effects of organization with the effects of particular program choices (or personnel). We have noted earlier the distortions that controlling program choices would have on determining the performance capability of organizational regimes. But allowing these variables to "float free" necessitates the longer time frame for evaluation, to allow the opportunity for initially poor program choices or designs to be recognized and corrected (or to permit possible deterioration of initially good choices).

The positive advantages of extending the time frame for evaluation follow directly from the definition of the critical time constants. Restricting evaluation to only a short period after start-up and before significant personnel turnover permits (likely ambiguous) measurement of only short-term productive efficiency and responsiveness of the organizational arrangement, under unusual "pioneer" conditions. Extending the time frame to cover the learning and correction time constant permits observations of the dynamic self-evaluating capability of an organizational regime. Covering the innovation cycle time would permit an even greater perspective or long-run dynamic behavior.

WHAT IS TO BE MEASURED?

Our earlier "production frontier" discussion suggests that several different types of measurements are necessary to evaluate organizational effectiveness. These include the cost and output levels of production in order to gauge production efficiency, and the match of output mix to the apparent preferences of citizens and consumers. As it is our intention here to consider aspects of evaluation uniquely relevant to organizational arrangements, we will not digress on the well-known, though no less vexing problems of output definition and measurement in public services. We will merely state here that quantitative data, representing variables which can be generally agreed to define service outputs, are often difficult to find (see Hinrichs and Taylor, 1969).

In addition to measuring production efficiency and short-term responsiveness to demand, the longer-run dynamic capability of the system must be measured. Thus, the contingencies of moving to a more efficient technology (frontier), or adjusting to production problems or changes in consumer preferences over time, must be recognized by measuring the *direction* and *rate of change* of outputs and costs *over time,* as well as absolute levels at a particular time.

Various pitfalls and hazards of experimental evaluation make it advisable to complement measurement of outputs and costs with other types of measurement. First, as we have mentioned, output variables may be difficult to define or to measure; thus, what we may be able to measure quantitatively may not truly correspond to our concept of performance. For example, reading and math scores neglect many other goals of public education, and keying on these indicators may warp evaluation conclusions. Second, our time frame for evaluation may of necessity be so constricted as to render very difficult the separation of program (and personnel) effects from the impacts of organizational variables. Third, evaluations in controlled social settings will more often than not be infeasible or imperfect in a scientific sense (Weiss, 1972). Hence, output and cost measurements may be confounded not only by the effects of internal (program and personnel variables, but also by external variables such as socio-economic setting and clientele characteristics.

There is no way to totally avoid these problems. To the extent that satisfactory output measurements and costs can be obtained over an adequate time frame under controlled circumstances, they should be so obtained. In circumstances lacking scientific control, or where time frames are constricted or output measures problematic, such information remains useful, but must be put in greater perspective since the cause and interpretation of performance changes can no longer be unambiguously derived. And since ideal experimental conditions are rarely achieved, other, "softer" data seem warranted as a complement in most if not all evaluations of organizational arrangements. Several different types of such "soft" information seem desirable. The first is a documentation (chronology) of various types of decisions and events including: (a) adjustments in operations designed to improve production efficiency, (b) changes in output mix and volume designed to respond to external (consumer) preferences, (c) changes in the type and quality of personnel, and (d) innovations in program or technology. Documentation of these occurrences would be made for the purpose of determining whether the kinds of events critical to successful organizational performance actually take place. Such documentation would not itself yield much information about the quality of the various decisions that triggered these occurrences. For this kind of insight, it seems necessary to look further into the *mechanisms* that exist for making decisions under different organizational regimes. In particular, to find out: (a) what events or "signals" provide the impetus for adjustment decisions; (b) what information is obtained by decision makers for basing adjustment decisions; (c) how does this information come to the attention of decision makers; (d) what controls are available for implementing the various levels of decisions and monitoring their impact; (e) what structural constraints interfere with implementation or monitoring. Such questions are intended to determine the existence and quality of the necessary "feedback" mechanisms of self-evaluation and correction (Young, 1971), and the "search and scan" mechanisms for innovation and change (Nelson, 1971), implied by organizational arrangements of different types.

Finally, it seems imperative to find out what drives decision-making from a motivational point of view. In particular, it is important to ask what economic, political, and bureaucratic incentives and pressures operate on decision makers to drive given decisions in one direction or another. Similarly, one must ask what sorts of motivations personnel tend to bring to their jobs, as influenced by the reward structure they face as well as the nature of the particular service itself. However, formulating direct questions to decision makers for the purpose of eliciting such information poses very difficult problems for the researcher (for examples of research successful in employing techniques for querying decision-making, see Kaufman, 1967; Hogarth, 1971). Perhaps what is ultimately necessary is an introspective analysis, in which the researcher, after developing a thorough understanding of the structure of the organizational arrangements and their implicit reward system, asks himself, "What would I do if I faced a given decision, and were the kind of individual tending to hold the particular decision-making position?"

In summary, the "process measurements" that we have just outlined are intended to develop several pieces of information useful for evaluating organizational arrangements:

(1) Are organizational mechanisms in place for sensing changes in demands and service performance?

(2) Do events (decisions) indicative of changes or adjustments in resource use, programs and technology, and service outputs actually transpire?

(3) Does the reward system implicit in organizational arrangements motivate and bias these decisions in directions that lead to improvements in performance?

As these kinds of information are developed, they may be juxtaposed with whatever "hard data" are available on outputs and costs to achieve as complete a picture of organizational performance and the merits of alternative arrangements as may be possible under circumstances of limited experimental control.

APPLICATION TO EXPERIMENTS IN THE
ORGANIZATION OF EDUCATION

The nature of our discussion here applies to a broad range of services, and experimental demonstration efforts, extending to education, medical and hospital care (see, e.g., Hardwick and Wolfe, 1971), legal services (e.g., see Hofeller, 1970), housing (see U.S. Department of Housing and Urban Development, 1973), and other functions. Our discussion here will be limited to one area—education—for two reasons. First, it is impossible in an article of this length to do justice to a wider array of functions; second, two of the most interesting and timely examples of experimental evaluation efforts focused on organizational change are taking place in elementary public school education—those in performance-contracting and in education vouchers.

In popular discussion, performance-contracting and voucher systems are proposed as possible alternative approaches to bringing the advantages of the private marketplace to bear on the delivery of education services. As such, both performance-contracting and voucher systems represent significant changes in the organization of elementary and secondary education. Performance-contracting introduces (private) educational contractors into the public school system under a reward system for eliciting supply response heretofore alien to public school teaching—financial profit based on measurable improvements in output performance. While this is an important change in the organization of educational supply, the demand structure for education is left relatively untouched—i.e., under performance-contracting, school districts and even individual schools interpret the collective will of the community much as they do under conventional public school organization. Under a voucher system, however, the organization of demand as well as supply is significantly altered, resting responsibility for demand articulation and leverage much more heavily on individual parents, and the burden of supply response on individual (public and private) schools.

PERFORMANCE-CONTRACTING

Performance-contracting is a system under which a school district enters a contractual arrangement with an independent firm to provide all or part of the educational services of a school or set of schools in the district. The contractor is paid a given sum for improving students' performances by certain increments or for bringing achievement up to certain prespecified levels; otherwise, the contractor is not paid. Hence, success or failure is measured by quantitative indicators, usually reading and math scores on standardized achievement tests, and translates into financial losses or profits to the supplier. In concept, during the life of the contract, the supplier is free within limits to select the methods of instruction and to experiment in an effort to foster achievement. The supplier is also free to move on if profits or the promise of profits fails to meet expectations after the contract is expired. On the other hand, the school district can choose not to renew the contract if it is dissatisfied and may look for alternative suppliers.

In practice, performance-contracting arrangements have been designed in a variety of ways, involving different kinds of payment formulas and contract durations, encompassing different proportions of the scope of a school's activities, and have been viewed in varying degrees of permanence as an ongoing or temporary supply mechanism.

While it is clear that performance-contracting represents an organizational change of a major sort, there are alternative views of the concept in circulation. First, since, in practice, educational contractors have emphasized innovation in instructional methods by favoring new techniques such as programmed learning, computerized instruction, audio-visual materials, and alternative ways of arranging classrooms and deploying instructional resources, educational performance-contracting itself is often (incorrectly) viewed an innovative "method of teaching." Or in the jargon of this paper, performance-contracting is seen as a new program rather than as an organizational change. Alternatively, performance-contracting is sometimes viewed as an institutional device for introducing programs (technologies) into the schools. In this view, performance contracts are seen as temporary arrangements, to be discontinued when the schools become capable of running the new instructional methods themselves. The latter, of course, is a legitimate use of the contract mechanism, but must be differentiated from the primary concept of *performance*-contracting as an ongoing organizational regime. In fact, the "performance" part of performance-contracting seems almost irrelevant to the use of this arrangement as a temporary mechanism to achieve programmatic change.

There have been two major evaluations of performance-contracting carried out by the Office of Economic Opportunity (1972a) and the RAND Corporation (1971) for the U.S. Department of Health, Education and Welfare. These two studies bear striking contrasts relative to the evaluation issues we are considering here.

The OEO experiment was an attempt at rigorous, scientific, quantitative evaluation; it exhibited the following outstanding features:

- eighteen school districts, selected as "representative" of different types of locales containing low-income, underachieving populations participated: four large urban, five small rural, nine middle-sized urban;

- groups of experimental and control students, contained in different schools, were selected in each of six grades; these groups were selected to meet criteria of skill deficiency, low income, and minority status;

- six private profit-making educational technology companies were selected to represent a number of different educational approaches and techniques. These companies each contracted with three of the eighteen school districts, widely varying in type, for the purpose of providing reading and math instructional programs;

- contracts between the technology companies and the school districts were written to reward the companies on the basis of reading and math achievements made by the experimental students, as measured by standardized test scores. The structure of the payment formula, identical for all the company-school district pairs, stipulated that the company would be paid a certain sum for every student reaching a minimum guarantee level, and additional sums for increments over this guarantee level, up to a maximum guarantee level. The particular prices and guarantee level amounts varied across the eighteen contracts, according to separate negotiations;

- the experiment took place over a one (academic) year time span;

- performance was evaluated by OEO by comparing the reading and math achievements of experimental and control students.

On the basis of results that showed no significant statistical differences between experimental students who had received instruction from the contractors, and the control students who had not, OEO concluded:

> The results of the experiment clearly indicate that the firms operating under performance contracts did not perform significantly

better than the more traditional school systems. Indeed, both control and experimental students did equally poorly in terms of achievement gains, and this result was remarkably consistent across sites and among children with different degrees of initial capability. On the basis of these findings it is clear that there is no evidence to support a massive move to utilize performance contracting for remedial education in the nation's schools. School districts should be skeptical of extravagant claims for the concept [Office of Economic Opportunity, 1972b].

The wording of this statement is guarded. However, the OEO conclusion has been widely interpreted as a strong statement of the failure of performance-contracting as an organizational concept.[2] What the OEO experiment actually showed was that several alternative instructional methods (programs in our terminology), operated by firms with a profit motive, failed to work in one year's time; but it proved very little else, for reasons related to the experimental design issues we have discussed.

First, it is clear that, assuming OEO was interested in testing performance-contracting as an organizational concept, it varied and controlled inappropriate factors. The programs—i.e., educational technologies—were selected for their variety and *fixed* for each contractor for the duration of the experiment. In fact, the OEO RFP states, "The purpose of this experiment is to evaluate the relative effectiveness of existing techniques, not to underwrite the development of new techniques" (Office of Economic Opportunity, 1972b).[3]

On the other hand, the principal organizational parameters—the payment formulas and types of suppliers (profit-making, nonprofit, and so on) were kept uniform throughout the experimental sample. The fact that such parameters might have made a difference is indicated by an incisive analysis by Peterson (1972), which showed that the particular structure of the performance contract can lead to differential treatment by contractors, of students at different levels of initial educational achievement. This suggests that changes in the payment formula could have important effects on contractor behavior and success in dealing with different groupings of students. In short, the OEO prescription is exactly the reverse of what would be needed to evaluate performance-contracting as an organizational arrangement—i.e., program variables should be left unconstrained while variations of organizational design should be tested.

Second, it is clear that one year is a very short time for evaluating organizational performance in education, especially where new technologies are introduced. Problems of start-up take weeks to work out, personnel turnover is virtually disallowed, major adjustments subsequent to mid-term performance measurement are hardly feasible even if modifications in program were allowed, and certainly there is no time for development, much less implementation, of innovations.

If only because a chief mechanism of the performance-contracting approach is the fact that contract *renewals* would be pursued by firms with confidence that they can succeed, while less sanguine or capable firms would be discouraged from rebidding, evaluation of the performance contract agreement would seem to require a multiyear time frame.

Finally, OEO's measurements were basically confined to quantitative output measures (reading and math scores). Little was learned (or at least documented) about the decision-making process by which contractors selected and manipulated resources and offered their services. In light of the other limitations of the design, such information would have been extremely useful for assessing the potential capability of the performance-contracting regime.

The RAND/HEW evaluations of performance-contracting were much less formal and rigorous than OEO's. Case studies were performed of demonstration efforts in six school districts: Norfolk, Virginia; Texarkana and Liberty-Elau, Texas; Gary, Indiana; Gilroy, California; and Grand Rapids, Michigan (RAND Corporation, 1971). Evaluation of these demonstrations had the following key features:

- studies were made of eight performance contracts operating in fifteen schools within six districts. The sample was chosen to provide "a diverse group of school districts and programs";

- selection of participating students varied by district, ranging from an entire school to categories of students with deficiencies in reading and math;

- seven private profit-making educational technology firms provided fifteen programs in the six districts. The scope of programs ranged from an entire educational agenda in the Banneker School in Gary, Indiana, to reading and math, to reading alone;

- most of the contracts called for payments per every student exceeding prespecified levels of reading and math achievement. These levels varied among the districts. One contract in Grand Rapids paid proportionally to student gains, without reference to prespecified levels of minimum achievement;

- four of the eight contracts were obtained by the districts through competitive bidding; the other four were sole source contracts. The contracts ran for one year, except for the four year contract in Gary. The evaluation by RAND was based on one year of operation in all cases;

- evaluation by RAND was not regarded as a scientific preplanned experiment but as a less formal series of (ex post) case studies; several dimensions of performance were considered including cognitive achievement, educational change and innovation, and the quality of evaluation activities undertaken by local education agencies, their contractors, and associated independent agencies. RAND's interest in local evaluation related to the use of such information for purposes of aiding and improving management decisions.

On the basis of its observations, RAND drew several conclusions about performance-contracting. On the performance level, it found no dramatic gains on standardized achievement tests, and widely varying but generally higher costs relative to conventional instruction, and concluded: "Unless further performance contracting programs achieve higher cognitive gains than past programs have, they will have to be justified on the basis of ancillary benefits such as curriculum development potentials" (Carpenter and Hall, 1971).

But perhaps more significant than its "output" findings, RAND made several acute observations regarding processes of decision-making under performance-contracting. For example:

- performance-contracting fostered a healthy emphasis on the student and his learning as a measure of program success [but] performance-contracting programs will probably continue to be narrowly focused because of difficulties of defining objectives in subject areas other than those involving simple skills or, in some cases, difficulties in measuring the attainment of objectives;

- performance-contracting is proving to be a useful research and development tool. People who are *not* a permanent part of the school system seem to be freer to implement radical changes in the classroom than are regular school personnel;

- performance-contracting does not seem to have generated large profits so far;

- performance-contracting has generated some follow-on contracts, only some of which tie fees to student achievement;

- performance contracts have enabled a number of firms to break into new markets and to receive publicity for their goods and services;

- established contractors tend to prefer other arrangements to performance-contracting, such as consultantships. Performance contractors will seek to convert performance contracts to other types of arrangements.

In assessing the RAND study as an exercise in evaluating organizational change, it is necessary to recognize the ex post nature of this inquiry. Cases (sample schools) were of necessity chosen from a few dozen ongoing independent performance-contracting demonstrations under way at the time. Thus, the benefits of preplanning and controlled experimentation were not available. One can only criticize on the basis of selection of sample cases. The selected cases *did* exhibit a variety of payment formulas, contract specifications, and bidding procedures. Overall, the apparent emphasis on organizational parameters in selection, rather than program technology are consistent with a study of organizational performance rather than of teaching method. However, it seems that RAND could have done more to exploit this variety in its analyses. What were the effects of different payment formulas? For example, did the "payment by proportional gain" formula in Grand Rapids result in the same kind of discrimination among student groups found by Peterson for the "threshold" payment formula used in Gary? Did the four contractors selected by sole source methods perform and behave differently from the four selected by competitive bidding? These questions were not addressed directly by RAND because of its "separate case studies approach." More emphasis on comparative observations would seem helpful.

RAND's evaluation covered a one-year time frame, and so is subject to the same criticism as the OEO effort on this score. An extended evaluation over three to five years would have allowed coverage of the long-run progress of the four-year Gary contract, as well as more extended observation of renewal and cancellation experiences with the other contracts.

As in the OEO experiment, the output (achievement score) results in the RAND cases proved relatively little about performance-contracting potential because of the limited time frame of observation, as well as, in this case, the noncontrolled nature of the evaluation. However, the observations of process that RAND recorded provide important insights, at least as valuable to the evaluation content as the output measurements. One obtains a sense of the directions educational supply would take over the long run, under a performance-contracting regime:

- curricula would be focused more and more on individual student achievement (as opposed to, for example, whatever else might be preferred by teachers) and would emphasize subjects where skills could be objectively measured;

- more emphasis would be given to introducing innovative methods in an effort to raise achievement scores and, hence, profits;

- if profits remained low—i.e., innovations continued to fail to produce significant improvements in student achievement—contractors would not continue to undertake performance contracts. Over the long run, one might expect educational technology firms to pursue contracts without performance incentives, and school districts to use contracting only to develop and implement (turnkey) desired curriculum (program) changes.

In summary, RAND's process measurements provide a likely scenario for what may be expected over time, in a regime of performance contracting. In the absence of a controlled experiment designed correctly for measuring the effects of *organizational* parameters over an adequate time frame, such information proves all the more valuable.

The principles for evaluating organizational change seem better appreciated in efforts to experiment with a proposal to reorganize public school education even more radical than performance-contracting—the education voucher system. But perhaps because of the major proportions of change it proposes and because of the structural limitations of experimenting with whole school districts at once, the opportunities for evaluating voucher proposals in a controlled scientific framework seem highly limited and the importance of maximizing evaluation output within these constraints seems all the more striking.

EDUCATION VOUCHERS

A system of performance-contracting allows school districts to choose among suppliers of educational services. The voucher system involves a more decentralized choice mechanism, vesting in the consumer (parent) the right to select the school his child attends.

Consumer choice is accomplished by distributing to parents "voucher certificates" of dollar amounts roughly equal to per student expenditures on education in the district, and allowing parents to "spend" these certificates at any one of a number of schools approved by the district for receipt of vouchers. The budgets of the schools are in turn determined by the voucher revenues attracted; hence, schools face financial incentives to respond to parents' educational preferences for their children.

A small but important literature has developed around the concept of education vouchers beginning as far back as Adam Smith (1937) and more recently Milton Friedman (1962), and continuing into other scholarly articles and papers preliminary to OEO efforts to begin experimental demonstrations (see LaNoue, 1972). The seminal work prepared by the Center for the Study of Public Policy (1970) develops a series of different organizational designs for voucher systems, and recommends a particular "regulated voucher system model" structured to overcome various objections to other designs. The recommended system for a public school district exhibits the following general attributes:

- an Education Voucher Agency (EVA), coincident with or a subdivision of the local school board and responsible for administration of the system, would receive all local, state, and federal education funds for the district and would pay money out to schools in return for voucher certificates;

- the EVA would issue vouchers to parents of every school-age child in the district, in an amount somewhat less than average per pupil expenditures in the district. Voucher values for students from low-income families would be supplemented by an additional amount;

- in order to become an "approved voucher school" eligible for cash vouchers, a (public or private) school would have to meet certain requirements regarding admissions policy, tuition, suspension and expulsion of students, accounting practices, and standards for curriculum and staff;

- families would designate their choice of schools for their children to attend. Schools would follow designated procedures for admitting students according to their choices;

- parents would submit their vouchers to the schools attended by their children, and the school would remit these to the EVA in return for money payments;

- the EVA would be responsible for assuring enough school capacity to accommodate all children, by helping to open new schools when necessary.

There is some distance between this conceptual model and the one actually implemented for experimentation in the Alum Rock School District in San Jose, California, in 1972. Yet the voucher system represents such a radical change in organization of elementary education that OEO received only a few serious inquiries from school districts to its financially generous offers for experimentation, and only Alum Rock agreed to cooperate with OEO even in a limited test of the voucher concept (Doyle, 1973). The Alum Rock voucher system design is restricted to public elementary schools and incorporates the following basic features (Alum Rock Union School District, 1972):

- parents of each participating child receive a voucher worth approximately the current average cost of educating a child in the Alum Rock School District. A second "compensatory" voucher is issued for disadvantaged students;

- six public schools participate in the experiment (initially). Each develops, under a single principal, two or more alternative, distinct educational programs;

- each parent selects an educational program and a school building for his child. Each child is assured of placement in his first choice program (but not school building);

- if a building is overapplied, additional capacity (e.g., by portable classrooms) is created wherever possible. If a program is overapplied, additional capacity for that program is created somewhere in the system;

- the budget of each program is determined by the voucher money attracted to it. Thus, programs can increase their budgets by tailoring their offerings to attract more students. Since voucher money is targeted to programs and not to schools, some degree of autonomy is implied for faculty or administration of individual programs;

- new programs are allowed to develop through various mechanisms, including: spin-offs from existing school programs, additional public schools opting to join the voucher experiment; community groups developing their own new programs; and outside groups proposing to provide new programs. In order for a new program to qualify for receipt of vouchers, it would have to be approved by the Education Voucher Advisory Committee, set up to advise the Alum Rock School Board.

The planned evaluation by the RAND Corporation (1972) seems to leave almost no stone unturned. Comprehensive data collection will cover a spectrum of output measures and changes over time; it will also process measurements, including occurrence of critical events and decisions such as curriculum changes, growth and decline of particular schools (entry into and exit from the educational market), attitudes and behavior of parents, students, school administrators, community groups, and the like, states of information, and structural parameters of the educational "marketplace."

Collection of such comprehensive information on process as well as performance seems especially necessary in the voucher demonstration context, for there is very limited opportunity here to introduce experimental controls. To begin with, there is only one demonstration site, and the achievement of this single demonstration should probably be viewed as a major political accomplishment.[4] Even if there were two or three experimental sites, the opportunities for controlling external variables would be highly constricted.

Two strategies for attempting such control might have been considered. First, *within* the demonstration district, there is the possibility of separating and randomly assigning students into experimental and control groups (as was done in performance-contracting by OEO). This requires setting aside one group of schools, parents, teachers, and students in the district to continue under the conventional system, while another very similar group participates in the experimental system. Unfortunately, such an arrangement would seem very cumbersome and problematic. There would inevitably be interactions between the experimental and control systems, in terms of bidding for educational resources, conflicts over district policies and procedures, and other possible competitive effects. All this would be added to the political problems associated with allowing one group of parents choice and another no choice, constraining schools from opting in or out of the control or experimental parts of the system, and the administrative burden of running two essentially different school systems in the same district. On the other hand, such an arrangement may be administratively feasible; in essence, the Alum Rock experiment comes close to this, as less than half the Alum Rock School District has participated in the experiment thus far, while the remainder has operated in a conventional mode. However, there was no random selection or control imposed in splitting the district into two parts, nor are there any formal plans to evaluate the two parts of the Alum Rock district on a comparative basis. In fact, long-range plans call for expanding the experimental part of the system, in large part by having additional schools from the conventional sector join the demonstration. While the latter would reduce the possibility of meaningful comparative evaluation, it of course would enhance the value of the demonstration as a test of "full blown" voucher system feasibility.

An alternative for controlling external variables would be to select several comparative school districts similar in socioeconomic make-up and educational experience to the experimental district, but geographically removed and operating under conventional organizational arrangements. Such matching districts could be used as controls, though they would serve imperfectly in this role because of the inevitable difficulties in achieving good matches between experimental and control districts. Still, the effort may be worth attempting, rather than depending on the general "wisdom and experience" available in conventional school systems, against which to compare experimental voucher system results.

In summary, the opportunities for formal control of external influences are very limited. Thus, voucher system evaluation must inevitably depend on data drawn from a non-rigorously controlled setting. Hence, it is especially important to focus on understanding the decision-making processes which drive the voucher system. The RAND evaluation plan seems to recognize this.

In addition to control of external variables, the issue of specifying and controlling (internal) program and organizational variables also arises in Alum Rock because of the ambivalent manner in which the term "program" is used. In particular, a "program" is defined both in terms of a particular instructional approach, e.g., "individualized learning" or "bilingual/bicultural" (for a list of available programs, see Alum Rock School District, 1973), as well as an organizational unit—i.e., a "minischool" within a school. (The latter use of the term "program" is, of course, in direct conflict with the terminology adopted in this paper.) As a result of the ambiguity in terms, potential problems arise in resolving (a) the organizational role of programs (minischools) versus schools, and (b) the rigidity with which programs (minischools) are required to adhere to their (initially) espoused instructional approaches.

The Alum Rock plan essentially leaves unresolved, the question of what actually constitutes the basic organizational unit of the Alum Rock system—i.e., the unit comparable to the firm in the private marketplace. Is it the minischool, which is administered by a separate faculty and whose budget is presumed (independently) determined by voucher funds attracted? Or is it the school, which contains several minischools and is administered by a principal who presumably has independence and authority on his own?

A system which is to operate on the market principle of a voucher arrangement could work in one of two ways. Either the basic organizational unit is the school, as a multiproduct (program) firm with the principal as its executive, *or* the basic unit is the minischool with an administration independent of the principal. In the latter arrangement, the role of the principal would be greatly diminished from its current form. In particular, his role might evolve into a building and resource management function, or an educational consultant function, without major policy decision-making content.

The danger in terms of experimental design is that the organizational structure of the demonstration system may not resolve itself in either of these two ways. In particular, an ambivalent situation could persist wherein principals preside over several semi-autonomous minischools which they fail to control but which they nevertheless impede by imposing arbitrary policies of "balance" in the size and variety of alternative programs within a school. The situation is recognized by those affiliated with the demonstration:

> The existence of individual minischools with their own goals and objectives and loyalties has created problems because the principals have felt an obligation to maintain an overall school integrity, and the internal conflicts between the staffs of the different minischools have created problems in maintaining an overall sense of the school [Levin, 1972].

Evidence from the first year of operation seems optimistic but inconclusive on resolving the situation. Experience varies among individual schools in the experiment, but there seems to be a tendency toward primary decision-making responsibility at the minischool level, with principals playing a coordinating and facilitating role. The minischools tend to be administered independently in collegial fashion by their separate faculties (observations based on conversation with staff members affiliated with the voucher project, and on Doyle, 1973). Still, the danger persists that the demonstration will develop according to a scenario wherein minischools are thwarted in their efforts to expand, contract, or vary their programs, while principals are similarly frustrated. Should this circumstance occur, the demonstration will not constitute a test of the voucher system model of organization as originally conceived, but of some less vital variant. And this contingency will have resulted from a failure to clearly specify at the outset the organizational parameters of the arrangement to be tested.

A second danger is the possibility of the demonstration evolving in such a way that organizational units (schools or minischools) are artificially restricted to particular kinds of instructional offerings. This could occur if programs eventually become rigidly associated with their instructional approaches rather than their organization identities; i.e., the demonstration system becomes viewed as providing parents choices among types of instructional approaches rather than sources of supply of educational services. In this circumstance, a tendency could develop to (centrally) constrain schools or minischools from making significant alterations in order to preserve the distinct instructional approaches (at the expense of organizational integrities). Or procedures for student placement might tend to be made on the basis of preference of instructional offering without reference to source of supply. The latter could occur, for example, if students selecting oversubscribed programs, as supplied by a particular school or minischool, are assigned to other (existing or especially created) suppliers with similar instructional offerings rather than allowing the original supplier to expand. In such contingencies, the market principle implicit in the voucher system concept would be violated and the demonstration would lose its significance as an organizational evaluation; rather, it would constitute no more than a limited test of free choice among alternative teaching approaches.

On the basis of the first year of the demonstration, it is hard to assess how serious these dangers might be. The schools and minischools seem not to have been highly constrained in their selection of instructional offerings; and adopted procedures seem to leave future modifications open to the discretion of these units, based on their ability to attract students. Furthermore, the procedures adopted for student admissions and determination of minischool capacities do seem to allow for a "marketlike" equilibration of consumer preferences with decentralized supply offerings, rather than central assignment on the basis of choice of instructional approach.[5] On the other hand, the problem of oversubscription of particular programs does not seem to have seriously arisen yet, certainly not to the point where major additions to individual minischool capacities (exceeding current building space) are required. Such a circumstance may provide the real test of how the system will be permitted to function over the long run.

So far, the Alum Rock experiment does not seem to have fallen into any of these various traps we have mentioned, but it is not yet out of the woods. Schools *or* minischools, as defined by an autonomous management, must be allowed to expand, contract, and modify instructional offerings freely, if a meaningful test of organization as opposed to programs, and of the voucher organizational concept in particular is to result. Vigilance seems required to keep to this course.

The Alum Rock demonstration is projected for an initial two-year period of limited experimentation. The first year was confined to six public schools, encompassing 3,900 students. The second year will expand the experiment to include thirteen schools and approximately 9,000 students, roughly half the Alum Rock School District. After two years, the project could be phased out if judged unsuccessful. Otherwise, consideration will be given to move to a "full" voucher system operated over an additional five-year period.

In terms of providing a meaningful test of the organizational concept of vouchers, it is doubtful whether the "transitional period" of the first one or two years will yield much more than a "debugging" of administrative procedure and a glimpse of ultimate organizational structure and behavior and educational performance. The key information will come from the extended five-year period, should the decision be made to continue the experiment.

CONCLUSION

The experiments and demonstrations in performance-contracting and education vouchers help illustrate the nuances involved in evaluating alternative organizational arrangements for providing public services. At the heart of the matter is the interaction between programs which define the technology of provision, and organizational arrangements which influence how well programs are selected, adjusted, and administered over time.

The key design parameters that evaluation planners may be at liberty to adjust in some degree are the selection of variables to be controlled and those to be systematically varied, the time span over which evaluation is made, and the selection of variables to be measured. In particular, we have argued that, within constraints of limited resources for evaluation and short time horizons for producing results to affect policy decisions: (a) organizational parameters should be systematically controlled and program variables allowed to float free; (b) that the time frame for evaluating organizational arrangements should be long enough to allow basic system adjustment behavior to take place; and (c) that measurements of the decision-making processes should be made to compensate for the limited opportunities for experimental control. The evaluation efforts in education that we have discussed succeeded to different degrees in achieving conformity with these principles.

It will be recognized, of course, that no discussion of this type can derive hard and fast formulas for designing organizational evaluations. The subject is too complex for that, and the realities of planning and implementing scientific evaluations in a social setting so very difficult to cope with. This is painfully evident in considering the difficulties evaluation planners have in adhering to the principles espoused here. Inevitable pressures arise for producing "hard (quantitative) results" in relatively short order, and there seems to be a natural tendency to focus on how the methods or technologies work rather than the influence of organizational arrangements. To counsel focusing on organizational parameters, collection of comprehensive information on process, and use of long evaluation time frames rubs against the grain of public policy makers and popular perspective. For this reason, evaluation planners much inevitably compromise—i.e., produce short-term performance information. This is no tragedy as long as the basic perspective is not lost. Our discussion here will have been useful if it succeeds in encouraging evaluation planners to strengthen future evaluation designs in the aspects we have highlighted, or if it educates laymen and public officials to the need for viewing efforts at organizational evaluation, in greater perspective.

The early experimental evaluations in the organization of elementary education should be recognized for their pioneering achievements. And it would only enhance their stature to have such efforts in future build upon their shortcomings as well as their strengths.

NOTES

1. The expression of social preferences will itself be shaped by the arrangements under which demand is organized (see the discussion on this point).

2. Press reports of the OEO evaluation were not nearly so conservatively worded as OEO's report. Quoting from an article by Rosenthal (1972): "The OEO's conclusions are thus likely to have wide impact and to strike a possible fatal blow to enthusiasm for the contracting approach." The article continues, "OEO's officials expressed disappointment today, even sadness, at their findings. 'We want it to work as much as anyone, knowing that we will have no solutions to teaching poor kids better,' Thomas K. Glennan, Jr., OEO's research director, said. 'But there is great value in learning which basket we should not be putting our eggs in,' he continued. 'Better to stop now, rather than wait until hopes–and spending–have become enormously inflated.' There have been earlier, more tentative indications that contracting did not accelerate achievement among the disadvantaged. But because of the extremely detailed and varied nature of its experiment, OEO's assessment is regarded as decisive."

3. In light of OEO's stated objective to test existing techniques, it is ironic that OEO has been criticized by the General Accounting Office for quite the opposite behavior that we are criticizing. Specifically, the GAO charged that "the companies were allowed to change their teaching approaches continuously during the experiment, and almost all used 'similar core materials' though each was supposed to have a distinctive program" (Washington *Post*, 1973; Joe Nay of the Urban Institute is thanked for bringing this article to my attention). If GAO's charges are correct, then our criticism of the experiment must be dampened. But, in fact, what may have happened is that OEO found itself *attempting* to constrain contractors from altering their methods, but failing, at least partially, in this endeavor.

4. According to Doyle (1973) and others associated with the experiment, success in Alum Rock bodes well for future demonstrations in at least three or four other sites in the near future.

5. Observations based on the "Voucher Information" packet provided by the Alum Rock School District, plus conversations with staff affiliated with the voucher project, the RAND Corporation, and the Center for the Study of Public Policy at Harvard.

REFERENCES

Alum Rock School District (1973) "Educational choices for your child."

――― (1972) "Transition model voucher proposal."

CAMPBELL, D. T. and J. C. STANLEY (1972) Experimental and Quasi-Experimental Designs for Research. Chicago: Rand McNally.

CARPENTER, P. and G. R. HALL (1971) "Case studies in educational performance contracting: conclusions and implications." Vol. R-900/1 HEW, December.

Center for the Study of Public Policy (1970) Educational Vouchers. Cambridge, Mass.: Harvard University.

DOYLE, D. (1973) "The San Jose educational voucher experiment." Washington Operations Research Council Conference on Urban Growth and Development, Washington, D.C., April.

FRIEDMAN, M. (1962) "The role of government in education," in Capitalism and Freedom. Chicago: Univ. of Chicago Press.

HARDWICK, C. P. and H. WOLFE (1971) "An incentive reimbursement/industrial engineering experiment." U.S. Department of Health, Education and Welfare Health Services Foundation, Blue Cross Association and Blue Cross of Western Pennsylvania, DHEW Publication (HSM) 72-3003, December.

HINRICHS, H. H. and G. M. TAYLOR (1969) Program Budgeting and Benefit-Cost Analysis." Pacific Palisades, Calif.: Goodyear.

HOFELLER, M. A. (1970) "An evaluation of the Office of Legal Services in Nassau County." American Institute for Scientific Communications, Hempstead, N.Y., October.

HOGARTH, J. (1971) Sentencing as a Human Process. Toronto: Univ. of Toronto Press.

KAUFMAN, H. (1967) The Forest Ranger. Baltimore: Johns Hopkins Press.

LaNOUE, G. R. (1972) Educational Vouchers: Concepts and Controversies. New York: Teachers College Press.

LEVIN, J. (1972) "Status report on voucher project." Memorandum to the Alum Rock Union Elementary School Board, December.

NAY, J. N., J. W. SCANLON, and J. S. WHOLEY (1971) "Benefits and costs of manpower training programs: a synthesis of previous studies with reservations and recommendations." Urban Institute Paper 2400-1, June.

NELSON, R. R. (1971) "Issues and suggestions for the study of industrial organization in a regime of rapid technological change." Yale University Economic Growth Center Discussion Paper 103, January.

Office of Economic Opportunity (1972a) "An experiment in performance contracting." Office of Planning, Research, and Evaluation, Washington, D.C.

——— (1972b) "An experiment in performance contracting: summary of preliminary results." Washington, D.C., February 1.

PETERSON, G. E. (1972) "The distributional impact of performance contracting in schools." Urban Institute Working Paper 1200-22, March.

RAND Corporation (1972) "Technical analysis plan for evaluation of the OEO elementary education voucher demonstration: technical dissertation." Santa Monica, California, February.

——— (1971) "Case studies in educational performance contracting." Vols. R-900/1-HEW through 900/6-HEW, Santa Monica, California.

RIVLIN, A. (1971) Systematic Thinking for Social Action. Washington, D.C.: Brookings Institution.

ROSENTHAL, J. (1972) "Learning-plan test is called a failure." New York Times (February 1).

SMITH, A. (1937) The Wealth of Nations. New York: Random House.

U.S. Department of Housing and Urban Development (1973) "Experimental housing allowance program: an overview." Office of Policy Development and Research, Washington, D.C., March.

Washington Post (1973) "School project of OEO scored in GAO report." (May 13).

WEISS, C. (1972) Evaluation Research: Methods of Assessing Program Effectiveness. Englewood Cliffs, N.J.: Prentice-Hall.

YOUNG, D. R. (1971) "Institutional change and the delivery of urban public services." Policy Sciences (December).

Part II

TOWARD A MORE SOPHISTICATED URBAN MANAGEMENT

5

Change and Innovation in
City Government

FREDERICK O'R. HAYES

☐ THE STRUCTURE AND ENVIRONMENT of municipal government varies widely in America among the states and regions and from
city to city. The principal difference is the division of authority
between the legislative body and the chief executive. The differences
are important. The mayors of New York and Detroit, for example,
are invested under strong mayor charters with authorities and
responsibilities that overshadow the limited role of both the
legislative mayors of Cincinnati or Minneapolis and the weak
executive mayors of Los Angeles and San Francisco. A process of
change and innovation under executive leadership consequently
carries a significantly higher probability of success in cities like New
York and Detroit, with strong mayor charters.

MUNICIPAL GOVERNMENT:
A LOW-CHANGE SYSTEM

Beneath the top decision-making level, American municipalities
are much alike in basic structure. This is a structure built in the first
quarter of this century. It is designed to ensure accountability of

public officials, to discourage corruption and to protect both decision-making and appointments from undue political influence. It has several characteristics.

First, nearly all positions in city government are covered by municipal civil service. Even in strong mayor cities, neither the mayor nor his appointed department heads typically have more than a handful of positions excepted from civil service requirements. In city manager cities, by custom or law, even department heads are often designated by internal promotion from within the department. In New York State, positions above entrance level must ordinarily be filled by internal promotion examination from among civil servants of the next lower rank in that job series and in that particular office or bureau. Open, competitive civil service examinations can be used only when the number of internal qualifiers is insufficient to fill the vacant positions. In New York City, the Chief Inspector of the Police Department, the Fire Chief, and the Chief of Staff of the Sanitation Department must have all entered the department at its lowest rung.

Rigidity begets loopholes, and it would be absurd to pretend that innovative mayors are so tightly bound by a system that offers them so little flexibility. But the system does place serious obstacles in the way of the introduction of new leadership in programs and areas where it is needed by the mayor. It also places a high premium in the administration of an aggressive and innovative mayor on manipulation of the system, and it forces the mayor to incur real risks of public and political censure in his efforts to maneuver his way through the civil service structure.

Even a good civil service tends to place a higher value upon stability and the maintenance of the current modus operandi than upon change and improvement. This is partially because the senior civil servants are the veterans and, often, the virtuosos of the system; they know how things are supposed to be done, and they are distrustful of changes in procedure and routines. Moreover, they are the survivors of an unusual group—people who take civil service examinations—and fewer and fewer of our young people of highest potential secure employment in this way.

A second characteristic is the use of the line budget specifying every position and its salary. New York City's government has more freedom than most since it uses only a modified line budget; changes in position distribution or salaries do not require legislative approval but can be done by the mayor. From the perspective of a department head, the difference may not be important. For every change in staffing pattern and every shift in resources from one object of expenditure to another, he must secure central approval of the mayor or his Bureau of the Budget. The line budget becomes another inertial force in the system, and an inducement to the department head to avoid difficulties by staying within his established budget structure.

A third characteristic is the role of the personnel department. Typically, the department head will find that any new position or change in an existing position must also be reviewed and approved by the Department of Personnel or the municipal civil service commission, adding another approval routine for changes of this character.

Fourth, he will find that, except for very small dollar amounts, contracts must be let by competitive bidding, or, as in New York, must have approval of the Board of Estimate. In some cities, although not in New York, contracts must all be approved by the city council whether or not they have been competitively bid. The department head may find that purchases must all be assigned to a central purchasing operation to assure lowest cost procurement. He may have, as well, to deal with an independent Comptroller, as in New York, who introduces his own requirements. An internal reorganization of his department may require council action or even an amendment to the City Charter.

These are all components of a central transactions review process aimed at requiring central review and approval of all changes from established routine and procedures. There are typically many more elements of the process. The above does not even touch on the special requirements associated with capital expenditures or the myriad ways in which city procedures must contend with state-imposed requirements. In New York, for example, shifts for policemen are specified by state statute and shifts for firemen by the state constitution.

To these structural elements, there has been added in recent years, in most of the large cities, the impact of collective bargaining and the unionization of civil servants. Most pronounced for teachers, policemen, and firemen, the impact of unionization is extending to all municipal employees. In New York, all but a handful of city employees are covered. No matter how the rules of the collective bargaining game are constructed, unionization makes the unions a factor in any proposed change in work organization or methods. In New York, the effects are extreme, reaching a stage described by some observers as "a system of public policy codetermination on the part of the unions" (Leavens et al., 1970: 621). On any significant change and many insignificant ones, this means that agreement must be worked out with the unions.

We might add to these factors the changing role of the organized public. In the last decade, community and various interest groups have operated with a determination and a militancy that has made many of them as formidable as the unions. Many changes in government are possible only if the terms can be negotiated with these groups. One observer has noted that proposals that a decade ago could be readily negotiated among traditional leadership cadres in a city now may be subject to veto by any one of a large number of citizen groups.

The purport of all of this is to indicate, first, that municipal government has been typically designed as a low-change system. It places special burdens and requirements on those who wish to change or improve the system. Innovations are possible, but only if they negotiate an obstacle course designed to assure that they are soundly based. An enormous amount of managerial energy must be devoted to negotiating this course by those who would make changes. For many, it is easier to live within the existing structure. The structural obstacles to change have now been reinforced by unionization of city employees and by the increasingly militant involvement of interested citizen groups.

Rarely is a municipal agency staffed at its top level with the numbers or talents required to take a large program of change through their complex of negotiations and justifications. Rarely is an agency, in fact, devoting the resources into investigating its problems or its opportunities for improvement that would produce a large and soundly based effort at innovation. This low level of investment in change and innovation is an accurate symptom of the political and public climate toward change in municipal government. The departmental commissioner and the mayor share basic and important constituencies that remain to be convinced.

The New York City government in 1966 had all these characteristics. The force of ingrained habit and the inertial stability of the system were strongly reinforced by the sheer scale of the city government. Scale lengthened the distance from the actors and operations to the ultimate decision makers. It could produce situations where the administrator of a hospital with an annual budget of $35 million could complain in public testimony about his inability to buy mop pails. And at the same time, far more important problems were beginning to emerge as the rapid pace of change in the world outside outdistanced the adaptation of the city's ponderous bureaucratic machinery.

POLITICAL LEADERSHIP

The inertial, earthbound characteristics of the municipal government system say something, per se, about the nature of the forces required to move it. The first is obvious. Movement does not occur as a result of casual or accidental intervention in the system; rather, it requires a deliberate and sustained effort to do so. Nothing less is likely to create a significant impact.

Effective and successful leadership is unlikely to arise within the bureaucratic system. With responsibilities divided, few actors among the senior civil servants occupy positions with the power and the capacity to make significant changes except on a modest scale and a gradual time table. Moreover, those with vested interests in the existing modus operandi are, unlike the senior civil servant, actors with political capacities, able to take their positions to a political arena. The very conditions surrounding inertia and stagnation in the bureaucracy demand correction through political, rather than administrative or bureaucratic leadership.

This is an important point, but one based on a realistic understanding of the basic situation. The possibilities of major reform and innovation within the bureaucracy are significantly reduced, under the best of circumstances, by the politicization and unionization of the employees who are most closely involved with change. When we add the frequent need to seek amendatory legislation or budget changes and the interest of external citizen groups in the maintenance of the status quo in many areas, political leadership is clearly critical to any significant program of change.

Underlying this perspective is the basic view that major change and innovation have become a legitimate political issue rather than simply an administrative problem. We should also recognize that change and innovation are a risk-taking venture, risks that career civil servants are ill-equipped by position and temperament to take. The circumstances of the current governmental structure demand external leadership.

Political leadership occupies an inherently important role in activities that involve significant risk-taking. Any effort toward change and innovation necessarily involves risks of different kinds. The first is the risk of disapproval and opposition by individuals and groups with a real or believed interest in the maintenance of the status quo. These groups have tended to become more important politically with employee unionization. The second major category of risk arises from the possibility that the planned change or innovation may be a failure. It may be a failure because success depended upon assumptions that were not realized. The assumptions may concern, for example, how program clients will react to a new method of new service, or prospective changes in the local economy, or the availability of skilled manpower, or how civil servants will respond to new directions. In all of these, there are finite probabilities that, in fact, the program planners' assumptions will not be fulfilled.

It is almost axiomatic that civil servants do not take—indeed, are ordinarily not supposed to take—significant risks. In fact, the political leader will usually prefer that changes involving significant risk be brought to him for decision. I would argue that the risk-taking characteristics of a policy of change and innovation have tended to increase during the last decade as a result of unionization, militancy, the politics of confrontation, and the declining public support of leadership figures.

Change and innovation are attempted and take place, of course, at varying rates, and it would be a misrepresentation to attempt to describe mayors or governors simply as innovative or noninnovative. There are, moreover, marked differences in the perspective of what is possible and what has to be done to make a proposed innovation feasible that differentiate sharply the approaches of two leaders with the same concepts of what ought ultimately to be done.

With all the qualifications, it seems clear that innovations in city government operations come during the administration of mayors with strongly favorable views toward innovation and with a willingness to take the political risks incident to innovation. John Lindsay was such a mayor, and the strong innovative thrust in government during his administration of the city was directly related to the Mayor's own orientation in this direction. Lindsay or someone like him is not a sufficient condition for a high level of innovation but such a mayor, except in rare cases, is a necessary condition.

I suspect that this stance in Lindsay is closely linked to many other of his characteristics. Lindsay ran for mayor knowledgeable about the problems of the city or their symptoms—dirty streets, crime, fiscal difficulties—but with no more than "white paper" sophistication about the running of the city government. He was truly an outsider, a Republican in a Democratic city, a Congressman with interests and responsibilities focused on the national rather than the local scene. His entourage as a candidate was almost devoid of anyone with experience in city government.

This meant that there were almost no areas of city performance where political commitments or obligations need moderate his criticisms. It meant that there were few cautionary voices saying that it cannot be done or that it would take years. With due allowance for the hyperbole of the political campaign, Lindsay entered office believing strongly in what he had said—that the city was a mess and that something could be done about it.

Part of it is in the character of the man himself. His predecessor, Robert Wagner, was an astute performer in this pluralistic world, a political man in the sense Theodore Lowi describes in *The Politics of Disorder*. Wagner understood the host of different interests that converged on any issue and the effort it required to produce a supporting coalition on any controversial issue. He was skilled in producing that coalition and cautious about any premature exposure of his position.

Lindsay had neither the perception nor the skills of the same calibre. But, more important, he was ideologically opposed to a process that, in his view, reduced forward progress to a snail's pace and perpetuated a system that failed to perform. This is part of a general viewpoint that despaired of the competence of the existing establishment and its modus operandi and saw the symptoms of its failure in New York's dirty streets as well as in Vietnam. This perspective in any man of good will is necessarily coupled with impatience. Lindsay, in these respects, is not unique. It is a perspective shared with Robert Kennedy in 1968, with Eugene McCarthy and with the earlier "sons of the wild jackass," the rebel Republican Senators of the West. What is uncommon is the application of that concept to the management of America's second largest government rather than to legislative policy-making.

In its application to government, this viewpoint brought Lindsay to distrust of the old—for it was not working—and hope, if not confidence, in the new—for it might work. It meant an a priori preference for new men, new ideas, new organization. It meant that the risks of innovation were heavily discounted because so low a value was placed upon the maintenance of the present. The process of accommodation and the yielding to political considerations were eschewed as a surrender of principle and as destructive of the hope for progress.

No one could retain these views in pristine form through eight years in the snake pit of the politics of the city of New York. Tempered and qualified as they were by "realism," however, the basic Lindsay thrust toward reform and reconstruction, change, and innovation survived through his entire administration.

I believe that the Lindsay perspective, with all its initial naivete and lack of realism, is basic and that the continuing thrust toward change and innovation in the city could not have been maintained without it. City Hall could have used more political deftness and, on some issues, superior outcomes could have resulted. But my guess—and it is little more than that—is that we can rarely, if ever, have the best of both worlds, that the perspective of the political man inevitably leaves numerous issues unconfronted and contro- versial problems undisturbed. On the other hand, the drive for innovation in a political environment will almost always be "unreal- istic" and lead to confrontation and conflict; the transitions will tend to be bumpy and difficult.

FULFILLING THE POLITICAL OBJECTIVE

Political leadership desirous, however passionately, of innovation and willing to take risks to do it does not necessarily result in successful innovation. The Lindsay administration, even after eight years, was still a patchwork quilt of success and failure. Programs and agencies managed with a sophistication almost unknown in municipal government still function side by side with others that have survived the same environment for eight years with little evident change. The basic difference is the quality of management, in the broadest sense of the word.

The basic lessons of the Lindsay administration for governmental change and innovation constitute, in my view, an argument for explicit as opposed to intuitive management approaches. This is a case, if you will, for the Harvard Business School approach to the management of large organizations. It means a heavy investment in the apparatus of management, in research and program analysis, in management information systems, in program development and design and in systems for the monitoring and control of implementation. It requires, as well, organizational structures and agency leadership compatible with this approach. From another perspective, we can say that a high rate of change and innovation requires a high investment in activities that generate change and innovation. The parallel to the business corporation with high research and development expenditures is directly relevant.

I speak of New York City, but the case has more general applicability, with qualifications. The qualifications are largely concerned with scale, for the benefits of what may be called modern management techniques increase with the scale of the operation. In a small town, a highway commissioner or a school superintendent may perceive directly most of the characteristics and quality of the operations he supervises. In New York and other large cities, the administrator or commissioner is, like Plato's people in the cave, dependent upon an indirect representation of reality. He must "see" his agency through the abstraction of statistics and reports. The quality of his vision will vary directly with the organization and focus of these abstract representations of his operations and problems; as the scale of the operation decreases, the value of direct observation, of insight, and of intuition acquire a more important place in management.

It should also be emphasized that these comments are chiefly applicable to managerial innovation, the major problems of the cities and their governments. The process of legislative innovation could well benefit with a greater attention to probable effects but, nonetheless, the simple power of an idea is substantially greater than it is within an executive and managerial structure. In the federal government, there is a far wider arena for legislative innovations and many for which there are not substantial problems of management or implementation.

In the sections which follow, I deal with five aspects of the process of innovation in the New York City government. These are: (1) the selection of administrators and commissioners; (2) the role of organization; (3) program and issue analysis; (4) the implementation of innovation; and (5) implications for the executive office of the mayor.

COMMISSIONERS AND ADMINISTRATORS

The men and women who headed the various city departments and administrations were absolutely crucial to Lindsay's capacity to achieve the improvements and innovations he wanted in city operation. No amount of attention by the mayor or his staff proved able to compensate in any appreciable degree for the deficiencies in the perspective or the capacities of the commissioner or administrator who ran the agency.

Administering any of the largest of the agencies of the New York City government is one of the most demanding public service jobs in the United States. By comparison, the secretary of a major federal department leads a sheltered and protected existence. The political heat can be intense. Any shortcoming in performance can produce an immediate attack on his leadership. He must defend himself against criticism, negotiate with the numerous interested groups and bodies exercising real power over his operations, and avoid repeated malperformance in areas with accumulated scar tissue. His knowledge about his own operations is usually limited. Information systems have been almost nonexistent. Even accounting support tended to be weak. Budget and personnel staffs have been small and narrowly focused.

For the most important agencies, Lindsay initially followed a blue-ribbon recruitment plan, usually an effort through a screening committee to find and get the best available professional in the country in the particular field. The effort was not without successes but the failure rate was high. Perhaps this is not surprising. The reputations of those enlisted were almost uniformly earned in quieter times, in a less critical environment, and, often, in endeavors that involved neither the managerial nor the political skills crucial in New York.

By the beginning of the second term, Lindsay was critical of the selection process and set up, within the Bureau of the Budget, a talent search operation to help find individuals with a greater capacity to lift the performance of the city departments. The talent search looked for individuals not simply with reputations as good managers but with a track record of significant change and innovation in the organizations they managed. Numerous prospects, otherwise highly credentialed, failed this test. The yield of successful innovators was small indeed and difficult to recruit.

The critical perspective helped, however, even though many appointments were made with no certainty that the desired characteristics would prove to be there. One result was a substantial discounting of professional expertise. Another was the increasing use of existing middle management talent, proven in staff capacities, chiefly in program analysis, but without demonstrated management capacities. In addition, the new appointees as a group were markedly younger than their predecessors. A large proportion were lawyers.

There were other differences. The new appointees came in with the benefits of a far more sophisticated view of the problems from City Hall and with no ambiguity as to the mayor's emphasis upon improvement and innovation. The appointment of deputy and assistant administrators was no longer left to the judgment of the administrators but received the same critical review and check out by the talent search. A high and explicit priority was placed upon the recruitment of supporting staff.

The results were, by no means, uniformly successful, but the batting average on appointments took a quantum leap upward. The key was not the recruitment of "better" men, but the recruitment of individuals with talents, skills, and viewpoints more compatible with the needs of the New York City government.

ORGANIZATION

The major organizational move of the Lindsay administration was the early proposal, largely implemented by the City Council, to consolidate nearly fifty existing departments into superagencies or administrations. No change made by Lindsay has been so extensively criticized. Yet, it was critical to the effort to introduce improvements and innovations into the government.

The superagencies were critical, because they established a base or home for the construction of innovation-oriented management structure. This structure did not exist in the departments and could not be created in any but the largest of them. Talent was too scarce and difficult to recruit to staff nearly fifty agencies and, moreover, would not often find the limited problems of a single small agency an attractive area of endeavor. The grouping into superagencies provided the necessary critical mass and program scope that was essential.

Furthermore, it introduced a new tier of management that was not buried in the day-to-day problems of program operation, as were the departmental commissioners. The administrators were in a position to give attention to the development of basic improvements in programs and their operations and of new approaches and programs.

In some cases, notably health, housing, and human resources, they were also able to tackle problems where related responsibilities were divided among departments. Common approaches were needed to the Departments of Health and Hospitals and to the employment programs of the Department of Social Services and the Department of Manpower and Career Development. The effective management of housing construction demanded a bringing together of responsibilities for relocation and property acquisition with financing and construction authority.

No reorganization is self-executing, and the Lindsay plan was no exception. Reorganization produced demonstrable results under administrators able to use the new structure and few benefits for those without that perspective and talent. Where it did work, the first symptom was the buildup of managerial staff, an expansion of overhead, if you will. The ultimate effect was a measurable improvement in the quality of operations and an expanded flow of program innovations. In most of the administrations, there are now signs of a favorable impact.

The Lindsay administration also ventured in a less tidy and structured manner into new organizational forms and instruments. The creation of a new Health and Hospital Corporation, a public benefit corporation to free the hospitals of city red tape, is the most important. This has yet to fully demonstrate its value, and, moreover, the original intent was greatly qualified by the compromises necessary to secure approval by the state legislature and the City Council. New nonprofit corporations were created for research, neighborhood poverty programs, for employment programs for ex-addicts and ex-criminals, for treatment of drug addicts, and for many other purposes. Various new construction funds and the Off-Track Betting Corporation also set operations in new, separate corporate public entities. These new forms and organizations require an evaluative essay in their own right. The lessons of the mixed experience to date indicate that there is a role for variant forms of separate organizations, especially for new and innovative programs, but that they demand a custom tailoring of organization to particular purpose and objectives. The municipal government is too constricted a vehicle to serve adequately all the objectives and programs for which it is responsible.

PROGRAM AND ISSUE ANALYSIS

Within the bureaucracies of the New York City government, the most visible change is the widespread development of staff units devoted to the analysis of issues and programs. They have had greater impact upon the city government than any other thrust by the Lindsay administration.

Until the last decade, there has been little attention in American government—whether federal, state, or local—to continuing efforts to evaluate programs in terms of their costs and effectiveness, and to identify possible alternative approaches superior by cost-effectiveness criteria. Even continuing surveillance aimed at effecting improvements in programs and their management has been uncommon. Over a period of time, often a relatively short period of time, programs are sanctified as part of the order of things and fall beneath the critical threshold of those responsible for their administration.

New York was in line with the national experience. There was almost no analytic staff in any of the city agencies. An attempt to build up such staff began, as it had with the federal government, as part of the Planning-Programming-Budgeting System. PPBS was likely to become a pro forma ritual without a significant capacity to feed the system with the results of program and issue analysis. In New York, unlike the federal government, the emphasis was placed upon program and issue analysis to the point where it represented almost the whole of PPBS.

The initial build-up took place in the Bureau of the Budget, with a new division of thirty to thirty-five program planners and analysts recruited largely from new degree recipients of the graduate schools and law schools. Most of the agencies proved conservative and distrustful of the analytic approach and skeptical of the capacities of the young and green staff which would do the analysis. The recruitment of analytic staffs and the introduction of the habit of analysis into the agency staffs proved painfully and disappointingly slow. The key was the administrator and, by the end of Lindsay's second term, most of the agencies had had administrators who bought the concept and recruited the staff.

The internal staff capacity was crucial, but doomed within any finite time period to be inadequate to the enormous program and issue analyses menu before the city. With a solid internal staff to identify priorities and supervise performance, it was possible to greatly expand the amount of analysis through the use of consultants and contract researchers.

Most important in this connection was the city's decision to establish as a joint venture with the RAND Corporation a new nonprofit research institution, the New York City Rand Institute, as a "think tank" devoted to the problems of the city and its government. The Ford Foundation has provided about $1.3 million to the support of the institute, and the city has contracted with the institute for studies at a level of about $2 million annually. It has performed in many areas of the city government, with its most outstanding work in the fields of housing and fire protection. The institute gave the city a genuine research capacity, insulated and protected from the daily pressure on internal staff to "fight fires" as hot issues arose. It has a staff with scientific and academic credentials beyond those of the city's own recruits, a needed complement to the internal analysis staffs.

Management and scientific consulting firms were also engaged. In program analysis, the work of the urban practice unit of McKinsey and Company was the most significant, extending into many different city agencies. The Vera Institute of Justice, emphasizing pilot projects more than analysis, was involved throughout the criminal justice system, including drug addiction treatment.

The Institute, McKinsey, and Vera relationships were built on the principle of continuity. The continuing relationship with the city meant that the consultants and researchers were relieved of the "study and run" syndrome that is the curse of that group. Ultimately, the outside researchers became almost as familiar with the peculiar program and policy topography of city agencies as the analysts within these agencies.

It would be difficult to identify with precision the city's expanding investment in program and issue analysis, but clearly it increased to somewhere between $7 million and $10 million annually with an impact that was both massive and extensive. Indeed, the literature on analysis in urban government is now dominated by the New York experience.

Public and academic attention tends to concentrate on techno-logical changes like the use of "rapid water" in the fire department or high-powered model-building and operations research. These are important, but the greatest role of analysis was simply bringing to light the basic facts of the complex reality of city programs. Analysis dissipated the anecdotal understandings of program operation that passed as information. It gave administrators, commissioners, and the mayor the capacity for the first time to see and understand what the city government was doing through its programs and activities. The objective was more to increase the effectiveness of city programs than to reduce costs, but a mere handful of the studies produced annual savings sufficient to pay the entire yearly bill for analysis.

IMPLEMENTATION

The hardest thing in the city is getting anything new done. The Lindsay administration faced the problem first in the lagging capital construction program, where the time required to complete a project was almost incredible. For the simplest buildings on the city's capital construction program—branch libraries, police precinct stations, and fire stations—the average time from site selection to construction contract award was a few weeks short of five years.

The problem arose from the division of responsibility among many actors—often both a client agency and a construction agency, sometimes a federal or state agency providing partial financing through grants, the Department of Buildings, the Bureau of the Budget, the Arts Commission, the Site Selection Board, the City Planning Commission, and sometimes others. Whatever the merits achieved by the extensive system of clearances, it left the construction agency with little sense of real responsibility for progress toward construction.

The administration began, early in 1967, to install a project management and information system to monitor and manage projects in the Department of Public Works. The system was extended, a year later, to all of the city's construction agencies. The system was developed and initially largely managed by consultants, Meridian Engineering, a small young firm, and, later, by a Meridian split-off, MDC Systems.

The system included a modified network structure, similar to but simpler than PERT or the Critical Path Method. A prototype schedule of progress milestones was laid out for every type of facility and adapted as necessary to the special situation of individual projects. Reports on project progress against the schedule were provided monthly or more frequently to all participants. The reports identified the reasons for project delays and the office and individual in which it was currently held up. A monthly management meeting was scheduled in each major functional area to straighten out lagging projects. The meeting rarely required high-level participation; most problems were resolved simply by distribution of the information.

The impact upon capital construction progress was nearly miraculous. The city's capacity to put buildings under construction increased from a little over $100 million at the beginning of the Lindsay administration to about $1 billion a year at its end, a massive increase even after discounting for the effects of inflation. A huge program involving over 2,500 major projects became administratively comprehensible and manageable. The capital budget was converted from a wish list to a realistic statement of projects which could be put under construction in the budget year. With construction costs rising during much of this period at one percent per month, the savings to the city from earlier construction were immense, probably several hundred million dollars.

Project management systems for major new programs and initiatives came later. The mayor had established a Policy Planning Council in the middle of his first year of office. The council, including the deputy mayor, the city administrator, the director of the budget, the chairman of the City Planning Commission and the mayor, concentrated during its first year on major policy issues supported by extensive analysis. Increasingly, it became clear that the fundamental problem was not right decision-making but implementation, getting the job done. A Project Management Staff was, accordingly, established under the PPC.

The Project Management Staff took on a number of projects of high mayoral priority. One was the enforcement of the new air pollution controls on both private and public buildings. Another was the program to substitute civilians for higher-paid policemen on desk assignments. Sometimes capital construction projects were included to provide the closer control that PMS could give. The new police headquarters building and the swimming pool program were of this character.

It was not easy. Schedules had to be worked out with the commissioner of the agency with primary responsibility. Sometimes preparing and monitoring schedules was not enough. Staff from PMS were detailed to help design an information system for the air pollution control program, to redesign the vehicle repair operation in the Department of Sanitation, and to help set up a new roving crew park maintenance organization. On at least one program—the expansion of the program for therapeutic residential treatment of drug addicts—the project management staff found it impossible to deliver.

But on most programs it worked remarkably well. Some agencies began to develop comparable staffs to manage a larger package of new departures and programs than the limited group of those with high mayoral priority.

Project management is probably the single most successful and most important innovation of the Lindsay administration. It addresses a problem common to all levels of American government —namely, the job of getting anything new done within a reasonable and predictable time period.

Project management is a dotted line overlay on the traditional hierarchic organization chart. A project coordinator cuts across the hierarchy to monitor the performance of the several organizational units with direct responsibilities. Even though limited to an informational and coordinating role, this process almost inevitably raises frictions with the line hierarchy and disrupts the normal structure of accountability. For the price paid, however, it offers the prospect of a vastly larger proportion of any group of projects actually getting done on time.

The system is also an important application of management by exception, concentrating attention and effort on improvements and innovation, discounting inferentially the need to devote managerial attention to the continuing day-to-day problems of agency and operation. This is important because a very large proportion of the attention of the mayor and his political apparatus should be on the changes, the new programs, the innovations which are part of the mayor's program and with which he is associated in the minds of the public.

THE EXECUTIVE OFFICE OF THE MAYOR

Elected political executives are often judged, by the public as well as by more informed observers of government administration, on their effectiveness as managers and administrators. The truth is that, from the President down to the mayors of our smaller cities, the office of the chief elected executive is not organized or staffed to manage anything but its own mail. The evolution of the Bureau of the Budget has provided staff support and a framework for budgetary and legislative decision-making but not for management of operations in any real sense.

The initiation of the Planning-Programming-Budgeting System began to extend the role of the Bureau of the Budget into problems of agency management. The bureau, in this role, was demanding that the agency give key program and management issues the analytic attention they required. It began to help agencies find and recruit analysts. It helped set up relationships with the New York City Rand Institute and with various consultants. It became a source of advice to the mayor on agency management and progress toward improvement. The change in roles was often subtle, frequently less than fully effective, but it represented a major shift in the outlook of the bureau, a shift more oriented toward the mayor's problems in managing the system and away from the traditional Bureau of the Budget concern with fiscal control.

In New York, the extension of a managerial outlook brought into being a number of new functions and activities designed to help the mayor do his job. Without any real design, these functions have ultimately come to rest in the Bureau of the Budget. Overall administration of a capital projects management and information system was not initially regarded as necessary. But, increasingly, it became evident that the integrity of the system required some central direction, and the temporary initial role of the Bureau of the Budget has become a continuing function.

The Project Management Staff in the Policy Planning Council was eventually folded into the Bureau of the Budget on the recommendation of the PMS director. He argued that the Bureau of the Budget alone could provide the support and muscle he needed to do his job with the agency. The talent search came to budget chiefly because the mayor wanted it done there—a symptom of the mayor's high regard for the quality of the Bureau of the Budget's recruitment of its own staff. It is now institutionalized as a division within the bureau.

The mayor added further to this consolidation of functions by transferring back to the bureau the management analysis functions given to the Office of Administration some fifteen years earlier. This marked the end of an effort to create within the Executive Office of the Mayor a positive managerial focus free of the traditional negative orientation of the Bureau of the Budget. The effort failed, because the bureau was powerful and the Office of Administration, without the budgetary responsibility, was powerless.

In 1971, the mayor initiated a citywide productivity program. This program, established for each agency, targets and schedules for performance in key areas. The program was, despite its name, not confined to matters of productivity or efficiency. It dealt also with other objectives—reducing backlogs, cutting response time to client requests or complaints, increasing police patrol on the streets, expanding drug addiction treatment programs, cutting back welfare rolls, and many others. The program was important because it established, in a far wider area than that covered by project management, a system of agency accountability to the mayor on performance. The staff responsibility must, of course, rest in the Bureau of the Budget because of the close links with other bureau operations and because this is the only institutional staff in the mayor's office with the capacity to do the job.

It remains to be seen whether the Bureau of the Budget can play as many different roles over a larger period of time as the new script requires. But it is clear that, at present, the mayor has, in the bureau's new activities, the managerial apparatus and the staff support that give him some real promise of actually being able to manage the city government.

CONCLUSIONS

No detached and objective observer would yet concede that New York is governable. But we have reached that point where it is possible to say that its government is manageable.

The management form is still in the process of evolution, but its dimensions and character are becoming clear. There are several key characteristics of the institutional development:

- first, organization into large agency complexes, whether superagencies or large departments, with the size and breadth sufficient to support a modern management structure;

- second, the build-up of program analysis and research capacity to levels immense by any prior standard in state and local government, supported by a permanent outside "think tank," as well as by the use of consultants;

- third, the parallel development of the specialized capacity to schedule and manage implementation in both new programs and innovations and in the city's capital construction program;

- fourth, the extensive use of new types of organization external to the city government proper for specialized activities and for new programs;

- fifth, the development in the Bureau of the Budget of a new capacity to support the mayor in his managerial responsibilities for the city government.

One effect of these changes on city operations has been to reduce the importance of the normal hierarchic channels of responsibility. The mayor or the administrators are both likely to turn not to the operating officials but to program planning units or even to the New York City Rand Institute or McKinsey and Company to detail a program of innovation. It is even more likely that they will place primary responsibility not on the operating official, but on the project management staffs to make sure things get done. These staffs have cross-cutting responsibility, ignore channels, and are placed outside normal channels. Some major programs are being operated entirely outside the city government structure. Over time, some of these functions will be more closely associated with the operating agencies. In sanitation, for example, an Office of Productivity staffed entirely by career sanitation officers implemented with great success a productivity program designed by an outside consultant. At the request of the top career staff, other functions, such as the design of new sweeper routes are being done by operations staff rather than by the industrial engineering group. These developments are marks of success—that the traditional operating units are interested in re-claiming functions, hitherto unexercised but properly theirs, of operations improvement and productivity. The new analytic and project management units must continue to exercise oversight and evaluation functions and to carry out analyses of broader scope and greater complexity.

The ultimate balance of responsibilities within this complex structure is not wholly predictable. The important fact is that the mayor has acquired from these changes, in their many different forms, a vastly enlarged capacity to actually manage the city.

REFERENCE

LEAVENS, J. M., D. BERNSTEIN, H. J. RANSCHBURG, and R. S. MORRIS (1970) "City personnel: the civil service and municipal unions," in L. C. Fitch and A. H. Walsh (eds.) Agenda for a City: Issues Confronting New York. Beverly Hills: Sage Pubns.

6

Systems Analysis in the Urban Complex: Potential and Limitations

GARRY D. BREWER

INTRODUCTION

□ AS A CONCEPT, SYSTEMS ANALYSIS is both very straight-forward and enormously complicated. To illustrate the former, it has been defined simply as

> a systematic approach to helping a decisionmaker choose a course of action by investigating his full problem, searching out objectives and alternatives, and comparing them in the light of their consequences, using an appropriate framework—insofar as possible analytic—to bring expert judgment and intuition to bear on the problem [Quade and Boucher, 1968: 2].

To illustrate the latter, such bêtes noires as rancorous invective (Hoos, 1968),[1] overblown promotion (Little, 1966; Dyckman, 1963; Steger, 1964), suspicion (Rosenbloom et al., 1971; Bacon, 1968), and narrow technical exploitation (Mesarovic and Reisman, 1972) have been generated.

AUTHOR'S NOTE: *My thanks go to Ralph Strauch and Willis Hawley for commenting constructively on an earlier draft, to Christine D'Arc Sakaguchi for trying (with only mild success) to teach me the difference between "which" and "that" and other imponderables of our mother tongue, and to Marjorie Roach for her usual skillful job of preparing various manuscripts.*

Let us try to clarify the concept of systems analysis by exploring both its potential usefulness and some limitations to its usefulness.

The concept defined above is not new; decision makers have sought systematic, useful, and creative answers since the days of the shaman and medicine man. What is new is the number, importance, difficulty, and profound implications of decisions confronting responsible officials everywhere, especially in our urban complex. In the urgency to resolve the problems necessary for making "good decisions," an urgency often bordering on desperation, the modern urban manager has turned increasingly to anyone who claims or even appears to have answers to his problems.

The plain fact is that no one really knows what to do; no one really understands many of the complex systems in which we all participate. We have learned that lesson the hard way, as we have undertaken one social program after another only to find the problems addressed were not resolved and, indeed, were frequently made worse by such interventions (Levine, 1968). This has led to a curious dilemma having a hopelessly circular quality to it: the more we act, the worse things may become, but the worse things become, the more we are tempted—indeed, forced—to act.

A basic question underlying this issue and guiding the following discussion can be simply stated: How can a complex society institute needed changes without great cost or extraordinary disruption?[2]

Creative imagination must be brought to bear on this matter, and the systems analysis concept has much to offer, as may be discerned in a careful reading of the definition quoted above.

- "Help a decision maker": This statement implies a demand and expectation that the client-analyst relationship will be based on shared experiences and rewarded according to the degree and kind of benefits resulting from the relationship; i.e., if the analyst does his job well, both he and the client benefit—and the converse.

- "Choose a course of action": This implies that one should adopt a manipulative, rather than a simply contemplative, attitude toward the subject matter, with manifest concern for the present and emerging future.

- "By investigating his full problem": This implies the importance of identifying and specifying the problem, selecting problem elements that are of interest and use to the decision maker, and comprehensive—rather than narrow, partial, or specialized—treatment of the problem.

- "Searching out objectives and alternatives: This implies the importance of creating new solutions and assessing them in light of the goals of the participants.

- "Comparing them in the light of their consequences": This indicates the need to predict, as well as possible, the most probable rewards and deprivations likely to accrue to the participants.

- "Using an appropriate framework—insofar as possible analytic": "Appropriate" means capable of satisfying and incorporating all the foregoing requirements; "analytic" stresses the need to use data responsibly.

- "To bring expert judgment and intuition to bear on the problem": This reemphasizes the client-analyst relationship and implies that each has an important and distinct role to play in understanding and solving the problem.

Other general and apparently straightforward concepts or frameworks have allowed—even invited—well-intentioned but spurious, partial application (for one very obvious example, see Lasswell and Kaplan, 1950).[3] Subtle but critical points are ignored or are misinterpreted (for a work with this as a main theme, see Gilmore et al., 1967); opportunistic converts leap on the bandwagon in hopes of cashing in on a good thing, with resulting discredit to both the convert and his newfound meal ticket. All the while, the problems for which the concept was devised remain, multiply, and demand attention.

To recite the limitations of a concept as general as systems analysis, even restricting them to applications in the urban complex, is to assess the failings of analysis and applied research in general. That being a practical impossibility, let us examine a few limitations that seem basic enough to shed light on the greater problem yet specific enough to indicate manageable and critical tasks that remain undone.

Though the emphasis in this paper is on the limitations of the systems analysis concept as it has been applied, my intention is fundamentally a constructive one: Where have we been; where are we now; where could we go; and what seems to be holding us up?

DATA DEFICIENCIES

Data suitable for conducting analyses are in desperately short supply in the nation's urban areas. Theoretical, technical-institutional, and conceptual deficiencies abound and severely limit efforts to understand and manage the problems of our cities. Let us look at just three of them: demographic information, urban management information systems (MIS), and social indicators.

THEORETICAL DEFICIENCIES: DEMOGRAPHIC DATA

Accurate information on the size, composition, and spatial distribution of a city's population is essential to efficient urban management (for an extensive treatment of this general topic, see Morrison, 1971). Urban decision makers operate with sketchy and usually outdated information about these matters.

Demographic information is required to set the level of city services to be provided; to determine the effectiveness of these services; and to identify future problems that might require action. Urban managers show scant understanding of which demographic features determine the demand for services and little appreciation of differences in the quality of urban services or of the effects of population on those services. And there is only the thinnest of social science knowledge about the relations between demographic processes and urban social, economic, and political change, nor is there a usable theory of governmental influence on those relations. Given the lack of knowledge about the sensitivity of demands to changes in population characteristics and about the production functions relating resources to measures of urban program objectives, it is not surprising that municipal information systems, where they exist at all, are unresponsive to the urban manager's needs to determine levels of service and service effectiveness.

To realize that present social science knowledge allows only the most limited identification of even basic data needed for effective urban management is a fundamental first step in understanding why systems analysis has not begun to realize its potential. In the absence of a basic understanding of elementary social relationships needed to identify, collect, and structure information files, efforts to construct data management systems have not borne fruit. Even if these theoretical limitations did not exist, technical and institutional factors have impeded the fruitful applications of systems analysis.

TECHNICAL-INSTITUTIONAL DEFICIENCIES: MIS

Municipal management information systems have been widely acclaimed as a means of improving the acquisition, control, and use of urban information.[4] But they have yet to show success in specific applications. In part, that is because MIS designers failed to consider urban decision makers' needs. In one early instance, not only were simple operational needs overlooked by the information system builders, but implementation of the system was ultimately made contingent upon "adoption by a city of major organizational, policy, and procedural changes" (Mitchell, 1966). Such institutional flexibility seldom exists, and in this respect creating a management information system differs little from any fundamental effort to innovate (Brewer, 1973b).

While each of the preliminary efforts to create and implement MISs is undoubtedly different in detail, there are many common features worth summarizing as a warning to unwary analysts and as a partial explanation for the limited utility of MIS so far. It is abundantly clear that, at some future time, analysts will depend heavily on these systems, but that time, unfortunately, seems quite remote.

- Large-scale, ambitous, comprehensive attempts to build information systems have a high probability of failure. The larger the scale, the greater the visibility and the more obvious the inevitable mistakes. The larger the scale, the greater the number, range, and vagueness of advertised end products, and the less likely their actual achievement. The greater the scale, the greater the perceived threat to an organization's integrity.

- It has yet to be demonstrated convincingly that a management information system enhances the relative political control or leverage of various decision makers.

- Research and technical specialists routinely justify their activities as "advances in the state of the art." That is, decision makers are seldom explicitly considered in the design of information systems.

- Insufficient attention has been devoted to the local context.

- Data collection and analysis are costly and have not become economically self-sufficient mainly because the information collected has not had demonstrable value to users.

- Research is needed to outline a master plan for MIS development and implementation. Continued funding of diverse systems may well be a waste of scarce resources better spent on one or a select few carefully controlled prototypes.

- Any information system has policy implications. The identification, collection, and analysis of data structure and simplify the world. The difficulty is to make these changes without disrupting ongoing processes unnecessarily—i.e., to be unobtrusive, but effective.

- Urban systems seem constitutionally subject to delay and reluctance to innovate. Destructive opposition is far easier than constructive action in the system.

- Urban governments are by and large ill-equipped to perform sophisticated data collection and management, not the least because they lack a stable complement of well-trained technical personnel.

- The definition and standardization of data have not been achieved.

CONCEPTUAL DEFICIENCIES: SOCIAL INDICATORS[5]

The concepts about social indicators are not very well defined, the uses of indicators are not well established, the catch-all notion of "quality of life" keeps intruding on rigorous and orderly measurement, and myriad obstacles—technical, conceptual, and definitional—impede the realization of the mythical social account.[6]

Uses. The uses of social indicators can roughly be divided into scientific and political, with a number of variations.

Scientific uses of social indicators include the generation of comparative information about different groups in a given geographic area with respect to an explicit time dimension. Included in this use are attempts to measure attitudes and values with respect to crime, pollution, housing, transportation, employment, ecology, and many other possible issues, both actual and potential. Many of these efforts have been thwarted by the lack of information in readily accessible formats for use in index and indicator construction. The result has been a plethora of special, one-time, unique surveys and not-too-successful attempts to adapt "surrogate" data to these purposes.

Political uses of indicators are not well established, but hints of such uses are provided in recent revelations about efforts to replace professional directors of the U.S. Bureau of the Census and the Bureau of Labor Statistics with nontechnical "friends" of the administration. With this in mind, the following list of potential political uses is rather easily generated: lobbying pressure; bases for claiming resources; ammunition in an adversary process; bases for organizational alliances and coalitions; symbols for persuasion; arguments for the mobilization of affected populations on certain issues. This incomplete list indicates the political potential inherent in the gathering of social indicators. It is only a matter of time before imaginative politicians realize that these uses might be turned to their own private advantage. When that begins to happen on an appreciable scale, the movement will in all likelihood blossom. Whether the benefits will accrue to the politicians or to society in general is uncertain.

Quality of life. Quality of life has been the rallying cry of many proponents and theorizers of social indicators, but it means many things to many people. The concept defies clarification and measurement.

At a minimum, we can distinguish between objective and subjective dimensions or components of the quality of life. The former is manifested by efforts to measure the condition of the environment, such as the kind and amount of pollution, the amount and quality of existing housing stock, and so forth. Assessed also are individual people's conditions, such as their status of health, level of education attained, the stability of their family, and so forth. In the second, or subjective dimension, personal experiences and attitudes are assayed: What are individual levels and kinds of frustration, satisfaction, aspiration, and perception?

The problem, simply stated, is the interconnection of the objective and subjective dimensions. Their relationships are not easily determined, and the forwarding of substitute or surrogate measures of the objective dimension, as if they truly represented the subjective, contribute little to the matter. Improvements in objective attributes do not necessarily relate on a one-to-one basis with improvements in subjective attributes, for instance. Real income, housing quality, the status of health, welfare, and pollution may all improve significantly, while at the same time such subjective dimensions as personal satisfaction with any or all of the objectively measured dimensions may decrease. Sweden in 1973 is a national case in point. By most objective measures, Sweden never had a better year, but indirect assessments of individual happiness[7]—i.e., the "quality of life," in a sense—did not show such improvement.

Summary of conceptual problems. Problems with social indicators have direct and serious implications for the construction and use of urban social data. And though those problems look circular—we have very little good analysis because there is so little usable theory; we have little usable theory because there is so little solid information and data; but there are very few decent data because there have been so few good analyses to generate theoretical insights that would tell which data to collect—progress will probably be made to break through the circle at various points over time. It is therefore important to understand what the present situation is to realize that the problem of doing good analysis is nearly as complex as the subject matter itself.

For example, underlying the subjective dimension of social indicators is a need to measure individual-level detail having to do with personal net-value assessments of wealth, power, well-being, affection, and so on. Furthermore, such assessments need a rich comparative and contextual basis, although comparison of what with what is not certain. Should the individual net assessments of value be compared against some absolute standard or goal, against an identical individual assessment at some earlier time, or against other groups in the population differentiated according to class, geographic location, socioeconomic status, or what? The answer to this, and many other basic questions, is just not known; in fact, we have only lately begun to pose the questions (these problems have been identified and discussed by Harris, 1973).[8]

There is an even more fundamental problem that limits the creative application of systems analysis to society's problems. Without precise and specific statements of goals—both for individuals and collectivities—attention in the social indicators movement has focused on the development, collection, and description of objective indicators. The basic problem, and rationale, for this preoccupation with measuring "things" is that we just do not know how to relate the objective and subjective dimensions well enough to do the measurement. One does not have to delve into problems as deep as the interrelationship of subjective and objective dimensions to run headlong into some serious difficulties with social indicators.

Problems of simple definition abound. It is difficult to distinguish between genus and species, and agreement about terms is subject to change. For example, the concept of alcoholism seems relatively straightforward, but drinking incidence, chronic alcoholism, and drunkenness are not the same things, and intermixing measures of each leads to serious questions of just what is being measured. The measurement of crime illustrates the second problem, that of changing definitions. There is no rational classification of crimes based on a set of properties that defines all crimes. Law often helps, but it is incomplete, varies from locality to locality, and is subject to change and reinterpretation. All three such properties in a concept being measured seem to indicate that the generation of stable, reliable, time-series information (a prerequisite for meaningful social indicators and hence useful urban analysis) is far from being realized.

The question of the level of detail or aggregation is often not well handled in discussions of indicators or in efforts to make and use the measurements. Most indicator information is aggregated at the national level; urban indicators, for instance, are mostly national ones, and thus there will probably be a powerful tendency to seek national solutions. But as any urban official will tell you—frequently with great indignation—*his* town is not the same as another, *his* problems are different and the solutions being forced on him by federal programs are not only inadequate but create more problems for him than they resolve. Most poignant is that the official is probably right on all counts.

Another major flaw in current activities relates to the objective-subjective distinction made earlier. There is very little individual-level information on perceptions, attitudes, expectations, and other concepts indicative of personal satisfaction. Instead, substitute measures are often used and misinterpreted as if they were measures of personal satisfaction. For instance, the divorce rate is often cited as a measure of satisfaction with the institution of marriage. However, we really do not know whether an increase in the divorce rate indicates a breakdown of the family as an institution as much as strains and frustrations with some other aspect of society. It has been hypothesized that high rates of divorce and family instability among minority groups are actually indicators of frustrated attempts to share in society's available power and wealth. If so, do divorce rates among rich, white suburbanites mean something quite different from what they mean for poor, black ghetto dwellers? Quite possibly, and the implications for the construction and use of stable social indicators are profound.

This calls attention to a technical problem—the need to standardize indicators so that time series can be generated that bear a close similarity to what is being measured in early and more recent times. The difficulty is that we cannot satisfy the statistical demand for stable definitions and reliable measures in a setting where the definitions are changing, where interpretations vary from group to group and time to time, and where differences in both absolute and relative terms from subgroup to subgroup of the same population are not even vaguely understood.

Other conceptually based technical problems plague the social indicators movement. Their variety is indicated in the list below.

- Social phenomena are not simple and require multidimensional treatment. A single index, by itself, is insufficient to represent most social phenomena.

- Quantitative elements tend to be more easily measured and hence overrepresented in social accounts and in social analyses.

- Reinterpretation of data collected for other purposes is seldom satisfactory.

- Patterns discerned in data change from one unit of analysis or analytic detail to another in response to different underlying patterns of expectation, identification, and demand. Hence, a simple correlation of percentage poor and a crime indicator may be quite high for a neighborhood, moderate for the city overall, and insignificant at the national level.

- Data often exist, but not in the public domain. Health information is particularly susceptible to this malady.

- Institutional biases contaminate many potential indicators. Data on social programs reflect administrative convenience more than the individual citizen's view of the situation.

A recitation of these problems is cause for both despair and hope: despair at how long the road is ahead and hope that we are beginning to piece together a map that might enable us to get from here to there with foreknowledge of some of the hazards along the way. We know that the terrain is enormously complex, although the nature and extent of that complexity are perceived in many different ways by those trying to find their way through it.

THEORY

THEORETICAL ORIENTATIONS

The analyst's general attitude toward his subject matter is crucial in determining how the subject will be perceived, analyzed, and interpreted (Deutsch, 1966). If one has a predilection for seeing linear, orderly, static, and rather simple patterns in the world—as a statistician or an actuary might—that suggests strongly the preferred methods and likely forms his abstracted views of the world will take. If, because of prior training, personal idiosyncracy, and the like, one looks at the world as composed of a small number of important elements configured in nonlinear, deterministic ways—as an engineer might—another set of methods producing a different world view will result.

The first problem orientation has been labeled *disorganized complexity,* where "[though] the number of variables is very large, and . . . each of the many variables has a behavior which is individually erratic, or perhaps totally unknown . . . the system as a whole possesses certain orderly and analyzable average properties" (Weaver, 1948). The second has been called *simplicity.* Weaver (1948: 538) supplemented these with a third orientation, *organized complexity* (for an adoption and application of this orientation to another substantive field, see Brunner and Brewer, 1971), "problems which involve dealing simultaneously with a sizable number of factors which are interrelated into an organic whole." Most social problems, including most of those confronting urban managers, conform more to the third orientation than to the first or second. This fact is important because the analytic tools developed along with Newtonian mechanics to manage *simple* problems and those created later to overcome statistical problems of *disorganized complexity* have not been matched for problems having properties of *organized complexity.* There are several implications for systems analysis (many implications are discussed in Simon, 1967; another recent and applications-filled source is La Porte, forthcoming).

Apologias about system complexity routinely appear in urban analyses, but with only the slightest sense of what is meant or implied by the concept. The following is typical:

> The scale and complexity of many urban problems, particularly social problems, are substantially greater than those customarily encountered by defense and space industries ... [and] the institutions responsible for coping with urban problems are not well suited for the management of change on such a scale [Rosenbloom et al., 1971: 19-20].

Just as there are discernible distinctions in general orientations toward subject matter, so are there differences in the approaches of individual systems analysts (compare the discussion of points made in the following section with Kuhn, 1964, and Hirschman, 1970).

Simplicity. In systems analyses exhibiting a simple orientation, two kinds of work and approach can be seen: mechanistic and utopian.

The mechanistic approach is predominantly associated with work produced by M. D. Mesarovic and many of his students and colleagues. It manifests a so-called multilevel or hierarchical-multilevel view of the world (Mesarovic and Reisman, 1972; Mesarovic et al., 1970). The view is not particularly novel and is directly linked with a classic problem of system decomposition. Decomposition implies some powerful assumptions about the context in which it is applied (Ando et al., 1963: 92-109),[9] and to the extent that these assumptions are untenable, the view and approach are, too, for the most part. Because of problems with recomposition (on the problem of recomposition, see Coleman, 1964: 444-447 f.),[10] data aggregation and information loss (see Orcutt et al., 1968),[11] and noncomparability of meaning for elements considered at different analytic levels of detail or resolution,[12] the application of the multilevel approach has been limited to simple engineering-like problems (Haimes and Macko, 1970; Matuszewski and Lefkowitz, 1970; Drew, 1972).[13] However, the approach is regularly being promoted as a way of managing urban complexity (Richardson and Pelsoci, 1972).

The utopian approach, based on a simple view of the world, is discernible in work spawned by Jay W. Forrester, who is, incidentally, an electrical engineer by training (see Forrester, 1969). This is not the place for a lengthy comment on the work itself; in a word, it reflects a simply viewed world (for a full critique, see Brewer and Hall, 1973). That is not a novelty either, being rooted in an ancient tradition documented in Reiner (1963) and interpreted in Boguslaw (1965).

Disorganized complexity. Examples of systems analyses that are oriented toward a view of the world as being in a state of disorganized complexity are too numerous to cite in full. Briefly, two representative subapproaches exist: the naive and the regressive.

The naive approach is well illustrated in Easton (1957, see also 1965), and more recently in Lineberry and Sharkansky (1971: 1-15 and generally). Extraordinarily complex systems are reduced to simple "models" having six to eight elements;[14] interactions among components are usually treated in reduced and abstracted form as simple, n x m, tables of correlation coefficients deduced from cross-sectional information. The number of relevant system elements is explicitly treated as being small but composed of many constituent parts whose "behavior" is reducible to simple mean values, which (if measured at all) are presumed to be derived from populations having regular statistical properties.

The regressive approach has been the special province of economists, and one extremely well executed, even exemplary, illustration of the genre is Muth (1969). It is impossible to tell where the analyst's technical skills end and his thorough substantive knowledge takes over. There is no substitute for knowing one's business thoroughly: better that an analyst have a minimal amount of data and an in-depth knowledge of the problem and context than tons of data and little or no substantive knowledge.

Organized complexity. There are relatively few examples of work that demonstrate a sophisticated understanding of the complexity of the urban context. Here again, however, two representative approaches are discernible: the utilitarian and the realistic.

A distinctly utilitarian flavor permeates Rosenbloom et al. (1971), a series of case studies from attempts to apply analytic techniques to urban management problems in several cities. These cases well document many of the practical difficulties encountered in the process, and in a no-nonsense, utilitarian, even hard-headed way. The cases themselves exemplify the mixed results obtained when some of the "standard" systems-analytic techniques are brought to bear on city problems. In Dayton, Ohio, for example, the weakness of existing social science theory from which quantitative models could be specified was only overshadowed by the poor quality of the urban data from which models could be specified and driven. In East Lansing, Michigan, the analysts were compelled to work for the "wrong" clients and were relegated to answering the "wrong" questions—a warning that the research environment may be as important as the research itself. And in New York City, the fate of the old systems analysis standby, Planning-Programming-Budgeting (PPB),[15] was found to hinge more on key personalities of uncertain tenure, the tendency of bureaucracies to distort innovations into often grotesque semblances of what was expected or had been realized in other implementations,[16] and the availability of a select few individuals possessing the required analytic skills (Bales, 1971), than on the pristine virtues or even the rationality of the PPB method itself.

The realistic approach is demonstrated by Lowry et al. (1972)[17] in their grappling with the organized (if somewhat perverse) and complex interaction of housing and welfare in New York City. That their work is not widely known indicates the relationship between client-analyst has been valued, for the moment, more than that between analyst and professional colleagues. Their study assesses the costs, results, and weaknesses of contemporary welfare housing policies in New York City. Alternative policies to improve the quality of housing for a fixed cost are created, and their various implications are estimated. Existing housing theory is relied on and enhanced by a rich assortment of methodological and logical exercises performed on a detailed, painstakingly assembled data base. Implementation strategies are laid out in sufficient detail and in plain English so that interested parties can make the informed choices required in the setting. Finally, the "multilevel" issue stressed in other systems-analytic work is treated realistically by subjecting the findings from New York City to experimental tests in several other settings to begin the hard task of piecing together a system view and policy on the national level (Lowry, 1971a, 1971b).

THE LIMITS OF THEORIES AND APPROACHES

Systems analysis is an idea general enough so that many different efforts end up being tarred with the same brush. It is a curse of general categories that a simplifying label often disguises underlying diversity. Such diversity is indicative of the creative latitude allowed the systems analyst in selecting his topics, structuring his problems, applying his methods, and formulating his results and recommendations. The quality of the product will vary with the judgment, knowledge, and skill of the analyst in carrying out each of those tasks (Novick, 1954).[18] However, this fact has been neglected, and the crucial importance of the analyst has been overshadowed by emphasis on methodological techniques, such as model-building and computer simulation (see Brewer, 1973c),[19] particular theories about the nature of the subject matter, such as the variations on the complexity theme already discussed,[20] and concern for existing, hard data to the virtual exclusion of important but hard-to-measure aspects.[21]

Formal model construction and interpretation, for instance, are basically judgmental (a point stressed by Quade and Boucher in their definition of systems analysis).[22] The approach adopted suggests methods for dealing with the subject matter but is no guarantee that valid results will naturally follow. Methods, as applied to problems characterized by their "organized complexity," have not yielded much more than insights and, occasionally, partial solutions. And those results owe more to the skill and judgment of the analyst than to the power or rigor of the method per se. Ignoring this, the methods-oriented analysts—e.g., many "regressionists"—behave as if the world were multivariate and Gaussian.[23] As one waggish colleague has remarked, "If the world were linear, George Dantzig [a foremost linear programmer] would be king."

This observation on the importance of the analyst's skill and judgment brings us to the more general consideration of how one selects items for analytic consideration (and, hence, judges the items not selected less important).

PRINCIPLES OF SELECTION[24]

SPECIALIZATION

The deadening effect of the human mind. People tend to be trained to think in narrow, linear, and sequential ways about life and problems, whereas very little in the world can be described by those adjectives. De Jouvenel (1967: 271) has stated the point well.

> We seem to have an innate liking for figures that are simple, regular, and symmetrical, which we look for in nature and society. But the history of scientific progress tells us of complex organizations found

precisely where we looked for simple order. . . . If the human mind had devised the world, it would be a dead world.

The deadening human mind is all too evident in many of the systems we have created (and analyzed) to serve urban citizens.

Professionalization. Professionals have been captured by their profession—its imperatives, perspectives, and tools. Political scientists, who are often very liberal in their general views about humanity and human behavior, become very conservative when it comes to specific details about designing, executing, and communicating the results of an analysis to other participants in the urban setting. They wear political scientists' glasses. Politicians wear glasses of a different refraction and tint, administrators wear yet another, and so forth.

To be a specialist means that one has concentrated on a limited field of knowledge. But there is a mismatch between the narrow and highly detailed information dealt with by a specialist, the somewhat more general and comprehensive information required by an urban manager, and the broad and grossly detailed information required by the average urban citizen. This fact has pronounced implications for the principles of selection employed in structuring an analysis, operating a city administration, or living in a city (Marney and Smith, 1972).

PRINCIPLES

One way of beginning to assess an urban analysis would be to identify which kinds of issues have been routinely considered and which participant's interests have been best served (for a clear statement of the issue, see Michelson, 1965). What topics, for instance, has an analyst selected for study, and what other topics have thus been omitted from the analysis?

Rationales. Many scholars and analysts use the stock rationale that their work contributes to science and to their professional specialties by generating knowledge and fresh insights.

Second, and less frequently, it is either acknowledged or can be inferred that selection has been based on calculations of expected rewards: increased status, prestige, income, security, and so on.

A third possibility is that the scholar-analyst has in fact worked in the interest of a societal objective, as indicated by conscious efforts to identify, define, rank, and then select research issues in a given context.

To discover what is involved in problem identification and problem selection, it is helpful to think about who constitutes "society"; that is, who are the relevant participants in a specified urban context and what do they consider to be their most urgent problems? The information sketched in Table 1 suggests roughly what might be involved.

Even the partial list of likely participants in the table forces one to face the fact that problem identification and selection are very complex and rich topics indeed. Not only are lists of likely problems different for each participant group, but the lists will differ from one location to another and over time for the same classes of participants. John Lindsay's list of salient issues for New York City was not the same as Nelson Rockefeller's; it is a safe bet that Mayor Beame's present and future lists of issues are different from those that Lindsay drew up. As obvious as that fact is, once stated, it is distressing to see just how little attention we analysts have given to it in the conduct of our affairs.

Implications. Assuming that the analyst chooses to be guided by societal objectives, his burden is to determine in numerous ways what the dominant issues in fact are. Because this is a heavy burden, much research and analysis today are rationalized more according to the imperatives of the individual scholar-analyst's discipline than by the imperatives of a given society at a given place and time.

When one arrays potential issues for study in this way, several important questions emerge. What portion of a given context's problem-issues has been selected for study and hence is judged to be more important (for a relatively concrete treatment, see Brewer, 1972: 17-26)? Whose preferences and biases are reflected in the items selected for study (Kemeny, 1960: 295-296)? Why have other portions of the problem space not been treated, and what are some likely results (Gould, 1967)?[2][5] To what extent do the selections represent the preferences and needs of the general society as compared with the preferences and needs of the individual analyst and his profession (Dewey, 1939)? And how are the implications that follow from this treated (Lasswell, 1971: chs. 5 and 6)?

These are important matters, for the selection act represents what our French colleagues refer to as a *déformation professionelle:*

TABLE 1
SOME URBAN PARTICIPANTS AND THEIR PERCEIVED PROBLEMS

Participants	Problems (partial lists, ordered according to imagined saliency)[a]
Administrators:	Finances, budgets, information.
Local	
Planners	Zoning, land use, environmental-impact statements,
School officials	finances, demographic trends, effectiveness,
Police	crime rates, personnel use, public image.
State	Similar problems on different levels of administration.
Federal	
Politicians:	Corruption, elections, taxes, inflation.
Local	
Mayors	Variants of these topics plus local issues.
Councilmen	
State	Variants plus state issues.
Federal	Variants plus federal issues.
Researchers:	Theory development, methodological application.
Social scientists	Problems change according to the time. For
Economists	instance, "hot" topics in economics ten years
Sociologists	ago were development and macro-theory;
Political scientists	today, other issues tend to dominate.
Applications specialists	
Operations researchers	
Planners	
Administrators	
Public	
Business	
Computer scientists	
Citizens:	Inflation, taxes, housing, schools, health, trust.
Economic subgroups	
Poor	Issues would have to be determined for each
Middle class	subgroup in specific locations at a specific
Rich	time.
Racial subgroups	
Political subgroups	
Demographic (age) subgroups	

a. Though these lists are purely speculative, such information could be determined by various measurement techniques: surveys of citizens' views, interviews of administrators, and so forth.

specialization in the interest of a sharpened and narrowed focus that prevents one from being concerned with all matters lying outside this focus (Barrett, 1958).[26] We all need to be reminded that any one narrow view alone is insufficient to understand and manage a complex subject; concentrating on the selection process is such a reminder.

DUE REGARD FOR CONTEXT

People generally have trouble thinking comprehensively (see Simon, 1969). Because they do, procedures that might improve understanding of the relationships between parts and wholes —emphasizing the interaction and change of past, present, and future events—should be explored and exploited.[27] Such procedures require a broad framework to indicate significant phenomena in a setting (the "principles of selection" issue) and to provide a tentative conception of the whole. Systems analysis, taken as a collection of procedures held loosely together by a comprehensive framework, satisfies many of these requirements—particularly in the hands of a competent analyst.[28]

LEVELS OF ANALYSIS

Different levels of analysis require different kinds of data, different analytic methods, different research modes, and so forth. Insufficient attention to such fundamental contextual matters has routinely limited the usefulness of analyses undertaken in the urban settings.

Because levels of analysis are regularly confused and mixed in the practice of analysis, it is useful to consider a scheme to differentiate them.

The following is only one possible way of delimiting representative levels of analysis:

- *Individual-level detail,* including studies of elite behavior, in-depth psychological studies of key individuals, and psychological profiles with extensive background information about the socialization of ruling and key decision makers.[29]

- *Small-group interactions,* including intraorganizational studies, comparative studies of related decision-making bodies, and crisis behavior (Crecine, 1969; compare Lipsky, 1970).

- *National-level analyses,* including national character studies and macro-process analyses.[30]
- *International relations,* comprising much of the literature so identified.[31]

Bearing in mind the difference in structure, process, and behavior at each level of detail would do much to clarify the matter.

If the properties at some higher level of analytic detail cannot be deduced from statements about constituent parts, and this seems to be true for most complex, organized systems, prediction will not always be possible. However, this strictly logical constraint does not foreclose on the obligation to do analyses that are more than good guesses. Emergent phenomena can be predicted using a consistent frame of reference, and not by simply extrapolating from a lesser to a higher level of detail. Emergent phenomena are always contingent on the total context in which they occur; however, because the context may be unique, prediction can be hazardous. Critical configurations essentially defining the context will never reappear in exactly the same fashion, although the likelihood of an event's recurrence is greater if it has happened "often" in the past. The basic problem is intractable because there is no sure means of demonstrating that the future will contain configurations analogous to those in the past.

All this implies that there is a great and continuing need to rely on multiple approaches to and perspectives on the subject matter, to use a variety of methodologies (extant and yet to be developed), and to conduct the analyses from several distinct but related levels of analytic resolution or detail. All these tasks should be done in as rich a contextual setting as possible.[32]

TIMING AND POLICY-MAKING

The probability of a predicted event's occurring depends on the state of theoretical knowledge, past historical routines, and the net interplay of values associated with participants at each of the pertinent analytic levels of detail.[33] Future events are, furthermore, subject to modification, deflection, or reinforcement by manipulative actions taken in the interim; such actions are, of course, what we commonly call policies.[34] Thus, to be concerned about policy is to be concerned about several related aspects of any specified context—e.g., operating goals, past trends, present structural relationships. To make assessments about the future—a prime objective of systems analysis—is to figure out what might happen in a specified context if nothing were done to intervene, if several plausible and feasible interventions were undertaken, or if only one option were taken.

CREATIVITY, SKILL, AND JUDGMENT

Devising those options, it is not generally recognized, is independent of the processes involved in analyzing the options (Levin and Dornbusch, 1973).[35] Relating actualities to possibilities involves one's creative orientation to the subject and mastery of the context, and anything that might be done to stimulate this creativity or to increase knowledge about the context is to be recommended. The analyst must move between contemplation of the narrowly specified details of a context and its fullest, most information-laden, and complex manifestation. This process and its results are more a function of the analyst's skill, knowledge, judgment, and creativity than of automatic projection, a point ill-appreciated by those who might be called mindless theorizers and insensitive data runners, to name only two cases of overspecialization.

In describing some of the basic limitations on the productive application of systems analysis, it has been necessary to operate on a fairly high and abstract plane. However, analysis takes place in the real world, which itself generates many limiting conditions. Those practical difficulties are worth considering next.

PRACTICAL DIFFICULTIES: IMPORTANCE
OF THE ANALYTICAL ENVIRONMENT

Why would a decision maker want to commit resources for the analysis of a system? What is in it for him? And what are the implications for the analyses that result?[36]

MOTIVES

If these simple-sounding questions are raised at all, and they seldom are,[37] there is often an implicit assumption that the motives underlying the willingness of policy makers to allocate scarce resources to conduct what often are very expensive systems analyses are basically the same as the motives that analytic specialists have. This assumption is usually incorrect, and it is important to delineate some plausible motives that may be guiding policy makers in buying analysis.

One motive might be to provide the appearance of using science; and in politics, where image is often treated as if it were more important than actuality, showing one's constituency that you are modern-thinking and management-oriented may be decisive in selecting systems analyses over intuition or other conventional decision-making techniques. I strongly suspect that many of those responsible in the federal bureaucracy for the millions of dollars spent, for example, on some of the large urban computer modeling projects and on the design and implementation of management information systems cared far less about the results or products than about receiving personal credit for having secured distinguished technical personnel to carry out the work.

Another possible motive concerns the large and expanding computational capacity in all levels of government. A good bureaucrat is not going to let a sizable computer investment and supporting staff sit idle. Excess public computer capacity may not be reduced by selling time on the private market, which suggests something about the kinds of analyses that are likely to be undertaken. It suggests the building of big analytic models, finely resolved in level of detail, models that require lots of time to process vast quantities of highly disaggregated information.[38] If one has a big machine and a lot of bright machinists, they must be kept busy.

Another reason for buying analysis might be simply to defer making a decision about a problem. Facing a general problem as complex as urban renewal, for example, a beleaguered decision-maker might think it smart to let a contract for a multimillion-dollar analysis that promises to resolve all problems at some vaguely specified future date. The bigger the project the better, because the bigger it is the harder it will be to deliver on time and the harder it will be for anyone to understand it.[39] If the decision maker is especially shrewd, he ensures that the delivery date for the final report falls when he expects to have moved elsewhere.

Pure and simple advocacy may primarily motivate a decision maker to enlist an analyst's assistance. This seems to happen more frequently in military applications than in urban ones because the military consumer is ordinarily technically more sophisticated than his urban counterpart. Advocacy would be evident motivation when a policy preference is clearly stated before the analysis is carried out; the analysis then "validates" that choice in subsequent resource claiming and adversary action.[40]

I do not rule out the possibility that a decision maker has a real problem to which he earnestly seeks a solution. I merely suggest that alternative motivations logically exist and must be considered.

CLIENT RELATIONS

The importance of client relations cannot be stressed enough. For technical specialists, talking to the client is often low on a list of priorities. The technician's "thing" is doing analysis. Who wants to be bothered by some ignoramus who neither understands nor appreciates the beautiful techniques, the theoretical elegance, or the rigorous treatment of data associated with a difficult analysis?

When the technician's penchant for minimizing client interaction prevails, what often results are analyses that, despite their elegance and so on, are not understood, are not relevant to the decision maker's actual needs, and hence are not used. Or client interaction may be limited to manager-salesmen specialists who may not adequately grasp the details of the work, with the result that client expectations become inflated beyond the capabilities of the analysis. Whether an analyst likes his client or not, he must interact, teach, and learn from him.[41]

The simple question of what the problem is, is often never asked; the answer is assumed. Without good client interaction, the answer is extremely hard to determine, for what is important is not the way the analyst perceives it, but the way the decision maker perceives it. Sadly, few decision makers are able to articulate their problem clearly enough that an analysis can be well or quickly undertaken to resolve the problem.

Even when the problem is determined, analysts seldom trouble to match the problem with an appropriate mix of methodologies; rather, the reverse happens all too frequently. One has a limited bag of methodological tools that he uses whether the problem is amenable to them or not. The investment of time and talent learning a certain technique predisposes an analyst to use that technique, regardless of the specific problem being considered.

This tendency is encouraged by current administrative procedures.

For instance, a request for proposal may prespecify the methods and approaches to be taken in solving the problem. Ideally, a client-sponsor should first hire good people and let them help formulate a statement of the problem, select the correct mix of analytic tools, and so forth. In a setting where good people are scarce (besides costing a premium) and shoddy work is commonplace, method-specifying procedures are understandable, if somewhat counterproductive.

A related question that is seldom asked is what an answer is. That is, what is an answer or an acceptable range of answers? Should it or they be sought or allowed to "fall out" of the analysis?

What the problem is and what an answer is are deceptively simple questions. They should be asked early in the analytic process, during the proposal or design phase. These questions assume mutual trust and respect between client and analyst. They presume that both parties can sit down, be honest with each other, decide what the problem is, and then ponder a few plausible answers. The question of what an answer is, though less often raised than the first, is just as crucial because it helps ensure that the selection process (discussed above) will operate to include, not exclude, needed information. Information that would be technically optimal is seldom politically or socially preferable; the question, if honestly posed, may present an analyst with a serious dilemma.

On the one hand, he can satisfy specialized professional norms by using readily available information and well-known methods to produce elegant, albeit irrelevant, analyses.[42] On the other hand, if, after considering the two basic questions, he decides that the available information[43] is not complete or appropriate enough to allow him to conduct the analysis, he may be forced to make a tough choice: (1) He can refuse to carry out the work on the ground that doing so will compromise his professional standards. (2) He can request additional resources (time and money) to collect the missing information before tackling the analysis—and thereby jeopardize the whole deal. (3) He can take on the job, do whatever is possible with the limited information within the contractual constraints, and make his recommendations in either vague and equivocating terms or in such profusion that one item on the final "shopping list" is bound to be the "right" answer—although it is the client's job to "pick the winner." Or (4) he can be honest, telling the client when and where information is deficient, doing his best to fill in the gaps as the analysis proceeds,[44] and standing by a select few recommendations at the project's end, for better or worse. Honesty is not always the best or the most prudent policy, unfortunately, and the fault is shared by client and analyst alike.

Political considerations have a way of intruding. What was an urgent problem when a large project began may have become less critical to the decision maker with the mere passage of time. Does the analyst trudge along, make serious compromises with the initial design of the project, forget the whole thing, resort to a "purely" scientific mode to satisfy professional colleagues, or what? What does one do when political fortunes,[45] whims, and fads[46] change, especially if one happens to be locked into a relatively long-term large-scale systems analysis?

BUSINESS AS USUAL

According to a cynic's definition of the validation of a completed systems analysis, validation is a happy customer. The ultimate validation is a follow-on contract. If the analyst makes his client happy enough that he receives more money to do more analyses, then the work may be thought to be validated. Stating the point is not the same as morally subscribing to it; it does appear, however, to be the way the world routinely works.

Business as usual cuts another way, and a client may find himself trapped by the analyst[47] —particularly in the case where large and complicated formal models or simulations have been developed.

Many large-scale models are not well documented (Brewer, 1973). The people who construct them may have a proprietary interest in keeping documentation hazy or even secret. When model-building analysts are asked about documentation, they routinely say, "It's proprietary; it's under development; we haven't had time to get around to it; we have a deadline to meet." One quickly discovers that there is no documentation or that the documentation is suitable for use only by the analysts who built the model in the first place. This may present some serious problems when the model user poses a new or slightly different question for his model to answer. Quite typically, the model will not be able to answer the question directly, an eventuality that the analyst provides for (often legitimately) in a follow-on contract to modify the model.

What options does the client-user have? He is probably not technically able to run the model; he probably does not have the documentation to allow someone else to do it for him; he has been trapped.

The onus is shared, however; the situation is not necessarily the result of devious or unscrupulous behavior on the part of the analyst. Bureaucratic agencies often have a high turnover of personnel knowledgeable about research or analysis. Development time for a large systems analysis may extend two or three years or longer, while the average tenure of individuals in an agency may be shorter, with the result participants at a project's end may not be the ones who initiated the work. In such a situation, documentation becomes all the more critical.

THE ANALYTIC PROCESS, AND
HOW TO APPRAISE IT

The logical phases in the life of an analysis include a proposal or design phase, a specification phase, a control phase, and a validation phase. An important part of validation is deciding whether the analysis is any good or not. To aid that challenging task, let us consider a few appraisal or evaluative questions.

THE PROCESS

The proposal or design phase relates to questions about who wants the analysis, what he intends to get from the analysis, and what is a reasonable statement of the problem and some answers to it. Is the work going to be used by a decision maker, or is it being done for different reasons? Specifically, what is the purpose of the work? If these questions are not answered adequately, it will be very difficult to evaluate the work later because evaluation depends on the basic purpose of the work.

Specification is the technical-theoretical phase. Are the problem elements identifiable? Can one select analytic elements that reflect societal objectives more than narrow professional ones? Can the elements be related systematically? Are there theories capable of informing these relationships? What data exist; what data have to be created, generated, adapted, and so on? As indicated above, the process may very well abort right here if the questions raised are honestly answered. Ideally, for many social problems, if one only got through the specification phase, realizing that there were no data and even less theory, and then decided to halt or greatly modify the process, it would still be an honest, even honorable effort. Reality is seldom ideal, the problems are not likely to go away, and someone is going to be making choices—analysis or not. One learns to do the best he can with what is available or close at hand.

The control phase pertains to good technical management of the data and analytic procedures used. It is then that the analysis is most likely to break down. Urban data do not exist in sufficient quantity, quality, or under sufficient control to allow much or very ambitious analysis. This judgment applies in spite of the large management-information system movement in the United States over the past fifteen years.[48] Nowhere do data exist to allow one to conduct large-scale systems analyses, such as building and running formal models.[49] No urban setting I know has its data under sufficient control to allow one to begin large-scale analyses without first taking time (often a great deal of time) to put the data into proper shape. If one does not have the luxury of time or is not overly concerned about control problems, there are some options—most of them undesirable. If one is a consultant on a contract cursed with rigid prespecifications and a nasty deadline, then one has to ignore all but the available information and hope that the intractable elements are read into the results by the decision maker. Or one may make every effort to estimate the missing information and warn the prospective user at every opportunity about problems of reliability and availability. Less ethical practitioners often resolve the matter by making the information up, fudging it, black-boxing it, or guessing at it; caveats hardly apply in this situation.

Validation is the fourth and final step in the process. It involves a host of questions that together constitute appraisal. Validation, if discussed at all, is usually reduced to narrow technical questions about adequacy and veracity of the analyst's formalizations. There is much more to the matter than mere technical detail, and one should consider what an expanded conception of validation might entail.

APPRAISAL

Certainly there is a technical dimension. What were the data requirements, and were they well and properly satisfied? If formal models were constructed, did they meet technical standards—e.g., were "mysterious mechanisms"[50] incorporated to enable the model to generate plausible outcomes, has the model been properly identified, have parameters been correctly estimated, and so on?

A theoretical dimension exists as well. What theories governed the analysis? If formal models were used, what theoretical assumptions were built into the models and how well do they match the assumptions held by the decision maker-client? Were the assumptions pertinent, and what would plausible alternative assumptions have looked like? Done to the analysis? Done to the recommendations?

An ethical dimension exists but has received scant attention. Whose view of the world is captured in the analysis (principles of selection)? Is it the bureaucratic technician's? The citizen's? The analyst's? Whose? What ethical possibilities have been constructed in the analysis? This is not an esoteric question, but one whose importance and neglect were stressed in the infamous body counts and indices of pacification so commonly used in our most recent war.

The intellectual and practical topics in these few ethical questions have only recently been addressed. This is probably because ethical specialists have been so preoccupied with their own profession's diversionary imperatives that they have been technically ignorant and because analytic specialists have been likewise so preoccupied that they have remained ethically naive (see Kaplan, 1963; Lindblom, 1959; Winter, 1969, 1966).[51] There is urgent need for a marriage of specialized ethical perspectives and technical skills; that marriage would yield great societal benefits. The problem now is to begin raising the ethical questions more often, clearly, and persistently.

Finally, there is the pragmatic dimension. Of what use is the analysis, and who cares? How has the analysis been used, and how does this use accord with the purposes set forth in the design or proposal phase? How much did it cost to conduct the analysis, and what are the direct (and indirect) benefits? Who is using the results of the analysis? At what level of the client's organization? In what ways? At what rate is the utility of the analysis depreciating? Should anything be done to update the work? Train in-house personnel? New studies? Forget it? And finally, is management any better, more effective, or more humane because the analysis was carried out?

CONCLUSIONS

Good analysis is hard to do. Given the severe limitations of data, theory, and methodological tools, there are more a priori reasons to expect urban analyses to fail than to succeed.

There are mixed motives for doing analyses. As there is great diversity in the urban complex generally, so is there diversity in the specific expectations, identifications, and demands of those who participate in, are responsible for, and must manage that complex.

Good analysts are a scarce resource. The skill, knowledge, and judgment of the analyst are of crucial, but apparently insufficiently appreciated, importance to the conduct of useful work. In an area more akin to art than science, the ratio of Picassos to house painters is small indeed.

The substantive context is enormously complex. As our usual piecemeal means to understand, and efforts to manage, the urban setting end up increasingly confused and self-defeating, whatever forces one to be more comprehensive should be adopted. In this regard, the general concept of systems analysis offers help, particularly because it encourages one to create practices and procedures that are more, not less inclusive.

NOTES

1. Hoos later expanded the ideas in this paper into a book-length indictment of the concept in application; however, this paper contains the flavor, if not the total substance, of the larger effort.

2. One promising technique is the social experiment, a technique whose limits and potentials are briefly explored in Brewer (1973a).

3. While the framework is there for the taking, the one who takes must read, understand, and apply the framework carefully.

4. The promotional literature is extensive and growing; Mitchell (1966) presents most of the general in-principle arguments normally encountered.

5. I have been greatly stimulated to consider the possibilities and limitations of social indicators by several German colleagues who are currently engaged in the design and implementation of regional information systems in the Federal Republic. While I appreciate the encouragement given particularly by Dr. H. Wolfgang Hartenstein of DATUM, e.V., Bonn-Bad Godesberg, the conclusions I have reached about the status and prospects of social indicators are solely my own.

6. A full-scale assessment is beyond the scope of this section; the interested reader is directed to Klages (1973) and Sheldon and Land (1972).

7. Worker satisfaction, alcoholism rates, divorce rates, suicide rates, and the like.

8. Harris and the author are currently working on an expansion and generalization of these related points.

9. This gives an idea of just how limited the approach is when applied to social science problems.

10. After doing an interesting exercise on this matter, Coleman is not so sanguine about the possibilities of recombining information assembled at lower levels of aggregation and getting back to what the "multilevelist" would call the "implications of higher-level problems."

11. There are quite different properties of information content at different levels of analysis, and it is a mistake to assume that one may disaggregate (or reaggregate) without taking these informational (meaning) differences into account for most social applications.

12. In gaming applications this is often referred to as the "levels of resolution problem." In other words, the whole is not necessarily equal to the sum of its parts in organized and complex social systems; it may be fundamentally different.

13. The number of published applications of the multilevel approach is very small, and, given all of its basic limitations, it will doubtless continue to be so.

14. The span of short-term memory has been found by cognitive psychologists to be around seven items or categories (Miller, 1967; see also Simon, 1969).

15. A standard source here is Novick (1965). Or see Lyden and Miller (1967) for a "sampler."

16. Brewer (1973b) discusses the problem of implementation in the Department of Defense from this point of view.

17. An earlier, extensive study of the overall housing market in New York set the stage for this analysis (Lowry et al., 1971).

18. Several successive issues of the *Review of Economics and Statistics* dealt with this.

19. This is a detailed description of the attempt to apply computer simulation to renewal planning in San Francisco and Pittsburgh; methodological and operational realities in the Department of Defense are treated in Shubik and Brewer (1972).

20. Wassily Leontieff (1971) made thoughtful comments on this; and Jean Rostand (1960) has treated the subject more broadly.

21. A statistician with years of experience in defense systems analysis, Ralph Strauch, has summarized shortcomings of working primarily with the quantitative elements of a problem and assuming that the nonquantified components will be handled exogeneously. "It is sometimes argued that such assumptions are made for 'analytical convenience,' and the results must, of course, be interpreted in a larger context. This argument would be valid if, in fact, the problems of interpretation in a larger context were regularly considered and assessed; but they seldom are" (Strauch, 1972: 7).

22. See above, the Introduction.

23. These arguments have been made at greater length by Strauch (forthcoming). I am indebted to him on several counts.

24. I have treated this matter at length elsewhere (Brewer, 1973d: 2-5, see also 1973c) but repeat the essentials of the argument here because it conveys many of the limitations of systems (and other) analyses in actual application.

25. In my opinion, a basic reason for the lack of success of various management-information systems relates to who structures (selects) the information in the system. To the extent that middle- and lower-level administrative personnel and computer specialists do so, to the relative exclusion of top decision makers and citizens, the information systems will be impractical for use in decision-making and irrelevant for use by citizens (Brewer, 1970).

26. Barrett has noted this problem and worked out some of its deeper implications.

27. Harold D. Lasswell (1971, most recently) continually reminds us of the crucial importance of the context in conducting social analyses.

28. Functionally equivalent frameworks exist and in some circumstances may be more

usefully employed—e.g., to replace the label "systems analysis," which has taken on sufficiently negative symbolic and affective connotations to render it obstructive to good problem-solving. The point is not so much what one calls it as that the minimal requirements are met.

29. The literature here, and in each of the noted levels of analysis, is extensive. Talbot (1967) is an example.

30. Dyckman (1960) treats the mixing of levels of analysis by example (see also Ham, 1965).

31. Two examples of the same topic, renewal, carried to the next higher level, are International Seminar on Urban Renewal (1958) and Abrams (1946). The increasing importance of so-called world or global cities should be noted here (von Laue, 1969).

32. Operational gamers and scenario writers, faced with the challenge of creating realistic, comprehensive, and playable scenarios, have known about these requirements for a long time. Only recently, however, have they begun to share this knowledge more broadly (for an excellent example, see DeWeerd, 1973).

33. Kaplan (1963) is a rich source of ideas on these points. Vickers has been long concerned about them as well (for a general discussion, see Vickers, 1973).

34. The manipulative concept is elaborated and contrasted with the traditional contemplative one in Lasswell and Kaplan (1950: vi and generally).

35. The article concludes with the following remark: "Solutions to social problems have to be created. They cannot be 'discovered' simply through positivistic investigations; the data in themselves do not yield policy choices."

36. As a guest of the Ministry of State for Urban Affairs of the Canadian Government in June 1973, I had occasion to explore these and related questions with the ministry's analysis staff. This chapter's discussion of practical limitations was generated in the course of those conversations.

37. In an otherwise competently done and useful paper, Bernstein et al. (1973) devote only minimal attention to these matters. The bulk of their discussion is technical. This is not to fault their effort, because it is rare enough that the basic questions have been raised, if only to show yet another result of the *déformation professionelle* associated with becoming a specialist.

38. This "suggestion" is borne out in fact. In a survey of operational military computer models, Martin Shubik and I encountered models having more than 200,000 separate computer instructions, one of which detailed the rate of fire of individual infantrymen in a division-sized operation (see Shubik and Brewer, 1972).

39. The bigger the analysis, generally speaking, the more grandiose the claims made in proposals to conduct the analysis. When resources are pooled from a number of research budgets to carry out an especially large project, the problem of working for many, frequently competitive, masters further complicates matters.

40. A cynic might even be able to find a relationship between the magnitude of an impending decision and the pounds of computer output thought to be necessary to back up that decision.

41. Explanations and justifications for this imperative are given in foregoing sections, particularly those dealing with world views, selection principles, and the need for different kinds of contextual information.

42. Professional norms do not always provide the kinds of guidance that one might expect them to (see Brewer, 1973e).

43. Information in this sense includes data, theory, and methodological components.

44. The relations among cost, analytic scope and content, and data collection are not generally appreciated. For instance, attitudinal data and other kinds of "soft" information often cannot be measured or can be measured only at great cost; it is time we face up to and respect this fact. There is no such thing as a cheap survey or a general-purpose urban model, and the sooner the clients are disabused of those beliefs, the better.

45. The net accomplishments of the New York City-Rand Institute, (NYC-RI) during the Lindsay years appear to have been positive; however, now what (Szanton, 1972)? The NYC-RI's most recent chief executive evidently made this calculation and answered the question by taking a job with the New York City School Board in the closing days of the Lindsay Administration.

46. In the mid-1960s, clients showed a predilection for large computer modeling projects and management information systems; by the late 1960s, this had been replaced by direct action ("throw some money on the problem") and esoterica (T-groups, sensitivity analyses, organizational development); 1973 marks the "Year of the Survey," if trends evident in the stream of requests for proposal emanating from Washington are indicative.

47. There is a principle of size or proportion at work here, most easily seen in the analogous field of credit. To default on a loan of $50,000 usually means bankruptcy for the debtor; however, to default on a loan of $50 million usually means that the debtor becomes a partner of his creditor. Likewise, cutting one's losses in the analysis business is an increasingly difficult matter the greater the investment in the analysis.

48. A major limitation of urban systems analysis is the virtual lack of suitable management information system support for these activities. While we have been developing MISs for over a decade, the development has been slow and the results have not been too exciting. What seems to be missing is a thorough, critical examination of these developments to sum up what has been learned so that the next decade's efforts may bear fruit faster.

49. In making this hard judgment, I use as a standard of comparison the excellent control that the military has over certain kinds of data used in their systems analyses. Information system design and management is one of the few areas where the experiences learned in the military can probably be easily adapted to the urban setting. I speak here of the technical issue of controlling the data, not the intellectual issue of constructing indices and social accounts discussed earlier.

50. This marvelous label was suggested by I. S. Lowry.

51. On June 12, 1973, an ad hoc advisory committee found the Tuskegee Syphilis Study of the U.S. Public Health Service to be "ethically unjustified." The committee also found that scientist-specialists should not be given such free rein in human experimentation, that there is no federal program to protect research subjects, and that a permanent investigation board should be established to regulate federally supported research involving human subjects. One wonders about the ethical dimension in analyses conducted by specialists who are *not* supposedly dedicated to the preservation of life and the relief of human suffering, if this is what results from specialists who supposedly are so dedicated.

REFERENCES

ABRAMS, C. (1946) The Future of Housing. New York: Harper.

ANDO, A., F. M. FISHER, and H. A. SIMON (1963) Essays on the Structure of Social Science Models. Cambridge, Mass.: MIT Press.

BACON, E. (1968) "American houses and neighborhoods, city and county." Annals of Amer. Academy of Pol. and Social Sci. (July): 117-129.

BALES, C. F. (1971) "The progress of analysis and PPB in New York City government," in R. S. Rosenbloom et al., New Tools for Management. Cambridge, Mass.: Harvard Univ. Press.

BARRETT, W. (1958) Irrational Man. Garden City, N.Y.: Doubleday.

BERNSTEIN, S. et al. (1973) "The problems and pitfalls of quantitative methods in urban analysis." Policy Sciences 4 (March): 29-39.

BOGUSLAW, R. (1965) The New Utopians: A Study in System Design and Social Change. Englewood Cliffs, N.J.: Prentice-Hall.

BREWER, G. D. (1973a) "Social experimentation and the policy process," pp. 151-165 in Twenty-Fifth Annual Report of the RAND Corporation. Santa Monica, Calif.: RAND.

––– (1973b) "On innovation, social change, and reality." Technological Forecasting and Social Change 5: 19-24.

––– (1973c) Politicians, Bureaucrats, and the Consultant: A Critique of Urban Problem-Solving. New York: Basic Books.

––– (1973d) "What's the purpose? What's the use?" RAND Corporation Paper P-5095.

––– (1973e) "Professionalism: the need for standards." Interfaces 46 (November): 20-27.

––– (1973f) "Documentation: an overview and design strategy." RAND Corporation Paper P-5052.

––– (1972) "Dealing with complex social problems." RAND Corporation Paper P-4894.

––– (1970) "A prototype office of human statistics: context, strategy, and recommendations." RAND Corporation Paper P-4439.

––– and O. P. HALL, Jr. (1973) "Policy analysis by computer simulation: the need for appraisal." Public Policy 21 (Summer): 343-365.

BRUNNER, R. D. and G. D. BREWER (1971) Organized Complexity: Empirical Theories of Political Development. New York: Free Press.

COLEMAN, J. S. (1964) Introduction to Mathematical Sociology. New York: Free Press.

CRECINE, J. P. (1969) Governmental Problem Solving. Chicago: Rand McNally.

DE JOUVENEL, B. (1967) The Art of Conjecture. New York: Basic Books.

DEUTSCH, K. W. (1966) "On theories, taxonomies, and models as communication codes for organizing information." Behavioral Sci. 11 (January).

DeWEERD, H. (1973) "A contextual approach to scenario construction." RAND Corporation Paper P-5084.

DEWEY, J. (1939) "Theory of valuation," in Volume 2 of the International Encyclopedia of Unified Science. Chicago: Univ. of Chicago Press.

DREW, D. (1972) "Applications of multilevel systems theory to the design of a freeway control system," pp. 156-175 in M. D. Mesarovic and A. Reisman (eds.) Systems Approach and the City. Amsterdam: North-Holland.

DYCKMAN, J. W. (1963) "The scientific world of city planners." Amer. Behavioral Scientist 6 (January/February).

––– (1960) "National planning in urban renewal." J. of Amer. Institute of Planners 26 (February): 49-59.

EASTON, D. (1965) A Systems Analysis of Political Life. New York: John Wiley.

––– (1957) "An approach to the analysis of political systems." World Politics 9.

FORRESTER, J. W. (1969) Urban Dynamics. Cambridge, Mass.: MIT Press.

GILMORE, J. et al. (1967) Defense Systems Research in the Civil Sector. Washington, D.C.: Arms Control and Disarmament Agency.

GOULD, P. R. (1967) "Structuring information in spacio-temporal preferences." J. of Regional Sci. 7 (Winter): 259-274.

HAIMES, Y. Y. and D. MACKO (1970) "Hierarchical structures in water resources systems management." IEEE Transactions on Systems, Man, and Cybernetics 3.

HAM, C. C. (1965) "Urban renewal: a case study in emerging goals in an intergovernmental setting." Annals of Amer. Academy of Pol. and Social Sci. (July): 44-51.

HARRIS, M. (1973) "An examination of social indicators: the relationship to the conditions of minority groups." University of Southern California Department of Urban Studies, July. (unpublished)

HIRSCHMAN, A. O. (1970) "The search for paradigms as a hindrance to understanding." World Politics 22 (April): 329-343.

HOOS, I. (1968) "A critical review of systems analysis: the California experience." Berkeley Space Sciences Laboratory Working Paper 89, December.

International Seminar on Urban Renewal (1958) Report: Urban Renewal. The Hague.

KAPLAN, A. (1963) American Ethics and Public Policy. New York: Oxford Univ. Press.

KEMENY, J. G. (1960) "A philosopher looks at political science." J. of Conflict Resolution 4 (September): 292-302.

KLAGES, H. (1973) "Assessment of an attempt at a system of social indicators." Policy Sciences 4 (September): 249-261.

KUHN, T. (1964) The Structure of Scientific Revolutions. Chicago: Univ. of Chicago Press.

LA PORTE, T. R. [ed.] (forthcoming) Organized Social Complexity: Challenge to Politics and Policy. Princeton: Princeton Univ. Press.

LASSWELL, H. D. (1971) A Pre-View of Policy Sciences. New York: American Elsevier.

——— and A. KAPLAN (1950) Power and Society. New Haven, Conn.: Yale Univ. Press.

LEONTIEFF, W. (1971) "Theoretical assumptions and nonobserved facts." Amer. Econ. Rev. 71 (March).

LEVIN, M. A. and H. D. DORNBUSCH (1973) "Pure and policy social science." Public Policy 21 (Summer): 383-423.

LEVINE, R. A. (1968) "Rethinking our social strategies." Public Interest (Winter): 88-92.

LINDBLOM, C. E. (1959) "The handling of norms in policy analysis," pp. 160-179 in M. Abramovitz et al., The Allocation of Economic Resources. Stanford: Stanford Univ. Press.

LINEBERRY, R. and I. SHARKANSKY (1971) Urban Politics and Public Policy. New York: Harper & Row.

LIPSKY, M. (1970) Protest in City Politics: Rent Strikes, Housing, and the Power of the Poor. Chicago: Rand McNally.

LITTLE, A. D. (1966) Community Renewal Programming: A San Francisco Case Study. New York: Frederick A. Praeger.

LOWRY, I. S. (1971a) "Reforming rent control in New York City: the role of research in policy-making." RAND Corporation Paper P-4570.

——— (1971b) "Housing assistance for low income urban families: a fresh approach." RAND Corporation Paper P-4645.

——— J. M. GUERON, and K. M. EISENSTADT (1972) "Welfare housing in New York City." RAND Corporation Report R-1164-NYC.

——— (1971) "Rental housing in New York City: II. The demand for housing." RAND Corporation Report R-649-NYC.

LYDEN, F. J. and E. G. MILLER [eds.] (1967) Planning, Programming, Budgeting: A Systems Approach to Management. Chicago: Markham.

MARNEY, M. and N. M. SMITH (1972) "Interdisciplinary synthesis." Policy Sciences 3 (September): 299-323.

MATUSZEWSKI, J. P. and I. LEFKOWITZ (1970) "Coordination and control in steel processing." IEEE Transactions on Systems, Man, and Cybernetics 3.

MESAROVIC, M. D. and A. REISMAN [eds.] (1972) Systems Approach and the City. Amsterdam: North-Holland.

MESAROVIC, M. D., D. MACKO, and Y. TAKAHARA (1970) Theory of Hierarchical, Multilevel Systems. New York: Academic Press.

MICHELSON, W. (1965) "Most people don't want what architects want." Trans-action (July/August): 37-43.

MILLER, G. (1967) "The magical number seven, plus or minus two," in The Psychology of Communication: Seven Essays. New York: Basic Books.

MITCHELL, W. H. (1966) "SOGAMMIS—a systems approach to city administration." Public Automation (April): 1-4.

MORRISON, P. A. (1971) "Demographic information for cities." RAND Corporation Report R-618-HUD.

MUTH, R. (1969) Cities and Housing: The Spatial Pattern of Urban Residential Land. Chicago: Univ. of Chicago Press.

NOVICK, D. [ed.] (1965) Program Budgeting. Cambridge, Mass.: Harvard Univ. Press.

––– (1954) "Mathematics: logic, quantity, and method." Rev. of Economics and Statistics (November).

ORCUTT, G. H. et al. (1968) "Data aggregation and information loss." Amer. Econ. Rev. 68 (September): 773-787.

QUADE, E. S. and W. I. BOUCHER [eds.] (1968) Systems Analysis and Policy Planning. New York: American Elsevier.

REINER, T. (1963) The Place of the Ideal Community in Urban Planning. Philadelphia: Univ. of Pennsylvania Press.

RICHARDSON, J. and T. PELSOCI (1972) "A multilevel approach and the city: a proposed strategy for research," pp. 97-131 in M. D. Mesarovic and A. Reisman (eds.) Systems Approach and the City. Amsterdam: North-Holland.

ROSENBLOOM, R. S. et al. (1971) New Tools for Urban Management. Cambridge, Mass.: Harvard Univ. Press.

ROSTAND, J. (1960) Error and Deception in Science. New York: Basic Books.

SHELDON, E. B. and K. C. LAND (1972) "Social reporting for the 1970's: a review and programmatic assessment." Policy Sciences 3 (July): 137-151.

SHUBIK, M. and G. D. BREWER (1972) "Models, simulations and games—a survey." RAND Corporation Report R-1060-ARPA/RC.

SIMON, H. A. (1969) "The psychology of thinking," in The Sciences of the Artificial. Cambridge, Mass.: MIT Press.

––– (1967) "The architecture of complexity," pp. 63-76 in L. von Bertalanffy and A. Rapoport (eds.) General Systems. New York: George Braziller.

STEGER, W. (1964) "Analytic techniques to determine the needs and resources for urban renewal action." Proceedings of the IBM Scientific Computing Symposium on Simulation Models and Gaming (December).

STRAUCH, R. (forthcoming) "A critical assessment of quantitative methodology as a policy analysis tool." RAND Corporation Report R-1423-PR/ARPA.

––– (1972) "Winners and losers: a conceptual barrier in our strategic thinking." RAND Corporation Paper P-4679.

SZANTON, P. L. (1972) "Analysis and urban government: experiences of the New York City-Rand Institute." Policy Sciences 3 (July): 153-161.

TALBOT, A. R. (1967) The Mayor's Game: Richard Lee of New Haven and the Politics of Change. New York: Harper & Row.

VICKERS, G. (1973) "Values, norms, and policies." Policy Sciences 4 (March): 103-111.

VON LAUE, T. H. (1969) The Global City. Philadelphia: J. B. Lippincott.

WEAVER, W. (1948) "Science and complexity." Amer. Scientist 36: 536-544.

WINTER, G. (1969) "Toward a comprehensive science of policy," in Houghton Lectures. Cambridge, Mass.: Harvard Univ. Press.

––– (1966) Elements for a Social Ethic. New York: Macmillan.

7

Muddling Through Analytically

JAMES W. VAUPEL

□ MY GUESS IS THAT the prospects for better individual decision-making depends not so much on the refinement of complex analytical techniques as on the *extension* of analysis to situations in which time is short and data are sparse. Consequently, this paper briefly describes and illustrates some analytical methods useful to busy decision makers. Part I of the paper discusses the range of decision-making methods, including decision-making by habit, intuition, incomplete analysis, and complete analysis. Part II explains how and why some simple quantifications may help a decision maker. Part III considers how a decision maker might decide how to make a decision.

THE RANGE OF
DECISION-MAKING METHODS

COMPLETE ANALYSIS

It is occasionally argued that decisions ought to be made on the basis of a method which I will call "complete analysis" and which is sometimes referred to as "comprehensive rationality" (Lindblom, 1959) or "systems analysis." Basically, this method requires the decision maker to:

(1) specify all possible decision alternatives,

(2) predict all possible consequences of each alternative,

(3) estimate the probability of each consequence,

(4) appraise the desirability of each consequence, and

(5) calculate which decision alternative yields the most desirable set of consequences.

As Lindblom (1959) has made clear, this method represents an ideal rationality that man will never attain because of limits of time, information, and intellectual capacity. For example, a complete analysis of setting the federal budget for cancer research would have to include, among other things, consideration of all the other ways this money might be spent (including spending the money on kidney research, educational research, raising military pay scales, and reducing tax payments) and all the implications of these alternatives (including the implications for the future of the Papacy, the profitability of oil wildcatting, the prospects for interplanetary travel, and the popularity of serial music).[1]

Since complete analysis is beyond human capabilities, decision makers have to "muddle through" by using habit, intuition, or some form of incomplete analysis. These three decision-making methods are discussed in turn below.

HABIT

Most decisions are not consciously made but rather result from some habit, automatic response, or reflex. Few individuals, for example, pause to think about whether to wear shoes to work, which pedal to push next while riding a bicycle, or which hand to use to shake hands. Indeed, when decisions are made by habit, it often is difficult to remember exactly which course of action was adopted. As William James (1892 [1968]) wrote, "Few men can tell off-hand which sock, shoe, or trousers-leg they put on first." Nonetheless, every man who wears socks, shoes, and trousers must, in effect, decide how to put them on.

INTUITION

Most of the decisions made consciously are made on the basis of a virtually instantaneous process which is commonly called "intuition" or "snap judgment" and which, for reasons explained below, is sometimes called "wholistic response." Most store managers, for example, decide whether to accept a customer's bank check on the basis of some intuitive snap judgment; many routine purchasing decisions are similarly made. You probably based your decision to begin reading this essay on snap judgment rather than on a detailed cost/benefit analysis, and you may decide whether to continue reading on the basis of an intuitive hunch rather than on a carefully reasoned justification. The words "intuition" and "snap judgment" are convenient to denote this kind of conscious (rather than automatic), direct, and immediate decision-making, but unfortunately they sometimes have derogatory overtones. In this essay, the words will be used merely to *describe* a very useful kind of decision-making process, and *not* to disparage this process.

INCOMPLETE ANALYSIS

Third and finally, some decisions are made on the basis of some analysis. The word "analysis" is derived from an ancient Greek word meaning "to break up," and in most dictionaries the primary meaning of the word is given as something like "separation of a whole into its component parts."[2] I will follow this usage in this essay and will use the word "analysis" to mean the breaking up or decomposing of a decision problem into two or more elements which are considered separately before the decision is made. "Complete analysis" requires decomposing a decision problem into *all* its ultimate elements and therefore involves, as indicated above, consideration of all possible alternatives and of all of the consequences of each of these alternatives. "Incomplete analysis," on the other hand, merely requires *some* decomposition. Both forms of analysis differ from intuition in that intuition is based on a single, wholistic response to the decision problem at hand rather than on separate consideration of two or more components of the decision problem.

Perhaps the simplest kind of incomplete analysis involves the following steps.

(1) One possible decision alternative is considered and on the basis of some intuitive judgment is assigned some very approximate value.

(2) Another possible decision alternative is compared with the first, and an intuitive guess is made as to whether it is better or worse than the first alternative.

(3) If the second seems better, it is chosen; if worse, the first alternative is chosen.

A dedicated TV viewer in a town in which only two stations could be received might use such a process to decide which program to watch. Similarly, a young couple might decide whose in-laws to spend Thanksgiving with on the basis of some simple analysis of this sort. If there were more than two TV stations or other viable alternatives for Thanksgiving, the procedure could be expanded into a series of successive comparisons.

A somewhat more detailed kind of incomplete analysis involves describing two or more alternative courses of action and then separately estimating both the likelihood and desirability of two or more consequences of each course of action. Consider, for example, the following hypothetical case study. George Barlow, mayor of Zenith City, had to decide whether to submit housing plan A or housing plan B to the city council for approval. Barlow made the decision by reasoning as follows:

> I can probably get Plan A through the council, but as for Plan B, which is more ambitious and more controversial, I think the chances are less than likely. However, I really prefer B over A; A is a good plan but B is a great plan—B would be better for the city and better for my reelection campaign too. And if I submit B and the council defeats me on it, that wouldn't be so bad—I'd look courageous and the council would look reactionary. Furthermore, the worst outcome would be if I submitted A and the council defeated me anyway. Nonetheless, as I said before, I probably can push A through and I really would rather have one of the plans passed than none at all. All things considered, I'll submit Plan A.

In making his decision, Barlow specified two courses of action and predicted in each case two consequences. Using the terms "probably" and "less than likely," he estimated the likelihood of each consequence. He appraised the desirability of the consequences as "good," "great," "not so bad," and "terrible." Barlow then apparently intuitively judged that "probably" "good" and "not probably" "terrible" was better than "less than likely" "great" and "more than likely" "not so bad," and therefore chose to submit Plan A rather than Plan B.

If Barlow wanted to do a less incomplete analysis, he might have considered some additional decision alternatives, such as submitting housing Plan C or submitting both Plans A and B, or submitting some modified version of Plan A, or waiting a month before submitting any plan. Furthermore, he might have calculated the probabilities of city council approval by separately considering how each council member would vote. Barlow might also have considered some additional, subsequent consequences whose probabilities might depend on what he did and how the city council reacted, such as the chances of getting a recreational facilities bill passed, the chances of being reelected, or the chances of winning a nomination to the U.S. Senate. Finally, Barlow might have separately considered the desirability of each outcome in terms of different objectives or different interests, such as the desirability of each outcome for Barlow personally, for Barlow's political party, for the poorer inhabitants of Zenith City, for Zenith City as a whole, for the United States as a whole, or even for the world as a whole. By further decomposing his decision problem in these ways, Barlow would move closer to the ideal of "complete analysis," but unless he considers everything, including the effect of his decision on the price of sirloin steak in Mongolia, his analysis will be incomplete.

QUANTITATIVE ANALYSIS

DECISION ANALYSIS

Barlow analyzed his problem in a verbal, nonquantitative way. A decision maker trained in the methods of "decision analysis" developed by Howard Raiffa, Robert Schlaifer, and others (Raiffa, 1968; Schlaifer, 1969) would probably structure the problem somewhat more formally and quantitatively. Suppose such a person —Anna Lisa Pedersen, for example—were mayor of Zenith City instead of George Barlow. Pedersen might begin her analysis of the decision problem by drawing a "decision tree," like the one depicted in Figure 1, that lays out the alternative decisions and likely consequences.

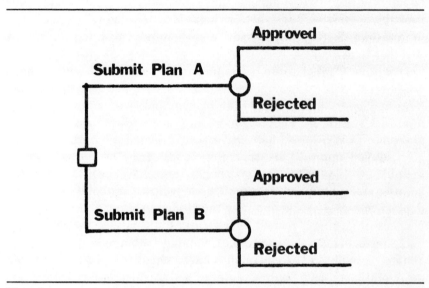

Figure 1.

Pedersen then would estimate the probabilities of the consequences, not by using terms such as "probably" or "less than likely," but by using numbers such as "80% chance" or "40% chance." These "subjective probability assessments" can be interpreted as follows. If Pedersen says there is an 80% chance that the council will approve Plan A, she means that the probability of approval is the same as the probability of drawing a red ball from an urn containing 80 red balls and 20 green balls. In particular, if Pedersen were offered two lotteries, one of which would pay $100 if the council approved Plan A and the other which would pay $100 if a red ball were drawn from an urn with 80 red balls and 20 green balls, she would value the two lotteries equally and be indifferent between them.

After subjectively assessing the probabilities of the various consequences in the decision problem, Pedersen would subjectively appraise the desirability (or "utility") of each of these consequences. In some kinds of decision problems, a decision analyst might express these figures in terms of dollars or some other unit of value, but in this instance, Pedersen probably would use what are known as BRLTs (an abbreviation for "Basic Reference Lottery Ticket," pronounced "brilts"). In using the BRLT system, Pedersen would assign a value of 1 to the best outcome (namely, approval of Plan B) and a value of 0 to the worst outcome (namely, rejection of Plan A). Outcomes of intermediate desirability would be given values between

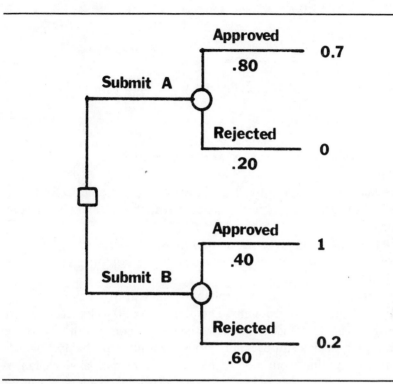

Figure 2.

0 and 1. In particular, Pedersen might give the "good" consequences a value of 0.7 and the "not so bad" consequence a value of 0.2. It is difficult to briefly explain how such BRLT values should be interpreted, but basically the value of 0.7 means that Pedersen considers the "good" outcome as being worth exactly the same as a hypothetical lottery that would give Pedersen a 0.7 (i.e., 70%) chance at the best outcome (namely, approval of Plan B) and a 30% chance at the worst outcome (namely, rejection of Plan A). Similarly, the value 0.2 for the "not so bad" consequence implies that Pedersen considers this outcome as being equal in desirability to a hypothetical lottery that yields the best outcome (approval of Plan B) with a 0.2 (i.e., 20%) probability and otherwise yields the worst outcome (rejection of Plan A).

Figure 2 depicts the decision tree with Pedersen's subjective probability assessments and subjective utility appraisals included.

Pedersen would now calculate the "expected value" of submitting Plan A as

$$80\% \times 0.7 + 20\% \times 0 = .56$$

and of submitting Plan B as

$$40\% \times 1 + 60\% \times 0.2 = .52.$$

Since .56 is greater than .52, Pedersen, like Barlow, would choose to submit Plan A.

THE ADVANTAGES OF THE DECISION ANALYSIS APPROACH

Pedersen might argue that her approach is preferable to Barlow's, at least for persons trained in decision analysis, for at least four reasons.

First, a decision tree is a useful way of keeping track of alternatives and consequences, especially in a complicated decision problem.

Second, quantification of probability and utility estimates often helps clarify decision problems. The phrase "an 80% chance," even if it only means "roughly an 80% chance," conveys much more information than a sloppy, ambiguous term like "probably," which can mean anything from a 51% chance to more than a 90% chance. Similarly a utility appraisal of 0.7, even if it is just a guess, is a more informative mode of expression than a phrase like "pretty good."

Third, given numerical values for probabilities and utilities, expected values can be readily computed, and the plan with the greatest expected value chosen. This is surely a more convincing decision criterion than arguing, as Mayor Barlow did, that "probably good" and "not probably terrible" is better than "less than likely great" and "more than likely not so bad."

Finally, as will be explained in the third part of this paper, numerical values for probability and utilities can aid a decision maker in deciding whether to expand his or her incomplete analysis by considering an additional factor or by gathering some additional information.

TAKING ADDITIONAL FACTORS INTO ACCOUNT

There are a number of additional factors Pedersen might want to take into account before deciding whether to submit Plan A or Plan

B. For example, Pedersen might think that the probabilities that the council will approve Plan A or Plan B depend on whether she approves the extravagant but popular recreation facilities program the council has just sent her. Furthermore, Pedersen might feel that her utility appraisals depend on what the different outcomes mean for her reelection chances, and she might therefore want to bring her reelection chances into the decision tree. In this case, Pedersen might expand her decision tree into the one shown in Figure 3. Given this analysis, Pedersen's optimal decision is to approve the recreation facilities program and then submit housing Plan B.

USING DATA

Another way Pedersen might extend her analysis would be to try to use some objective data to help estimate the probability and utility values. For example, Pedersen might separately consider the probability that each of the councilmen will vote for Plan A (or Plan B). The past voting records of the councilmen might provide some clues as to how each councilman will vote on the housing plans. To get a better idea of the effect of her decisions on her chances for reelection, Pedersen might commission a poll of some sample population in Zenith City. To better estimate the consequences of Plan A or Plan B for the residents of Zenith City, Pedersen might do (or have a subordinate do) an analysis of the various costs and benefits associated with the plans, including the net costs of the plans to the city in both the short and the long run.

INCOMPLETE ANALYSIS AS DECOMPOSED INTUITION

By taking some additional factors into account and by using some objective data to estimate probability and utility values, Pedersen could make her analysis a little more complete. Nonetheless, as indicated above, no analysis can ever be made perfectly complete (or even nearly complete). Thus, Pedersen's analysis, like any other analysis, will have to ultimately rest on a number of intuitive judgments.

First, intuitive judgments must be made as to which factors and decision alternatives to consider in an incomplete analysis of a decision problem. In the Zenith City case, for instance, Barlow and

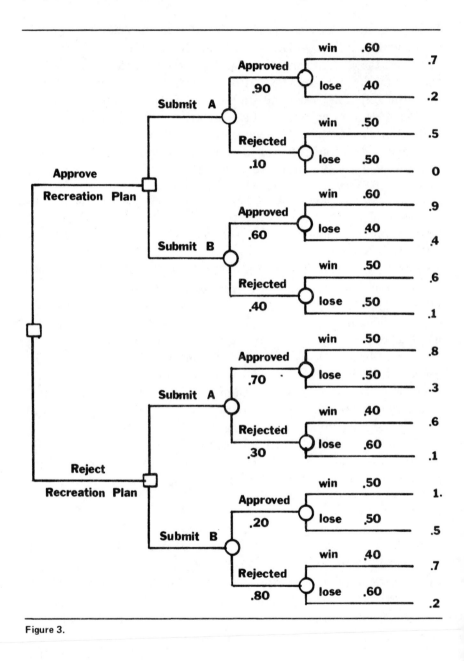

Figure 3.

Pedersen only considered two alternatives and a very few possible consequences; they left out of their analysis the possibility of submitting Plan C, compromising on a watered-down version of Plan B, logrolling with Councilman Smythe on his school construction proposal, the chances for a Senate candidacy, the effect of the plans on land-values in Zenith City, and even the effect of the plans on crime rates in Manitoba.

Second, intuitive judgments must be made about the probability values used in an incomplete analysis of a decision problem. This is the case even when objective data are available to help estimate the probabilities. Objective data necessarily concern past events, but the probabilities to be assessed concern future events. Thus, at the very least, intuitive judgments must be made as to whether the future is sufficiently similar to the past. In most cases, it is clear that there are no very exact precedents for future events, so that the conclusions of a statistical analysis of past data must be combined with some subjective guesstimates about the future in order to assess the probabilities of future events.

Third, intuitive judgments must be made about the utility values used in an incomplete analysis of a decision problem. As with probability estimates, this is the case even when objective data are available to help appraise utility values. At a minimum, an intuitive judgment must be made as to whether some objective index, based perhaps on net dollar costs or dollar profits or responses on a questionnaire or number of lives saved, adequately measures the desirability of the outcomes. In most cases, the desirability of the outcomes will depend on some considerations that cannot be objectively measured. For instance, a decision maker's appraisal of the desirability of an outcome may depend on his or her judgment about the morality or justice of the outcome.

In short, all incomplete analyses are utlimately based on a collection of intuitive, subjective judgments about which decision alternatives to consider, about which outcomes to consider, about the assessment of probabilities, and about the appraisal of utilities. In a sense, then, decision-making is and, since complete analysis is beyond human capabilities, probably always will be, more "art" than "science." Consequently, experienced decision makers who have developed effective habits and intuitions through extensive on-the-job training will, other things being equal, outperform novices.

This does not imply, however, that the systematic decomposition of a decision problem or the statistical analysis of objective data is foolish or futile. Some incomplete analysis (i.e., decomposed intuition), based in part on objective data, almost always yields a decision that seems better to the decision maker than his or her initial, wholistic, intuitive judgment. Other things being equal, decision makers who understand the techniques of decision analysis and statistics will tend to make better decisions than decision makers who rely solely on routine responses, hunches, and snap judgments. Various experiments psychologists have performed to test alternative modes of decision-making confirm the usefulness of analysis and data.[3]

In short, while the quality of a decision maker's decisions will always depend on the decision maker's wisdom and experience, the quality of these decisions can be significantly improved through the intelligent use of the various techniques of systematic analysis.

DECIDING HOW TO DECIDE

DECISION-MAKING DECISIONS

Decisions can be made in a variety of different ways ranging from automatic responses and intuitive judgments through the simplest forms of verbal, incomplete analysis to more complex and quantitative forms of incomplete analysis that begin to approach the infinitely complex, infinitely time-consuming systems analysts' ideal of comprehensive rationality. Consequently, before a decision maker makes a decision, the decision maker must, in effect, decide how to make the decision.

This decision-making decision has three major dimensions. First, the decision maker must decide how much time and other resources to spend making the decision. Second, the decision maker must decide how to divide the allocated time between the task of gathering information and the task of thinking (i.e., gathering information from his or her own brain). Third, the decision maker must decide how to divide the allocated time among the tasks of designing alternative courses of action, predicting the likelihood of various outcomes, appraising the desirability of these outcomes, and finding the optimal course of action. Deciding how to decide is clearly a complex decision problem.

If most decisions are made by intuitive judgments, an even greater proportion of decisions about how to decide are made by intuitive judgments. Only in a relatively few instances—such as the case of a decision about whether to undertake an expensive experiment or sample survey before reaching a final decision—are decisions about how to decide commonly made by detailed analysis. In virtually all cases, third- and higher-order decisions (i.e., decisions about how to make decisions about how to make decisions about . . .) are based on intuitive judgments. Thus, another reason why a decision maker cannot be comprehensively rational is that at some point a decision maker must start thinking about a decision without first stopping to think about whether to think about the decision.

It makes sense, however, to spend some time analyzing the general problem of how to make decisions, as some general rules may be helpful to decision makers in judging whether and how far to decompose a decision problem. The approach I will take in developing these rules is not normative, but rather what Howard Raiffa (1968) has called "prescriptive." That is, if *you* are faced with a decision problem and if *you* are persuaded by the discussion that follows, then *you* might want to adopt some of the suggestions made. My claim is not that the discussion defines "rational" behavior, whatever that is, but merely that some decision makers may find the discussion useful.

WHY ANALYSIS?

The case for making decisions on the basis of analysis rather than on the basis of habit or intuition rests on the simple proposition that, in most cases, analysis results in "better" decisions. By "better," I merely mean that after the analysis has been done, the decision maker prefers the decision indicated by the analysis rather than the decision he or she would have made before the analysis. That is, analysis is useful to a decision maker because analysis frequently leads the decision maker to change his or her mind and adopt a new course of action that seems, to the decision maker, superior to the course of action that otherwise would have been taken. Of course, if analysis indicates the same decision that was indicated by habit or intuition, this decision can be adopted. It thus appears at first glance that decision-making by analysis is superior to decision-making by habit or intuition and, indeed, the more complete the analysis, the better.

Loosely speaking, a second argument in favor of analysis is that analysis may be useful to a decision maker in convincing others that some particular decision is the best one or was the best one that could have been made despite the bad consequences that followed it. Thus *after* having made a decision, a decision maker may want, as part of his or her decision strategy, to do some window-dressing analysis. However, to the extent that the aim of analysis is to reach a decision, rather than to justify a decision, the case for analysis depends solely on the probability that the analysis will produce a significantly better decision than would have been produced by habit or intuition.

THE CASE FOR HABIT AND INTUITION

The case for basing decisions on habit and intuition rests on the fact that the process of analysis costs the decision maker some of his or her time, or the time of subordinates, or the money to hire assistants. In a sense, the best decision maker would be one who could automatically make optimal decisions without wasting any thought, time, or other resources decomposing decision problems or gathering data. As William James (1892 [1968]) urged:

> The more of the details of our daily life we can hand over to the effortless custody of automatism, the more our higher powers of mind will be set free for their own proper work. There is no more miserable human being than one in whom nothing is habitual but indecision, and for whom the lighting of every cigar, the drinking of every cup, the time of rising and going to bed every day, and the beginning of every bit of work, are subjects of express volitional deliberation.

While it is perhaps hard to imagine the misery of a human being who deliberates the lighting of every cigar, anyone who has lived a while in a foreign country probably has experienced the dizzying exhaustion that sets in when old habits no longer suffice and decisions must be made about whether to shake hands, what to eat next, and how to interpret a gesture or a grimace.[4]

The advantage of habit and intuition over analysis is perhaps most clearly demonstrated by the fact that "education," in large measure, involves the acquiring of habits and intuitions that save time by allowing an individual to rapidly make certain decisions. That is, the bulk of a person's education (or "training") is aimed at enabling the person to get away with *not* analyzing a large variety of commonly occurring decision problems. Otherwise,

> the whole activity of a lifetime might be confined to one or two deeds. . . . A man might be occupied all day in dressing and undressing himself; the attitude of his body would absorb all his attention and energy; the washing of his hands or the fastening of a button would be as difficult to him on each occasion as to the child on its first trial; and he would, furthermore, be completely exhausted by his exertions [James, 1892 (1968)].

Moreover, just as the progress of an individual depends on the development of habits and intuitions, so progress in human knowledge occurs when hard-earned flashes of genius became so routine and commonsensical that it is possible to use them without rederiving them.

The savings in time produced by making decisions by habit or intuition is important because decision makers have so little time. In fact, time is so scarce that all of us, either consciously or unconsciously, in large part blindly accept the status quo and never consider vast numbers of potential decisions that might alter our lives for the better. Descartes, for example, consciously made the decision *not* to question the religion and customs of seventeenth-century France so that he would have the time to analyze a few philosophical issues. And *you* may never have considered whether parents should be allowed to determine the sex of their future children, whether an innovative health care program just implemented in Abilene should be adopted by your community, whether to quit your present occupation, move to Montpelier and become a landscape architect, or whether higher taxes should be levied on individuals who are born smarter, stronger, or more attractive than the average. If you could relegate more routine decisions to the effortless custody of habit and make more nonroutine decisions on the basis of swift, intuitive judgments, you would have more time to consider these and other decisions. Of, if you prefer, you would have more time to play tennis, read mystery thrillers, listen to music, or whatever else you like.

It thus may seem, on second thought, that decision-making by habit or intuition is to be preferred to decision-making by analysis and, indeed, habit is to be preferred to intuition since habit is automatic and immediate, while intuition wastes a moment or two of conscious attention.

THE pd $>$ c RULE

The flaw in this line of reasoning is, of course, that habit or intuition may not produce the best decision. That is, as discussed above, analysis in nearly all cases will yield a decision that will seem to the decision maker to be at least as good as the decision indicated by habit or intuition. Clearly, the benefits of analysis must be balanced against the costs of analysis to determine whether an analysis should be done and how complete it should be.

The general rule, as stated, for example, by John Rawls (1971), is that "we should deliberate up to the point where the likely benefits from improving our plan are just worth the time and effort of reflection." In other words, a decision maker should begin or continue to analyze a decision problem only as long as the expected costs of (further) analysis are less than the expected benefits. The expected benefits of analysis depend on two factors: (1) the probability that the analysis will yield a different decision and (2) the difference between the values (i.e., "expected utility") of this decision and the decision that otherwise would have been made. If the probability is p and the difference in values is d, then the expected benefits of the analysis equal p times d. If the cost of the analysis is c, then the net benefits of analysis will be p times d minus c. Thus, a decision maker should begin or continue to analyze a decision problem as long as the decision maker thinks that pd $>$ c (i.e., that p times d exceeds c).

Because deciding how to decide is itself a decision problem the pd $>$ c rule can also be derived by doing a decision analysis. Basically, the decision maker faces two alternatives: (1) no further analysis and (2) further analysis. The first decision has one outcome—namely, adoption of the decision which is currently thought best. Let the expected value of this outcome be designated by v. The second decision has two possible outcomes: (1) the decision maker reaches a better decision and (2) the decision maker does not reach a better decision (i.e., does not change his or her mind about what to do). If p is the probability that the decision maker will reach a better decision, if d is the expected difference between the value of this better decision and the old decision (which has a value of v) and if c is the expected cost of further analysis, then the decision tree can be drawn as shown in Figure 4. Decision 2 ("further analysis") should be chosen over decision 1 ("no further analysis") if

$$p \cdot (v+d-c) + (1-p) \cdot (v-c) > v$$

which reduces to the formula

$$p \cdot d - c > 0,$$

or, equivalently,

$$p \cdot d > c.$$

It might be noted that a simplified form of the $pd > c$ rule is the rule of thumb known as "satisficing" (Simon, 1957). Basically, "satisficing" involves choosing the first decision alternative that satisfies the decision maker's goals. A decision maker who satisfices is in effect making the judgment that once a satisfactory course of action has been uncovered, the probability of finding a significantly

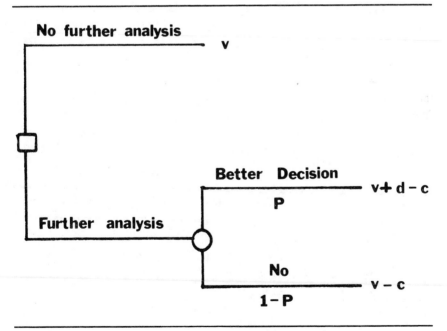

Figure 4.

better course of action is so low that the cost of further analysis is not worthwhile.

How can the pd > c rule be used? For a decision maker used to thinking in terms of this rule, the rule may become "second nature" and consequently be either semi-consciously or unconsciously employed in the making of most decisions. Conscious and deliberate use of the rule will sometimes be helpful in deciding whether to make some incomplete analysis a bit more complete. That is, a decision maker might use the rule to decide whether it is worthwhile to consider an additional course of action, gather some additional information, or decompose some probability or utility according to some additional factor.

In most such cases, the decision maker will probably use the rule in a rough, verbal way. In particular, p might be estimated by phrases like "fair chance," "less than likely," or "virtually certain"; d might be estimated by phrases like "hardly any difference" or "substantially better"; and c might be estimated by phrases like "little costs" or "very expensive." For example, Mr. Barlow, in considering whether to consider a third housing plan—Plan X, say, that he might submit to the city council instead of either Plan A or Plan B—might have reasoned as follows:

> It's hardly likely that I would actually decide to submit Plan X
> rather than Plan A or Plan B. In any case, Plan X, at best, would be
> only marginally better than Plan A. Besides, it would cost me a lot
> of time to figure out how good Plan X really is. Therefore, I won't
> worry about Plan X.

Thus, in effect, Barlow estimated p as "hardly likely," d as "marginally better," and c as "a lot of time," and decided not to consider Plan X because he judged "hardly likely" "marginally better" to be not worth "a lot of time."

In cases where such verbal analysis is not persuasive, a decision maker may want to assign approximate, subjective values to p, d, and c. And in some cases, a decision maker may want to do a formal decision analysis (complete with decision tree and, if possible, the use of objective data) of whether or not to extend his or her decision analysis. Such formal decision-making decision analysis is most commonly used in cases where a decision maker has to decide whether to gather some additional information which might significantly help him or her but which is expensive to gather. Consider, for example, the decision problem faced by a politician who is considering running for some office. A sample survey would help the politician assess his chances, but sample surveys are costly. Therefore, before making his decision about whether to run for the office, the politician may first want to do a decision analysis of whether to purchase a sample survey.

SENSITIVITY ANALYSIS

If a decision maker has done a decision analysis of a problem and has assessed the probability values and appraised the utility values, he or she can use these numbers to get some clues about what the expected benefits (i.e., value of $p \cdot d$) of further analysis are likely to be. One useful technique is known as "sensitivity analysis." Basically, sensitivity analysis involves calculating how much any probability assessment or utility appraisal would have to change to make a difference in the indicated decision. In the mayor's problem, for example, the optimal decision would shift from "submit Plan A" to "submit Plan B" if

- the probability of A being approved were less than 74% (rather than being 80%),
- the probability of B being approved were greater than 45% (rather than being 40%),
- the utility of submitting A and getting it approved were less than 0.65 (rather than being 0.7), or
- the utility of submitting B and getting it approved were greater than 0.27 (rather than being 0.2).

Knowledge of the sensitivity of a decision to changes in probability or utility values may be helpful in deciding whether to engage in further analysis. In particular, in the mayor's problem, since fairly small changes in her rough guesstimates could change her decision, Pedersen might decide the expected benefits of further analysis are worth the expected costs and therefore proceed to decompose the problem a little further.

EVPI CALCULATIONS

Another technique that can provide a decision maker with some clues about whether further analysis will be worthwhile involves the calculation of the "expected value of perfect information" or "EVPI." Suppose that in the mayor's problem Pedersen could determine for sure whether Plan A would be approved or rejected by the city council. How much would this "perfect information" be worth to the Pedersen? Since Pedersen now thinks that there is a 80% chance Plan A will be approved, it follows that she thinks that there is a 80% chance that this perfect information will tell her that Plan A will be approved. In this case, the perfect information will be of no value, as the perfect information will not change the mayor's mind about what to do: as before, Pedersen will submit Plan A. On the other hand, there is a 20% chance that perfect information will indicate that Plan A will be rejected. If Pedersen knew this, she would submit Plan B. As indicated in Figure 2, Pedersen appraised the value of submitting Plan A and having it rejected as 0, while she calculated the expected value of submitting Plan B as 0.52. Thus, if Pedersen knew that Plan A were going to be rejected, she could expect to gain 0.52 value units by submitting Plan B instead of Plan A. Consequently, the expected value of perfect information about whether Plan A will be approved or rejected will be given by the product of 20% (the probability the information will result in a better decision) and 0.52 (the expected difference in value this better decision will make), or $0.2 \cdot 0.52 = 0.104$.

The expected value of perfect information is useful since the expected value of any partial or imperfect information must be less than this value. In other words, the EVPI puts an upper bound on the value of p $\cdot$ d. Thus, if the cost of gathering *imperfect* information seems greater than the expected value of *perfect* information, it will probably not be worthwhile to gather the imperfect information. For example, Pedersen may be able to get a better estimate of the probability that the city council will approve Plan A, but if she thinks that this better estimate will cost more than the equivalent of 0.104 utility units (the expected value of perfect information), then she should not attempt to get this better estimate.

SOME IMPLICATIONS OF THE pd > c RULE

The pd > c rule does not imply that some particular form or style of decision-making is superior to others: the determination of p, d, and c depends on the context of the decision problem and the subjective guesstimates of the decision maker. In some cases, the rule may indicate that the decision should be made on the basis of habit or intuition; in other cases, on the basis of a simple analysis; and in other cases, on the basis of a lengthy and detailed analysis. And just as different ways of making decisions may be best for different kinds of decision problems, so different styles of decision-making may be appropriate to individuals in different situations.

Consider, for example, the different situations of a manager (or "policy maker"), staffer (or "policy analyst"), and academic (or "policy researcher"). While managers tend to allot considerable time to decision-making, staffers tend to allot even more time, and academics probably allot the most time. On the other hand, managers tend to consider a large number of decisions, staffers many fewer decisions, and academics the least number of decisions. Thus, managers spend the least average time per decision and therefore must rely on snap judgments and truncated analyses to a greater extent than staffers and to a far greater extent than academics. Of course, a busy manager does not have to allocate a short amount of time to each decision he makes: a fairly long amount of time might be spent analyzing a few decisions in detail. To the extent that time is spent on a few decisions, however, less time will be left for the remaining decisions. Thus, a manager who carefully analyzes some of his decisions will have to rely on snap judgments about the rest of his decisions (or about the adequacy of his subordinates' recommendations concerning these decisions).

Since the skills required to make good snap judgments, and truncated analyses are probably different from the skills required to make good detailed analyses, people trained in analysis may not necessarily be particularly good managers and policy makers. Although this proposition deserves empirical checking, it may be that the educational preparation of managers and policy makers should emphasize the back-of-the-envelope calculations, "soft" statistics, "quick-and-dirty" analysis, the science of delegation (including the science of asking probing questions), and the logic of drastic simplification.

In addition to the distinctions between the decision-making

contexts of managers, staffers, and academics, a distinction might be drawn between the decision-making contexts of rich decision makers and of poor decision makers. For two key reasons, a rich man may tend to rely more on habits and intuitions than an equally rational man who happens to be poor. First, the rich man probably has more decisions to make, as he has a greater amount of money to decide how to spend. Second, the rich man may place a higher value on his time, thus raising the cost to him of spending time on decision-making. Unless the benefits of spending time on decision-making increase to compensate for this increased cost, the rich man will tend to allocate less time to decision-making. For example, a wealthy man may choose to spend less time making decisions and more time on some more pleasurable and, presumably, expensive activity, such as fishing for marlin in a yacht off St. Thomas, perhaps while "drinking Brazilian coffee, smoking a Dutch cigar, sipping a French cognac, reading the New York *Times,* listening to a Brandenburg Concerto and entertaining his Swedish wife—all at the same time, and with varying degrees of success" (Burenstam Linder, 1970).

In sum, the rich man may have a larger number of decisions to make but nonetheless decide to allot less time to making those decisions. Hence, if a rich man and a poor man are equally rational, the rich man may tend to rely more on habit and intuition than the poor man. One implication of this hypothesis is, as Staffan Burenstam Linder (1970) argues, that increased prosperity (i.e., economic growth) may shift decision-making styles toward greater reliance on snap judgments and short analyses.

A large variety of distinctions other than those between managers, staffers, and academics and between rich men and poor men might be drawn, but hopefully these two examples sufficiently illustrate the basic point of this section. The point simply is that the amount of time a decision maker should spend on a decision problem depends on the tractability of the problem, the importance of the problem, and how busy the decision maker is. More precisely, decision makers should follow the pd > c rule and spend time on a decision problem only as long as the expected benefits of doing so exceed the expected costs.

NOTES

1. An amusing "Test for Systems Analysts" was published in *Modern Data* (August 1973). The question on "economics," for example, was: "Develop a realistic plan for refinancing the national debt. Trace the possible effects of your plan in the following areas: Cubism, the Donatist Controversy, the wave theory of light. Outline a method for preventing these effects. Criticize this method from all possible points of view."

2. See, for example, Webster's *New World Dictionary* or Webster's *Third New International Dictionary*.

3. My colleague, Gregory Fischer, suggested the following references: Goldberg (1970); Miller et al. (1967); Moscowitz (1972); Slovic and Lichtenstein (1971).

4. I am indebted to David Good for this example.

REFERENCES

BURENSTAM LINDER, S. (1970) The Harried Leisure Class. New York: Columbia Univ. Press.

GOLDBERG, L. R. (1970) "Man vs. model of man." Psych. Bull. 73: 422-432.

JAMES, Wm. (1892 [1968]) "Habit." The Writings of William James, edited by J. J. McDermott. New York: Modern Library.

LINDBLOM, C. E. (1959) "The Science of 'Muddling Through.' " Public Administration Review 19 (Spring): 79-88.

MILLER, L. et al. (1967) "JUDGE: a value-judgment-based tactical command system." Organizational Behavior and Human Performance 2: 329-374.

MOSKOWITZ, H. (1972) "R & D manager's choices of development policies in simulated R & D environments." IEEE Transactions on Engineering Management 19: 22-30.

RAIFFA, H. (1968) Decision Analysis: Introductory Lectures on Choices Under Uncertainty. Reading, Mass.: Addison-Wesley.

RAWLS, J. (1971) A Theory of Justice. Cambridge, Mass.: Harvard Univ. Press.

SCHLAIFFER, R. (1969) Analysis of Decisions Under Uncertainty. New York: McGraw-Hill.

SIMON, H. (1957) Models of Man. New York: Wiley.

SLOVIC, P. and S. LICHTENSTEIN (1971) "Comparison of Bayesian and regression approaches to the study of information processing in judgment." Organizational Behavior and Human Performance 6: 649-744.

Part III

DECENTRALIZATION

8

Service Delivery and the
Urban Political Order

DOUGLAS YATES

☐ AFTER A DECADE OF PROTEST and demands for participation and community control, urban government appears to be entering a new era. Now that the "urban crisis" has been discovered, debated, and in some quarters dismissed (Banfield, 1970), government officials and academic analysts alike have increasingly come to focus on "service delivery" as the central issue and problem of urban policy-making.

This shift from "crisis" rhetoric and dramatic solutions to the discussion of service delivery is in itself highly interesting. As compared with hopeful plans for community control, the quest for improved public services is a more modest and limited urban "solution." In particular, it represents a new emphasis on everyday urban problems and on the capacity of government to perform basic functions. In this sense, the service delivery orientation hits closer to the lived experience and expectations of urban residents than the previous, highly generalized desire to "save the cities." For one thing, it is difficult to see how a government can solve its dramatic problems if it cannot solve its routine ones. For another, the idea of an "urban crisis" was and is an abstraction that soars above the less dramatic, but more fundamental failure of government to meet those daily service needs that shape the quality of life in urban neighborhoods. In short, the desire for improved service delivery and urban management brings urban policy-making back to the street level and back to the roots of city government: the world of "street-level bureaucrats" (Lipsky, 1969)—teachers, policemen, firemen, social workers, and garbage men.

Viewed from this perspective, the problem of urban management today is that, instead of asserting, as we used to, that there is no Democratic or Republican way to clean the streets, we are now asking whether government is capable of cleaning the streets at all. That we should have moved from a confident expectation to an open question about the city's capacity to deliver basic services indicates that the service delivery issue cuts very deeply. Indeed it raises fundamental questions about the structure and functioning of the urban political and administrative system.

My contention in this paper is that service delivery problems are the most fundamental urban problems and that they reflect deeper structural problems in the urban political system. More precisely, my argument is that improved management and service delivery depends not so much on the introduction of new money or new efficiency experts but on the relationship between the structure of public service institutions and the structure of citizen demands. If this is true, the student of urban management must focus on citizen demand for public services, the bureaucratic organization of service delivery, and the point of intersection between the two: the street-level relationship between citizens and public employees.

This mode of analysis provides an explanation of why recent urban "solutions" did not work that is very different from other, more familiar views. The most familiar recent explanations of policy failures are that not enough money was spent, that there were too many disjointed programs, that there was too little citizen participation (or too much), that policies were misdirected, or that the problems being attacked could not be solved with available instruments of public policy (Banfield, 1970). Whether these explanations stress the economic, social, or intellectual foundations of urban policy, they all have in common a tendency to ignore the machinery of government and existing political forces in the city. They view policy initiatives in an administrative and political vacuum. Careful attention is paid to the design of policy instruments but very little attention is paid to the setting of policy innovation and implementation. Put another way, these explanations ignore the "black box" of political and governmental process and thus produce a detached input-output model of urban policy-making.

How is it then that the political and governmental system of the city determines the success (or failure) of urban policy-making, in general, and service delivery, in particular? What happens to policy initiatives in the "black box" of city government and politics?

My view, simply put, is that the new urban policy or "solution" is injected into a political and administrative system that is fragmented to the point of chaos. Further, there is no coherent administrative order to implement and control new policies. What exists instead is an extreme pluralism of political, administrative, and community interests which produces what I would call "street-fighting pluralism." And in this context, the likely fate of the new policy initiative is that it will be ripped apart in the street fight between rival political interests.

At first glance, it might seem that this political fragmentation is a constant in the American political system and that pointing it out here provides no special understanding of urban government. I would argue, however, that the fragmentation entailed by "street-fighting pluralism" is indeed a distinctive feature of the urban system and for the following reasons. First, in most large cities, the system of political representation is weakly developed and provides little articulation of citizen interests (Alford and Lee, 1968). In particular, city councilmen typically are part-time officials who are underpaid and understaffed and rarely involved in significant policy-making. Urban political representation is also fragmented by overlapping city, county, state, and national constituencies. Who services the interests of urban residents and whom does the urban resident look up to for help: his councilman, state assemblyman, state senator, congressman, or senator? At one time, the great machines provided political linkage between urban representatives. But with the demise of the machines in most cities, urban representatives are now apt to produce rival and often warring sovereignties.

What makes this political fragmentation especially important is that it is mirrored at the street level by a deep fragmentation of citizen interests. At higher levels of politics, citizen interests are articulated by organized and functional pressure groups—labor unions, veterans' associations, agricultural lobbies, and the like. But at the street level, there is a paradox in citizen interest articulation that impedes and fragments collective action. On the one hand, as consumers of city services, residents are dealing with the most tangible and visible kind of government outputs—outputs that impinge on their personal lives every day. Thus, the relationship between government and urban residents in service delivery is constant, salient, and tangible: all of which we might expect would stimulate citizen demands on end action against government. On the other hand, citizen interests and demands are fragmented by the very nature of urban public services. That is, because urban services are personal, direct, and locality-specific they are—in terms of both delivery and citizen needs—highly divisible. In terms of delivery, urban services, in contrast to pure public goods like national defense, can easily be "divided"—allocated differentially to different groups of citizens. Indeed, one major source of service complaints against the city is that x block or neighborhood does not receive its fair share of public services.

More important, urban residents have very different needs and demands for urban services. While citizens have a relatively undifferentiated need for national defense or postal service, urban residents have particular locality-specific needs for services like police, fire, education, and garbage collection. In the first place, neighborhoods differ in their demands as between services. Some neighborhoods will be satisfied with police and fire protection but discontent with garbage service, while other neighborhoods will have opposite needs. In short, demands by individuals and neighborhoods for a particular service differ both qualitatively and quantitatively. In terms of quality, differences in community structure strongly affect the precise nature of service demands. In terms of quantity, economic, social, and physical factors (e.g., population density and age of housing) affect the amount of fire protection and garbage collection required in an area. A low-income neighborhood with a large number of drug addicts will have different demands for police protection than will a commercial strip, an entertainment area, or a neighborhood of middle-income homeowners or those with a high proportion of elderly people or college students. Moreover, demands for police service may vary dramatically between different groups within such neighborhoods. In low-income, high-crime neighborhoods, for example, older residents will often demand much tougher law enforcement against young "toughs" while young residents, who like to hang out, will demand less police surveillance and less law enforcement. The same difference may exist between students and other residents in a college district. In these cases, qualitative differences in the demand for services complicate the determination of what kinds of laws should be enforced or not enforced, what kinds of crime are feared, what time of day police protection is most needed, and what kind of "public place" residents wish to maintain.

In addition to neighborhood differences, service demands will often differ on a block-by-block, household, or individual basis. Many urban service problems affect a very small public, and the solution of one problem is entirely independent of the solution of others. The defective traffic light, the broken catch basin, the abandoned car, the after-hours bar, the rubbish fires in a vacant lot, the broken park benches and swings, the noise and fumes from a small factory, and the addict meetingplace are all problems that affect a small group of urban residents intensely and other residents not at all. In short, because the delivery of and the demand for urban services is so locality-specific, one resident may be satisfied with service delivery and his neighbor highly discontent. For the fact that garbage is collected regularly at one home does not mean that it has been collected down the block or around the corner. More important, according to this logic, it does not matter to an urban resident that urban services are delivered well elsewhere. This gives him no material or symbolic satisfaction. Rather, his concern and demand are that services be provided to him, his family, his house, and his block. It is for this reason that citizen demands for urban services are deeply fragmented.

Given the fragmentation of citizen interests, myriad service demands tend to grow up from the street level spasmodically—almost randomly—when a new service problem develops or when residents decide they have to do something about a long-standing problem. And, given the fragmentation of political representation, the natural response of angry citizens is to bring their protests to the mayor: to fight city hall (Lipsky, 1970). In fact, it is precisely the constant barrage of citizen demands and the resulting skirmishes between city hall and the neighborhoods that give the idea of "street-fighting pluralism" its distinctive meaning and substance.

If the mayor becomes in this way the focal point of the city's fragmented political system, he is not, however, in a structural position to provide coherent policy-making and service delivery. In part, this is so because the mayors of most large cities lack the formal powers to control the administrative system that delivers services to urban residents. His formal power over service delivery is shared with independent elected officials, with independent, or semi-independent departments, boards, and commissions (boards of education, police departments, and city planning commissions). Moreover, the urban executive's control over policy-making and service delivery is limited because it is shared among city, state, and federal administrators of public services. The mayor may be blamed for health service problems in the neighborhoods, but the nature of those services is obviously controlled to a large extent by the policies and programs of the Department of Health, Education and Welfare, and by parallel state bureaucracies.

Further, the mayor lacks control over the urban administration because of the fragmentation of that system and its own street-level character. On the first point, even though many service problems are interrelated, the various service delivery departments tend to function as separate feudal baronies. Departmental subunits and related departments that are involved in service delivery are so numerous in most cities that the administrative route from policy innovation to implementation is, even with the help of "project management," a labyrinth. This proliferation of administrative units and subunits constitutes a pluralization of government that reflects past ad hoc attempts at specialized problem-solving and contributes to the city's street-fighting pluralism. For when city government tries to gather together the various administrative pieces that control a given service delivery sector, it finds a plethora of administrators who have a vested interest and a putative sovereignty over policy-making. As a result, any attempt by the mayor to redirect or reorganize service delivery will lead him into complex disputes between administrators over "territorial" rights and responsibilities and thus into intense bureaucratic conflict.

Moreover, this interagency fragmentation and pluralization of subunits is compounded by certain fragmenting dislocations in the hierarchical structure of service delivery bureaucracies. As police, fire, sanitation, and education bureaucracies grew larger and more complex, "downtown" administrators sought to rationalize and gain control over their dispersed street level bureaucracies by centralizing power in central headquarters. The primary effect of centralizing administrative power was to weaken the power and authority of district officials and street-level supervisors (such as police captains, school principals, and sanitation foremen) who in the past had direct control over and responsibility for service delivery at the point of contact with urban residents. This shift in authority to downtown thus created a dislocation of control and accountability in service delivery hierarchies that was reinforced by the distinctive role of street-level bureaucrats—policemen, teachers, and sanitation men. Since urban services, unlike many other public services, involve a direct and often personal relationship between public employees and citizens, it is the street-level bureaucrats who must make constant, on-the-spot, personal judgments and, in so doing, determine the nature and quality of urban services. The wide discretion of street-level bureaucrats has frequently been noted (Wilson, 1968: 7), and it is indeed a crucial structural determinant of inconsistent, incoherent, and fragmented service delivery. Although the police commissioner or the school suerintendent may try to lay down broad-gauged policies and administrative practices, the foot soldier out on his own on the beat, in the garbage truck, or in the classroom determines whether central policy is innovated. And it is almost certain that, even if they were followed, policy directives would lack the subtlety and detail to provide workable operating guidelines for a policeman dealing with an ambiguous or delicate arrest or for a teacher trying to respond to different student needs and problems. Thus, given the independence and discretion of street-level bureaucrats and the weakness of field supervision, urban bureaucracies have precious little administrative control over service delivery at the crucial point of contact between city and citizens.

What this means is that service delivery takes place in a highly decentralized administrative "marketplace" with many different consumers and producers trying independently to strike a bargain on a wide range of goods and services. The diffuseness of this street-level service market is further increased by the inherent subjectivity of judgments about the quality of urban services. What, after all, constitutes satisfactory police, fire, or garbage service? In the case of police service, city government is often faced with conflicting citizen demands and values. The subjectivity of citizen demands is readily apparent here and for obvious reasons. Consider a more concrete service like garbage collection, the effectiveness of which we might expect could be more precisely and clearly measured. But even with garbage collection, the impact and effectiveness of the service rendered are highly arguable. Here, as with most other urban services, the problem in citizen evaluation concerns criteria and parameters of assessment. If a particular block gets as many pickups as any other block, but the block is still not "clean," the residents are unlikely to be satisfied with the service. If the same block gets more pickups than other blocks but still is not clean, how should residents view their service? What about residents on whose blocks pickups are reduced because the city feels the area can be kept up with only two pickups a week? In all these cases, the quantity (or level) of service is not, from a citizen view, a satisfactory indicator of the quality of service (the cleanliness of the block). And to add another complication, what about the block that is in fact kept clean but where citizens are angry because pickups are occasionally missed or are irregular? Here citizen evaluation is not even a function of the overall quality of service but of expectations concerning the manner of its delivery. If it is hard enough to have city and citizen agree on what counts as an acceptable schedule of pickups per week, it is that much harder to decide on what counts as "clean." Is the block clean if it is well cleaned except on weekends or during snowstorms or after truck deliveries or when cars are illegally parked on the street? Is it clean if the street is immaculate and the sidewalks are not? Or when the streets and sidewalks are clean but vacant lots and back alleys are not? Quite obviously, the answers to these questions depend in the first place on citizens' subjective appraisals—on their own tastes, circumstances, and expectations. A block that is clean enough for a teenager who uses the street as both eating place and playground may not be clean enough for a homeowner or a shopkeeper.

In addition, the appraisal of service quality is complicated by differing conceptions of the parameters of government's (and in this case the sanitation man's) responsibility and capability. The source of this problem is that garbage collection, like most urban services, quickly becomes intertwined with other service delivery problems and with underlying neighborhood conditions which cannot be altered by ordinary public services. On the first point, the sanitation department often cannot deliver its services properly if abandoned cars and illegally parked cars are not removed by the police; but citizens are likely to blame them anyway for service deficiencies that result. Equally, even though garbagemen responsible for daily collection are not responsible for bulk pickups or the cleanliness of vacant lots, the fact that the latter service problems persist will likely affect residents' appraisal of the department's effectiveness in keeping the streets "clean." More important, the garbageman's success in doing his job is critically affected by the kinds of disposal facilities provided by landlords, by population density, and by the social and economic character of their "beat," but these determinants of "cleanliness" are beyond his control.

In short, the subjectivity inherent in citizen demands, and with it, the difficulties in setting performance standards, establishing clear responsibility, and evaluating service delivery, reinforce the deeply rooted pattern of fragmentation in both the supply and demand for service delivery. Lacking workable standards for supplying and evaluating urban services, both citizens and street-level bureaucrats are, to a large extent, left to their own personal judgments and to whatever piecemeal accommodations can be worked out between the "servers and the served." Put another way, in this service delivery context of disparate expectations and fuzzy evaluations, it is every man for himself, and service delivery becomes a series of atomistic encounters between citizens and public employees.

To summarize my argument, because of the pervasive fragmentation in the structure of both citizen demands and urban institutions and because of the "street-fighting pluralism" that results, existing mechanisms for establishing public control over service delivery have become very weak. And if the sources of this weakness are structures, it follows, I think, that the solutions to the service problem must lie in a restructuring of the present service delivery system.

THE EVOLUTION OF THE
URBAN POLITICAL SYSTEM

As we have seen, the fragmentation and street-fighting pluralism that characterize city government exist in large measure because of intrinsic atomizing forces in the urban polity. But these patterns also have deep historical foundations. They have developed over time as a result of the evolution of the urban political system. That is, the present governmental structure has been built by accretion. New additions have been added to the edifice layer on layer by earlier urban political actors dealing with different demands and pressures and trying to correct quite different deficiencies in the structure they inherited.

Like most historical political systems, the urban polity began with relatively simple governmental structure and with limited functions and pressures on it. In the colonial and pre-Civil-War periods from 1760-1840, American cities were literally and figuratively "closed corporations" (Griffith, 1927). They were literally so in the sense that only a small percentage of residents—namely, property owners—were allowed to vote, and local governing bodies were typically controlled by a small, self-perpetuating group of "city fathers." American cities were figuratively "closed corporations" in the sense that the early city fathers had little to do or even discuss beyond "minding the store" and preserving harmony in the body politic. More than anything, urban politics in this period must have resembled the elite-dominated, status-quo-oriented village politics described by Vidich and Bensman (1960) in *Small Town in Mass Society*. This was the Yankee era of striking social homogeneity, of crude public facilities, and of rag-tag public services (such as the night-watch and volunteer firemen). But even in this uncomplicated and undramatic period of governance, the roots of fragmentation were being sowed. As the physical city grew, the strongly centralized but passive caretaker government did little to establish public control over the expanding city. As a result, both the city and its public institutions and services grew piecemeal. Newly developed areas received services and facilities as the need arose, and the consequence was pervasive geographical fragmentation of public control. In addition, just outside the boundaries of the central-city areas, new cities and villages sprang up as urban settlements extended into farmland. This is, of course, the same pattern that we encounter today with the proliferation of suburban governments in metropolitan. The difference is that, by comparison to the present, this early urban political order was Lilliputian, involving numerous tiny polities without the population to demand extensive services or the governmental machinery to provide them. And everywhere, this was a world of silent government, of city fathers running their cities and villages as closed corporations.

After 1840, this early political order was transformed by the impact of massive immigration from Europe. The social impact of immigration is self-evident; a relatively homogeneous community became not only diverse, but divided. The great machines of the nineteenth century were in large part the produce of this ethnic division between Yankees and immigrants. And, as such, they represented a waging of the social struggle by political means. As is well known, the urban machine provided both material and symbolic benefits to the new immigrant population—patronage, an ombudsman-style linkage between city and citizen, and ethnic recognition. More important, at least in political terms, the urban machine emerged as one possible solution to the structural problems presented by the earlier political order. In the first place, the order presided over by the city fathers lacked political roots in the urban community and thus provided little in the way of political and administrative communications between city and citizen. Indeed, a political order that was remote and out of touch with street-level needs in the early nineteenth century clearly lacked the ability to adapt to the new service demands of an unfamiliar population. The machine was, in this sense, an experiment in political adaption, an attempt to forge new channels of communication between city government and its citizens. Equally important, the machine provided a solution to the progressive fragmentation of the political and administrative structure. The purpose and effect of the great machines was to pull together the scattered pieces of political power and forge a new, coherent system of public control (Merton, 1957: 72). If this system, as it developed, was increasingly characterized by corruption and graft, it was, where successful, an enormously powerful instrument of centralized control. And in its most dramatic incarnations, in Tammany Hall, Jersey City, and Kansas City, to name three machine cities, the new system led to political monopolies, in the form of boss rule that established an entrenched structure of political control from city hall to the street corner.

The relationship between the machine order and the organization of service delivery is a little-explored subject that has important implications for the development of urban administration. As is well known, the urban machine provided personal services to immigrants needing assistance and mediation in dealing with a distant and unfamiliar government. And today, in the absence of the machine, the need for a functional equivalent to the political linkage between citizens and ward leaders has led to the creation of neighborhood service centers, little city halls, and ombudsman experiments. Further, the machine provided far more substantial benefits to a smaller group of residents: patronage jobs in a rapidly expanding public sector. Since the public sector grew as a result of the need to govern and deliver services to a dramatically expanded and diversified population, it can be said that the machine was at first the creation of immigration and then the creator of a new political order based on the impact of immigration on the structure of government. Put another way, immigration provided the political opportunity, and the machine provided the political innovation to capitalize on that opportunity. Immigration and the machine thus exist in a special historical relationship. For the capacity of the machine to deliver the level of service and jobs necessary to establish extensive political control depended on the immigration-produced transformation of the social and governmental order. By contrast, according to this analysis, a political innovation like the machine could not be made in a period when the size of government was either fixed or contracting.

In addition, the success of the machine as a mechanism for public control depended on the nature of the services that were demanded and supplied during its period of growth. The machine flourished in the course of providing personal services and favors (which established political communication, loyalty, and electoral support) *and* large-scale works (which generated large amounts of capital, produced highly visible benefits to the community, and conduced to the centralization of power and public control).

The new technology of traction, bridges, water supply, tunnels, subways, and the like created large bond issues, contracts, and expenditures. In turn, when controlled by the machine, these public works projects provided jobs in construction, administration, and maintenance, windfall profits for "honest" grafters who knew where the city was going to build, money in the city hall till (not all of which was spent on the projects), and the ability to reward political allies in the granting of government contracts. Quite simply, the task of building the physical structure of the city carried with it powerful political resources in the form of both capital and operating expenditures that could easily be translated into political money.

The construction of large-scale public works also gave the machine a highly salient political product and thus the appearance of successfully delivering services to urban residents. By contrast, the delivery of social services typically lacks the salience, concreteness, and visibility in impact that characterizes public works projects, and for that reason delivering social services is likely to be less persuasive political advertising for city hall.

Perhaps most important, the construction of large-scale public works required centralized coordination and administration—a function that the great machines were only too happy to provide. Whereas the simpler public services—such as street paving—of an earlier period were locality-specific and could be provided on a fragmented, ad hoc basis. Bridges, sewer systems, and traction were, in economic and political terms, natural monopolies and could not easily be planned and administered under a fragmented system of public control. In this structural sense, the impact of machines and that of large-scale public works were complementary, for both had the purpose of tying the city together, articulating its many parts into a more coherent whole. But, as with its relationship to immigration, the strength of the machine was that it harnessed itself to those forces that were transforming the structure of the city, and used those forces as its essential political resources in establishing public control.

In sum, the machine gained and centralized political power by providing both direct personal services and large-scale public goods. However, in important ways, this thrust toward political central-ization did not eliminate the existing fragmentation of the service delivery structure. Although the machine sought to establish an intricate system of political links and obligations, it was little concerned with the administrative organization of public services. It was important to the machine's control that police commissioners as well as police captains be dues-paying members of the political organization, but the machine was not particularly concerned if the commissioners had little administrative control over their captains in the field. In fact, the slight evidence that exists on urban public institutions in the machine age suggests that district officials and field supervisors ran their subunits like small feudal barons and made their own special arrangements with local citizens and politicians as to how services would be delivered and as to which would be enforced or not enforced against which group of citizens. This system of dispersed policy-making and standard-setting may indeed have allowed public employees to respond to local demands and mores, but it also continued and, in fact, deepened the fragmentation of service delivery created in the earlier era.

The reform reaction to the machine's political order is well known. The reformers' response to the machine's power and corruption was to break up the system of centralized power and establish strong administrative control over urban government. But there was a sharp conflict between these two purposes which had serious implications for the organization of service delivery. Re-formers were interested in establishing two quite different kinds of public control—one negative and one positive. The negative form of public control was a desire to prevent monopolies of political power, and, as in anti-trust, the solution was to deliberately fragment power by parceling it out to independent boards and commissions (Kaufman, 1969). The idea was to frustrate the political larceny of the machine by locking power up in a series of separate governmental safe deposit vaults. But having achieved this kind of negative public control, it was structurally impossible for reformers to achieve the second, positive form of control—coherent administrative control over urban government and service delivery. For, having divided power in the hope of taming it, there was no way to simultaneously achieve stronger and more coordinated public control of urban bureaucracies and service delivery. Rather the political order of reform added new political fragmentation to the existing administrative fragmentation in the city.

The reform movement was clearly successful in some cities in disrupting the machine, but it had little impact on bureaucracies and thus on the organization of service delivery. Indeed, the reform movement's only remedy for the daily problems of service delivery was the introduction of the merit system and, with it, the creation of a civil service. However, this policy reform was an instrument for controlling personnel recruitment and promotion, not an instrument for reorganizing the operating structure of city departments.

The coming of the New Deal also had a deep impact on the urban political structure. Much has been written about how New Deal social programs and services displaced the personal services provided by the machine and thus undermined the machine's political position. Be that as it may, the proliferation of federal social programs had another important and undeniable effect on urban government. The creation of new social programs at the national level led to a pluralization of the service delivery structure at the city level. And this meant that a new layer of sprawling, loosely coordinated agencies was added to the sprawling, loosely coordinated administrative structure that already existed. In terms of the administrative control of city government, the net effect of New Deal social innovation was to compound the problem of fragmentation that, by now, had become a historical plague on urban management and service delivery.

In addition, the growth of federal involvement brought with it a further division of policy-making in the design of urban services. While in the past the states were the cities' limited partners and overseers in organizing urban services, the emergence of the federal government created a more complex partnership and, indeed, a three-ring circus of shared and competing public control.

In the urban political system that developed after 1940, many of the patterns that existed earlier were repeated with slight variations, and there were also several unsuccessful attempts to reverse the existing pressures toward fragmented administration and service delivery. The city of the 1940s and early 1950s was a quieter and less-pressured place than it had been before and was to become later. In broad historical terms, this was an unfamiliar interlude between European immigration and depression, on the one hand, and massive black migration, on the other. In this relatively static city whose citizens were undoubtedly more concerned with international relations than with public services, the process of bureaucratization in urban institutions grew apace. The problem of service delivery, as it existed at this time, was how to consolidate and control the chaotic service bureaucracy. In this context, the apparent solution was to increase administrative control in "downtown" headquarters and to strengthen the power of mayors through strong mayor charters. The attempt was to create sufficient centralized power to counteract the powerful centrifugal forces in the governmental system. The result was formal-legal efforts at centralization (charter revision) and the continued buildup of central office administration. Had the city remained relatively static, these measures might have proved successful. But the suspicion remains that these attempts at central bureaucratic control had the effect of building yet another layer of administration on top of the existing system and thus rendering the structure of city government more intricate and cumbersome.

In any case, by the middle of the 1950s, it was becoming obvious to many mayors and urban analysts that social and economic pressures were growing dramatically and that city government lacked the capacity to deal with them. The city's housing stock, built during the period of immigration and rapid growth, was now old and decaying—and with it the main public facilities—schools, public transit, and even the water system. More important, the migration of large numbers of poor blacks, Puerto Ricans, and Chicanos from rural areas placed heavy demands on existing social services and contributed to the rapid growth of central-city slums.

Urban renewal was the federal government's initial response to this new awareness of a deepening "urban crisis." Although the renewal program in no way solved the problem of spreading slums and blight, it did have a powerful impact on the structure of urban government in many cities. This impact can be seen most clearly in cities like New Haven and Philadelphia, where renewal programs were most fully developed and where, as a result, the politics of renewal were most pronounced. In these cities, renewal led to an increased centralization of political power and to the rise of new "bureaucratic machines" serving as instruments of central control (Wolfinger,

1974). In the first place, renewal provided a new source of patronage and capital expenditures that translated into potent political resources in the hands of energetic mayors like Richard C. Lee of New Haven (Lowe, 1967; Dahl, 1961). In addition, because renewal involved large-scale projects requiring coordination and control, renewal provided structural support for entrepreneurial, centralizing administrators just as earlier public works had for political bosses and parks development did for Robert Moses in New York, to take another well-known example. It is entirely possible, of course, that Mayor Lee could have won election without renewal, but it was renewal that provided the dramatic, highly visible, and tangible issue that allowed him to dominate a city for almost two decades. Also, it was renewal that generated the staff expertise and patronage jobs required to build a bureaucratic machine in Redevelopment Agencies. Thus, in New Haven and other cities, redevelopment agencies harnessed the centripetal pressures set in motion by renewal planning and became the dominant force in city administration.

The rise of urban regimes advancing ambitious renewal programs (and armed for this purpose with federal funds and new bureaucratic resources) had important consequences for the delivery of ordinary public services. For the emphasis on bricks and mortar could not help but divert attention from social services, and the highly centralized, project-oriented renewal bureaucracies inhabited a political world far removed from the ordinary concerns of residents at the street level. In this period, big projects drove out small ones, and large-scale thinking about the future of the city drove out small-scale thinking about the delivery of particular services in particular neighborhoods.

In short, the political order based on urban renewal strengthened the power of mayors and established the power of redevelopment bureaucracies. But it did nothing to reduce the gap in political communication between city and citizen, and it did not reduce the fragmentation in the structure of service delivery. For the new centralization and coordination of policy-making in the redevelopment sector did not spill over into other service delivery sectors, such as police and education. However strong executive action might have been in the renewal arena, it was not aimed at the task of restructuring the organizational setting in which police captains, school principals, and sanitation foremen worked. And, in any case, the structural barriers separating the different feudal baronies in service delivery were sufficiently strong to allow business to proceed as usual in schools and police stations.

The reaction to the rather imperial character of urban renewal is well known. But in the early 1960s, local residents began to protest against disruptive renewal plans which they had no voice in determining (Davies, 1966). Equally, citizens began to demand that city government address itself to the quality of its low-skilled, poorly educated, nonwhite immigrants in the center city. But also, as we have seen, the demand was an evolutionary reaction to a political system preoccupied with bricks and mortar. And finally, the demand for new and improved social services stemmed the street-level view that public service bureaucracies were unresponsive, uncoordinated, and unaccountable. Thus, by this time, the problems of fragmentation and remoteness—the latter created by swollen central offices—had come to roost. And thus, too, the lack of effective public control over service delivery had become a highly charged political issue.

If we understand the development of the urban political system as a reactive, evolutionary process in which adjustments designed to correct old problems also bring about new and unforeseen problems, we can easily see how the protest and community action movements of the 1960s were at once apparently logical solutions to the problem of unresponsiveness in service delivery and at the same time powerful contributors to the syndrome of ever-deepening fragmentation in urban government.

The "community revolution" (Bell and Held, 1969), rooted in the civil rights movement and nourished by the war on poverty, brought large numbers of new participants into the urban political arena and changed the relationship between city and citizen in many low-income urban neighborhoods. At the least, the various strategies of citizen participation in the late 1960s had the effect of creating a voluble neighborhood voice able to articulate complaints about public services and corresponsiveness of service bureaucracies. This voice, in turn, had the effect of establishing a communications system between citizens and city agencies that had been largely absent since the heyday of the machine and the wardheeler. The problem with this form of political communication was that it was adversary and largely one-way. For community action was, in the final analysis, a strategy for fighting city hall, the schools, the housing authority, and the police. Indeed, community action was designed by its original architects in Washington to provide creative conflict—to shake up remote and sluggish bureaucracies. Of course, the targets of protest in the existing government system were not without resources in dealing with protest groups from "the community." Aside from straightforward delay and avoidance tactics, urban administrators under community pressure could always fall back on the fragmentation of the service delivery system by passing the buck to another agency or by relegating community demands to their own bureaucratic labyrinths (Lipsky, 1970).

In short, interaction between citizens and city government typically evolved into a ritualized game of shadowboxing in which the political energies created by community action were largely consumed by the constant sparring with government. In addition, the shadowboxing game, as played in government, was a closed and hollow one. Instead of producing administrative efforts to deal with fundamental problems of service delivery, it channelled administrative energy into the business of defending the government—of dodging or parrying community demands and protests.

The community revolution also contributed to the fragmentation of administration and service delivery. For, with the development of protest groups and poverty programs, thousands of new neighborhood institutions were erected in American cities. Some community organizations disappeared as quickly as they grew up, and many of them were able to sink only very shallow political roots in their communities. In all, the result was a diffusion of unrelated storefront organizations across urban neighborhoods. And since funding for community action typically provided for a few attractive staff positions in each organization, new but well-entrenched feudal barons were quickly added to the already crowded baronial structure in city politics.

Unlike urban renewal, the community action program created strong centrifugal pressures on urban government. Whereas renewal sought to restructure the city, viewed as a whole, community action was designed to improve urban life in particular neighborhoods. Because community action sought a multifaceted, locality-specific approach to service delivery, it also contributed to the growth of a diffuse anti-poverty bureaucracy at the city hall level.

Finally, the growth of poverty programs brought with it a new federal and state involvement in urban policy-making and thus further complicated and divided administrative power, authority, and responsibility for social service delivery. The federal government played a prominent role through the Office of Economic Opportunity in the development and evaluation of community anti-poverty programs. And, somewhat later, many states entered the anti-poverty arena, some creating departments of community affairs designed to focus and coordinate the services and programs provided through state government.

The decline, if not the final demise, of community action has stirred a heated debate about the cause of the program's failure. Whatever other factors may be pointed to, I would place primary emphasis on two kinds of fragmentation that have emerged out of the evolving urban administrative system. Viewed in this perspective, the problem of community action is no different from the more general problem of service delivery: both are reflections of structural problems in the urban political order. The first kind of fragmentation is that which exists both between different citizen groups at the street level and between citizens and city employees. The second and equally important kind of fragmentation is that which exists between federal, state, and city government. In addition, as we have seen, the fragmentation at the street level and in the federal system is compounded by the weakness of the basic control mechanisms in urban bureaucracy.

During and immediately after the discordant experience of community action, may other experiments were tried out as cities continued to search for answers to their service delivery problems. On the one hand, community control was advanced as a dramatic and immediate solution to the perceived remoteness and unresponsiveness of city government. On the other hand, in New York and elsewhere, the attempt was made to control service delivery by creating superagencies which placed a new layer of even more centralized administration on top of the existing structure. Also, in some cities, the role of budget bureaus and analytical staff units were strengthened as another way of achieving central control. So, too, the Model Cities program rose and fell as an instrument of coordinated local planning and, more recently, revenue-sharing has been implemented—in large part, as a way of consolidating administrative control in city hall.

However, as with earlier urban solutions, these experiments have not resolved the structural problems in the urban government that frustrate the development of an effective service delivery system. Given the disappointing record of recent urban innovations, urban management remains very much in flux, and control over service delivery is up for grabs.

With the benefit of considerable historical hindsight, we can draw several conclusions about the evolution of the urban political order and its service delivery system. In the first place, the evolutionary process has been marked by a sequence of governmental reactions to problems created in large measure by previous reactions. In this sense, the development of urban management and service delivery has traced a dialectical course between problem-solving and problem-creation. Nevertheless, there are two important constants in this zig-zag pattern of development. First, the urban political system has proved its adaptiveness in the face of massive social transformations. Both the machine and the "community revolution" were ingenious and largely spontaneous inventions of sidewalk politics designed to accommodate new pressures on the system. At the same time, despite its adaptiveness and ingenuity, urban administration has never come to terms with the steady growth of fragmentation in the urban system, and, in fact, the various new adaptations have often had the unforeseen consequence of deepening the syndrome of fragmentation.

Today there are new actors in the struggle for public control —public service unions, minority group activists, leaders of defensive working-class neighborhoods, reformers, and law and order candidates. But none of these actors has the political strength to reorganize service delivery simply by force of political will. More generally, given the social and economic diversity of urban neighborhoods, it is hard to see what kind of cohesive new political order could be established to overcome the present political fragmentation of the city. The numerical dominance of the immigrant population provided the basis for the great machines, and the dominance of white, working-class voters provided the political support in the 1950s for strong mayors like Daley and Lee. But today, the political makeup of the city is itself highly pluralistic: a melange of low-income neighborhoods (with their own ethnic and economic divisions), defensive working-class neighborhoods, growing areas of upwardly mobile homeowners and pockets of upper-middle-class reformism. In this context, the men at the top in urban government, be they active or passive, will have trouble establishing effective political control over their cities. Entrepreneurs like Lee, or Allen of Atlanta, are likely to be replaced by frustrated crusaders like Lindsay, Stokes, and Gibson. And confident bosses from Tammany Hall and men like Daley, or Whalen in Jersey City, are likely to be replaced by cautious political brokers like Robert Wagner of New York, Perk of Cleveland, Stenvig of Minneapolis, and Guida of New Haven. This distribution of and change in leadership styles in city hall may be represented as shown in Figure 1.

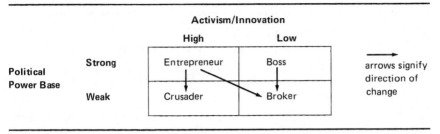

Figure 1: MEN IN CITY HALL: LEADERSHIP STYLE AND DIRECTION OF CHANGE

Given the difficulty of establishing a new political order, the mayor, city manager, or top-level administrator must therefore find structural solutions to structural problems. On the one hand, the fragmentation of citizen demands at the street-level means that programs and services are immediately pulled apart by myriad competing interests. This diminishing effectiveness of service delivery in a context of proliferating demands is the main effect of street-fighting pluralism, and it constitutes one main reason why city governments cannot solve their service problems. The problem can be represented as shown in Figure 2.

The other structural weakness in service delivery stems from fragmentation in policy-making between four levels of government: federal, state, city, and neighborhood. This fragmentation affects both the coherence of administration and the allocation of services and fiscal resources. The structural problem here is an administrative pluralism that cuts the program and resource pie into so many pieces that the impact of government in any given neighborhood is greatly diluted. Put another way, the present structure of intergovernmental

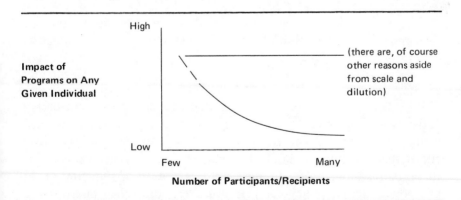

Figure 2: THE IMPACT OF SOCIAL PROGRAMS ON URBAN NEIGHBORHOODS: PROBLEMS OF SCALE AND DILUTION

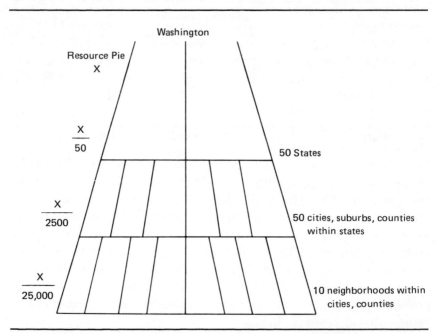

Figure 3: THE POLITICS OF FEDERALISM: THE PIE-CUTTING PROCESS

policy-making on service delivery involves a "trickle down" process—in which, at present, programs and resources are channeled in at
least 25,000 different directions as they move from Washington to
the street level. This pie-cutting syndrome, which is further reinforced by revenue-sharing, can be represented as shown in Figure 3.

URBAN SOLUTIONS:

A STRUCTURAL APPROACH

If urban management is frustrated by mutually reinforcing
patterns of fragmentation rooted in the structure of urban government, what solutions can be applied to urban "problems"? As a first
cut at this question, it should be clear that the toughest problem for
the urban administrator to deal with is that of fragmentation in
intergovernmental policy-making. For, here, the urban manager has
little control over the behavior of counterparts in Washington and
the statehouse. More than that, the urban policy maker is not only

controlled by decisions made a higher levels of government, but he is also subject to the uncertainty, ambiguity, and changeability of federal and states policies. Of course, there is a structural solution to this aspect of the service delivery problem, but it is not within the power of local political actors to bring it about. That solution would be to establish administrative instruments for systematic, coordinated decision-making on urban issues in the Congress, in the Executive Branch, and in state government. In addition, such a structural solution would entail new methods and mechanisms for intergovernmental planning and coordination (Sundquist, 1969). To an extent, the revenue-sharing dimension of the "New Federalism" addresses this structural problem by shifting public control of urban policy to city halls. Nevertheless, with or without revenue-sharing, federal and state government will inevitably make legislative and administrative decisions which have a critical impact on urban delivery. And unless these decisions—for example, new programs, funding levels, and administrative standards and procedures—are articulated in an explicit way, the problem of intergovernmental fragmentation in service delivery will not diminish.

Looking beyond the essentially "incontrollable" federal arena, we reach the problems of fragmentation at the city hall level and at the street level which the urban administrator can address directly and for which he may find viable structural solutions. At the city hall level, the central need is for policy planning that will chart and monitor the workings of service bureaucracies, illuminate the interconnections between different facets of the service delivery system, seen as a whole, and allocate resources equitably to different communities. As an example of this function, the policy planning performed by the Budget Bureau and other analytical units in New York City has increased in scope and effectiveness in the last five years. But this broad-gauged system analysis, however important, does not get to the heart of the fragmentation problem.

At first glance, it might seem that the structural answer to the fragmentation of departments at the city hall level is to build new devices for interdepartmental planning and coordination. Indeed, there is a clear need for policy coordination. And the idea of city hall service cabinets dealing with the interfaces between service bureaucracies has been discussed and implemented in some cities. But despite the apparent good sense of coordinating services at the city hall level, this solution is at best insufficient and at worst may be yet another attempt to strengthen the wobbly service delivery structure by building a neat new administrative penthouse on top. As in engineering, the edifice and its new penthouse are likely to be undermined by the weakness of its foundations. If this is so, the task of restructuring urban service delivery must begin with the foundations of service delivery: the street-level relationship between the servers and the served. Structural solutions at this level involve both mechanisms for citizen participation and communication between employees and citizens on the one hand, and administrative relations within and between district-level service bureaucracies, on the other.

Citizen participation has taken many forms in the last decade and there is no need here to discuss the strengths and weaknesses of the various approaches. My contention is that the kind of citizen participation that will improve service delivery is that which articulates and aggregates citizen demands and forges a link between citizens and public employees. How precisely can these objectives be achieved through citizen participation? One very simple and undramatic mechanism for aggregating citizen preferences is the block association or the neighborhood association. In many cities, these associations have grown up spontaneously and function as "homemade" instruments of collective action. These associations do not require elaborate administrative structures. Members meet in a church or a neighbor's living room, and the organizations are based on concrete common concerns and one built around existing social networks in the neighborhood. As such, they are easily able to identify, aggregate, and set priorities among local service problems and needs. In addition, they are easily able to monitor and evaluate the performance of policemen, firemen, and garbagemen. Most important, there appears to be a crucial economy of scale in citizen organization at the block or subneighborhood unit. Given our earlier analysis, we would expect that, in larger groups, the diversity —indeed, the fragmentation—of citizen demands and problems would frustrate attempts at interest articulation and aggregation. In short, given these virtues, the block and subneighborhood associations provide government with a viable communications link in the neighborhood. If this is true, the question becomes how to stimulate the development of block associations and how to design the link between block associations and city government.

On the first point, one way for government to stimulate the formation of block associations is to pay block leaders a small stipend to work as service monitors. With existing block groups, it may well be an adequate incentive to neighborhood residents that participation produce faster and more responsive service delivery. On the second point, the necessary condition for forging a city-citizen link is that block leaders have direct access to the neighborhood service administration in the police station, the sanitation garage, and the school. My reason for expecting that this form of citizen participation will succeed where so many other approaches have failed is that, instead of creating an adversary relationship, this monitoring and evaluation role is based on mutual self-interest. The street-level bureaucrat requires information and feedback on his services and neighborhood problems if he is to increase both his control over service delivery and the responsiveness of his services. The citizen, on the other hand, needs an open communications channel to the service bureaucracy if he is to record local needs and complaints and then get any reaction.

More generally, the block association represents one way to create an ombudsman structure in service delivery and thus establish direct communication between the citizen and the city government. There are other ways to do this. District-level officials could appoint citizen ombudsmen or create neighborhood service councils. At present, the trouble with police precinct councils and many school councils is that they lack a specific task, responsibilities, and accountability. They often exist as forums for casual conversation and for improving public relations. But if their task was to monitor and evaluate service delivery and if they served as effective conduits for citizen complaints, they would have a substantive role to play in service delivery. Moreover, the creation of a cooperative, service-oriented relationship between city and citizen is one way to curb the protracted and frustrating fights over power and control that have characterized many experiments in citizen participation. Many cities have begun to experiment with ombudsmen, and many have also created neighborhood service centers and little city halls as communications centers between public employees and citizens (Washnis, 1972). In terms of the implicit model of city-citizen communication proposed here, it is crucial that neighborhood centers not only be administrative outposts, but that they involve local residents as service monitors and ombudsmen.

Taken together, all these mechanisms of citizen participation, if fully developed, could bring about an entirely new communications system in street-level service delivery. At the lowest level, the monitoring function could be elaborated to the point where every block would be covered by a citizen "reporter," and, indeed, such a system has already been developed in one neighborhood government experiment in New York. At a higher level, community service councils could be developed that would be concerned with all services (and their interrelations) in the neighborhood. Such a mechanism would have the additional virtue of providing a central switchboard for the fragmented communications system that has developed with the proliferation of community organizations.

To this point, the discussion of an ombudsman model for citizen involvement in service delivery has one clear weakness: without a discussion of complementary structural changes in service bureaucracies, it is like the sound of one hand clapping. At present, the administratively centralized but still fragmented service bureaucracies are in no position to deal effectively with ombudsmen or departmental service councils, much less with interdepartmental citizen service councils. The structural solution to the bureaucratic part of the service delivery problem is administrative decentralization within departments and the creation of coordinating devices at the district level. The reason for administrative decentralization is simple: if local administrators do not have decision-making power and accountability, there is no point in gearing up citizen involvement at the district level. The only sensible strategy for citizens would still be to fight city hall or central headquarters. Equally important, if public control over service delivery is not established at the point of contact with citizens, there is no reason to think that services can be controlled or improved at all. The reason for establishing administrative mechanisms at the street level to coordinate the work of different departments is that, as was noted above, the closer one gets to the street level, the more apparent it is that many service problems are multifaceted or interstitial. If these premises are correct, administrative power in service bureaucracies should be pushed down to the school, precinct, and firehouse level, and interdepartmental service cabinets should be created at the neighborhood level. The service cabinet strategy has been launched in New York City, and it has brought about the beginnings of administrative coordination and communication at the street level.

In sum, my proposed structural solution to the urban service delivery problem is to build a new communications system from the bottom up rather than trying to reform the system from the top down (as has been the intent of many experiments in scientific urban management). The heart of this decentralized system would be a joint bureaucratic service cabinet—citizen service council structure which would involve citizens and public employees in focused attempts to solve concrete service problems. It would avoid the long-range planning exercises, general discussions about grievances, and power struggles that have changed the name of decentralization from a panacea to a dirty word in many cities. Below the service cabinet—service council structure would be a network of ombudsmen, block associations leaders, and service monitors who would provide the detailed information about needs and problems that the service delivery system presently lacks. Above the neighborhood cabinets and councils, in this model, would be a policy planning system in city hall concerned with productivity, project management, and citywide allocation. And, on this conception, the administrative role of higher-level government would focus on the design of service programs and on interdepartmental and intergovernment cooperation, but not on the administrative details of service delivery. Put another way in this system, administrative control of service delivery would be focused at the street level, and higher-level policy-making would be concerned with new policy development and with resource allocation.

Finally, one might naturally wonder what chance such a decentralized system of service delivery would have of being implemented. My answer to the feasibility issue is at once optimistic and pessimistic. On the one hand, one can be optimistic about the system's feasibility because it does not require massive new appropriations—in fact, it would require little in the way of new resources. In addition, in architectural terms, the proposed system is a simple one. It does not involve large-scale administrative overhead or the creation of new bureaucracies. What it does require is a rewiring of the communications system in urban government, a rearrangement of existing pieces in the urban jigsaw puzzle. On the other hand, one must be pessimistic about the willingness of entrenched political interests to permit a restructuring of administrative power within bureaucracies and to welcome increased citizen participation. Perhaps paradoxically, the creation of a decentralized service delivery system would require strong action by chief executives throughout the American system to bring it about.

REFERENCES

ALFORD, R. and E. LEE (1968) "Voting turnout in cities." Amer. Pol. Sci. Rev. 62 (September): 768-814.

BANFIELD, E. (1970) The Unheavenly City. Boston: Little Brown.

BELL, D. and V. HELD (1969) "The community revolution." Public Interest 16 (Summer).

DAHL, R. A. (1961) Who Governs? New Haven: Yale Univ. Press.

DAVIES, C. J. (1966) Neighborhood Groups and Urban Renewal. New York: Columbia Univ. Press.

GRIFFITH, E. (1927) Modern Development of City Government. London: Oxford Univ. Press.

KAUFMAN, H. (1969) "Administrative decentralization and political power." Public Admin. Rev. 29 (January/February).

LIPSKY, M. (1970) Protest in City Politics. Chicago: Rand McNally.

——— (1969) "Toward a theory of street-level bureaucracy." Presented at the annual meeting of the American Political Science Association.

LOWE, J. (1967) Cities in a Race with Time. New York: Harper & Row.

MERTON, R. K. (1957) "The latent functions of the machine," in R. K. Merton, Social Theory and Social Structure. New York: Free Press.

SUNDQUIST, J. L. (1969) Making Federalism Work. Washington, D.C.: Brookings Institution.

VIDICH, A. and J. BENSMAN (1960) Small Town in Mass Society. Garden City, N.Y.: Doubleday Anchor.

WASHNIS, G. (1972) Municipal Decentralization and Neighborhood Resources. New York: Praeger.

WILSON, J. Q. (1968) Varieties of Police Behavior. Cambridge, Mass.: Harvard Univ. Press.

WOLFINGER, R. (1974) The Politics of Progress. Englewood Cliffs, N.J.: Prentice-Hall.

9

Decentralization for Urban Management: Sorting the Wheat from the Chaff

ANNMARIE H. WALSH

INTRODUCTION

☐ DEBATES ABOUT URBAN DECENTRALIZATION often beg the questions they purport to answer. Reasonable citizens agree that it is paramount to stop deterioration of neighborhoods, to ease localized fiscal burdens, and to create a sense of efficacy among the city's peoples. How do we judge, however, the likelihood of these goals being furthered by one or another reorganization plan. What changes in programs and politics can reasonably be expected to flow from rearranging formal government structure? Precisely what is it we are trying to decentralize and with what effect?

This chapter attempts to lay out a framework within which decentralization proposals might be realistically assessed.

AUTHOR'S NOTE: *Substantial portions of this chapter previously appeared in* The City Almanac *VII (June 1972), published by the Center for New York City Affairs, New School for Social Research. This chapter attempts to outline a framework for analysis, and for this purpose, citations are kept to a minimum. The reader interested in sources and further reading should see* Selected References: Decentralization and Neighborhood Government, *published by the*

A statement that was directed to decentralization in industry some fifteen years ago rings true relative to government today:

> It has become something of a fad to have a decentralization program—or at least to talk about decentralization. . . . Unfortunately, much of the talk and writing includes vague generalizations, misconceptions and propositions that are only partly thought out [Smith, 1958].

This situation reflects a broader failure, the failure of scholars and practitioners to develop systematic, empirically based methods for designing and evaluating government organization. Traditional reform principles have long been under attack. But no acceptable set of concepts has taken their place. Now many reformers are simply undergoing conversion from the faith of Executive Centrism to the faith of Community Decentralism. The sentence with which several recent discussions have opened reflects a kind of mystical fatalism: "Decentralization is an idea whose time is come." That statement is part of a romantic dialectic of reform and disillusion that has cruel impact on public expectations. It is time to insist, rather, on responsible data gathering and hardheaded analysis as a basis for organizational design.

Any government reorganization is costly, using up public funds and energy, heating up political conflicts, and raising public expectations. We live with the results for decades. It is crucial, therefore, that hard thinking, practical research, and broad consultation take place before city governments are launched into uncharted waters.

There are three prerequisites for sensible evaluation of any decentralization plan: (1) a definition of decentralization must be made in terms of clear, practical meaning; (2) the goals sought must be identified; (3) evidence must be gathered that the decentralization planned would contribute to those goals. This chapter explores the definition and goals of decentralization and the need for evidence to predict its likely results.

Office of Urban Policy and Programs, Graduate School of CUNY (33 West 42 Street, New York, New York 10036). Substantial parts of this chapter previously appeared in my "What Price Decentralization in New York? (City Almanac 7 [June]) and "Lessons from London for Governing New York City," prepared for the New York City Commission on State-City Relations (April 1972). The City Almanac *is published by the Center for New York City Affairs, directed by Henry Cohen.*

THE MEANING OF DECENTRALIZATION

To decentralize means, literally, to reverse a centralization process. The term implies that one has determined by empirical research or experience that something is too tightly centralized and has designed a set of changes predicted to reduce that concentration.

Decentralization is, then, a broad concept, with application to business, land use, even biology. We are here concerned with its application to urban government. Even thus narrowed, it is an idea of many faces that must be more focused to be implemented.

JURISDICTION AND POWER

Two common meanings can be extracted from current discussions:

- To decentralize is to reduce the scale on which public services are operated.
- To decentralize is to redistribute effective political power more widely among people and groups.

Discussion of governmental decentralization has tended to blur these two basic issues: geographic scale and political power. Changes in geographic scale are frequently claimed—in some cases altogether speciously—to be efforts at political decentralization; these claims divert attention from real redistributive issues.

The geographic scale on which a particular operation is managed may have some effects on its costs, its quality, its equity, its character. These will vary with a specific set of factors: local politics, the distribution of resources needed for the activity, the distribution of people who consume the service, its impact on other people, requirement for local variations and special knowledge, and economies or diseconomies of scale. All these are factors relevant to determining the scale for any given function, whether it be neighborhood, city, region, state, or nation. Cogent arguments have been made, for example, for federalizing the financial and regulatory aspects of public assistance. The public assistance consumption market crosses state and local boundaries (potential welfare clients migrate). Resources are unevenly distributed. Public assistance is a function that involves redistribution and, therefore, requires jurisdiction sufficiently broad to take in an average scatter of wealth levels. There are severe externalities (e.g., suburban zoning adds to the city welfare load). At the same time, a strong case can be made for localization of the operation of social services—family counseling, home nursing, day care, and so forth.

Another illustration of scale factors is water supply development. As urban growth puts pressure on localized systems, regionalization of water source development and distribution is needed. The geography of resource distribution and consumption market has compelling effects on organizational choices for water supply. Hence, there are some practical methods for evaluating the geographic scale on which such an operation should be organized.

The second, and more significant meaning of decentralization —distribution of power—requires separate consideration. Political values and power goals are relevant to this consideration. So are the abundance and distribution of governmental resources, such as leadership, information, executive and technical manpower, and fiscal capacity.

These two aspects of decentralization—the geographic and political—can vary independently of each other; that is, operations on a local scale can be accompanied by local control, central control, or a mixture of both. The example of water supply is a case in point. Where it has been decided that water supply should be managed at the metropolitan level (a resolution of the geographic issue), this function has been assigned in different parts of the world to an elected metropolitan government, to an association of local politicians, to a national or state regional office with explicit metropolitan jurisdiction, or to a public corporation governed by a board representing one or all levels of government. Each alternative represents a different resolution of the political issue.

Changes in the geographic scale of an operation and changes in the ultimate distribution of power, although interrelated, are conceptually separable. The distinction is crucial both for devising strategies for modification of urban government and for assessing the effect that given modifications are likely to have on existing patterns of behavior.

The possible combinations of geographic scale and influence distribution are myriad and should be considered with some breadth of mind. Local power can be exerted through large-scale organization. The U.S. Congress, for example, is notable for its response to local power centers. Representatives are accorded the right to veto certain decisions affecting their areas (appointments and public works).

Conversely, geographically localized government may be subject to the dominant influence of centralized political parties, centralized interest groups, or state and federal officials. British borough councils, for example, follow the lead of Labour or Conservative Party policies on some school issues that are hotly battled over in local boards and councils of U.S. cities.

Small jurisdictions often show very narrow internal patterns of political control. Most available studies of local politics in the various community action programs and in towns throughout the United States show low rates of voter participation. *Where oligarchic or factional patterns of local politics persist, we would actually narrow political power distribution by decentralizing the scale of operations.* Hence, two aspects of decentralization as defined—geographic scale and distribution of power—vary separately, often in delicate counterbalance.

SERVICES AND FUNCTIONS

There is yet another set of distinctions that must be handled clearly to know what we are talking about when discussing decentralization. What activities and roles are being considered for broadened influence or reduced scale of operation?

For the most part, what industry has termed "decentralization by product line" is not feasible today in city government. Business can usefully consolidate all the activities it takes to turn out one product or service, delegating sufficient power to that product division to get the job done expeditiously. In government, this would entail consolidating all functions (planning, legislation, financing, budgeting, purchasing, contracting, distributing, and so forth) relating to a single public service or product (e.g., water supply, public housing, sanitation, law enforcement, health services). Some reorganization in this direction could have significant payoffs in city government, but there are distinct limits because of intergovernmental patterns. All major public goods and services are dependent on a complex intergovernmental production system. We could consider, for example, decentralization of housing code enforcement. That alternative must then be judged in the context of rent regulation and housing finance remaining in other spheres. Many of the financing and regulatory functions for locally managed services are vested in state and federal government, out of the reach of city government reorganization. State legislatures frequently legislate elaborate details of local administration.

Every pattern by which these labors are divided sets up a network of intergovernmental interdependencies that can expedite or choke the system; that can counteract or support meaningful decentralization. No sensible evaluation of a decentralization plan can be made without careful exploration of this dimension. We must ask: Who will be planning, financing, budgeting, producing, allocating equipment, distributing, and receiving public demands and complaints, relative to each service involved? How many agencies will it take to actually get the task accomplished? Without answers to these questions, one cannot assess whether a planned change would expedite or stalemate action, would bolster or hamstring local power to get things done. Simply creating a localized jurisdiction is meaningless. Its power and predictable tendencies depend on its place in the intergovernmental system and its functional abilities.

GOVERNMENTAL RESOURCES

A third dimension is crucial to judging decentralization; that is the distribution of resources, human and fiscal. The world offers plentiful examples of decentralized authority that is counterproductive because resources are scarce. Neighborhood governments, for example, can have little effect if they cannot hire competent staff and if they command financial resources grossly deficient relative to public needs and expectations. The wealth of our nation is concentrated in its metropolitan areas. As long as huge chunks of it are sheltered by suburban municipal boundaries, neighborhood government in the inner city may be a very frustrating experience.

To summarize the definitional path trodden thus far, decentralization entails changing the structure of government in order to broaden political influence or reduce the scale of operations of specific activities. One of those changes does not necessarily lead to the other. Furthermore, the results of the change depend directly on what place the shifted activities will have in the aggregate pattern of shared action necessary to produce a governmental service; on how dependent the local jurisdiction would be on other agencies to approve decisions, provide finance, take complementary action, and so on. The results of the change also depend directly on the distribution of fiscal and manpower resources. Very scarce resources are difficult to decentralize. Nor can resources be decentralized if they are not centrally controlled to begin with, if they are unevenly distributed among local portions of the urban area. In that case, reducing the size of operational jurisdiction will accentuate uneven distribution of power over resources.

To assess the extent to which any plan to set up districts or subcity governments would actually produce political decentralization, one must consider what roles in public service production would be controlled by the smaller geographic jurisdictions; how dependent they would be on decisions of other governments; what resources they would realistically command; and whether the power structure relating to those roles and resources would, in fact, be broad or narrow.

THE AIMS OF REORGANIZATION

One of the continuing difficulties of making sensible judgments about reorganization proposals is that few of them state their aims in anything but vague and sweeping terms. Public acceptance of this failure can be explained by a kind of frustrated cynicism that characterizes current attitudes toward government. Things are so bad, the reasoning goes, that any shakeup has good odds of being an improvement of some sort.

This is an appealing idea, but reorganization is not free. It is costly, and it can have very serious consequences for the future of city politics and of city life if it creates neighborhood governments that then find themselves powerless to do anything significant about the important problems of their constituents (drugs, jobs, housing quality, law enforcement, and so on), and are rife with internal conflict that weakens the community in the larger political arena.

Hence, we must clarify what we wish to accomplish by reorganization. The goals that are implied by reform rhetoric and by the obvious problems of city government sort out into the following categories:

- *Participation:* to provide the city's people with the opportunities to exercise real power over their urban environment, and to communicate with decision makers when they have specific problems.

- *Equity:* to improve the distribution of public goods and services, assuring that various parts of the city get their fair shares.

- *Adaptability:* to facilitate the adaptation of the mix of services and the way in which they are delivered to the varying desires and sensitivities of people in different parts of the city.

- *Community:* to give institutional expression to a sense of community and neighborhood pride.

- *Finance:* to slow the spiral of city government costs, easing the local tax burden, and increasing the funds available for priority needs of the city's peoples.

- *Management:* to improve the operations of government by speeding and expediting action, facilitating coordination among interrelated functions, and encouraging imagination and innovation.

- *Accountability:* to make public employees more loyal to the policy goals set by elected officials and improving procedures for calling them to account for low productivity, insensitivity, or graft.

Some of these goals are more feasible than others. Some of them have little to do with decentralization under present conditions. Analysis of the way the status quo operates is absolutely essential to determining whether any decentralization plan would further these goals.

More important, if these are in fact the ultimate aims of reorganization, then we should start our review of city government with them and work back to specific types of structural changes that we might devise to achieve them. To start, instead, with our minds made up on one structural alternative is to beg the question entirely. There is no reason to assume a priori that "neighborhood government" is more likely than any other innovation to produce the improvements sought.

Let us now look more closely at these goals and at the kinds of evidence that must be garnered to decide on appropriate forms of reorganization.

PARTICIPATION

Participation is an important end in itself. The first part of that goal is enhancing the opportunity for individuals and local groups to influence the urban environment. This can be pursued in several ways. Decentralization of political power is essential to it. A wider distribution of power within citywide politics would clearly enhance the influence of the groups that are weaker to begin with. This, of course, is not easy to achieve. There are many ways of influencing the distribution of local power: community organization and the development of local institutions; making decision processes more visible; cutting down on the direct contacts between bureaucracy and established interest groups; increasing voter registration and minority candidacies; strengthening the city legislature; and revising electoral methods. These are all possibilities, with, of course, uncertain results.

Geographic decentralization that establishes subcity governments may or may not contribute to this end. The experience with poverty councils, Model Cities councils, and urban local school boards, for example, is not encouraging. Narrow voter participation and factional politics have been common. Moreover, the allocation of government powers and resources to potential district governments is crucial in this respect. If neighborhood governments are not given a sufficiently broad range of roles and resources to have some real impact on the state of housing, on the safeness of streets, of the health of children—then they would not convey *power over the urban environment* to the people even if they operated in thoroughly populist fashion.[1]

Another aspect of participation is improving the access of the citizen and small groups to points of formal decision-making, wherever they may be. Patterns of communication in and out of government are important from this point of view. Information is a vital part of effective influence. Insofar as officials responsible for certain matters are known and accessible, they feel more pressure to respond to reasonable complaints and demands.

There is not much to be gained in this respect simply by returning to the "long ballot," adding to the list of councils and elected officials that the voting public can keep little track of. Multipurpose districts of city administration may be more able to make this improvement only *if* responsible officials in them are both equipped with the power to act and easily identifiable and accessible to the general public.

The operating styles of executives and top officials are also relevant. Channels to city hall open and close with informal executive habits. Moreover, the casework function of the legislature can be crucial; the U.S. Congress is far better equipped, for example, to intercede with the bureaucracy for its constituents—even for individuals who reside thousands of miles from Washington—than are most city councils. Congressmen and their staffs spend more time dealing with specific problems of their constituents than considering national legislation.

Finally, the habits and attitudes of various sectors of the population are important variables. The vicious circle is well known by now: people who have traditionally not had access to power often do not seek it.[2] Without community enthusiasm and community organization, "neighborhood government" can operate, in effect, as a conservative extension of centralized "establishments."

Finally, the district boundaries will have significant effects on how power is redistributed in any decentralization scheme. The long history of the gerrymander had demonstrated that, from the point of view of any particular group, it is folly to judge the scheme before the lines are drawn.[3]

EQUITY

The distribution of public goods and services to various parts of the city may be easier to alter in the short run than the distribution of power, but broader power distribution is probably necessary to sustain equity in the long run. The normative dimensions of what constitutes a fair share of the city's goods and services, of how to redress long-standing inequities, of the degree of redistribution desirable, are beyond the realm of organizational analysis. These fundamental commitments are basic to the equity goal, and our society has not fully made them. It is not that we do not know how to equalize urban outputs. In fact, there is no evidence of political commitment to redistribute from high-income, high-voting districts to low-income, low-voting districts.

Assuming willingness to seek at least some quantitative measures of minimally fair distribution, however, an information system is required to permit determination of existing current patterns of distribution. Districts could be identified for which each department of government could develop expenditure/per-capita data and other measures of output. This is absolutely essential to permit areas of the city to determine what share they *are* getting. There is some evidence from a recent analysis in New York City (1972) that the results might diverge considerably from popular impressions.

District budget-breakdowns and district budget-bargaining opportunities can be developed within the framework of existing city governments. In the long run, this kind of process might contribute significantly to political decentralization. Neighborhood governments with their own elected officials and their own budgets are likely to contribute to this goal only if they have substantial resources and expenditure powers. If not, then improvements in citywide budgeting and distribution systems would have greater effect on net equity of service outputs.

There are broader dimensions to the equity goal as well. Distribution of nonservice outputs is one; of jobs, of contracts, of consulting roles, and so forth. Political control is often sought as a means to patronage, and shifts in central distribution of patronage tend to alter pressures for local control.

Distribution of tax burdens is on the input side of the equity formula. Regional equalization, state and federal urban aid formulas, and the operation of the property tax are key issues on that side.

ADAPTABILITY

Adaptability of government activities to unique community conditions and desires is frequently called for. To assess its importance, we need to survey the needs and preferences of various subdivisions of the city.

The adaptability of city government might be enhanced by making departmental operating procedures more flexible, by geographically decentralizing a good deal of managerial discretion to competent executives assigned to district offices, and a whole range of internal government reforms that are suited to cutting down on red tape, predecision, and rigidity. Whether elected local governments with their own budget and legislative powers are needed for this purpose depends on the character of the society.

In theory, localized representative government is more politically efficient for translating popular demands into government action under certain conditions. These conditions are (1) that the local society has greater internal consensus on the public goods and services it desires than does the society of the larger jurisdiction; (2) that the preference structure of the local society differs substantially from that of its neighbors; and (3) that the activities involved do not have substantial spillover effects on people of other locations. In sum, small local government has the potential for superior responsiveness if and only if the locality is a fairly distinct, relatively homogeneous social unit with some functional independence from its neighbors. One need only look at a small Mississippi town to understand the unresponsiveness to some groups of localized government that does not meet one or more of these conditions.

Little coherent study has been made of whether the neighborhoods of our large cities in fact have distinctive preference structures relative to the mix of government services. It may be, on the contrary, that local populations want the same things as those in neighboring jurisdictions—only more of them (less of them in the case of "undesirable" installations like welfare housing). In that case, the equity goal takes precedence. The crux of responsiveness in that situation is not neighborhood government but broader access to central means of resource allocation. This brings the subject back to the need for fundamental political decentralization—i.e., wider distribution of effective influence, with or without use of geographically smaller jurisdictions.

COMMUNITY

Giving institutional expression to local communities is a traditional value in American politics. The term community connotes a social unit with common values, internal ties of friendship and organization, and group identity. Decentralization plans frequently propose "community" or "neighborhood" governments where the existence of a neighborhood is, in fact, questionable. Heterogeneous areas of 300,000 residents, for example, are far larger than neighborhoods in which playgrounds, shopping, and direct contacts are shared. But they are smaller than the area in which a family's welfare is determined—where jobs are found, where future residential choices lie, where major political decisions will inevitably continue to be made.

Again, empirical inquiry is required on the stability and perspectives of urban subpopulations in order to design plans that would really give jurisdictional expression to "communities." The experience of the community action efforts of the 1960s should also be carefully reviewed in order to come to a conclusion as to how to go about encouraging the development of community identity and joint effort. It may prove more important to provide funds and equipment for hundreds of small neighborhood institutions than to create large urban districts with limited efficacy.

FINANCE

There is no automatic relationship between political or geographic decentralization and government costs, except insofar as economies or diseconomies of scale are identified with respect to the particular activities that might be shifted (for the beginnings of a rare effort at empirical analysis of political or geographic decentralization and government costs, see Merget and Shalala, 1973). Data on economies of scale are somewhat inconclusive, but they do indicate that subcity district governments would probably have to depend heavily on central purchasing and on contracting for central services (resembling the governmental system in Los Angeles County).

An important point that requires emphasis is that the literature reflecting comparative analysis of local fiscal data shows *no stable correlation between the size of government and its costs.* Density, poverty, levels of services provided, and the results of the collective bargaining process are the major determinants of local government cost increases; none of these would be seriously affected by geographic decentralization per se.

Per capita local government expenditure is higher in New York City, for example, than in any other portion of New York State. But it has been rising fast throughout the state in the last seven years, and the second-ranking group is local government in Nassau County—a collection of villages, towns, and small cities.[4]

On the other hand, any major reorganization has immediate costs. Geographic decentralization of government structure entails the monetary costs of new capital plant, of multiple planning, budgeting, and executive staffs. It will be important that these be estimated fairly, so that the public can judge with at least a seat-of-the-pants, cost-benefit calculation.

The most important financial impact of creating new governments, however, is the resultant rise in government activity. New pressures for services, new expectations, new officials seeking to make their mark—all add up to substantial increases in aggregate expenditure. This phenomenon largely accounted for the thirty percent increase in property taxes in Greater London in the five years following reorganization there.

Hence, any realistic reorganization plan must confront the need to make additional funding available if any of the improvements sought are to be effected. It is hard to see how this can be done without coping with the issues of federalizing public assistance and establishing some form of state or regional fiscal equalization, allowing the poorer parts of the metropolis to tap resources where they are concentrated. This has been a crucial aspect of reorganization in European cities, and *structural decentralization in U.S. cities without regional fiscal reform would have token results at best.*

MANAGEMENT

Better management is an important target of reorganization. As city dwellers are all too aware nowadays, political influence is of limited value if the governmental structure cannot deliver the services we demand of it. If government performance were satisfactory, many people in the city would undoubtedly prefer not to be bothered with direct participation.

One method of improving management in both government and business is to delegate some decision-making to lower levels of organization, at the same time concentrating executive powers at both central and lower levels of management in order to define goals and priorities, to fix responsibility for whole programs, to coordinate related activities, and to supervise employees. This kind of approach involves geographic decentralization in that some decision-making and supervision would occur in smaller-scale subdivisions of administration. It also may contribute substantially to political decentralization, insofar as the patterns of decision-making within the bureaucracy become more visible and accessible to public influence.

The advantages to be sought are several: coordination; better distribution of the management load; multiplying sources of initiative; linking decisions to on-the-spot opinion and information; stimulating leadership; reduced red tape and multiple handling to speed and simplify management and to decrease the points at which action can be stymied; and, above all, increased productivity in public operations.

The trick is to do all this without so multiplying the sources of conflict or adding to the power of less-than-competent personnel as to be counterproductive. Designing reorganization for these purposes requires close examination of the operating procedures within city departments, study of the practical problems of establishing multipurpose service districts, analyzing needs for strengthened middle management; and reexamination of the whole maze of checks and counterchecks of personnel and budget administration that cut into management prerogatives.

Governmental decentralization puts a squeeze on human resources that must be resolved if any of these management aims are to be achieved. Nothing can be gained from delegating decision-making or supervisory authority to lower levels of organization unless there are high-caliber managers to act at that level. There are economies of scale to attracting staff and executive talent. Persuading a man who aspires to work in the mayor's office to sign up for a planning staff in a peripheral district is not simple. Serious consideration must be given to increasing the supply of honest, energetic and qualified executive personnel to meet the increased demand generated by decentralization. This requires, for example, consideration of lateral entry and special recruitment.

Finally, research on the behavior of administrative hierarchies and practical experience tell us that much real power in administration is buried in the lower levels. The boss's command is only effective when the man down the line chooses to carry it out. To harness activity to coordinated executive decision-making and community political expression requires design of a structure of incentives strong enough to alter the actual behavior of the bureaucracy. This brings us to perhaps the most important target of reorganization: accountability.

ACCOUNTABILITY

One of the central issues of modern democracy is how to engage the loyalties of a large, tenured, and specialized bureaucracy to the goals and day-to-day preferences of elected officials, managers, and the public. The effectiveness of political executives is vital to this aim. It should be clear to all newspaper readers that the network of restrictions that hamstring executives in city government has not succeeded in the purpose for which it was designed—to eliminate graft. Corruption has proven more adaptable than leadership. This implies reexamination of state and local regulations governing civil service, auditing, purchasing, contracting, and performance supervision, to recreate the kind of concentrated management responsibility that is utilized in city governments abroad and in industry at home.

Thorough analysis is required of state civil service regulations and related provisions to determine their influence under current conditions. Similar assessment of the "working condition" portion of collective bargaining agreements is needed, and special recruitment programs such as New Careers should be surveyed. The unionization of management in city government raises complex questions relating to the pattern of loyalties in administration and to the means of applying performance standards. If we are interested in heightened accountability, we must examine a number of developments to determine who represents management in city government and what their prerogatives and responsibilities are.

Moreover, the political function of the bureaucracy has too long been overlooked in American political literature and folklore. This fact is reflected in superficial distinctions between "administrative decentralization" and "political decentralization" that are frequently made by state and city officials and the press. "Political decentralization," it is said, takes place when elected councils or full-fledged municipalities are created. "Administrative decentralization" is merely delegation of authority within a hierarchy. This distinction is made on purely formalistic grounds and is not relevant to the aims we have discussed. Administration *is* political. The scope and complexity of government being what they are today, it is inevitable that vital *decisions are made and unmade by the bureaucracy.* Making government responsive and accountable would be a far greater contribution to real *political* decentralization, as we have defined it (broader distribution of influence in society, at all geographic levels) than creation of councils, if the councils are to be elected by a minority of residents and have command of very limited resources. The foggy distinction between "administrative" and legislative decentralization is misleading and formalistic. Decentralization of either administrative or legislative institutions entails two things that are not always related in simple fashion—the redistribution of power, and the geographic dimensions within which power is exercised.

The opposition to community control generated by the organized city bureaucracies has strengthened its appeal to those interested in making those bureaucracies more accountable. To achieve increased accountability, reorganization in this direction should be carefully designed. Accountability of a highly organized bureaucracy with its own citywide associations and citywide bargaining agreements, with rights grounded in state law, manned by many who are not residents of the city, is not likely to ensue from the mere creation of elected district councils. The administrative branch of government has long had more power in many U.S. cities than the legislative branch. Meaningful reorganization must, therefore, focus on changes in the behavior of the bureaucracy, and on the sources of its power in statutes, collective bargaining agreements, and interest group alliances.

SUMMARY

Two kinds of conclusion emerge in relief from this discussion of the aims of reorganization. First, the ends usually cited for decentralization depend, for achievement, on a whole range of other means as well. Geographic decentralization is only one of many alternative changes in the structures and procedures of government that should be considered to achieve greater participation, equity, adaptability, community, stronger finance, improved management, and accountability in urban government. Some of the important alternative areas for possible reform are outside the scope of city charters; for example: regional finance and statewide equalization; civil service, collective bargaining and management supervision; party structure and election procedures; internal departmental operations and information systems; community attitudes and organization; discrimination in and out of government.

Second, this discussion yields an agenda of research that is prerequisite to design of any decentralization plan that is likely to have reasonably predictable results. We need to know something about the stability and preference structures internal to a sample of a city's "neighborhoods." We need to assess past experience with community action programs, local school boards, and borough management, looking closely for evidence of their impact on service delivery and political participation. We need operations research directed at existing service delivery systems. Study of the motivations and incentives of various levels of the bureaucracy is key. We need to review the intergovernmental distribution of power for selected services and issues that are of concern to the neighborhoods. To what extent are key chunks of power over personnel, service quality, site selection, and so forth insulated from decentralization in the state government? We need to assess city service output and quality by neighborhood. Finally, realistic assessment of the costs of reorganization and the fiscal needs of any new governments should be an integral part of any proposal. These kinds of research will not yield definitive answers; reorganization is still more art than science. But answering these questions as best we can in limited time will at least make our guesswork "educated."

EXISTING DECENTRALIZATION PLANS

The discussion thus far presents a framework for evaluating concrete decentralization proposals. Do they define the phenomenon they are attempting to institutionalize? Is it mainly a change in geography or in power distribution? Do they spell out their aims and present evidence that the plan presented would contribute to those aims?

ILLUSTRATION: NEW YORK CITY

One highly publicized proposal for "a new structure of city government" was released in 1972 by the State Study Commission for New York City (Costikyan and Lehman, 1972).

Its recommendations were, briefly, as follows. Districts of 200,000-300,000 population should be established within the city of New York. Each district would be headed by an elected council and an elected or appointed executive. These district governments might have the following functions: street cleaning, community hospitals, street maintenance and lighting, local parks and recreation, local "nonviolent" law enforcement (not regular police functions), garbage collection, "local" health services, social services (not including the major one—public assistance), housing management and maintenance (not rehabilitation or construction), and code enforcement (not code approval). The report did not identify where the district boundaries would be.

The report's plan for the central-city government would establish commission-type government, in effect. Citywide executive power would be shared by mayor and policy board (including the current borough presidents, renamed county executives). The five county executives would be the supervisors of district governments within their jurisdictions. In effect, then, three tiers of authority would be established within the city itself: district governments, county executives, city government.

According to the report's recommendations, the city's comptroller would have the power to "surcharge" district officials who, he finds, have "misspent" funds. The policy board would have the power to transfer functions between the city and district governments, and to completely "supersede" district governments where "appropriate." The legal meanings of "surcharge," "misspend," and "supersede" where "appropriate" are difficult to predict, but certainly hint at enormous powers to be vested in the comptroller and policy board.

The policy board would be given budget and planning staff. Prior to preparation of the district budgets, the city policy board would devise a formula to allocate among the districts a fixed percentage of city revenues; thirty percent of all city revenues was the illustrative figure used. Apparently, the allocation formula battle for the budget would be an annual (if not year-round) event. The board, not the mayor, would propose the ultimate city budget—a combination of the district and central budgets. The local districts would also have their own budget offices.

The arguments presented by the report in support of its recommendations are diffuse, but they seem to cluster around three major fallacies: the Myth of Neighborhood Heritage, the Myth of Centralized Power, and the Myth of Omnipotent Structure.

THE MYTH OF NEIGHBORHOOD HERITAGE

The report and its proponents proposed to "return power to the people in their neighborhoods" (Logue, 1972); "that there may once again be communities in New York City within which people . . . take part and know that their participation is meaningful" (Costikyan and Lehman, 1972: 1). In fact, however, power never did reside in general population groups within the neighborhoods of our big cities; much less were there cohesive communities as large as 200,000-300,000 people. The image of neighborhood power has cultural roots in American ideology (our yearning for town-meeting society) but it has little place in twentieth-century urban political history. Local political party organizations once had considerable power, but this power tended to consolidate at the county level in New York. And that is where power is likely to migrate under the commission's proposal for New York City. In any case, that pattern of party machine was not a model of grass-roots government we would want to return to without serious thought.

THE MYTH OF CENTRALIZED POWER

The Myth of Centralized Power maintains that a decade of reform to centralize power in the chief executive, and in comprehensive budgeting and policy processes associated with him, succeeded in New York City. In other words, the proponents believe that power over city government is now highly concentrated. But, it is held, our experience with centralized power is disappointing. Thus it is proposed that we should undo that centralization.

On the contrary, one can more convincingly argue that efforts to centralize power never succeeded in the first place, and one cannot undo what does not exist. Serious students and avid practitioners of politics in many large cities of the United States find that real power is highly fragmented. Despite his impressive formal authority, there is very little that the mayor of New York can do without complex, lengthy, and uncertain bargaining. What we have is a high degree of decentralization, not along geographic lines, but along functional lines. Power is fragmented among specialized pockets of single-purpose interest groups, entrenched bureaus, corporations, public employee associations, local, state, and federal officials. To impose geographic decentralization on top of this functional fragmentation, without any real analysis of where power lies and how it might be shifted to neighborhoods or anywhere else, is just adding a whole new category of organizations to an already debilitating fray. The real challenge is to shift from decentralization of power among specialized functional interest groups to decentralization of power among a broader range of general publics (neighborhood publics, if that proves appropriate and feasible).

THE MYTH OF OMNIPOTENT STRUCTURE

The most disturbing aspect of current discussions of decentralization is the claims they make for vast social, economic, and political changes that will somehow automatically flow from altering the structure of government.

After a general disclaimer that it presents no panacea, the New York Task Force report asserts, for example, that creation of its admittedly limited neighborhood governments will do the following: cope with the alienation of citizens, slow the growth of local government expenditure, give us more efficient government, make social services more humane, probably halt much housing deterioration, and produce more intimate relations between residents and the police. All these assertions can be contradicted by available information about the root causes of the conditions cited. The structural changes suggested in the report are at best marginal to these problems, and in some instances may be negative.

Let us look closely at a few of these assertions.

(1) Citizen alienation. Apathy toward local government is common and shows no correlation with the size, or for that matter the formal powers, of local government. It is a chief concern in Greater London where only thirty-five percent of the eligible voters turn out for elections in the thirty-two boroughs, and a sample survey showed that a substantial portion of Londoners did not even know the name of their borough.

Electoral participation drops with size of jurisdiction in New York State and elsewhere in the United States. Neighborhood politics in New York is factional and unstable. Many neighborhood corporations have experienced the lack of media coverage and of civic habits that dooms efforts to develop broad neighborhood interest.

New York's neighborhoods are varied, and the Task Force report has not analyzed them. Such information as there is indicates that some neighborhood governments could be controlled by minority factions or middle-income groups within them; others would rouse broad interest only when defending themselves from intrusions from without, such as low-income housing or regional schools.

(2) Holding down costs and making government efficient. The argument is made that government for population groupings of 300,000 or less has been shown to be more efficient and less costly.

This misinterprets the literature on optimum city size and the available analyses of the determinants of local government expenditure. *It is the nature of the urban settlement, not the size of its government, that has had some demonstrated impact on costs and efficiency.* Particularly, density, incomes, and certain social conditions associated with large-scale urbanization affect expenditure.

The Task Force report cites Albuquerque, among other medium-sized cities, as comparable to the proposed district governments. Bedford-Stuyvesant, however, will not become remotely comparable to Albuquerque simply by endowing it with a municipal subgovernment.

(3) Halting housing deterioration and humanizing social services. The structural changes recommended in the report are extremely remote from these problems. Two major studies of the causes of housing deterioration and abandonment in New York (Sternlieb, 1970; Kristof, 1970) show that housing code enforcement and managment—which the Task Force report proposes to decentralize—are marginal factors at best. Replacing absentee ownership with resident ownership of dwelling units would be more to the point.

Moreover, it is hard to see how social services could be made "vastly more humane" without confronting needs for new policies on comprehensive minimum income, full employment, and child care that can only start with the federal government.

(4) Police-community relations. The Task Force report predicts that redrawing precinct boundaries will produce more "intimate relationships" between local residents and the police. The police department itself would not be decentralized. It is difficult to believe that such simple solutions as changing precinct boundaries and creating local councils with no enforceable police power would fundamentally alter social relations with the police, any more than geographically decentralizing social work and building inspection would resolve the basic housing and welfare problems.

Actually, in New York, a central police commissioner (Murphy) made a serious effort to tailor police services to the neighborhood, and to consolidate sufficient supervisory power within the force to break down personnel resistance to changes. It was the mayor who sought a civilian review board and a fourth platoon. It was a citywide commission that revealed patterns of corruption that have a great deal to do with failure to slow the growth of narcotics abuse.

Neighborhood governments would not have the clout to compete with centralized police associations like the PBA. And they might reduce the clout of the mayor and the police commissioner sufficiently to be left at their mercy.

In summary, the Task Force did not answer the questions essential to realistic reorganization. People probably do not care whether units of 250,000, one million or eight million are used to elect officials to oversee police activity. They want effective, honest, and fair-minded police—and more of them. How to achieve this functional result should be a starting point for devising reorganization strategy.

(5) Costs. Finally, crucial costs of reorganization are not considered in the Task Force report. The highest costs may be in time and energy needed to cope with intensified conflict. One of the major functions of the political system of a large city is to manage conflict. The Task Force proposals do not purport to create self-sufficient neighborhood governments in New York. As proposed, the districts would compete for funds, programs, and policies to come from federal, state, and city sources. Two kinds of conflict might be intensified: ethnic and intergovernmental.

Intensifying intergovernmental conflict and red tape would produce an enormous waste of political and executive energy. Every year the neighborhood governments would battle over distribution formulas for a portion of the city budget, which the city would feel it could not spare.

Moreover, the powers to do the simplest tasks in the neighborhood would remain scattered throughout the intergovernmental system. State legislation was required, for example, to permit removal of license plates in order that local police and sanitation officers could dispose of abandoned cars in a neighborhood. The report does not recognize this debilitating fragmentation of power.

The Task Force report calls for "local personnel management." It is the case, however, that nearly all issues traditionally considered to be in that rubric are now handled by collective bargaining agreements and state civil service regulation. Examination of these was not on the Task Force's agenda.

In summary, although its stated aims are easy to agree with, this plan demonstrates no reasonable likelihood of achieving them. The issues it raises remain under consideration by a charter review commission.

ILLUSTRATION: LONDON

One of the supporting arguments used by the New York State Task Force and others is the contention that major European cities are decentralized. (In fact, in terms of the definition given here of political decentralization, they are more centralized urban political systems than is common in the United States.) Particularly cited as a model is the 1963 reorganization of London. In fact, if one looks at all the aspects of decentralization described in the first part of this chapter, the evidence shows that London reorganization to consist in at least as much centralization as decentralization.

The London Government Act of 1963 created a metropolitan government for Greater London; changed the boundaries of local government jurisdictions within Greater London; and shifted some functional responsibilities between layers. The boundary changes and the functional reassignments resulted from compromise between those who argued for *larger* authorities to increase "administrative efficiency" and those who defended the representative qualities of *small* local governments.

Prior to 1963, Greater London contained nearly one hundred units of local government scattered on two levels. Most of the jurisdictions dated from the end of the nineteenth century. After 1963, Greater London included only thirty-three units of local government, most of them twice the size of their predecessors.

Before reorganization, the upper layer included London and Middlesex Counties, and three independent "county-boroughs." Parts of four other counties also fell within what is now Greater London. The lower layer included: twenty-eight "metropolitan boroughs" and the "city corporation" within London County, and fifty-four other units of local government ("municipal boroughs" and "urban districts") in the remaining counties. All totaled, then, the governmental population consisted of nine units in the upper tier and eighty-three in the lower tier, plus numerous special authorities.

The London Government Act consolidated these jurisdictions into one unified regional government—the Greater London Council, and thirty-two lower-tier units—the London boroughs. In other words, in terms of boundary change, the reform continued a long tradition of two-tier government and tended in the direction of geographic centralization by consolidating and increasing the size of jurisdictions on both levels. For example, the "City of Westminster" had a population of 95,000 before the change, 244,000 afterward; the suburban towns of Feltham, Heston-and-Isleworth, and Brentford-and-Chiswick were consolidated into the single borough of "Hounslow."[5]

For the most part, public services that had been provided by the eighty-three municipal governments prior to reorganization were transferred to the thirty-two consolidated borough governments. And for the most part, services provided by the six counties were transferred to the consolidated Greater London government.

The net result of preceding history and the reorganization is metropolitanwide management of water supply, almost 600 miles of main roads, drainage and waste disposal, fire services, traffic management, mass transport, major parks, and related services. Education and building construction control remained consolidated in the inner London boroughs, as they had been in London County. Some functions are subject to fairly intricate patterns of shared responsibility between the Greater London Council (GLC) and the boroughs; this is the case with planning, housing, and management of refuse, roads, and parks.

Hassles between the GLC and boroughs over housing have jeopardized decentralization planned for that sector. The 1963 act contemplated the decentralization of the housing powers that had rested with London County after a transition period, but to 1972 the Greater London government had retained most of those powers, and there is serious doubt as to whether the boroughs will take them over.[6] Further powers in the area of planning and construction control may also be transferred to the Greater London Council in order to strengthen the planning function, which has not developed satisfactorily. Greater London Council powers over mass transport were expanded in 1970.

Finally, some services that had been provided by the counties were decentralized to the boroughs in 1963. These were limited health and welfare services, and schools in the suburban county of Middlesex. School decentralization in the inner city was rejected. The kinds of health and welfare services that were decentralized approximate what we call social work, home nursing, and institutional care (homes for the aged and children), plus environmental health inspection. In general, that decentralization worked satisfactorily. However, it is now proposed that home nursing and other "personal health" services that had been decentralized in Greater London in 1963 be transferred to area health boards under the National Health Service. The national government recently required functional consolidation of the social work and children's services within single social services departments of each local government.

British analysts have concluded that expenditures on decentralized social services increased very rapidly following 1963, partly because of service expansion. (Expenditures on centralized services also increased.) Another result has been greater diversity in the quality and quantity of the decentralized services. These variations correspond more to varying professional and political pressures among the boroughs than to variations in need according to an official review (Royal Commission on Local Government in England, 1968).[7]

The most striking conclusion that could be drawn from direct, informed comparison of London with U.S. cities is that our local governments may be suffering from the sheer strain of too many demands on them. The overwhelming proportion of the New York City budget, for example, is channeled into services that neither the London boroughs nor their predecessors provided. The national government takes care of many of the public needs in London that we rely on city government to fulfill. Police and public assistance are cases in point. The metropolitan police have been a national service since the early nineteenth century.

Table 1 lists many of the functions performed by New York City government and indicates their counterpart source in London.

To make a comparative evaluation, then, picture New York City government relieved of most community development and poverty programs, public assistance payments, police services, higher education, collective bargaining for the uniformed services, jails, property assessment, hospitals, drug control, and manpower problems. And picture it engaged in local politics almost devoid, so far, of racial pressures and ethnic competition. In addition, even the activities that *are* locally managed in London are closely supervised by national officials. Conflicts over building site selection, educational curriculum, and so forth, that tear U.S. cities apart, are quietly resolved with departments of the British national government. London's planning and building decisions can be appealed to a national cabinet member. Local government proposals for dealing with slum clearance are submitted to the Minister of Housing and Local Government. The British government has approval powers over even local park changes, and placement of waste disposal facilities in Greater London.

There are crucial differences between London's boroughs and U.S. urban neighborhoods. The London boroughs were drawn so that they would correspond as closely as possible to traditional municipal

TABLE 1
**NEW YORK CITY FUNCTIONS AS PERFORMED IN LONDON
BY LEVEL OF GOVERNMENT**

Functions	New York City Government	British National Government	Greater London Government	London Boroughs
Community development	x			
Public assistance (income support)	x	x		
Social work	x	x		x
Parks	x	x	x	x
Recreation	x		x	x
Police	x	x		
Roads	x	x	x	x
Highways	x	x		
Traffic	x		x	
Education	x		Authority responsible to the GLC	Suburbs only
Collective bargaining	x	x	For white-collar workers only	
Consumer affairs	x	x		
Jails	x	x		
Sanitation	x		x	x
Water supply	x		Separate metropolitan board	
Assessment	x	x		
Fire	x		x	
Environmental health	x			x
Hospitals	x	x		
Housing	x		x	x
Addiction services	x	x		
Manpower training and placement	x	x		
Youth services	x			x
City planning	x		x	x

units. The boroughs were shaped to encompass relatively stable populations with community identity, developed institutions, and experienced officials. The twin foundations of whatever political decentralization there is within London are tradition and social restraint. These characteristics complement the unifying forces of party government and of tight national supervision.

Each borough and the Greater London Council is governed by the majority party in power within it—a coalition that rarely breaks down over local issues. Local elections are contests between party tickets. Political campaigning for them rarely shows any signs of ethnic conflict. Neither does it manifest public involvement in substantive local issues. Public debates over appointments, school management, or public housing sites are rare and subdued. For example, community groups have not interfered with the complete authority of school principals over curriculum and organized activities in the schools.

These characteristics contrast sharply with conditions in New York: factional politics and transient coalitions; ethnic, economic, and emotional divisions among and within its neighborhoods; and its lack of stable, grass-roots community identity or governments to build on.

By several measures, the politics of New York are decidely more open and participatory than the politics of Greater London. In the 1960s, voter participation in elections for borough councils and for the Greater London Council ranged from thirty-five to forty-five percent of the eligible electorate, compared to seventy-five percent in British national elections and seventy to seventy-five percent in New York's mayoralty elections.

Perhaps more important is the absence in London of most of the channels of participation taken for granted in New York. Nominations to political office are made by the party organizations in London; there are no direct primaries. Press, radio, and television coverage of local issues is relatively scanty. Referenda are not used. The public, for example, does not vote on bond issues; instead, local governments must obtain loan sanctions from the national government.

Greater London and its boroughs are governed by unpaid, part-time, elected councilors, and by specialized committees appointed by the council to oversee specific government departments. Important policy decisions are made in the committees, which are closed to public and press. Public hearings are rare.

The bureaucracy dominates local government in London. There are no political executives, as we know them. The government department heads are professionals who do not shift when a new council is elected or when party control changes. Many of them are appointed, not by elected officials, but by the town clerk, the head bureaucrat in the borough. Elected officials and committee members rely upon the town clerk and the chief officers of the government for information and advice; they do not have sizable staffs of their own.

Apparent public satisfaction with government performance in

London is partly due to decentralized authority *within* the bureaucracy, and to performance standards with management perogatives, permitting, for example, dismissal for inefficiency or ineffectiveness.

Decentralized authority within administration can be quickly illustrated with the examples of schools. The principal of *each individual school* makes pedagogical policy, determines curriculum, selects textbooks, and hires and assigns teachers for that school. He does this within the framework of budget and policy laid down by the suburban borough council's education committee or the Inner London Education Authority, and by the inspectors from the National Education Department. His immediate authority does allow him, however, to respond quickly and sensibly to local needs, without referring every move up the line to a central headquarters.

In summary, to avoid the Myth of Omnipotent Structure, we must look to a broad range of political, social, and program differences as well as to organization in order to compare systems. Other attributes of urban management in London and other major European cities deserve close scrutiny. They utilize metropolitan jurisdictions that encompass city and suburbs, as well as local urban governments; they have fiscal equalization schemes and national finance for income support; they have national systems of health care and massive public support for housing (the majority of new housing built in Greater London is built by governments).

Finally, realistic consideration of the London model must include review of the performance of London government. In this, we are fortunate that the British analysts have attempted some objective evaluation. According to one of the principal proponents of the reforms (William Robson, writing in Rhodes, 1973), "It is now clear that the London Government Act of 1963 contained several major defects which have seriously impeded the benefits which the reorganization could yield. These relate, in most instances, to a failure to confer sufficient powers on the Greater London Council and an excessive insistence on the independence of the London boroughs." This assessment is made mainly by criteria of management and finance. Participatory democracy was neither a priority goal nor an evaluative criteria in London.

CONCLUSION

Students of urban government agree on at least one point: existing systems are unsatisfactory. It is the thesis of this chapter that diagnosis and prescription for reorganization must be comprehensive if they are to have any likelihood of improving the actual behavior of, and popular attitudes toward, government. This means considering not merely geographic decentralization, but also political decentralization—dispersal of power. It means considering a broad range of structural reforms, not merely creating neighborhood governments, but also helping community organizations, strengthening city councils, re-creating management prerogatives, reforming regional tax and welfare systems, improving budgetary and operational information systems, eliminating discrimination in employment, and so forth.

Without complementary changes of other types, creation of district governments that could not fundamentally improve urban life and that further insulate middle- and upper-income portions of urban society from the city's problems would be an extremely conservative reform.

It is clear that city governments in the United States attempt to provide far more goods and services that they have been able to deliver fairly and effectively; and far more than *city* governments abroad try to provide. Some activities need to be shifted up to state or federal levels. Some may usefully be shifted down to district offices or district governments. Both *upward and downward shifts* must be considered together.

The advantages for better management of shifting some activities downward could be considerable if central decision-making and coordinating powers are sufficiently strengthened to avert escalation of debilitating conflict and to lend coherence to government activity overall. Hence, for the aim of better management, the details of how district and citywide powers would intermesh are crucial.

Decentralization of political power would not follow automatically from establishment of district governments. It is an erroneously *assumed* advantage of geographic decentralization. Other types of reform might contribute far more to this end. Each group interest in the city needs to examine closely any given proposal for "decentralization" to determine pragmatically whether it would gain or lose by it.

NOTES

1. Rapid disillusionment is a common public reaction to reforms that fail to enhance power over the environment. One of the key advantages of community corporations is their ability to start with achievable aims and build community interest from practical results (see, for example, Gifford, 1970).

2. Discrimination affects people's sense of self-efficacy, which in turn affects likelihood to participate in political activities. Income also appears to be associated with participation rates in some studies. The point is that the variables of participation are several, and in certain circumstances the mere establishment of district governments would be cathartic. Alan Altshuler (1970) points out that community control in the ghetto would have mixed effects on ghetto political organization.

3. Neighborhood government may give black, Chicano, or Puerto Rican groups majority status only if the lines are drawn around largely segregated neighborhoods, for example.

4. This example illustrates the tradeoffs among goals of any reorganization. As Alan Altshuler (1970) has pointed out, efficiency has not been given a high priority in state and local politics. Suburban governments represent the choice to purchase insulation, homogeneity, and small scale at the costs of inefficient and inequitable use of taxes. Similar choices may be made in the city, but with scarce fiscal resources the costs may be less tolerable.

5. There was and is no central-city government in the London scheme. What is known as the "City of London" is an historical anomaly: one square mile presided over by a Lord Mayor, sovereign by assignment of the Crown by annual payment of six horseshoes and some tools. Its residential population is under 5,000. London County was considered the inner city. It included 28 municipalities (metropolitan boroughs) with populations ranging from 21,000 to 339,000. After reorganization, this "inner London" portion corresponding to London County includes 12 boroughs, ranging from 192,000 to 329,000.

6. The 1963 act called for transfer of public housing stock from the Greater London Council to the boroughs. Initial proposals to carry this out in 1970 were withdrawn from Parliament after objections were brought to the attention of the minister. After independent inquiry, transfer of 44,500 dwellings out of a total of 253,500 was approved. Two large boroughs refused transfer.

7. The more recent and comprehensive evaluation of London reorganization is Rhodes (1973: 123). It concurs with the above evaluation and concludes: "The principal advantage gained for the health and welfare functions from the new structure has been to create order over the whole Greater London area out of what had been a multiplicity of large and small authorities."

REFERENCES

ALTSHULER, A. (1970) Community Control: The Black Demand for Participation in Large American Cities. New York: Pegasus.

COSTIKYAN, E. N. and M. LEHMAN (1972) "Restructuring the government of New York City." Task Force Report on Jurisdiction and Structure of the State Study Commission for New York City.

GIFFORD, K. D. (1970) "Neighborhood development corporations: the Bedford-Stuyvesant experiment," in L. C. Fitch and A. H. Walsh (eds.) Agenda for a City: Issues Confronting New York. Beverly Hills: Sage Pubns.

KRISTOF, F. A. (1970) "Housing: economic facets of New York City's problems," in L. C. Fitch and A. H. Walsh (eds.) Agenda for a City: Issues Confronting New York. Beverly Hills: Sage Pubns.

LOGUE, E. J. (1972) Testimony before the Scott Commission, February 2.

MERGET, A. E. and D. E. SHALALA (1973) "Dollars and cents: the fiscal implications of decentralization in large city governments." Presented at the National Conference on Public Administration, April.

New York, City of (1972) Municipal Expenditures by Neighborhood. New York: Office of Administration.

RHODES, G. (1973) The New Government of London: The First Five Years. London: Weidenfeld & Nicolson.

Royal Commission on Local Government in England (1968) The Lessons of the London Government Reforms. London: HMSO and London School of Economics and Political Science.

SMITH, G. A., Jr. (1958) Managing Geographically Decentralized Companies. Boston: Harvard Business School.

STERNLIEB, G. (1970) The Urban Housing Dilemma: The Dynamics of New York City's Rent Controlled Housing. New York: Housing and Development Administration.

10

Decentralization: Fiscal Chimera or Budgetary Boon?

JONATHAN SUNSHINE

□ DECENTRALIZATION IS VERY MUCH IN VOGUE as a medicine for the ills of large cities. Its enthusiasts see it producing improved delivery of services and making government more sensitive, accountable, responsive, and accessible to the city's citizens.[1] Skeptics are more inclined to believe that whatever improvement of services decentralization produces—if it does indeed produce any— will be solely the result of a Hawthorne effect; that decentralization is a symbolic (rather than substantive) solution to urban problems, assuaging people through the appearance of activity rather than with real improvements; and that much of big-city mayors' enthusiasm for it derives from the new neighborhood-level organizational bases it can give them or from the possibility it affords them for diverting citizen wrath from themselves onto district officials.

Whatever their evaluation of decentralization, discussions of it have stressed political, administrative, and social issues to the relative neglect of fiscal ones (for a review of the literature and an extensive bibliography, see Shalala, 1971). Fiscal considerations apparently have seemed less important, less interesting, or both. This paper attempts to redress the balance by demonstrating that there are

AUTHOR'S NOTE: *The author gratefully acknowledges the support of Harvard University's Center for International Affairs. Opinions expressed herein are not necessarily those of the Office of Management and Budget, the Center for International Affairs, or any other organization with which the author has been or is associated. An earlier version of this paper was presented at the 1973 annual meeting of the American Political Science Association.*

interesting fiscal questions in decentralization and by showing the importance of fiscal politics—that is, politics centered about fiscal issues—to the probable degree of success of the decentralization movement. To show that there are interesting fiscal issues, it will discuss the question of how, under a decentralized system, resources ought to be allocated among districts. Undoubtedly, when push comes to shove, politics will bulk large in the solution of this question. But, meanwhile, we with the time to think ahead, with a claim to expertise in public policy, and with hopes of having a salutary effect on the course of public affairs through professional recommendations, should be seeking rational answers to the resource allocation problem. Second, to show the importance of fiscal politics, it will review some current developments in that field which probably will have a major effect upon the extent, if any, to which decentralization is implemented and upon the form it will take.

The argument will generally remain at an analytical level with the points it makes being applicable to the decentralization of most cities and most functions. In the interest of concreteness, however, the discussion will turn frequently to New York City—the locus of the most widely publicized decentralization activities, and the example best known to the author. Similarly, several illustrations from varied functional areas are included for the benefit of those readers who prefer less abstraction or who have special interests or competences in particular functional areas.

HOW TO FINANCE DECENTRALIZATION

To the first question first: As policy experts, what system should we recommend for apportioning funds among the districts into which a large city is to be divided?

DISTRICT SELF-SUFFICIENCY

Discussions of decentralization often suggest its goal is the closeness of city hall and the responsiveness to citizens that (presumably) already obtain in suburbs. Small size is taken to be the wellspring of these virtues—the typically greater education, wealth, and general political efficacy of suburbanites is usually neglected. Thus suburbs are held up as models for decentralization. Let us then analyze first the suburban model. Self-finance will be the rule. For all decentralized functions, each district will have to pay, by itself, in full, the entire cost of any activity it undertakes. Correspondingly, it will have no responsibility for the costs of any activities its neighbors undertake willy-nilly, whether it likes them or not. Every tub on its own bottom!

This sounds like a marvelous formula for fiscal responsibility. A district that wants something has to pay for it, so it will undertake only activities that are worth their cost. Similarly, inefficiency will not be tolerated in district operations, since the district itself pays for the waste involved. In contrast, to take the New York example, current New York City operations are seen as ones in which all eight million residents pay for a neighborhood's activities, so that the neighborhood has little incentive to worry about benefit/cost ratios; and in which inefficiency is tolerated because the vast bulk of its cost falls on someone else's shoulders. How accurate these perceptions of New York or other large cities are, why nonbeneficiary neighborhoods do not now object to wasteful projects, and whether medium-sized cities will behave any less wastefully are questions that ought to be answered before accepting the purported virtues of the suburban model. However, even on its own terms, this model has two major deficiencies, one of equity and one of efficiency.

The former is well known to students of local governments and their problems: Both needs and resources are unevenly distributed among communities, needs with little or, more likely, a negative relation to resources. While it makes much sense to require the community of average needs and average resources to cover its own costs, few would not want to require the high-resources/low-needs community to contribute some of its resources to other areas, or to give some outside help to the low-resources/high-needs community. The high-resources/low-needs community is exemplified, in the existing suburban system, by a town which has a large shopping center within its boundaries, and derives lucrative revenues from it. These revenues typically result predominantly from the patronage of nonresidents; and it is not only academics, but also these neighbors, who have asserted that high-resources/low-needs communities should be required to aid less-well-off areas. In large cities, similar overprivileged areas will result from district self-financing. In New York, for instance, the district that encompasses Lower Manhattan and the one that includes Midtown Manhattan will both be in a position to tax a tremendous quantity of real estate and earnings. The reverse type of community, one with little tax-generating ability and a large proportion of "problem" residents, is less familiar in the suburbs, but is the category into which city slum areas fall.

By creating such grossly underprivileged districts, a self-financing decentalization plan in most central cities would exacerbate the needs-resources disparity which is already a major problem of metropolitan areas (for an empirical analysis of the point with respect to Buffalo, Rochester, and Syracuse, see Callahan and Shalala, 1969). At present, central cities can at least smooth out disparities among neighborhoods, and they often have a concentration of commercial or industrial activity (although in many cases this is dwindling) which somewhat compensates for the concentration of needs they face.[2] District self-financing would eliminate these tempering forces and would create extremes of poverty—and probably extremes of wealth—greater than those already extant in the metropolitan area.

If local self-financing were adopted, dynamic forces would probably make the situation even worse. Consider the relative tax levels in two self-financing areas, one with lesser resources per unit of need than the other. To maintain equal service levels, the former will need higher tax levels than the latter. (Indeed, if disparities are significant, the usual result has been that the less fortunate jurisdiction combines inferior services with higher taxes.) Both businesses and mobile individuals will have an incentive to locate in the more fortunate community, since they will then get more services, pay lower taxes, or both. Differential movement will enhance the tax base of the already more favored community,[3] which will in turn exacerbate disparities and (hypothetically, at least) speed up the differential migration.

This dynamic worsening of disparities is again a problem which has already been seen to be serious with respect to central cities versus suburbs, but which would probably be intensified if self-financing districts within the central city were adopted. The large size and central location of the city deter some firms and persons from abandoning it, whereas merely to abandon one of its neighborhoods carries far fewer negative consequences.

Besides these static and dynamic equity arguments against district self-financing, there is an argument, well known to economists, that it is not even efficient. Economic theory indicates that marginal cost pricing—in decentralization terms, paying for district purchases with district money alone—leads to efficiency—in this case, a proper mix between governmentally provided and privately purchased goods and services—only if there are no externalities—i.e., if the buyer gets for his money the full and exclusive enjoyment of whatever he purchases. If someone else gets some of the benefits without paying for them, the price per unit benefit that the purchase pays will be higher than the true price (the purchaser get for himself fewer benefits than his purchase brings in total), and the purchaser will therefore buy less than he would were he paying the true price. Now many governmental services have spillover effects (know technically as positive externalities) of just this sort. For instance, if one district invests in parks, residents of surrounding districts are likely to make some use of those parks and—by common sense as well as economic logic—ought to pay some of their costs. Similarly, spending on housing code enforcement or urban renewal is likely to enhance the value not only of neighborhood property, but also that of property in surrounding districts. And education increases the value of an individual (a value measurable for many purposes in increased earnings or increased taxes paid), but people frequently move beyond the neighborhoods in which they grow up, so that the community which pays for their education does not reap its benefits.

The appropriate treatment for positive externalities, again according to economists, is for higher levels of government to finance the activities in question to the extent that persons outside the local district benefit from them. The problem of disparities in the needs/resources ratio points in the same direction—to a role for the city at least, and probably for regional, state, and federal governments also, in the finances of decentralized districts. At this point, then, we seem to have "solved" the problem of financing decentralized districts to the point of deciding that the most obvious alternative, self-finance, is undesirable; that higher levels of government must retain a role; and that disparities in the relationship of resources to needs are a major problem which ought to be central in determining an allocation formula. This "solution," however, still leaves many issues unsettled, as an examination of the options open under it will show.[4]

RESOURCES

Two formulae developed in the field of education finance fit the parameters of this solution and are sufficiently simple and attractive to merit first attention. One, proposed by the Fleischmann Commission (New York State Commission on the Quality, Cost, and Financing of Elementary and Secondary Education, 1972: ch. 2), calls for complete financing by higher levels of government. In education, revenues will be raised entirely by the state and distributed to localities on the basis of an equal per pupil allocation. (Disadvantaged students, however, are weighted to count as one and one-half pupils.) Districts can spend these funds as they please; local autonomy is thus preserved. Indeed, the Commission argues that local autonomy is enhanced by its plan, since local officials can stop worrying about raising money and can concentrate entirely upon the operation of their system. The strength of this proposal, its authors stress, lies in its meeting a very fundamental goal of education —equality of opportunity. That is indeed an important objective; the allocation formula is easily applied to the decentralized districts into which a large city might be divided; and so we have a promising candidate for a solution to our problem of finding an appropriate allocation formula.

However, full higher-level funding (citywide funding, in our case) has serious problems as a general allocation formula—that is, one intended for district activities in general, not merely for education. First, it eliminates local choice about the *level* of spending on a given activity. The Fleischmann Commission defends this as a virtue in the field of education, arguing that it is as unfair that a child be penalized throughout life by a poor education because his parents' neighbors *choose* to spend little on education as it is that he similarly suffer because his neighbors happen to be poor and cannot finance adequate education (New York State Commission on the Quality, Cost, and Financing of Elementary and Secondary Education, 1972: 2.45, 2.46). Perhaps this is a sound argument in education, which is singularly a capital investment in future generations.[5] But it hardly accords with the general spirit of decentralization, or with the strong case for local choice about spending levels in most other areas of service. Should a district not be permitted, to take an example at the other end of the spectrum, to choose between spending heavily on public recreational and cultural activities, or doing little and allowing its residents instead to obtain their rest and recreation privately (or for that matter, spend their income on other types of items)? Similarly, should decentralized districts not be free (within rather broad limits) to spend more or less on fire services, letting residents choose to pay higher or lower insurance premiums and take different risks of losses? And should they not be free (again within limits) to decide how clean they want their streets to be, perhaps putting their dollars into other objectives they value more?

Second, while full funding from citywide sources would solve the problem of disparities in resources from district to district, this Fleischmann Commission approach does not get into the problem of measuring needs, which also vary from district to district. Again, in the field of education, there may be a simple solution. The number of pupils, it could be widely agreed, is a decent approximation of the need for educational resources—especially if some special weighting for disadvantaged children is included. (Note, however, that the Commission does not present evidence for its 1.5 weighting for the disadvantaged. Since disadvantaged students are a problem that is both serious and significantly varying from district to district, the allocation formula, even for education, could be seriously inadequate.)[6]

In other fields, however, the measurement of need is generally a more difficult issue. In the police field, for instance, crime data are notoriously poor, indices debatable (the FBI crime index, for instance, includes no white-collar crimes and has been denounced as racially and culturally biased), and measurements are all based on a situation in which police resources are already at work and thus (presumably) are affecting the measurements. (For instance, a district may have a low measured crime rate because it is saturated with police, who have significantly reduced the crime rate. But if resources are then reallocated, based on measured crime rates, and the district gets few resources in keeping with its good current statistics, it may quickly undergo a serious worsening of measured crime.) Here the "full city-level funding" formula only indicates where money should come from, not how it should be divided among districts. Similarly with fire protection—needs depend in a complex fashion on characteristics of the population and the structures in each district, and needs must be assessed before funds raised citywide can be allocated among districts.

In summary, while full funding by higher levels of government is an apparently attractive and simple formula for decentralization finance and one which does meet the problem that resources vary from district to district, it does not give districts a choice in how much they wish to spend, nor does it indicate what to do in the face of district-to-district disparities in needs. In the field of education, whence this formula comes, neither deficiency may be serious, but in other areas they are.

The first of these deficiencies can be rather neatly dealt with by power equalization, another concept pioneered in the field of education finance. Power equalization seeks to give communities a choice in how much they spend while eliminating the disparity in resources which has so far made local financing incompatible with many notions of equity. It does so by setting up a system of subsidies and surcharges administered by a higher level of government so that for any given tax rate, all school districts, whether on their own they are rich or poor, will have in hand an equal sum per pupil to expend. For instance, the system might give every district seventy dollars per pupil for every mill of property tax.[7] Should the property tax in a given district actually raise less than this amount, the higher level of government will make up the difference. Should the tax raise more than this amount, the excess must be contributed to the higher level of government for redistribution to other districts. Thus, a district's innate wealth ceases to determine the level of services it can finance with any given tax rate: With an equal tax effort, each district can maintain an equal level of service. Again, this is an attractive system, readily applicable to financing decentralization in large cities. It preserves the resource-disparity-overcoming feature of full citywide funding, while also permitting local choice about service levels.

But it too has drawbacks. Like full centralized funding, power equalization does not give guidance on how needs should be measured, a problem which must be solved before it can be meaningfully implemented.

Besides this issue of needs, power equalization tends to push into the open a host of questions about what is the proper measure of resources. If one finances activities with a property tax, one set of districts (the property-poor ones) will be net beneficiaries. Income tax finance would make another set of districts (the income-poor ones) the net gainers. Sales tax finance would give a third result. The first two options are particularly likely to lead to conflict. Since property taxes are widely felt to be regressive[8] and income taxes to be progressive, different strata of the community may line up behind these two financing modes. For many large cities, the question of which tax instrument to use hardly arises, since they are by law restricted almost entirely to the property tax. However, this dependence on a single tax is decreasing, and, particularly in New York City with its very wide variety of taxes, the debate could become heated.

A closely related issue is whether one should make the power equalizing system progressive. For instance, one system[9] gives school districts (approximately) seventy dollars per mill up to $1,000 per pupil and thereafter only forty dollars per pupil per mill. Such progression, it may be argued, discourages excessive spending and compensates for the regressive nature of the property tax.

These issues of appropriate tax base and structure, it should be noted, are not unique to power equalization. They are more or less standard problems in public finance that exist—at least theoretically—in any tax system. However, the fiscal politics of adopting a power-equalization-based decentralization system is likely to bring them into the open with unusual force.

A third problem with power equalizing is that it can be difficult to comprehend, particularly if progression is introduced. Indeed, a member of the staff of the Wisconsin gubernatorial task force which formulated the two-rate system described just above, reports that task force members (largely the usual group of high-level public and professional representatives) in some cases subsequently indicated they had not understood the fiscal implications of what they were doing, and would not have approved of it if they had known. Again this is a problem in fiscal politics, not a technical issue, but it deserves mention here as an important practical difficulty in power equalization.

Finally, power equalization suffers from the fact that its impact on central-level finances is uncertain. To see this, assume that the tax level/expenditure ratio is set to what would actually be raised on a citywide average. For instance, returning to the examples, assume that each mill of property tax in New York City in fact generates seventy dollars per pupil on a citywide basis, and that the system for districts is therefore set up using this figure. Now if rich districts (those with an above-average property/pupil ratio) generally enact higher tax rates than poor districts, the surplus they generate will exceed the subsidy poor districts must be given, and the citywide budget will show a net gain. If, conversely, poor districts generally enact higher tax rates than do rich districts, then subsidies they need will exceed the surpluses the richer districts produce, and the central treasury will suffer a net drain. Either outcome is quite possible,[10] and the uncertainly involved may act as a strong deterrent to the adoption of power equalization.[11]

The foregoing discussion of full centralized funding and of power equalization has shown them promising in dealing with the problem of resource disparities, but inadequate in their failure to deal with the needs side of the equation. The next section seeks to overcome this weakness by developing an analysis of needs.

NEEDS

The most obvious approach to needs is to assume they are proportional to population. This is a simple, easy-to-understand formula—in a full centralized funding system, it translates into equal per capita grants for each district; in a power equalizing system, it means that the subsidies and surcharges must be such that a given tax rate generates an equal yield per capita in each district. It is embedded in many existing intergovernmental aid programs. And it generally effects a redistribution from richer to poorer areas, since taxes paid per capita are higher at higher incomes, even for regressive taxes like the property tax, while the yield to districts is on a strict per capita basis. Finally, this view of needs has been the basis upon which expenditures are usually analyzed. For instance, those few studies of district-by-district expenditures which have been made for New York City report their findings on a per capita basis (New York City Administrator's Office, 1972; Temporary State Commission to Make a Study of the Governmental Operation of the City of New York, 1973: ch. 5).

Unfortunately, however, there are few needs for which population is an accurate measure. Consider the following example to see how it misleads. The City Administrator's office analyzed New York City expenditures by neighborhood using expenditure per capita on each function as the data for comparison. Far the highest expenditure for fire—one much above any other district—appears in Lower Manhattan. Crazy? Especially since the buildings there are not wooden firetraps, but fireproof metal, glass, and concrete structures? Not at all! There are many, many office buildings and extremely few residents in the Lower Manhattan district. Thus an expenditure per cubic foot of building space (or per square foot of floor space, or per dollar of structure value, if these are more appropriate measures) that is quite low is divided by very few caputs and yields a high per capita expenditure. The data mislead one into thinking that fire costs in this area are, by some reasonable comparative standard, very high.

Much the same holds for per capita police expenditures. They are very high in Lower Manhattan, although this is not a high-crime area. Again the explanation lies in the large amount of property there is to protect, with the sparseness of residents further inflating the per capita figure. (Besides protecting property, direction of the area's heavy daytime traffic and protection of its very large daytime working population presumably also consume a good deal of the police manpower assigned to this district, although neither con- tributes to its resident population, which is the divisor of the ratio in question.) These examples should serve to demonstrate that some city services are directed not to people, but to things, or to a complex mixture of things and people, so that population is a poor measure of the need for them.

Even for those activities which are solely services to people, population is usually not a good measure of needs. Consider welfare payments. Here there is a fairly clear notion of what meeting needs means—it is to provide eligibles with benefits according to a payments schedule. While there is debate about appropriate eligi- bility rules and payment levels, it is clear that, for any reasonable values of these parameters, meeting needs will result in per capita expenditures which differ widely from district to district. Recrea- tional and cultural expenditures are perhaps the only ones for which allocating funds according to population would be a good approxi- mation to meeting needs.

Since regarding needs to be proportional to population is unsatisfactory, another approach is necessary. To develop one, let us take some services whose objectives are fairly clear and on which data are reasonably good, and then try to figure out how one would want to allocate resources. Consider, for instance, that part of police activity which deals with crime (as opposed to traffic, parades, visiting dignitaries, and so on) and abstract from most statistical problems by assuming there exists a satisfactory and agreed upon crime index. Despite the sanitized and simplified nature of this example, there are several competing criteria by which resources might ideally be assigned.

First, one could make a good case that funding should be allocated according to the magnitude of the problem—if crimes are twice as numerous in one district as in another, the first should get twice as much money for police. This is a readily understood criterion; and it is one that can be easily linked to either a full central funding or a power equalization system. Under the former, funds are allocated directly, with each district getting so many dollars per crime. (Strictly speaking, the allocation will be so many dollars per crime index unit. Presumably some crimes will be weighted more heavily than others in the index.) Under the latter option, the subsidies and surcharges are set so that dollars obtained per unit of tax rate are proportional to the crime index—i.e., to take the two districts again, the first should get twice as many dollars per mill of tax as the second.

Note, however, that the crime index used should be that which would be found in the absence of any police—otherwise the data will be distorted. For example, consider again an area with a high innate crime rate which, thanks to very heavy policing, has an actual crime rate no higher than another, less troubled neighborhood. Use of "with-police" crime rates would suggest equal resources be assigned to each, whereas, in response to the underlying "without-police" crime rate, there are already more police in the first district, and it is fairly clear one wants to keep things that way. Now without-police crime rates may be difficult to know with any accuracy (although rough relative rates may be obvious), and few would want to make the experiment that would disclose them—pulling all police out of the city to see what would happen to crime rates in each neighborhood.

Another problem with this allocation formula is that perfectly reasonable and only slightly varying definitions of what constitutes a problem can lead to radically different district-by-district allocations. An education example will serve to show this difficulty.

Consider a remedial reading program and two districts with pupil reading levels distributed as in Figure 1. One district is perfectly average—the median reading levels is right "on grade." The other district is somewhat subnormal—its median reading level is one grade below the norm.[12] If we see the problem with which remedial reading should deal as "reading levels below average," then half the children in the first district are problem readers, about five-sixths of those in the second are, and the second should get one and two-thirds the remedial reading resources (per child) that the other one does. However, it makes more sense to say "*By definition,* half of all children read below the norm for their grade. Thus reading a bit below the norm is nothing to worry about; it's being seriously below it that should be of concern."[13] If, then, one takes "reading one or more grades below the norm" as the problem, about one-sixth of the children in the first district are problem readers, half those in the second are, and the second should get resources in the ratio of 3 to 1. However, there is nothing sacred in the one-grade figure. Particularly by high school, one might well want to argue that it is only lagging two or more grades behind the norm that is a deficiency serious enough to warrant remedial help. By this criterion, about 2.3% of the children in the first district and 16% of those in the second need help, and resources should be assigned to the districts in the ratio of approximately 7 to 1. Other criteria give other results. For instance, using three grades down as the cutoff point calls for an allocation of about 18 to 1.

Any remedial or "help the underprivileged" program will show this strange phenomenon. The appropriate allocation of resources among districts will be sensitive to the precise definition of those in need and the more restrictive that definition, the more concentrated the allocation will be in the worse-off districts.

Given these problems involved in a "resources proportional to needs" formula, perhaps the economists' favorite advice—equalize marginal productivity—should be followed. Marginalist economics can get a bit technical, but its injunction is basically the familiar managerial principle of putting additional resources where they will do the most good. The virtue of the "equalize marginal products"

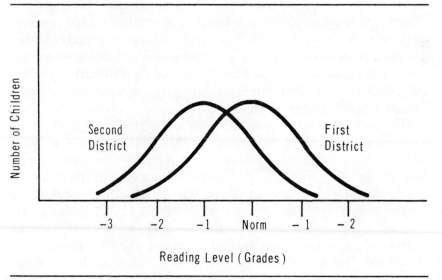

Figure 1.

formula is that it is the most efficient solution to the resource allocation problem—that is, to take the police example, it produces the greatest reduction in *citywide* crime for a given amount of police resources. (If resources are not allocated so that marginal products are equalized, citywide crime can be cut by transferring resource units from districts where the amount each unit reduces crime is less to districts where each reduces it by more.) However, marginalism has a serious political drawback. Besides indicating where to put additional resources, it directs that *existing* resources should be removed from other places and shifted to high marginal product areas—a transfer not likely to be welcomed by resource-losing districts.

Again, there is a difficult data problem—to know how changes in police resources affect the level of crime. The difficulty is compounded by the fact that the precinct-level experiments so far conducted indicate that a major effect of "beefing up" one precinct is to increase the crime rate in surrounding areas. Nevertheless, the necessary data are probably more readily obtained than those needed for the preceding formula. After all, one can realistically experiment to determine the effect of modest changes in the level of police resources in different districts.

Other problems with this allocation criterion are more serious. For one, it is relatively difficult to explain, and the distribution of resources it calls for, although ascertainable by trial and error, is not likely to be known in advance. (One would have to know the marginal product of police resources at a variety of total resource levels in each district.) Political decision makers are unlikely to trust the experts and buy this "pig in a poke"—they will know neither how much resources they will be getting for their district, how that will compare with what other districts get, nor what the resulting level of security is going to be.

The most serious criticism of marginalist allocation is that, for some important public functions, it does not at all accord with society's basic objectives. Education is probably the best case in point. Learning[14] most likely is a function not only of the resources which government devotes to the formal education process but also (among other things) of abilities, experiences, and attitudes generated by a person's sociocultural milieu. These other factors, of course, are generally less favorable among the "disadvantaged" population. Therefore, the marginal product of the nth dollar per pupil (i.e., the gain derived from spending one more dollar per pupil when n−1 are already being expended) will generally be greater for advantaged than for disadvantaged children. Learning among the former is abetted by the background factors mentioned, while learning among the latter is hampered by them. To equalize marginal products—which, remember, gives the efficient solution (the greatest total learning for a given amount of educational resources)—then requires that more resources per student be devoted to advantaged than to disadvantaged children. Although this is indeed the way to get the most education for a given resource outlay, and although it is the model that the American higher education system largely follows, it is contrary to the dominant concept of equity in primary and (probably) secondary education. The purpose of this part of the education system is generally regarded to be to impart to every person the same quantity (or at least some basic minimum) of learning. To do so requires more, not less, resources for disadvantaged pupils, and this inefficient but equity-serving allocation is generally (or at least proportedly) followed.

The marginal productivity criterion *is* the one to follow if your goal is to get the "biggest bang per buck" citywide *and* you are perfectly indifferent to the distribution of benefits among districts. However, given that politics under decentralization is likely to become very much a contest among districts to get their (self-perceived) "fair share" vis-à-vis the other districts, the political future for marginalist allocation is not promising.

A third possible approach to needs is to distribute resources so that, upon their employment, the problem they attack is reduced to an equal level in all districts. In the police example, for instance, this would mean concentrating forces in "bad" areas so that their crime rate was reduced to that which obtains in good areas. In education, this is the equality of opportunity model—put resources into the ghetto schools until the children who go through them reach achievement levels equal to those of the best upper-middle-class districts. And, in sanitation, it means distributing resources so that the streets in all neighborhoods are equally clean.

This is a readily understood criterion and one whose goal is quite attractive. Its achievement can easily be tested post hoc, although prior planning would be uncertain and a solution would probably have to be reached through trial and error, since knowledge of production functions in public services is generally poor. Once the necessary information was developed, this criterion could be linked to a power equalizing system, giving districts all the flexibility that implies. Under power equalizing, subsidies and surcharges would be worked out so that for each district a given tax rate gave it funds to achieve the same services output—crime rate, street cleanliness, educational achievement level, and so on—as any other district with an equal tax rate.[15]

Probably the biggest problem of the "reduce problems to equal magnitudes" criterion is that it is extremely redistributive. It is, for instance, hard to imagine reducing ghetto crime levels to those of upper-middle-class areas, even with a tremendous concentration of police resources. Or, perhaps, this result might be achieved if police resources in wealthier areas were greatly reduced and the situation there allowed to deteriorate. (With sufficient encouragement in the form of differential policing levels, "common" criminals might learn to take a subway or bus and ply their trade in wealthier neighborhoods rather than in the poor and crime-ridden areas where most of them live.) Similarly, the resources needed to bring the performance of ghetto schoolchildren up to those of upper-middle-class children must be enormous—if the task is indeed possible—although again equalization might more readily be achieved if heavy "leveling downward" of good areas were permitted.

It is hard, however, to envisage the middle class of large cities not fighting tooth and nail in the political arena against any extensive leveling down. And, indeed, the growth in New York of private street patrols and of nonpublic schooling suggests that, if public services fail, middle classes can and will resort effectively to private institutions to prevent leveling down. Thus, reduction of problems to an equal level seems an allocation principle unlikely of full implementation, although the attractiveness of its goal is sufficient that one might expect some funds to be distributed on the basis of it.

CONCLUSIONS

What conclusions may be drawn from the preceding study of allocation formulae? First, both resources and needs vary in an irregular fashion from district to district, and both must be considered in allocation decisions. Second, on neither side of the equation is any one system entirely problem-free. A number of competing formulae have attractive features. Third, on the needs side particularly, different criteria seem appropriate for different functions. For instance, the efficiency criterion of equalizing marginal productivity fits higher education, while primary education is built around an equal achievement for each student model that strongly counters efficiency criteria, and recreation funds are possibly best allocated on an equal expenditure per capita basis. Since the expert can give no one clear and simple answer, the likely outcome is a complex compromise—rather like that in federal revenue-sharing—which incorporates multiple criteria at once. Moreover, different solutions may well be adopted for different government functions.

A further consideration makes a general-purpose compromise formula, similar to that of revenue-sharing, an even more likely result. All the approaches discussed so far (with the exception of district self-financing) have assumed a separate formula, separate calculations, and separate funding for each function decentralized. This assumption is not unreasonable if only two or three functions are decentralized. However, if the scope of decentralization is much greater, it is more realistic to expect that all functions will be funded through one formula or through a very limited number of formulae. In this case, with a formula having to allocate funds to multiple functions, formulae will probably consist of a basic per capita allocation plus perhaps two or three non-program-specific correction factors—such as per capita income, per capita real estate value, own-source revenues, unemployment rate, or percentage of population below the poverty line—intended to compensate for variations in needs and resources.

FISCAL POLITICS OF DECENTRALIZATION

We turn now from relatively theoretical questions of how funding *ought* to operate in a decentralized system to a more applied study of politics of fiscal questions that are likely to affect the course and results of decentralization.

GENERAL EFFECTS OF THE
FISCAL POLITICS OF DECENTRALIZATION

Most of the allocation decision questions raised in the first half of this paper are ones which currently face big cities, although this fact is little recognized. All of the needs-side issues are there, since cities now provide services to many neighborhoods, and there exist (even if they are not discussed) possibilities of changing the distribution of services. To take but one example, police operate in precincts, and the allocation of manpower among precincts is a decision little different from those involved in designing an allocation system for police under a decentralized plan.

Furthermore, in jurisdictions which have some choice of tax instrumentalities, the resources side also presents many choices that have to be made and which can have widely varying effects on the city's neighborhoods. For instance, extending New York City's sales tax to food is a possible major revenue measure (estimated yield, $100 million) that would most heavily hit poor areas; cranking upward the high end of the personal income tax, on the other hand, would have almost all its effect on wealthy neighborhoods. Although cities cannot—as decentralization might permit—have the same tax collected at different rates in different parts of the city, the choices open to them do provide plentiful opportunity for shifting considerable tax burdens from one set of neighborhoods to another.

Despite these opportunities for choices, "non-decision-making" is the dominant style of operation. There are, for instance, a few systems analysts paid to worry about how fire-fighting resources should be distributed in New York City, but it is certainly not an issue that surfaces in public or political discussion. Even more telling, crime is a major theme in the political arena, yet despite the fact that the New York *Times* periodically runs front-page research articles on its distribution, the allocation of police resources among neighborhoods is not an issue that gets discussed publicly. (Indeed, even the systems analysts New York City employs to work on police department issues do not seem to do much along these lines.)

Why do distributional issues not enter the political arena? The answer, I would argue, is in large part that the current structure of the political system directs attention away from them. Information on the amount of a given service any particular neighborhood gets, or the amount it gets compared with other neighborhoods, just is not to be had. To develop it requires a major research effort.[16] Allocation formulae, if they exist at all, are prize examples of *arcania bureaucratica*. More often, distributional decisions are made on an ad hoc basis within the bureaucracy, with no clear locus of responsibility and no clear criteria.

Decentralization would change this situation in several ways. First, districts would be defined, giving people a uniform frame of reference for their comparisons. Second, district budgets would (probably) be developed and enter the domain of reasonably available information. Third, claims that a district was not getting its fair share of services or was paying more than its fair share of taxes would probably become a stock in trade of district leaders seeking to advance the well-being of their constituents. Thereby information (or misinformation) on relative levels of taxes and services would be forced to the attention of the politically interested public. Moreover, the competition among districts to be considered shortchanged would probably stimulate the formulation and advocacy of competing allocation criteria, each district lining up behind criteria which showed it badly off.

Thus, decentralization probably would touch off a continuing debate of considerable magnitude about how taxation should operate and how public services should be distributed.[17] Those who believe that government is improved by deliberate, conscious decision-making and the open airing of issues will welcome this development.

However, it has drawbacks. For one, cities currently perform a great deal of redistribution—an amount which, however, is not much in the public consciousness since the sums involved are not currently pulled together in any neat fashion. District budgets would change that.

A rich district—take the "silk-stocking" East Side of Manhattan as a paradigm—would find itself making tremendous tax contributions: New York's income tax is progressive; the per capita market value of residential real estate is high in well-to-do neighborhoods because of the more spacious and better-quality dwellings their residents inhabit, and real estate taxes paid are even higher since new property is typically assessed closer to market value than are older structures.

On the expenditure side, rich districts would discover that the cost of the city services they receive is low. Buildings are relatively new, well-kept, and fireproof, and their inhabitants are relatively careful, so fire expenditures are low. Similarly, the resident criminal population is low. This translates into relatively low police expenditures, despite the concentration of valuable items to steal, since criminals work for the most part close to home even if there are richer targets elsewhere. (To be accurate, one should refer to the population of perpetrators of those forcible crimes with which the police department mostly concerns itself. The stock manipulators, price fixers, competition restrainers, and their ilk *do* concentrate in wealthier areas.) The affluent have relatively few children and, in New York City at least, send a large proportion of these children to private schools for both higher and "lower" education. Thus public spending on education in these areas is also low. Finally, they include practically no welfare or Medicaid recipients, and their consumption of public hospital services is negligible, which means they get next to nothing out of some of New York's largest expenditure categories.

In short, rich districts pay much, much more in taxes than they get back in public services.[18] Poor districts—take a slum as the paradigm—stand in the reverse position. They make low tax contributions and are very heavy consumers of all the city services in the preceding list.

Now, as decentralization makes this situation more and more clear and puts a dollar sign on what it is costing the rich to aid the poor, the better-off areas of cities are likely to become increasingly unwilling to bear the burden, and cities will be forced into heavy cutbacks of their redistributive programs. In New York, open admissions to CUNY and the massive system of municipal hospitals might be the first to be cut, since they are chosen rather than mandated programs (welfare falls into the latter category) and are "extremist"—"avant garde," if you prefer—in that duplicates are hardly to be found elsewhere in the country. Their elimination would cut redistribution by literally hundreds of millions of dollars. In many cities, the better off might force a reallocation of police, fire, and sanitation spending that would alter the distribution of these services in a regressive direction. Perhaps liberals should not be so enthusiastic about decentralization as they usually are.

A second drawback of decentralization arising out of fiscal politics is the possibility of increased divisiveness and polarization in the city. Proponents of decentralization have usually anticipated the opposite result—cynically put, that decentralization will take "the heat" off city officials and make district officials into the "them" who are blamed for everything. However, the preceding paragraphs have indicated how the well-to-do are likely to become more fractious under decentralization as a consequence of the fiscal position they find themselves in. At the same time, the debate about allocation criteria that I foresee may magnify the dissatisfaction of poor areas. These neighborhoods probably will develop output-oriented standards and point out that, although all citizens are supposedly equal, city services in fact are distributed so that the poor suffer more crimes, more fires, and more dirt and garbage in public places; are less educated by the schools, and the like. Meanwhile, various middle strata, motivated by the allocation debate to find criteria according to which they are being short-changed, may also find ample cause for dissatisfaction.

In short, the fiscal politics of decentralization may well give every neighborhood in the city first, an argument that it is being short-changed by others; second, political leaders who, both for electoral purposes and to maximize the fiscal well-being of their district, will keep harping on this argument; third, the facts and figures with which to make the case; and fourth, a set of structural units—decentralization districts—around which polarization can take place. This is a sobering prospect for those who have foreseen in decentralization a panacea for the tensions of the big city.

Reflection on generalized decentralization has led to these rather gloomy conclusions. What are the lessons to be learned from specific service areas in which there are developments in fiscal politics that promise to have a significant effect upon decentralization? Education will be discussed first, followed by health and welfare.

EDUCATION

In New York City, a system of decentralized education is already in operation. Yet, from the fiscal standpoint, there is surprisingly little devolution to this decentralization. Each district has resources allocated to it by the central Board of Education, rather than being allowed to decide what level of spending (and taxing) it will pursue. Moreover, although the allocation consists basically of a lump sum—so that districts hypothetically might, à la Fleischmann, have a choice about what to spend their money on—budgetary tightness plus centrally negotiated contracts which specify maximum class sizes and work loads and otherwise generally restrict flexibility combine together to leave districts very little discretion (Wytock, 1973). Reality is different from rhetoric—or, in more popular terms, money rather than mouth shows where convictions currently lie.

Money—or, more accurately, fiscal politics—also offers clues to the probable future course of school decentralization. The most important trend in educational finance in the last few years has been to a larger role for states vis-à-vis localities. The impetus has been varied. Court decisions holding that district-to-district resource disparities violate equal protection constitutional clauses (Serrano v. Priest, 1971, is the most famous), legislative fears that other such rulings are impending, and legislative response to the growing burden of the local property tax have been prominent motivations.

States have generally started out with the intention of providing financial aid only and have avowed their intention to stay out of policy, in accordance with "the time-honored American tradition" of local autonomy in education. (The Fleischmann Commission's claim to be enhancing local autonomy is typical.) However, he who pays the piper finds it difficult to ignore the tune, especially when its price is increasing rapidly. State legislatures are generally suspicious of big cities and are often controlled by conservatives with relatively anti-spending values who are aghast at the size and growth of large-city education budgets, particularly when teacher's unions get into the picture and accelerate the rate of increase.

To take New York as an example, the supposedly autonomy-supporting Fleischmann Commission, in the later sections of its report, recommends state-set salary scales and statewide collective bargaining (New York State Commission on the Quality, Cost, and Financing of Elementary and Secondary Education, 1972: ch. 13). The state legislature met in special session in the summer of 1973 to limit teachers' pensions (as well as those of all other public employees in the state). And the idea of statewide collective bargaining is very much in the air. All this is further along than increased state aid, on which there has been relatively little activity since the Fleishmann Commission report appeared. The picture for higher education in New York City is much the same. Albany is committed to paying half the cost. The governor, in turn, is demanding a new board of directors for CUNY, which he will control, control over tuition charges, and other increases in the state's decision-making role. The commitment to a large financial role is clearly seen by state-level officials as entailing an enlarged role in policy-making.

In general, then, an increased financial contribution from states seems to be leading, despite some initial protestations to the contrary, to increased state control over education. Indeed, where taxpayer displeasure at rising local property taxes is the primary cause of state action, increased state control may precede, or occur in the absence of, an increased financial contribution.

One implication for decentralization is obvious. That which becomes a state-level function—and more and more is moving this way—cannot be a prerogative of decentralization districts. Ironically, then, just as decentralization is appearing, developments are under way which guarantee that decentralized districts will never have the range of powers that citywide school systems or (perhaps more relevantly) suburban systems have enjoyed.

However, for at least two reasons, this need not count in the negative column. First, groups fearful of what powerful local boards of education might do have been a major source (in New York City, far the dominant source) of opposition to decentralization. If enough important functions (and the right ones) are removed to the safety of state control, this source of opposition may dwindle away. That could greatly speed the spread of decentralization in education.

Second, with an increased state role, a system becomes conceivable in which only the state and the decentralized district are in the education business, and the large city drops completely out of the picture. From the fiscal standpoint, this possibility becomes realistic if either full state funding or power equalization is implemented. Indeed, even the less radical step of improving state education aid formulae so as seriously to curtail existing inequities could do the job. Such a two-tiered system might be favored by decentralization advocates. Under it, the decentralized district could wind up with more power than it would have had if the state's role had not expanded. The elimination of a citywide educational system, with the accretion of powers to the district that would result therefrom, might well outweigh power lost to the state. Those who deplore the dead hand of 110 Livingston Street (New York City Board of Education central headquarters)—and they are many— probably would, on reflection, prefer such a two-tiered option to more conventional decentralization plans. In addition, the greater simplicity of a two-tiered system as compared to a three-tiered one, and the resemblance of the former to the familiar and much-lauded suburban model, might make for better political salability.

In short, the fiscal politics of education point to a greater role for states, thus limiting the powers that could be devolved through decentralization. However, this need not hinder decentralization. Indeed, it may aid decentralization, if one is willing to contemplate decentralized districts of limited powers.

HEALTH AND WELFARE

Health and welfare (a category including publicly provided medical services, medical insurance, social services, and transfer payments to the needy) is another field in which there are major ongoing developments in fiscal politics that are likely significantly to affect decentralization. Here the stimulus to change has been skyrocketing costs, and the primary response has been for each level of government to try to hive off its share of costs onto the next higher level.

By now it seems clear that in a few years the federal government will take over many programs. Aid to the aged, blind, and disabled was changed from a federal-state-(and generally) local mix to a fully federalized system (which, however, the states may supplement) at the beginning of 1974. Bills "in the hopper" on welfare reform and national health insurance will reduce or eliminate the role of states and localities in these lines of business—and this is true of both the proposals of the incumbent administration and those of its opponents.

Again, then, it is obvious that—at least in a few years' time—there will not be much left to decentralize. Both the setting of eligibility and other standards and the administration of the system are likely to be federalized (or at least moved up to the state level)—and therefore not amenable to neighborhood decentralization—in the areas of health insurance and transfer payments to the needy.

Nor is the short-term prospect, at least in New York, more favorable to decentralization. Albany has been upset by the rapid growth of welfare payments, since it must meet a fixed fraction of the total bill. Its immediate strategy (as ͵sed to the longer-term hope of getting the feds to take over ͵ompletely) has been to criticize violently local welfare administration, intimating the costs would be much less if the rolls were not swelled by ineligibles. The states' welfare Inspector General, in his review of local eligibility determinations, consistently has come up with the finding that there have been errors in some 30% of the cases.[19] The governor's current proposal is to take over administration of welfare while still requiring localities to pay their 25% of total costs—just the reverse of what a decentralizer would propose. Yet New York City will probably support this plan—or rather, will oppose it verbally, proclaiming the excellence of its own Human Resources Administration, but will not really work at stopping it—since it is a promising first step toward getting the state to take over local costs.

While not identical, the short-term picture in other states is similar: Localities are trying to shed their share of welfare costs, state governments are being receptive in varying degrees, and the possibility of decentralization is disappearing. Much the same holds for Medicaid. Thus, neither in the long term nor in the short term, is there much prospect for decentralization of transfer payments to the needy or of governmentally provided health insurance.

In both fields, this is a severe disappointment to some critics of the existing system. They have attacked what they regard as an unresponsive administrative style, a hostile eligibility determination process, and inadequate benefit levels; and have often seen in decentralization a possible remedy to all three. Indeed, there has been a belief that even if benefit levels are externally determined, decentralization could help to overcome their inadequacy through a sufficiently sympathetic interpretation of the complex rules which define just which payments and special allowances each case is entitled to.[20]

In contrast to the outlook for health insurance and transfer payments, the future may augur well for decentralization of publicly provided health services and (perhaps) social services. To take the former first, public provision of health care in the United States has classically existed because the poor could not pay for care and society was unwilling to allow them to suffer total neglect. A limit to the scope of public clinics and hospitals was therefore imposed by the fact that they could derive little in the way of fees from the services they rendered and had almost entirely to be tax supported. The general public did not want the poor to have no treatment facilities, but it certainly was not going to tax itself much for their benefit, nor provide them with high-quality services.

The development of a health insurance system tends to undermine this limitation. Just as private physicians have found they can collect from the aged and poor, thanks to Medicare and Medicaid, and so have become more willing to treat these groups, so city hospitals and clinics find themselves suddenly able to collect from their patients. They have become potentially self-supporting, and when they do become financially self-sufficient, they should be able to expand to medically (or bureaucratically) optimal size, rather than being limited by tight-fisted burghers.

Existing health insurance programs do not get them this far. There is still significant dependence on central financing, so fiscal constraints continue strongly to limit the scope of public health care facilities.[21] Now the various national health insurance plans under consideration—and again this is true of both administration and Democratic proposals—promise major improvements in this respect. Under them, it should be feasible for decentralization districts to establish those health care facilities which each community feels are needed, and for these to be self-financing, or close enough thereto, for fiscal constraints to cease being a significant problem.[22]

On the other hand, a good insurance program might also have a reverse tendency to undermine publicly provided health care facilities. To some extent, the absence of private facilities in poor areas has been a consequence of the limited ability of the poor to pay for services. Insofar as this factor—rather than prejudice, indifference, and fear—has kept private physicians and hospitals from adequately serving the ghetto, health insurance may bring the private sector back in and thereby curtail opportunities for public facilities.

In social services, the positive prospects for decentralization are less clear, although certainly extant. The main opportunity arises from the fact that federalization of welfare payments is likely to lead to a "decoupling" of social services from income maintenance. Although the federal government probably will take over income maintenance, it quite possibly will not want to hire an army of social workers and take on the provision of social services. This could leave the provision of these services as a relatively free-floating activity, up for grabs by decentralization districts or even by private social agencies. If an opening for new providers does occur, community pressure and professional opinion among social services personnel will probably combine to push toward decentralization. On the other hand, the need to provide social services to each of millions of welfare cases might well lead to a decision to dispense social services through the income maintenance bureaucracy, or through a parallel large-scale organization, rather than to tackle the problem of checking to make sure that every one is being cared for by one of a large number of separate agencies.

In short, the current trend of fiscal politics in the health and welfare field is clearly toward federalization of programs, thereby eliminating the possibility of decentralization to community districts. Even in the period before the federal government takes over, the dominant trend is likely to be toward an increased administrative and financial role for the states, and away from any decentralization to the community level. Health care facilities constitute the one probable exception to this pattern. Here the financial independence made possible by national insurance promises to assist the development of community-level public facilities, particularly in poorer areas where the supply of private facilities is inadequate.

CONCLUSIONS

An examination of fiscal politics indicates the prospects of decentralization are less rosy than generally is believed. Extensive decentralization is likely to bring under public scrutiny allocation decisions currently made by "non-decision-making." A gain in rationality therefrom is questionable, given the politicization and district self-serving that will probably develop around these issues. Moreover, a reduction of redistributive services and, contrary to common expectations, an increased polarization in the city are also likely consequences of widespread decentralization.

In some major areas of city services, such as police, fire, and sanitation, there are no major trends in fiscal politics likely to affect the course of decentralization. In education, a greater financial role for states is likely to entail a greater policy-making role for them; but decentralization, although of a limited character, may oddly enough be abetted thereby. In health and welfare, fiscal politics is leading rapidly toward federalization of programs. This will in general preclude decentralization, although health care facilities may be an exception. Again, then, fiscal politics indicates that the prospects for decentralization are relatively poor.

NOTES

1. This formulation comes from Costikyan and Lehmann (1972: 12). This is the report of the task force on jurisdiction and structure of the Scott Commission, the New York State commission charged to study the government of New York City.

2. For instance, New York City, like most central cities, has a net in-commutation of workers—in the New York City case, roughly a half-million people—whose earnings it can tax. Furthermore, over seventy percent of the commercial property in the state is located in New York City.

3. It is assumed, as generally has proved true, that mobile businesses and individuals are tax-base-enhancing. Zoning and discrimination, rather than innate mobility differences, may well account for this result.

4. The rest of this paper abstracts from the issue of externalities which, theoretically at least, is readily solved by a system of matching grants.

5. If valid, though, it would certainly dictate full federal funding of education, for why should a child be penalized if his state of residence happens to choose to skimp on education? New Yorkers probably would not be keen on this, since they currently spend very heavily on education (highest in the nation, some 50% above the average, and considerably more than other rich, industrialized states) and thereby help to maintain their leading position in the state per capita income sweepstakes.

6. The commission's own observation that physically handicapped children often need classroom teacher-pupil ratios around 1 to 5 rather than the approximately 1 to 30 of normal classes suggests that the 1.5 weighting may be seriously inadequate. Given this figure on the handicapped and given the failure of so many compensatory programs, it may well be that severe sociocultural handicaps also can be overcome only with miniscule-sized classes, and that the appropriate weighting for the disadvantaged is more like 5 than 1.5.

7. This example is based on a Wisconsin program (see the Final Report of the State of Wisconsin Governor's Task Force on Educational Financing and Property Tax Reform).

8. Professional opinion among economists now disputes this belief, but what matters here is not academic opinion but what the general public believes to be true.

9. This is the form the Wisconsin program actually took.

10. A priori, one might expect poorer districts to have lower tax rates since they will spend a smaller fraction of their income on "luxuries" other than food, clothing, and shelter—such as education. On the other hand, they historically have had higher tax rates, despite lower per pupil expenditures, because of much lower per pupil tax bases. And this historical pattern might be continued.

11. A state could be fairly certain to achieve a net gain from enacting power equalization by setting the yield/tax ratio lower—to $60 or even $50 per pupil per mill. However, this "profiteering" might be politically untenable, and the size of the net gain would be uncertain.

12. For convenience, a normal distribution with standard deviation = 1 grade level has been assumed.

13. Many professional educators—and even more so, classroom teachers—are not aware of this point and do worry if any of their charges are below the norm. (Goldenberg, 1973).

14. "Learning" is used here as a general term for the intended output of the educational system. No judgment about the appropriateness of that output is implied.

15. The mechanics would be complex, but understanding could be easy. For each service, a two-column table would be constructed in which one column listed tax rates and the other indicated the output-measured level of performance that each tax rate would buy. Such tables would be unusually simple to understand, and would encompass what the general public needed to know.

16. For instance, New York City Administrator's Office (1972) and Temporary State Commission to Make a Study of the Governmental Operation of the City of New York (1973), which do contain information of this sort, are both major research works.

17. Support for this conclusion is provided by the fact that, when the City Administrator's neighborhood expenditure study appeared, a discussion of just this sort was for a while a major activity of local political clubs.

18. Note that this comparison is on a dollar-for-dollar basis. An equity comparison —based on what wealthy neighborhoods can afford to pay as compared to other areas and upon their service needs as compared to those of other areas—might well indicate that wealthy areas come out as gainers, not losers.

19. This figure has held in upstate investigations as well as ones of New York City. Note, however, that the bulk of purported errors are misclassifications—cases receiving more or less than the Inspector General finds them entitled to. He also finds eligibles categorized as ineligible, as well as the bogeyman of ineligibles classified as eligibles. Thus, although he claims errors in thirty percent of classification determinations, rectification of these errors would reduce welfare payments by nowhere near thirty percent. This fact, needless to say, has not been widely advertised.

20. The welfare Inspector General's findings, discussed in the previous note, suggest that this belief is realistic. There apparently is considerable scope for discretion in interpreting the rules, or at least sufficient confusion about them that bending them under the pretense of confusion should be easy. The existence of leeway—or confusion—is attested to by the fact that the inspector finds a mix of errors in both directions, not evidence of one-way "welfare-chiseling."

21. For instance, only about half the budget of the New York City Health and Hospitals Corporation, which runs the city hospitals, is currently derived from payments for services.

22. The Nixon Administration's health strategy runs strongly along these lines—provide adequate insurance and get out of direct funding of delivery facilities, such as neighborhood health centers.

CASE

SERRANO v. PRIEST (1971) 5 Cal. 3d 584, 487 P.2d 1241.

REFERENCES

CALLAHAN, J. and D. E. SHALALA (1969) "Some fiscal dimensions of three hypothetical decentralization plans." Education and Urban Society 2 (November): 40-53.

COSTIKYAN, E. N. and M. LEHMANN (1972) Re-Structuring the Government of New York City. New York: Praeger.

GOLDENBERGER, E. P. (1973) Personal communication. Simmons College, Boston.

New York City Administrator's Office (1972) Municipal Expenditures by Neighborhood. New York.

New York State Commission on the Quality, Cost, and Financing of Elementary and Secondary Education (1972) Report.

SHALALA, D. E. (1971) Neighborhood Governance: Proposals and Issues. New York: American Jewish Committee.

Temporary State Commission To Make a Study of the Governmental Operation of the City of New York (1973) New York City: Economic Base and Fiscal Capacity. Syracuse: Syracuse University Maxwell School Metropolitan and Regional Research Center.

WYTOCK, D. (1973) Personal communication from the Lead Planner for Education, New York City Budget Bureau.

11

Community Control and Governmental Responsiveness:
The Case of Police in Black Neighborhoods

ELINOR OSTROM
GORDON P. WHITAKER

☐ CREATION OF NEIGHBORHOOD-SIZED GOVERNMENTS within large American cities has been proposed as a way to increase the responsiveness of municipal officials to their local constituents. Police are among those officials often thought to be least responsive to citizens. Black citizens are among those constituents cited as least satisfied with the performance of local police and other public officials. Because of the controversy surrounding neighborhood police service to urban black Americans, this area is particularly appropriate for inquiry into the effects of community control.

This study compares one big-city- and two neighborhood-sized departments in terms of the police services they provide to residents of similar areas. Since community control experiments have not yet

AUTHORS' NOTE: *The authors are appreciative of the financial support provided by the Center for Studies of Metropolitan Problems of the National Institute of Mental Health in the form of Grant 5 R01 M1 19911-02, by the National Science Foundation, Grant GS-27383, and by the Afro-American Studies Program at Indiana University. We want to thank Shelley Venick and*

been instituted, it is not possible to examine directly the consequences of reducing the scale of large police jurisdictions. Similar
neighborhoods served by different-scale jurisdictions within a single
metropolitan area do provide the opportunity to assess compara-
tively some of the probable effects of community control.

Neighborhood police service is only one of the municipal activities
for which community control has been proposed. Additional
research on the effects of size of jurisdiction on the quality of other
urban public services is also needed. This study investigates two
questions: (1) Is community control conducive to greater governmental responsiveness? (2) Does community control create obstacles
to the effective provision of public services? The study examines
some specific problems which have been suggested as arising due to
community control. It evaluates the extent to which those problems
have developed in the small communities studied. It also assesses the
relative responsiveness of the small-scale and the large-scale governments. The evidence presented here deals with police services in five
neighborhoods of a single metropolitan area, but the findings can,
with caution, suggest parallels in other places and for other urban
public services.

THE NEED FOR MORE RESPONSIVE
POLICE AGENCIES

. Protection of property and person is more desperately needed by
the poor than by the rich. While the poor have less to lose, they feel
the consequences of loss more acutely. The respect police accord
those of high social status is likewise important to those who have

*Susan Thomas for their conscientious work as research assistants, and Dennis
Smith for his considerable help and criticism. A portion of this paper was
presented at the 1971 Annual Meeting of the American Political Science
Association in Chicago, September 7-11, 1971. Data were obtained from the
National Opinion Research Center, Study N-506. Each suburban sampling area
in the NORC frame for this study was coded as either a separately incorporated
suburban municipality or as an unincorporated suburban place using the 1967
Census of Governments definition of municipalities. Only respondents living in
separately incorporated municipalities are included in the category of white-
suburban reported above.*

suffered the indignities of lower status in an affluent society. However, black Americans charge that they continue to receive inferior police protection and suffer more abuse from the police than do the majority of Americans. The stance of many black citizens toward the police has shifted from "resentment to confrontation" (Fogelson, 1968). The resentment has been based on many charges related to the unresponsiveness of the police—police brutality, police corruption, lack of police protection in the ghetto and the lack of effective mechanisms for protest and remedy (National Advisory Commission on Civil Disorders, 1968; Hahn, 1971; Campbell and Schuman, 1968).

In many instances, the resentment appears to be widespread. In a recent comparative study, for example, TenHouten and others found that two-thirds of the respondents living in ghettos agreed with the statement that "police rough up people unnecessarily when they are arresting them or afterwards" (Ten Houten et al., 1971: 236). Resentment against police also appears to be based on experience. In a study of fifteen cities, using data obtained from four different sources, Rossi and Berk found that ghetto residents' grievances concerning police reflect the reality with which they live. Police brutality as a salient local issue was related to the existence of more abusive police practices, less responsiveness on the part of a local police chief to black grievances, less knowledge by the police of local black residents, and more personal experience by blacks of police abuse (Rossi and Berk, 1970: 122-125; see also Lieberson and Silverman, 1965). Aberbach and Walker (1970b: 1212) found that among the blacks they interviewed in Detroit during 1967, personal experiences of police mistreatment were negatively associated with political trust and that individuals with low levels of political trust were more likely to be able to imagine a situation in which they would riot. In general, they found that attitudes of political trust were not mere reflections of an individual's basic personality or of background factors. Their most important explanatory variables were "those which arise from the workings of the social or political system" (Aberbach and Walker: 1970b: 1214).

Police responsiveness varies by neighborhood. In their study of the large Denver police force, Bayley and Mendelsohn (1969: 114) found:

Ethnicity is a primary determinant of the amount and kind of contact people have with the police. Within ethnic groups there is by and large no association between age, sex, and class and whether an individual has been stopped and arrested or has called the police for help or talked over difficulties with them.

The differences in contact which Bayley and Mendelsohn find related to ethnicity are largely related to the neighborhood in which individuals live. Denver, like most large American cities, is residentially segregated. Bayley and Mendelsohn learned from their surveys of officers that "police do carry certain predispositions into their contacts with minority people, especially in minority neighborhoods, that can produce a double standard in enforcement behavior" (Bayley and Mendelsohn, 1969: 166). A similar dynamic may also be at work in Milwaukee, Seattle, and Detroit where blacks also report more unfavorable contact than whites with police and where neighborhoods are generally racially homogeneous.

Jacob, in his study of black and white Milwaukee neighborhoods, found that "the general reputation of the police in black ghettos has become so bad that good experiences do not bring correspondingly good evaluations" (Jacob, 1970: 72). He identified this phenomenon as "neighborhood culture" and considered it "one of the intervening variables between experiences and perceptions." Similar findings in Seattle were explained as a "contextual effect" whereby "persons in sub-communities subject to relatively high probabilities of arrest, develop less positive attitudes toward police whether they themselves have been arrested or not" (Costner et al., 1970: 46). The phenomenon which these authors identify as "neighborhood culture" and "contextual effect" appears to be a reflection of the lower levels and poorer quality of service provided to citizens living in black neighborhoods. That is, the low evaluations black citizens make of their police reflect the unresponsiveness of the police serving their neighborhoods.

Trends found in specific large cities have also been established nationwide. In a survey conducted for the President's Commission on Law Enforcement and Administration of Justice in 1966, the National Opinion Research Center administered a nationwide survey including several questions asking respondents to rate their police services. Analysis of the data from that survey shows that black residents of large center cities consistently rated police services lower than white residents of either center cities or incorporated suburbs. As shown in Table 1, black center-city respondents were less likely than white respondents at all income levels to rate their police as respectful, as paying attention to complaints, as giving protection to the people in their neighborhood, or as being prompt. Wealthier white respondents living in center cities tended uniformly to rate the police higher than did poorer white respondents. White respondents living in independently incorporated suburbs tended to rate the police higher at all but the lowest income level than did white center-city residents. The ratings of police services by black respondents were *not* positively associated with income levels as were the ratings by white respondents. In fact, black respondents of higher income levels tended to be less likely to give high ratings to police than black respondents of lower income levels. Thus, the criticisms

TABLE 1
Rating of Police Services by Black Center-City Respondents,
White Center-City Respondents, and White Incorporated
Suburb Respondents Controlling for Income[a]

	Income Levels of Respondents				
	Less than $3,000	$3,000 to $5,999	$6,000 to $9,999	$10,000 & Higher	n
Percentage Giving Highest Rating to Police for:					
Being respectful to people like themselves					
Black—Center City	40	28	33	25	256
White—Center City	69	59	67	74	877
White—Incorporated Suburbs	62	65	78	77	430
Paying attention to complaints					
Black—Center City	36	18	18	8	242
White—Center City	47	49	54	64	823
White—Incorporated Suburbs	50	56	57	70	387
Giving protection to the people in the neighborhood					
Black—Center City	23	17	22	9	241
White—Center City	50	42	53	55	829
White—Incorporated Suburbs	50	58	60	72	412
Promptness					
Black—Center City	30	16	24	8	249
White—Center City	50	42	53	55	803
White—Incorporated Suburbs	47	56	63	73	379

a. Data obtained from the National Opinion Research Center, Study N-506. Our thanks to Patrick Bova for assisting us in working with these data. Only respondents living in separately incorporated municipalities are included in the category "White-Suburban" reported in the table.

of police by black citizens are not restricted to the poor, but are shared by all segments of the black urban population.

Complaints related to police brutality and harassment coupled with complaints of insufficient police protection have seemed somewhat paradoxical to some observers. However, it would appear that the practice of "preventive patrolling" utilized by some police forces simultaneously increases the resentment of residents and diverts police manpower from other activities such as answering calls and investigating the many crimes which do occur in the ghetto (Hahn and Feagin, 1970). Across the country, victimization rates for blacks are higher than for whites at all levels of income for serious crimes against the person (Ennis, 1967: 30). Black residents living in cities of over 100,000 population were considerably more likely to cite a need for self-defense when asked: "Do you think that people like yourself have to be prepared to defend their homes against crime and violence, or can the police take care of that" (Feagin, 1970: 799)? Thus, the simultaneous criticism of too much and too little policing may be valid. Police seem to be failing to serve the residents of many black neighborhoods in U.S. cities.

Increasing attitudes of confrontation have become all too obvious. While some confrontations between black citizens and police have occurred within the institutional settings provided by elections, courts, and review boards, many have occurred in the streets. Street confrontations have occurred particularly in larger cities in which the proportion of black citizens has increased significantly. Riots, assaults on officers, and the stony hostility or taunting jibes which often greet policemen are all reflections of hostility between black citizens and their police. Confrontations on the street reflect the absence of opportunities for confrontation through the regular institutions of government. The failure of established governmental institutions to be responsive to black citizens' demands for effective and impartial police services represents a serious threat to domestic peace and order.

ALTERNATIVE STRATEGIES FOR
ENHANCING RESPONSIVENESS

PROFESSIONALIZATION AS A REMEDY

Two remedies are frequently recommended for reducing the overt antagonism, mistrust, and hostility toward the police by many black citizens. These are similar to those often suggested as strategies for enhancing the responsiveness of government generally. One remedy involves increasing the "professionalization" of public servants. According to James Q. Wilson (1968a: 175), a "professional" police department is one "governed by values derived from general, impersonal rules which bind all members of the organization and whose relevance is independent of circumstances of time, place or personality." Professional departments are said to have attributes which include the following:

- recruitment on the basis of achievement,
- equal treatment of citizens,
- negative attitudes toward graft both within the force and in the community,
- commitment to training of generally applicable standards,
- bureaucratic distribution of authority.

In communities served by professionalized departments, law enforcement may be stricter, but it is thought to be more equally applied to all groups than in communities served by nonprofessional departments (Wilson, 1968a). Reliance upon brutality as a means of social control is thought to be less within such departments than in nonprofessional departments.

However, tensions between black citizens and the police have not lessened in the cities with police departments described as highly professionalized. Two of the departments most frequently characterized as "professional," Oakland and Los Angeles, have also been observed to take strong punitive actions against blacks (Skolnick, 1967; Jacobs, 1966). More "professional" recruitment, training, and authority structure does not necessarily entail equal treatment of citizens. Even advocates of professionalization recognize the "limitations of professionalization especially when it is used to rationalize the employment of preventive patrolling and the other extraordinary tactics which transform the Negro ghettos into occupied territories" (Fogelson, 1968: 247). Police professionalization may have served more to insulate the police against external criticism than to reduce the level of discrimination by police against black citizens.[1] James Q. Wilson (1963: 201)—a firm advocate of police professionalism—has argued that "professionalism among policemen will differ from professionalism in other occupations in that the primary function of the professional code will be to protect the practitioner from the client rather than the client from the practitioner." Thus, "professionalization" may in fact decrease rather than increase a police department's responsiveness to citizen needs and preferences.

Paul Jacobs vividly describes resistance to any meaningful review procedures by the Los Angeles Police Department prior to the Watts riot. One of its basic strategies was to demean civil rights groups and others calling for outside review. In the department's 1964 *Annual Report,* for example, the charge was made that

> the detractors of law enforcement stepped up their pervading accusations of police misconduct and pleas for an independent review of police practices in an attempt to create an atmosphere of apprehension, predicting that the streets of this city would also become an arena in which the issues of the civil rights movement would be settled [cited in Jacobs, 1966: 99].

During the same year, 121 complaints were lodged with the police department concerning the excessive use of force. Only 21 were sustained. However, in none of the 21 cases where charges were sustained did the officer charged receive the penalty associated with the use of excessive force. Officers were allowed either to resign without penalty or to receive a lesser penalty (Jacobs, 1966: 98-99).

At least one early champion of the "professionalization" remedy has recently reversed his position. Burton Levy, after a two-year period of intensive observation of police departments across the country for the U.S. Department of Justice concluded that recruitment, training, and community relations efforts did not seem to have a significant impact on police practice.

> The problem is not one of a few "bad eggs" in a police department of 1,000 or 10,000 men, but rather of a police system that recruits a

significant number of bigots, reinforces the bigotry through the department's value system and socialization with older officers, and then takes the worst of the officers and puts them on duty in the ghetto, where the opportunity to act out their prejudice is always available [Levy, 1968: 348].

Professionalism as a remedy for the problems of resentment and hostility toward the police among black citizens would appear to have serious limitations. It provides no leverage for blacks to demand improved service and is thus an inadequate device for institutionalizing confrontation.

COMMUNITY CONTROL AS A REMEDY

A second general proposal to alleviate the growing tension between black citizens and government is community control (for the best in-depth overview of the issues involved, see Altshuler, 1970; see also Shalala, 1971). In the case of police services, proponents argue that reducing the size of local police jurisdictions and bringing the jurisdiction under the control of the citizens living in the community served will increase responsiveness of police to the preferences of citizens. Under a more responsive institutional structure, police would be expected to provide services needed by community residents, thus increasing citizens' satisfaction with police services.

However, community control has been strongly questioned as an effective reform strategy. Sherry Arnstein (1969: 224) has summarized some of the most frequently articulated arguments against community control in the following overview:

> Among the arguments against community control are: it supports separatism; it creates balkanization; it enables minority group "hustlers" to be just as opportunistic and disdainful of the have-nots as their white predecessors; it is incompatible with merit systems and professionalism; and ironically enough, it can turn out to be a new Mickey Mouse game for the have-nots by allowing them to gain control but not allowing them sufficient dollar resources to succeed.

Let us briefly examine these arguments against community control and the responses in favor of such a system.

Separatism. The first argument is that community control supports racial separatism. Given existing patterns of residential segregation, the population of local communities would be more racially homogeneous than the population of citywide areas. Once boundaries were drawn, it is argued, the tendency toward homogeneity of communities would increase as citizens scurried to move out of areas where they were in a minority. The result might be that police forces in each type of community would be much more oriented to abusing members of the minority race in that community than now occurs in the big city.

Proponents of community control reply that segregation is a fact imposed on black citizens by the unwillingness of white citizens to allow integration in any meaningful form (Spear, 1967; Tauber, 1968). Community control would not appreciably increase the amount of segregation and racism currently in existence—it would give to those who had been denied open access to housing a greater opportunity to control what happens in their own neighborhood. There is no evidence that blacks controlling their own areas would be more racist in orientation. Aberbach and Walker (1970a) found that eighty-eight percent of the black residents of Detroit interviewed in 1967 preferred to have the "best trained police, no matter what their race" patrolling in Negro neighborhoods rather than "Negro police only." Interestingly enough, of the whites interviewed, twenty-two percent (as compared to twelve for blacks) thought that Negro police only should patrol in black areas.

Balkanization and economies of scale. The second argument is that community control creates balkanization of public services and is more costly and less efficient. This is an old argument repeatedly presented by advocates of metropolitanwide governments. Advocates of metropolitan government recommended the *elimination* of most of the currently established units of local government in metropolitan areas (Zimmerman, 1970).[2] Metropolitan reformers assume that large economies of scale exist for all public services and thus urge the creation of one or a few large-scale public jurisdictions to serve an entire metropolitan area. Those associated with this movement argue that decreasing the size of police agencies and increasing their number within a particular metropolitan area would increase the costs of service and lead to grave problems of coordination among diverse agencies. The sheer presence of a large number of local units is frequently cited as evidence in and of itself that coordination of efforts among such a multiplicity of juris-dictions cannot be accomplished. Coordination within a single large jurisdiction is presumed to be more easily accomplished than cooperation among many jurisdictions.

Proponents of community control have argued that economies of scale do not exist for such services as police and education and that, consequently, community control may not lead to an increase in the cost of local services. Large-scale agencies could continue to provide such services (which do benefit from large scale) as transportation, water, sewage and to help provide some of the financing for smaller units within the larger unit (Meltzer, 1968; Mayer, 1971). Just as large units may be more effective and efficient in the provision of certain services, smaller units may also be more advantageous for other services. Police services such as neighborhood patrols and emergency aid appear to be of this type. Furthermore, in many situations in the United States, a number of disparate public jurisdictions are able to coordinate efforts through joint agreements, contracting, and distinct distribution of authority (Ostrom, Tiebout, and Warren, 1961; Bish, 1971; Warren, 1966; Bish and Warren, 1972). Community control would enable blacks in the center city to have the personalized, small-scale service provided today to whites in the suburbs (Ferry, 1968). Suburban residents have vigorously fought against being included within large, metropolitanwide govern-mental jurisdictions. Why should residents of the center city be the only ones who cannot have small-scale public agencies responsive to the particular needs of their communities (Rubenstein, 1970; Babcock and Bosselman, 1967; Press, 1963)?

Lack of participation. The third argument against community control is that local decision-making within small communities is more "undemocratic" than that within larger units (Kristol, 1968; Perlmutter, 1968). Critics point to the low turnout of voters and the ineffective bickering among "poverty representatives" in many of the early community action programs. Because of the relative homo-geneity of an individual community, they also argue that there would be less challenge to local leaders who may be more demagogic than leaders of large, heterogeneous city governments. The intimacy of the local community, furthermore, may lead to corruption and lack of uniform enforcement practice (Wilson, 1968b; Prewitt and Eulau, 1969; for a different argument, see Rossi, 1963: 12).

If black citizens have genuine control concerning local affairs, however, participation levels may be expected to increase (Gittell, 1968). While participation in many programs in the past has been low, it is unreasonable to expect high participation in newly organized arrangements whose potential benefits may be quite nebulous. Many programs have used "participation" as therapy rather than as a means to enable local people to exercise substantial control over events affecting them (Mogulof, 1969; Arnstein, 1969). People do not learn to participate actively or constructively in a short time period. If meaningful control were placed in the community, individuals would begin to learn that it was worthwhile to participate and how to participate more constructively.[3] Once community control was established, the effect of having local public officials sympathetic to the needs and aspirations of local citizens would decrease the general level of alienation among black citizens living within the ghetto of a typical large American city.[4]

Amateur public servants. The fourth argument is that small, community-controlled police departments would be less professional. It is assumed that a relatively large department is needed to be able to afford adequate salaries, good training facilities, and sufficient levels in the bureaucracy to achieve meaningful advancement for ambitious young personnel (Altshuler, 1970: 39). It is frequently argued that small departments cannot attract as qualified employees as can large departments. Such personnel, employed in specialized, hierarchically controlled departments are seen as necessary to improved police service.

Proponents of community control argue that many of the consequences of "professionalization" have been to keep blacks from obtaining jobs due to irrelevant educational requirements or middle-class, biased examinations (see also Baron, 1968).[5] The establishment of less-bureaucratized forces with police officers living in the community they serve and sympathetic to the life style of the residents is seen as a benefit rather than a cost.[6] Career opportunities can be pursued among small jurisdictions by lateral movement as occurs in many school districts rather than relying on vertical movement in a single bureaucracy. Neighborhood agencies can be expected to be less effective in providing specialized units for criminal investigation but more effective in providing police patrol services to the neighborhood (Ostrom, Parks, and Whitaker, 1973). A centralized police force could continue to provide specialized police services for the entire city.

Lack of financial resources. Finally, it is argued that community control may be a futile strategy if significant reallocation of resources is not also accomplished at the same time. Impoverished areas would remain just that—impoverished areas. Once separated from the rest of the city, black citizens would find it difficult to obtain from white citizens living in separate jurisdictions the resources needed for effective programs. Community control might prove to be a cruel joke. Those in "control" would not have sufficient resources to be able to accomplish their goals (see Altshuler, 1970: 53-54). Consequently, the long-run consequences of community control might be further bitterness, disillusionment, and alienation among black citizens (Aberbach and Walker, 1970a: 1218).

There is, however, considerable doubt that extensive redistribution in favor of the poor does occur in larger political units. In a study of American cities which had adopted one of the reform measures leading toward greater consolidation, Erie et al. (1972) found no tendency toward redistribution of wealth among elements of the populations within reformed metropolitan institutions. Moreover, many of the needs of poor areas are not solved by the mere infusion of more economic resources. Even if more funds are available, the services provided by the larger government may not suit the affected community. More effective service depends upon fitting public services to the particular needs of a community. Milton Kotler describes the deliberation of a community corporation in a poor neighborhood of Columbus, Ohio, concerning medical services. Doctors were proposing "fancy new clinics with interns rotating the work day by day." However, the people in the neighborhood corporation "said no, they didn't need anything as elaborate as a big clinic. What they needed was a night doctor. . . . Neighborhoods like this need doctors who work on a different schedule" (Kotler et al., 1968: 16). If the views of the professionals had prevailed, more money would have been spent, but the people living in the neighborhood would not have been as satisfied with the type of medical service provided. Many (but, of course, not all) of the problems of the ghetto relate to the need for services tailored to residents' own needs (Itzkoff, 1969).

Finally, the financing of services in a public jurisdiction does not always have to come entirely from the area itself (Ostrom, Tiebout, and Warren, 1961). Redistribution formulas by which larger units provide some of the funds for smaller units are used by both state and federal governments. Effective organization of the local community may enable sufficient pressure to be brought at metropolitan, state, or federal levels to achieve further redistribution of resources. Such resources could then be utilized in a way responsive to the preferences of local residents in various types of areas rather than as a result of decisions made by a single set of officials for all areas.

AN EVALUATION OF THE ARGUMENTS
RELATED TO COMMUNITY CONTROL

Several studies have been undertaken to evaluate the warrantability of the arguments for and against community control. Most of these studies have focused on white, middle-class neighborhoods. When the performance of relatively small police jurisdictions (serving under 20,000 people) is compared to the performance of relatively large forces (serving 200,000 to 450,000 people) serving similar white, middle-class neighborhoods, the smaller jurisdictions were found to produce higher levels of output at similar or lesser costs (see Ostrom and Whitaker, 1973; Ostrom, Parks, and Whitaker, 1973; Ostrom, Baugh, Guarasci, Parks, and Whitaker, 1973; IsHak, 1972). An additional study examined a range of police departments located throughout the United States. The 102 departments included in this study served jurisdictions ranging in size from 10,000 to 8 million. When the levels of output of these police departments are compared, smaller departments are found to produce equal or higher levels of service for similar or lower expenditure levels (Ostrom and Parks, 1973). Thus, contrary to the arguments against community control, larger departments do not appear to provide higher levels of service. Therefore, the small scale implied by community control does not necessarily entail loss of effectiveness or efficiency.

Local control of the police has not been a salient issue for most middle-class, white citizens. Several reasons can be stated for this lack of saliency:

(1) Wealthier white citizens are often well served by large-scale police jurisdictions and thus tend to be more satisfied with large-scale jurisdiction than black citizens or poorer white citizens living in the same city (see Table 1).

(2) White citizens—living in large cities—who are dissatisfied with the services provided by their police, or for that matter with any other aspect of local government, can move to a different jurisdiction relatively easily (Tiebout, 1956).

(3) White citizens at all income levels except those under $3,000 income, living in separately incorporated suburban jurisdictions, appear to be receiving higher levels of service than either white or black citizens living in center cities (see Table 1).

White citizens appear to have the opportunity to receive the kinds of urban public services they desire either by moving to a jurisdiction which will provide them or by exercising a greater voice in the articulation of their demands for service (Orbell and Uno, 1972).

However, most black citizens and members of other minority groups are excluded from these options. They are prevented from moving to smaller, suburban jurisdictions where public services are subject to more direct control by residents. Black citizens can find housing primarily in the most crowded sections of central cities and rarely in suburban jurisdictions. They rarely have effective channels for articulating their service demands to big-city governments (Parenti, 1970). Neither are black citizens able to compete effectively for control of police policy in large cities. Consequently, many black demands have been focused on "decentralization" of large-scale police forces already serving central cities.

Given the concentration of black citizens in most large cities and the low percentage of blacks in most suburban cities, it is extremely difficult to locate adequate research sites to examine the consequences of increased levels of local control for black citizens. There are, however, several independently incorporated black communities located in the Chicago metropolitan area. A small study was recently undertaken to evaluate the consequences of community control for the residents living in two separately incorporated black communities by comparing the police service they receive with that provided to residents of matched neighborhoods within the city of Chicago.

THE AREAS STUDIED

The villages of Phoenix and East Chicago Heights, Illinois, are both small and poor. The population of Phoenix in 1970, according to official census figures, was 3,596, while village officials feel that the population was closer to 5,000. The official census figure for East Chicago Heights was 5,000. Village officials feel that at least 2,000 East Chicago Heights residents were missed in the official census. In 1970, the median family income in Phoenix was $7,600 while that of East Chicago Heights was $6,750. The median value of homes in Phoenix in 1970 was $15,900 and in East Chicago Heights was $16,000 (Illinois Regional Medical Program, 1971). Whenever socio-economic rankings of the municipalities surrounding Chicago have been published, these two villages have always been among the lowest five municipalities (see De Vise, 1967; Illinois Regional Medical Program, 1971; Chicago *Sun Times,* 1972).

Each village is governed by a six-person Board of Trustees and an independently elected mayor and village clerk. All village officials are black. The ratio of village residents to members of the Board of Trustees is less than 1,000 to 1. Other than the full-time village clerk, all other elected officials serve in a part-time capacity. The mayor and village trustees all hold other jobs and attend to village affairs during the evenings and weekends. However, village officials spend almost all their "free" hours working for the village. The level of volunteerism is high in the villages. Tasks such as clearing snow and salting roads are performed by the trustees along with village citizens who have volunteered to help. Community projects such as painting or repairing a public building are frequently organized on a voluntary basis with a community cookout scheduled for relaxation after the work is completed. Both communities are served by volunteer fire departments.

The police forces in the villages are quite small. The size of each force fluctuates considerably. At the time of the study, however, Phoenix employed four full-time and fifteen part-time officers. East Chicago Heights had six full-time and five part-time officers. Part-time police officers were paid at the rate of $1.60 per hour. Full-time officers received approximately $400 per month. Police-men in the villages received little formal training. Both chiefs and some officers have received training at police institutes run by the state of Illinois. Training within each department is provided by the more experienced officers.

The villages face a perplexing problem with regard to training. When they have provided funds to send a regular patrolman to a police training program, they have frequently lost the patrolmen within a short time to one of the surrounding municipalities which pays higher salaries to police officers. Both villages find that their best police officers are frequently lured away after they gain experience on the village force. Consequently, there is a high level of turnover on both police forces. Inexperienced individuals who gain experience in a small police force and demonstrate proficiency in police work are able to follow better career opportunities by seeking employment in other jurisdictions.

Each village has two or three radio-dispatched cars. However, police cars are out of operation for relatively long periods of time due to the high costs of repair. The lieutenant in one of the villages usually drives his personal car (which he has equipped with a radio at his own expense) in order to reduce the operating expenses of the village department.

Both villages cooperate with neighboring villages when extra help is needed in any of the south suburban municipalities. They rely upon the Cook County Sheriff for investigative services and laboratory work when needed.[7]

Financially, both departments have extremely limited resources. The police department budget for each community is approximately $40,000 per year. One of the villages has had a long-standing reputation as a speed trap and, until recently, traffic fines provided most of the financing for its police department.

The sample areas within the city of Chicago were selected to match as closely as possible the socioeconomic characteristics of the independent villages. Some factors which affect police service were thus controlled through the use of a most-similar-systems research design (see Prezeworski and Teune, 1970). A comparison of respondents in the Chicago neighborhoods to those in the independent communities is shown on Table 2. A major socioeconomic difference between the two types of sample areas relates to housing patterns. Thirty-seven percent of the respondents living in the independent villages reside in public housing. All these respondents live in East Chicago Heights in two-story, low-density public housing units. It was not possible to find within Chicago, with similar public housing units, a neighborhood which was not greatly dissimilar to the independent communities on most of the other socioeconomic factors. One Chicago neighborhood chosen did have a large low-rent apartment complex within it. Residents of public housing may generally rate public services less favorably than nonpublic housing residents. Consequently, the presence in our sample for the independent communities of a large number of public housing residents biases that portion of our sample downward with regard to citizens' evaluations of services received.

TABLE 2
BACKGROUND CHARACTERISTICS OF RESPONDENTS IN THE
TWO TYPES OF NEIGHBORHOODS (in percentages)

	Two Independent Communities	Three Chicago Neighborhoods
Age of Respondent		
16-20	10	16
21-30	16	18
31-40	32	30
41-50	20	15
51-65	10	12
Over 65	11	10
(n)	(213)	(294)
Sex of Respondent		
Female	62	64
Male	38	36
(n)	(213)	(294)
Husband's Occupation		
Professional-Managerial	13	12
Clerical-Sales	9	14
Craftsmen-Foremen	14	18
Semi-skilled	26	25
Unskilled	14	14
Retired	14	12
Unemployed	10	5
(n)	(118)	(156)
Ownership of Housing		
Buying home	48	73
Renting	15	27
Public housing	37	0
(n)	(201)	(280)
Length of Residence		
Less than 2 years	16	20
2-5 years	13	27
6-10 years	17	23
More than 10 years	54	37
(n)	(209)	(290)
Education		
8 years or less	33	18
Some high school	26	27
High school graduate	29	32
Some college	9	15
College graduate	2	6
(n)	(198)	(274)
Number of Dependent Children		
No dependent children	16	25
1 or 2 dependent children	26	32
3 or 4 dependent children	26	22
5 or 6 dependent children	18	13
More than 6 dependent children	14	8
(n)	(198)	(274)

Chicago is governed by a strong, independently elected mayor and a city council of 50 members. The mayor of Chicago dominates the city council as well as the executive departments, including the police. The ratio of Chicago residents to members of the Chicago City Council is more than 65,000 to 1.

The Chicago Police Department is one of the most modern, best trained, and best financed departments in the country. The force had over 12,500 men at the time of this study. Patrolmen received from $9,600 to $12,000 per year, depending on years of service. The department conducted extensive training programs, including in-service instruction and a thirty-one-week cadet program for recruits. The proportion of blacks serving on the Chicago police force was substantially less than the proportion of black residents in the population of Chicago. While blacks made up approximately forty percent of the Chicago population, approximately twenty percent of the patrolmen were black; eight percent of the detectives were black; and four percent of the lieutenants were black (Jackson, 1970; see also Baron, 1968).

Chicago is divided into twenty-one police districts, each with its own station. The Englewood station is located within one of the neighborhood areas included in this study. Thus, residents of that neighborhood have somewhat more immediate access to police than those of the other two study areas inside Chicago. However, all radio cars in Chicago are controlled by a central dispatch office. In terms of telephone access to police, all three Chicago neighborhoods are quite similar due to this central dispatching. The Chicago Police Department has highly specialized units to handle a variety of investigative and support activities. The total budget for the Chicago Police Department during 1970 was $190,922,514. Expenditures for police services in the three Chicago neighborhoods have been estimated at $1,720,000 (Whitaker, 1971). Thus, over fourteen times as much was spent on policing each of the three Chicago neighborhoods as was spent by the villages for local police services there.

Given the relative similarity of the sample areas but large difference in financial resources allocated to police and the differences in the training of the personnel employed by the different types of police departments, one would expect the Chicago Police Department to provide a much higher level of service to residents than would the village police departments.

LEVELS OF POLICE SERVICE PROVIDED

In general, citizens living in the independent communities received equal or higher levels of service than residents of similar neighborhoods in Chicago. As we have discussed elsewhere, there are no generally agreed upon methods for measuring police output (Ostrom and Whitaker, 1973; E. Ostrom, 1971). Because of our interest in services provided to citizens, we have utilized survey methods to obtain two types of indicators of police output. The first type of indicator is the police-related experiences which respondents have had. Levels of criminal victimization and the quality of a variety of police actions are assessed in this way. The second type of indicator consists of citizens' evaluations of service levels. In eight items, citizens were requested to evaluate various aspects of police service. Five additional items were included to obtain respondent's evaluations of local government in general.

Citizens' reports of the police services they have received are summarized on Table 3. For four of the indicators, service levels are reported to be quite similar (tau less than .10). For the other three indicators, village respondents are more likely to indicate higher levels of service (tau greater than .10). A similar pattern is seen for citizens' evaluations of the equality of police services provided by their local forces. As shown on Table 4, police services are judged similarly by respondents on four of the indicators. On the other four indicators, village respondents are more likely to give high ratings than are Chicago respondents. The service levels reported in this study are much lower than those reported in white neighborhoods of Indianapolis and Grand Rapids. The pattern is the same, however; residents of independent communities report services of similar or higher levels across a large number of indicators when compared to residents of similar central-city neighborhoods. In no case do residents served by large police departments report higher levels of service on any indicator. This finding is particularly surprising in the Chicago area because of the substantially smaller amount of funds devoted to the village police forces. The experiences of these five neighborhoods cannot be generalized to all black neighborhoods, but the study does provide evidence which bears upon arguments for and against community control.

Separatism. The first objection raised to community control is that it would encourage racial separatism in American cities. If

TABLE 3
COMPARISON OF SERVICE LEVELS RECEIVED

Indicators of Police Services Received	Independent Communities	Chicago Neighborhoods	Tau for Complete Table
Percentage of Respondents:			
Reporting that they were not victimized during preceding 12 months	75	74	.01
(n)	(195)	(276)	
Reporting they do not stay at home because of fear of crime	58	45	.13[a]
(n)	(205)	(276)	
Receiving high levels of police followup to reported crime[c]	59	46	.13[b]
(n)	(32)	(48)	
Calling on the police for assistance not related to victimization	19	24	−.05[b]
(n)	(193)	(269)	
Reporting police arrival in less than 5 minutes	60	48	.11
(n)	(30)	(44)	
Reporting effective police assistance[c]	95	94	.01
(n)	(36)	(62)	
Reporting fair treatment when stopped by own police force[d]	77	70	.06
(n)	(35)	(97)	

a. $p < .001$
b. $p < .05$
c. Coded "effective" when respondent indicated that police handled the matter, police gave emergency aid or police solved problem.
d. Includes respondents indicating that they were treated nicely or in a fair manner.

community control were to encourage racial separation, one would expect residents of the two independent black villages to have a *stronger* preference for black officials than respondents of the three Chicago neighborhoods. However, residents of the independent black villages were neither more nor less likely to prefer black officials. Twenty percent of the black respondents preferred black officials whether or not they lived in a separately incorporated community served by black officials.

Nor did a higher proportion of respondents in the independent communities express strong racial identity (Mitchell, 1973). Strong racial identifiers living in the Chicago neighborhoods were, however, extremely negative in their views toward the legitimacy of local

TABLE 4
COMPARISON OF CITIZEN EVALUATIONS OF
POLICE SERVICES (in percentages)

Evaluation Indicators	Independent Communities	Chicago Neighborhoods	Tau for Complete Table
Percentage of Respondents:			
Evaluating local police-community relations good	46	44	.01
(n)	(181)	(254)	
Evaluating local police response time very rapid	26	25	.01
(n)	(144)	(197)	
Believing local police do not accept bribes	37	21	.18[a]
(n)	(110)	(157)	
Agreeing they have some say about what police do	49	47	.01
(n)	(196)	(264)	
Agreeing that local police have the right to take any action necessary	61	38	.24[a]
(n)	(192)	(270)	
Agreeing that redress is possible for police mistreatment	67	66	.01
(n)	(176)	(270)	
Agreeing that local police treat all equally according to the law	46	18	.31[a]
(n)	(105)	(181)	
Agreeing that local police look out for the needs of the average citizen	56	36	.21[a]
(n)	(180)	(264)	

a. $p < .001$

institutions. On the other hand, strong racial identifiers living in the independent communities were more positive in their support of local institutions than were medium or weak racial identifiers (Mitchell, 1973). Given an opportunity to live in a separately incorporated black community, those with strong racial indentification appear to become supporters of regular political institutions rather than antagonists.

Balkanization and economies of scale. Opponents of community control frequently assert that small units of government are most costly due to their failure to realize supposed economies of scale in the provision of public services. In the Chicago study, however, the independent communities did not spend more for police protection than was spent by the city of Chicago in policing the neighborhoods studied. In fact, expenditures in the independent communities were much lower. Each independent community spent approximately $40,000 in support of its local police department in 1969. During the same year, the Chicago Police Department, according to our estimates, incurred expenditures averaging over $500,000 for each of the neighborhoods investigated. Similar or better services appear to have been provided by the smaller communities for about seven percent of the cost of the service provided by the larger police department.

Further, with regard to cooperation and coordination, there is considerable evidence of cooperative efforts between the smaller police departments and other local police agencies. Emergency mutual aid arrangements exist between the small black communities and some of the neighboring white communities. The Cook County Sheriff's Department, a large-scale agency with overlapping jurisdiction, provides a number of back-up and technical facilities for the two villages and many of the other small Cook County municipalities.

Lack of participation. The third objection to community control relates to a fear that small communities will be more undemocratic and their officials less responsive to the preferences of citizens than leaders in larger communities. Our findings indicate the opposite in the area studied. As shown on Table 5, village residents were more likely than Chicago residents to agree that citizens can get satisfaction from local officials. Village residents also were more likely to believe that local officials were interested in their neighborhoods. These findings are consistent with our findings that more village residents rated their police as responsive than did residents of the city of Chicago. Residents of villages are somewhat less likely to believe that local elections make a difference. Some might argue that this finding indicates a lack of willingness to participate in local elections and thus demonstrates that small-scale governments are less democratic. On the other hand, this finding may reflect the higher level of homogeneity in the villages and a belief that village government will be responsive regardless of electoral outcome.

Amateur public servants. With regard to the argument concerning professionalism, we did find that, by most standards, the village

TABLE 5
COMPARISON OF CITIZEN EVALUATIONS OF
LOCAL GOVERNMENT (in percentages)

	Independent Communities	Chicago Neighborhoods	Tau for Complete Table
Percentage of respondents agreeing that:			
Citizens can get satisfaction from talking with local public officials	53	33	.21[a]
(n)	(170)	(256)	
Who gets elected to local office makes a difference	66	76	−.15[b]
(n)	(195)	(279)	
Citizens can do something to prevent local corruption	37	39	−.03
(n)	(186)	(264)	
Citizens can influence the way the town is run	60	54	.04
(n)	(178)	(261)	
Local government is interested in their neighborhood	46	31	.13[c]
(n)	(182)	(265)	

a. $p < .001$
b. $p < .005$
c. $p < .05$

police would not be called professional. Because of their limited resources, the villages paid policemen very poorly. The average salary of patrolmen was under $2.00 an hour. Many of the police officers were part-time policemen and held full-time jobs elsewhere. The chief of police in each village was paid approximately two-thirds the salary of an *entering* patrolman in the Chicago Police Department. Officers were poorly trained and equipped. However, despite the obvious handicaps under which the village police pursued their duties, the citizens which they served rated police services as good as or better than similar citizens being served by the highly professional Chicago Police Department. Rates of criminal victimization and quality of police activity followed the same pattern. One would hardly argue that police service in the two villages could not be improved by increased training and higher salaries paid to officers. However, "professional" police without some means of relating to the people they serve do not seem to be more effective than even very "nonprofessional" police who are subject to community control.

Lack of financial resources. The final argument raised by opponents of community control relates to redistribution of resources. The two villages studied are very poor and are forced to rely very heavily on their own limited tax base. One of them has found it necessary to enforce traffic laws aggressively on a national highway within its jurisdiction to augment the funds available to the police department. Limited redistribution does occur as a result of the services provided by the Cook County sheriff, but it does not appear to be very great.

A potentially more important source of redistribution are the revenues which the villages receive under a recently enacted Illinois statute which returns a small percentage of the state income tax to incorporated communities. In personal interviews, both village mayors stressed the importance of the small additional source of revenue from this state source to the operation of the village government. If greater financial support were available from county, state, or federal funds, the villages could improve police pay levels, training of patrolmen and equipment.

At the time of this study, the village governments found themselves in the strange position of not qualifying for most state and federal aid. They are both too small and too poor. Most grants offered by the Law Enforcement Assistance Administration, for example, are directed to police departments of medium and large cities whose budget exceeds a certain minimum level. The *only* federally controlled grant available through LEAA funds to the villages of Phoenix and East Chicago Heights in 1970 would support studies leading toward consolidation of their police forces with neighboring communities. While these police forces have been able to work out cooperative mutual aid arrangements with their white neighbors, consolidation with a neighboring police force is not a politically viable solution. The two villages are physically separate, and joint grant proposals have been refused due to lack of contiguity. Considerable redistribution could be accomplished with only minor changes in state and federal policies to open up opportunities for grants and other funds to small, very poor communities.

However, redistribution of resources, itself, is not sufficient to bring about responsive police services. It appears that considerable resource redistribution is currently occurring within the city of Chicago. More resources are probably being devoted to policing in the black neighborhoods studied than are derived in revenue for such purposes from these areas. Residents of these neighborhoods, however, find police services no better and police somewhat less responsive than do village residents *despite* the much greater difference in resources devoted to policing.

A POLICY RECOMMENDATION

Police effectiveness depends, in part, on police understanding the nature of the community being served and police openness to suggestions, criticism, and complaints. This is particularly true of the kind of police services citizens receive in their own neighborhoods. Community control appears to be one way of enhancing the possibilities of citizen-police communication, thereby increasing both citizen support of police and police responsiveness to citizen preferences.

This policy alternative has already been adopted by those living in independently incorporated, small communities. Perhaps some of the problems of our largest cities might be more effectively dealt with by employing a similar remedy within their jurisdictions. There is no need for the elimination of the large city or its police department. Many police problems are city-, state-, or nationwide. Such police problems require a diversity of relatively large-scale jurisdictions. Moreover, some specialized police services can be provided better by larger-scale units. Communications and records, laboratory facilities and specialized investigatory details may be more economically provided by larger, citywide units.

Locally controlled police agencies could be established within the boundaries of a larger police jurisdiction to serve the particular needs of the large city's diverse neighborhoods. While many observers have assumed that overlap of jurisdictions was in and of itself wasteful and to be avoided, overlap of jurisdictions may be necessary to deal simultaneously with problems of varying scale (V. Ostrom, 1971). The United States has a number of federal agencies existing concurrently with a number of state and local agencies. Just as some police problems are only city- or statewide in scope, others extend only to a single neighborhood. Furthermore, the FBI and other police agencies with broad jurisdictions must rely on local agencies for specific and detailed knowledge of particular areas if their efforts in specific places are to be effective. Those at the top of many large-city police departments may have grave difficulties in getting an accurate picture of local conditions within their districts. Few patrolmen in our largest cities live in the districts which they serve. Many large departments rotate personnel among precincts on a regular and short-term basis. Often the familiarity of officers with a specific area and the residents being served is viewed as something to be avoided. Rotation is frequently justified as a means to avoid corruption. Interestingly, we found that citizens of the small black villages were less apt to indicate that their police took bribes than were black residents of the Chicago neighborhoods where patrolmen are frequently rotated.

The problems in obtaining an adequate knowledge about local situations have lead several large-scale police departments to experiment with local commander systems and other arrangements to decentralize administrative control of neighborhood patrol forces. While this reform may increase direct supervision of patrolmen in the field and may lead to more effective coordination of their efforts within neighborhoods, it may be expected to decrease the responsiveness to citizens of patrolmen serving these areas. Beat commander systems further isolate the patrolmen from formal accountability to the general public through regularly established channels. Local control of the police would involve the establishment of formal structures of accountability to the public being served as well as direct internal supervision of patrolmen on the job.

An effective means of establishing local control of the police in large cities might be to set up neighborhood districts to handle a variety of locally confined public problems. Such units would require some means for public selection of officials and the authority to levy local taxes and establish local ordinances. Such districts, of course, would function within the context of larger city, state, and federal governments and be subject to the laws of the larger jurisdictions. Officials and ordinances in the smaller jurisdictions would be submitted to the scrutiny of the courts at all levels. With regard to the police in particular, citywide regulations could enable a local patrolman to pursue a fleeing suspect anywhere within a broad jurisdiction. The large police department serving the city as a whole would be available for technical assistance and specialized investigation in all areas of the city. Citywide forces could also be utilized to supplement the needs of any local area in times of emergency. Redistribution to the poorer neighborhood districts within the large city could be provided from citywide as well as state and federal sources.

Community control of police may, thus, provide an institutional framework for the effective expression of black citizen demands for impartial police service. Calls for protection and respect cannot be expected to disappear. Black citizens have come to regard equal governmental treatment as a right. Professionalism alone does not appear to provide sufficient controls so that police will be responsive to their needs for protection and respect. Community control places that responsibility on the people themselves and provides them with mechanisms by which to exercise it.

NOTES

1. Harlan Hahn (1971: 385), in reviewing the effects of police professionalism, indicated that "the trend toward professionalization also has been occasioned by an increased sense of expertise and self-esteem among police officers and by a corresponding deprecation of external influences. The intense opposition of police forces to proposals for civilian review boards during the 1960s, therefore, appeared to be consistent with some of the major tenets of professionalism. . . . Basic to the impetus for professionalization was a prevalent belief among police officers that they should be freed from outside interference and that they should be allowed to pursue their central mission—the prevention and control of crime—using their own methods and judged by their own criteria."

2. See Ostrom (1972) for a description of the theoretical structure implicit in the metropolitan reform movement. See Piven and Cloward (1967) for a discussion of the effect of metropolitanwide governmental units for black citizens.

3. See Marshall (1971) for a review of the findings from several studies on the amount and type of participation in poverty programs during the late 1960s. She argues that those among the poor who have participated in the programs studied "exhibit increases in political education, information about how the system works, and in organizational skills and feelings of political efficacy" (Marshall, 1971: 473).

4. Anthony Orum (1966) argues that lower-class blacks are more likely to belong to organizations than lower-class whites and that voting rates for blacks of all classes have risen since 1952. Marvin Olson (1970) has also found that, when socioeconomic status is controlled, the rates of participation of black citizens are higher than those for white citizens and the rates of participation have been increasing between 1957 and 1968.

5. In a survey of police serving predominantly black areas in thirteen American cities, Groves and Rossi found black policemen to perceive the people in the area they are serving to be less hostile than did white policemen serving the same areas (Groves and Rossi, 1970: 732). They conclude that "most of what a white policeman anticipates from black citizens is determined by factors other than the actual level of hostility in a city. A good deal of the perceived antagonism appears to be a projection of the policeman's own fears and prejudices—although a high level of acquaintance with community residents, leaders and other individuals tends somewhat to mitigate a highly prejudiced policeman's projection of his hostility" (Groves and Rossi, 1970: 741).

6. James Q. Wilson (1968a: 190) describes the officers of a highly professionalized police department serving "Western City" in the following manner: "The city in which they

now serve has a particular meaning for only a very few. Many live outside it in the suburbs and know the city's neighborhoods almost solely from their poilce work. Since there are no precinct stations but only radio car routes, and since these are frequently changed, there is little opportunity to build up an intimate familiarity, much less an identification with any neighborhood. The Western City police are, in a real sense, an army of occupation organized along paramilitary lines."

7. After this study was completed, the informal cooperation between villages was formalized through the establishment of the Suburban Mayors' Planning Group, involving the seven communities of Harvey, Markham, East Chicago Heights, Dixmoor, Chicago Heights, Phoenix, and Robbins. The first project undertaken by this group was a joint proposal by the seven communities to the Illinois Law Enforcement Association to establish a cooperative crime prevention program. (The proposal was, incidentally, turned down after extended negotiations because none of the communities involved had sufficient financial resources to meet the matching requirements of the program.)

REFERENCES

ABERBACH, J. D. and J. L. WALKER (1970a) "The attitudes of blacks and whites toward city services: implications for public policy," in J. P. Crecine (ed.) Financing the Metropolis: Public Policy in Urban Economies. Beverly Hills: Sage Pubns.

——— (1970b) "Political trust and racial ideology." Amer. Pol. Sci. Rev. 64: 1199-1219.

ALTSHULER, A. A. (1970) Community Control: The Black Demand for Participation in Large American Cities. New York: Pegasus.

ARNSTEIN, S. R. (1969) "A ladder of citizen participation." J. of Amer. Institute of Planners 35: 216-232.

BABCOCK, R. F. and F. BOSSELMAN (1967) "Citizen participation: a suburban suggestion for the central city." J. of Law and Contemporary Problems 32: 220-231.

BARON, H. M. (1968) "Black powerlessness in Chicago." Trans-action 6: 27-33.

BAYLEY, D. H. and H. MENDELSOHN (1969) Minorities and the Police. New York: Free Press.

BERGER, C. J. (1968) "Law, justice and the poor," in R. H. Connery (ed.) Urban Riots: Violence and Social Change. New York: Random House.

BISH, R. L. (1971) The Public Economy of Metropolitan Areas. Chicago: Markham.

——— and R. WARREN (1972) "Scale and monopoly problems in urban government services." Urban Affairs Q. 8: 97-122.

CAMPBELL, A. and H. SCHUMAN (1968) "Racial attitudes in fifteen American cities," in Supplemental Studies for the National Advisory Commission on Civil Disorders. Washington, D.C.: Government Printing Office.

Center for Governmental Studies (1970) Public Administration and Neighborhood Control. Washington, D.C.

Chicago Sun Times (1972) "The ranking of 200 suburbs in status list." (August 15): 12.

CLARK, T. M. (1970) "On decentralization." Polity 2: 508-514.

COSTNER, H. L., R. O. HAWKINS, P. E. SMITH, and G. F. WHITE, III (1970) "Crime, the public and the police." University of Washington. (mimeo)

DE VISE, P. (1967) Chicago's Widening Color Gap. Chicago: Interuniversity Social Research Committee.

ENNIS, P. H. (1967) Criminal Victimization in the United States: A Report of a National Survey. Washington, D.C.: Government Printing Office.

ERIE, S. P., J. J. KIRLIN, and F. E. RABINOVITZ (1972) "Can something be done? Propositions on the performance of metropolitan institutions," in L. Wingo (ed.) Reform of Metropolitan Governments. Washington, D.C.: Resources for the Future.

FEAGIN, J. R. (1970) "Home defense and the police: black and white perspectives." Amer. Behavioral Scientist 13: 797-814.

FERRY, W. H. (1968) "The case for a new federalism." Saturday Rev. 30 (June 15): 15.

FOGELSON, R. M. (1968) "From resentment to confrontation: the police, the Negroes and the outbreak of the nineteen-sixties riots." Pol. Sci. Q. 83: 217-247.

GITTELL, M. (1968) "Community control of education," pp. 63-75 in R. A. Connery (ed.) Urban Riots: Violence and Social Change. New York: Random House.

GROVES, W. E. and P. H. ROSSI (1970) "Police perceptions of a hostile ghetto: realism or projections." Amer. Behavioral Scientist 13 (May/June): 727-744.

HAHN, H. (1971) "Local variations in urban law enforcement," in P. Orleans and W. R. Ellis (ed.) Race, Change and Urban Society. Beverly Hills: Sage Pubns.

--- and J. R. FEAGIN (1970) "Riot precipitating police practices: attitudes in urban ghettos." Phylon 31: 183-193.

HALLMAN, H. W. (1971) Administrative Decentralization and Citizen Control. Washington, D.C.: Center for Governmental Studies.

Illinois Regional Medical Program (1971) Chicago Regional Hospital Study. Chicago: Chicago Association of Commerce and Industry.

IsHAK, S. T. (1972) "Consumers' perceptions of police performance: consolidation vs. deconcentration; the case of Grand Rapids, Michigan, metropolitan area." Ph.D. dissertation. Indiana University (Bloomington).

ITZKOFF, S. W. (1969) "Decentralization: dialectic and dilemma." Educ. Forum 34: 63-69.

JACKSON, J. (1970) "On the case." Chicago Daily Defender 4 (Weekend Edition for August): 22-28.

JACOB, H. (1970) "Black and white perceptions of justice in the city." Law and Society Rev. 6: 69-90.

JACOBS, P. (1966) "The Los Angeles police." Atlantic Monthly 218 (December): 95-101.

KOTLER, M. et al. (1968) "Table talk/finding the city." Center Magazine 1 (May): 14-17.

KRISTOL, I. (1968) "Decentralization for what?" Public Interest 11 (Spring): 17-25.

LEVY, B. (1968) "Cops in the ghetto: a problem of the police system," pp. 347-358 in L. H. Masotti and D. R. Bowen (eds.) Riots and Rebellion: Civil Violence in the Urban Community. Beverly Hills: Sage Pubns.

LIEBERSON, A. and A. R. SILVERMAN (1965) "The precipitants and underlying conditions of race riots." Amer. Soc. Rev. 31 (December): 887-898.

MARSHALL, D. R. (1971) "Public participation and the politics of poverty," pp. 451-483 in P. Orleans and W. R. Ellis (eds.) Race, Change and Urban Society. Beverly Hills: Sage Pubns.

MAYER, A. (1971) "A new level of local government is struggling to be born." City (March/April): 60-64.

MELTZER, J. (1968) "A new look at the urban revolt." J. of Amer. Institute of Planners 34 (December): 255-259.

MITCHELL, M. (1973) "Racial identification and public order in black communities." Indiana University Department of Political Science Studies in Political Theory and Policy Analysis.

MOGULOF, M. (1969) "Coalition to adversary: citizen participation in three federal programs." J. of Amer. Institute of Planners 35: 225-232.

National Advisory Commission on Civil Disorders (1968) Report. Washington, D.C.: Government Printing Office.

NIE, N. H., G. B. POWELL, Jr., and K. PREWITT (1969) "Social structure and political participation: developmental relationships." Amer. Pol. Sci. Rev. 62: 361-378.

OLSON, M. E. (1970) "Social and political participation of blacks." Amer. Soc. Rev. 35: 682-697.

ORBELL, J. M. and T. UNO (1972) "A theory of neighborhood problem solving: political actions vs. residential mobility." Amer. Pol. Sci. Rev. 66: 471-489.

ORUM, A. M. (1966) "A reappraisal of the social and political participation of Negroes." Amer. J. of Sociology 72: 32-46.

OSTROM, E. (1972) "Metropolitan reform: propositions derived from two traditions." Social Sci. Q. 53 (December): 474-493.

––– (1971) "Institutional arrangements and the measurement of policy consequences in urban areas." Urban Affairs Q. 6: 447-475.

––– and R. B. PARKS (1973) "Suburban police departments: too many and too small?" pp. 367-402 in L. H. Masotti and J. K. Hadden (eds.) The Urbanization of the Suburbs, UAAR VII. Beverly Hills: Sage Pubns.

OSTROM, E. and G. WHITAKER (1973) "Does local community control of police make a difference? Some preliminary findings." Amer. J. of Pol. Sci. 17 (February): 48-76.

OSTROM, E., R. PARKS, and G. WHITAKER (1973) "Do we really want to consolidate urban police forces? A reexamination of some old assertions." Public Administration Rev. 33 (September/October): 423-432.

OSTROM, E., W. BAUGH, R. GUARASCI, R. PARKS, and G. WHITAKER (1973) "Community organization and the provision of police services." Sage Prof. Paper in Administrative and Policy Studies 03-001.

OSTROM, V. (1973) The Intellectual Crises in American Public Administration. University: Univ. of Alabama Press.

––– (1971) The Political Theory of a Compound Republic. Blacksburg, Va.: Center for the Study of Public Choice at Virginia Polytechnic Institute.

––– C. M. TIEBOUT, and R. WARREN (1961) "The organization of government in metropolitan areas: a theoretical inquiry." Amer. Pol. Sci. Rev. 55: 831-842.

PARENTI, M. (1970) "Power and pluralism: a view from the bottom." J. of Politics 32: 501-530.

PERLMUTTER, N. (1968) "We don't help blacks by hurting whites." New York Times Magazine (October 6): 31.

PIVEN, F. F. and R. A. CLOWARD (1967) "Black control of the cities: heading it off by metropolitan government." New Republic (September 30): 19-21; (October 7): 15-19.

PRESS, C. (1963) "The cities within a great city: a decentralist approach to centralization." Centennial Rev. 7: 113-130.

PREWITT, K. and H. EULAU (1969) "Political matrix and political representation: prolegomenon to a new departure from an old problem." Amer. Pol. Sci. Rev. 62: 427-441.

PREZEWORSKI, A. and H. TEUNE (1970) The Logic of Comparative Social Inquiry. New York: John Wiley.

ROSSI, P. H. (1963) "The middle-sized American city at mid-century." Library Q. 33: 3-13.

––– and R. A. BERK (1970) "Local political leadership and popular discontent in the ghetto." Annals of Amer. Academy of Pol. and Social Sci. 391: 111-127.

RUBENSTEIN, R. E. (1970) Rebels in Eden. Boston: Little, Brown.

SHALALA, D. E. (1971) "Neighborhood governments: rationale, functions, size and governmental framework." Prepared for the National Consultation on Neighborhood Government, Institute of Human Relations, New York.

SKOLNICK, J. (1967) Justice Without Trial: Law Enforcement in Democratic Society. New York: John Wiley.

SPEAR, A. H. (1967) Black Chicago: The Making of a Negro Ghetto, 1890-1920. Chicago: Univ. of Chicago Press.

TAUBER, K. E. (1968) "The problem of residential segregation," in R. H. Connery (ed.) Urban Riots: Violence and Social Change. New York: Random House.

TenHOUTEN, W., J. STERN, and D. TenHOUTEN (1971) "Political leadership in poor communities: applications of two sampling methodologies," pp. 215-254 in P. Orleans and W. R. Ellis (eds.) Race, Change and Urban Society. Beverly Hills: Sage Pubns.

TIEBOUT, C. M. (1956) "The pure theory of local expenditure." J. of Pol. Economy 64: 416-424.

WARREN, R. (1966) Government in Metropolitan Regions. Davis: University of California, Davis, Institute of Governmental Affairs.

WHITAKER, G. P. (1971) "Urban police forces: size and scale in relation to service." Ph.D. dissertation. Indiana University.

WILSON, J. Q. (1968a) "The police and the delinquent in two cities," pp. 173-196 in J. Q. Wilson (ed.) City Politics and Public Policy. New York: John Wiley.

——— (1968b) Varieties of Police Behavior. Boston: Harvard Univ. Press.

——— (1963) "The police and their problems: a theory." Public Policy 12: 189-216.

ZIMMERMAN, J. F. (1970) "Metropolitan reform in the U.S.: an overview." Public Admin. Rev. 30: 531-543.

Part IV

REORGANIZATION

12

Organizations of the Future

CHRIS ARGYRIS

□ PUBLIC AND PRIVATE ORGANIZATIONS are coming under increasing attack. The consumer public is becoming alarmed at the costs to maintain organizations and the resulting inadequate services and low-quality products. Whereas a decade ago, the nation as a whole expressed a high degree of confidence in the effectiveness and credibility of organizations, during the last three years the confidence has markedly decreased (Argyris, 1973).

There has resulted an understandable clamor to change organizations in order to get them to become more effective. Before we can change organizations effectively, we need to know why they are becoming less dependable.

In examining social life, the first observation that can be made is that organizations of all types and varieties are in trouble: business firms, trade unions, churches, schools, local governments, governmental business, and so on. If organizational deterioration is occurring across the range of organizations, then the causes must be beyond the specific goals of organizations. Thus, if an overemphasis on profits is an important cause, then why should churches and schools be in trouble?

Perhaps it is governmental pressure (local, state, or national) that may be a cause. But this reason seems rather weak, since there are many organizations having difficulties that are not significantly interfered with by government. Indeed, in some cases, such as local community government, the major governmental intervention is probably to pay the bill.

It is also difficult to imagine that the causal factors for organizational deterioration in such a wide variety of organizations lie in their environment. The great variance in environments would imply that special factors affect individual organizations. Such an explanation, if valid, will have little action value because all that would be asserted is that there are no general patterns or basic causes.

SOME CAUSES OF ORGANIZATIONAL DETERIORATION

There are some factors that are common to most public and private organizations, and they are related to the way they are designed. Organizations, for the most part, tend to be designed on the basis of similar managerial theories. Organizations follow the imperatives of traditional scientific managerial and bureaucratic theories. They tend to have some variant of the pyramidal structure; they tend to define work in specialized terms, and they tend to use managerial controls (budgets, reward and penalty systems, production or service records) that follow the theory implicit in the pyramidal structure.

There is a growing body of literature that is exploring the causes of organizational deterioration. Basically, these scholars tend to see some of the causes (but not the only causes) as stemming from the oversimplified view of man assumed by traditional organizational designs and the lack of attention paid to the way people actually deal with one another (interpersonal relations) or in groups (group dynamics) while making decisions, plus the realities of intergroup rivalries and organizational norms that support these activities (Argyris, 1964; Bennis, 1966; Golembiewski, 1969, 1967a; Herzberg, 1966; Katz and Kahn, 1966; Likert, 1967; McGregor, 1960; Marrow et al., 1967; Maslow, 1970; Tannenbaum, 1968; Whyte, 1969).

Recently the writer has attempted to summarize some of the findings of this group regarding human and organizational problems at the lower and upper levels (Argyris, 1971).

CHARACTERISTICS OF THE
LOWER-LEVEL WORLD

(1) Work is highly specialized and fractionalized; it is broken down to the simplest possible motions. It is assumed that the easier the work, (a) the greater the productivity, (b) the shorter the training time needed, (c) the greater the flexibility for interchangeability of the worker, and (d) the greater the satisfaction of the employee because the less the frustration or responsibility that he will tend to experience.

(2) Responsibility for planning the work, defining production rates, and maintaining control over speed is placed in the hands of management and not in the hands of those actually producing.

(3) Responsibility for hiring employees, issuing orders, changing work, shifting employees, evaluating performance, and defining and disbursing rewards and penalties is vested in top management.

The exact degree to which these assumptions or premises are followed varies with the organization. However, we may hypothesize that, to the extent that organizations attempt to follow the consequences of these premises, they will tend to create a work world for the lower-level employees in which the following conditions will be true:

(1) Few of their abilities will be used. Those abilities that will be used will tend to be the ones that provide more limited potential (in our culture) for psychological success (such as finger dexterity and other motor abilities). The abilities more central to self-expression and psychological success, such as the cognitive (intellectual) and the interpersonal abilities, will tend to be utilized minimally.

(2) The employees will tend to experience a sense of dependence and submissiveness toward their superior and feel that he has little control over the crucial decisions about his organizational life.

(3) The worker will tend to experience a decreasing sense of responsibility and self-control because he knows that someone else will tell him what to do, how well and when he ought to do it, how much he ought to perform, whether or not he has performed adequately, and so on.

(4) The more rigidity, specialization, tight control, and directive leadership the worker experiences, the more he will tend to create antagonistic adaptive activities. However, the employees are limited in creating adaptive activities that are antagonistic to the system, for

they are subject to dismissal or reprimand. Consequently, they increase their risks as the adaptive activities become more antagonistic. These activities could be resolved by institutionalizing some of them, by developing a new organization that has equal power—a trade union, for example. Or employees can resolve the issue by withdrawing psychologically so that the frustration and stress are not too incapacitating. The exact solution will probably vary from one organization to another and within each organization at different stages of development.

(5) The probable forms of the informal activities are:

- absenteeism
- turnover
- aggression toward the top
- apathy, indifference, gold-bricking
- trade unions
- increasing demand to relate compensation to the degree of dissatisfaction, tension, and stress experienced on the job
- market orientation
- alienation

The existence of these informal activities tends to upset and frustrate management. In most cases, management tends to respond by increasing its control over the people through tighter administrative procedures, stricter edicts, and a liberal sprinkling of personal advantages (human relations programs, courses in economics) to make the employees accept their situation. Unfortunately, the actual impact is almost the opposite. The workers tend to feel even more frustrated, psychologically failing, and in conflict. Thus, there is a feedback to close the loop and develop a self-maintaining system.

(6) Administrative controls are a critical component of modern systems. The underlying assumptions of these managerial controls are that (a) management (through some staff experts) plans and controls human effort; (b) the control of human effort is manageable by logic and systematically developed by relatively quantitative techniques; and (c) this latter is achieved by the use of the principles of exception, which means behavior is monitored (on a sample basis) and investigated when it deviates from the plan.

These assumptions tend to create control systems which provide for little influence by the people whose behavior is being controlled (except if they cheat and distort the information they provide). Moreover, the quantitative techniques used rarely mirror the reality experienced by the employee; indeed, they are not designed to do so. A good managerial control provides information that management needs; it does not show what the employees experience. Finally, control systems tend to create a fear of being trapped by logic and the principle of exception.

In short, managerial controls tend to create group rivalries, force groups to think of their own and not the other's problems, reward an overall point of view rarely, and place groups in win-lose situations in which they are competing with each other for the scarce resources.

(7) The underlying philosophy of most managements tends to be to appoint leaders whose styles are similar to, and consonant with, the impact of the pyramidal structure and the administrative controls. The predominant pattern tends to be what Likert calls "production-centered," McGregor calls "theory X," and Argyris calls "directive leadership."

Leaders are urged to be strong, to drive, sell, pressure, and coerce employees to increase their productivity and loyalty. They are charged with getting the facts and controlling the problem-solving, decision-making, and implementation processes. They are also charged with evaluating the performance and the individuals and groups. Research has shown that such leadership results in a decrease of the individual's experience of psychological success and essentiality, and of the probability that he will give valid information to those above or to his peers. Such leadership also tends to increase the frequency of destructive win-lose interdepartmental rivalries.

Moreover, the members become minimally involved in the health of the system (that is the boss's responsibility) and rarely take the initiative to explore and correct their group processes. If the group is effective, it will primarily be the responsibility of the leader.

Finally, systems tend to produce norms, policies, and practices that create binds for the lower-level employees. Most of these binds have been produced by the so-called human relations fads. If one examines the human relations programs designed by management to better human relations, one sees that they produce some contradictory messages. On the one hand, management tells its people they are important, to be trusted, and to be respected; yet management designs programs in which it takes the responsibility to rejuvenate, motivate, and inspire the workers. Management says it is ready to make changes in order to increase the effectiveness of the systems, but until recently the major changes it has been willing to make are changes in employee personnel. Little or no significant changes have been made until recently (and this in a very few firms) in the way jobs are designed, managerial controls are used, leadership is used, and rewards and penalties are distributed. Another frequently documented finding shows that managers tend to emphasize the importance of participation and democracy, yet their actual behavior rarely permits true participation; rather, it tends to encourage a "pseudo-participation."

CHARACTERISTICS OF THE
MANAGERIAL WORLD

(1) The higher up they go in the organizational hierarchy, the more individuals are able, if they wish, to alter the system. They are not as bound by the structure, technology, and control systems as are the members at the lower level. Assuming as we have from the outset that the members have the necessary technical competence, we see that the biggest barriers to change at the upper levels are the interpersonal relationships, the group and intergroup relationships, and the system's norms. These barriers to effectiveness are rarely potent when the system is dealing with routine, programmed, and nonthreatening information. The barriers become especially difficult to identify and overcome when the system is dealing with innovative, nonprogrammed, and threatening information.

(2) Why do administrators tend to create interpersonal, group, and intergroup relations that inhibit system competence precisely when it is most needed? Part of the answer to that question is that they are programmed from an early age with interpersonal incompetence (Argyris, 1968). The cause of the incompetence lies in the fact that most people who aspire toward positions of power and are educated in our schools are taught to internalize the values of the engineering-economic-technological world which dominate our lives. The writer's research indicates that there are three such identifiable values. They are:

(a) The significant human relations are the ones which have to do

with achieving the organization's objective. In studies of over 265 different types and sizes of meetings, the indications are that executives almost always tend to focus their behavior on getting the job done. In literally thousands of units of behavior, almost none show that men spend some time in analyzing and maintaining their group's effectiveness. This is true even though in many meetings the group's effectiveness was bogged down and the objectives were not being reached because of interpersonal factors. When the executives are interviewed and asked why they did not spend some time in examining the group operations or processes, they reply that they are there to get a job done. They add, "If the group isn't effective, it is up to the leader to get it back on the track by directing it."

(b) Cognitive rationality is to be emphasized; feelings and emotions are to be played down. This value influences executives to see cognitive, intellectual discussions as relevant, good, workable, and so on. Emotional and interpersonal discussions tend to be viewed as irrelevant, immature, and not workable.

As a result, when emotions and interpersonal variables become blocks to group effectiveness, all the executives report feeling they should *not* deal with them. For example, in the event of an emotional disagreement, they would tell the members to get back to facts or to keep personalities out of it.

(c) Human relationships are most effectively influenced through unilateral direction, coercion, and control, as well as by rewards and penalties that sanction all three characteristics. This third value of direction and control is implicit in the chain of command and in the elaborate managerial controls that have been developed within organizations.

(3) The impact of these values upon the interpersonal relationships within an organization is to create a pattern of behavior which may be identified as pattern A. The development of pattern A may be summarized briefly as follows.

To the extent that individuals dedicate themselves to the value of rationality and getting the job done, they will tend to be aware of and emphasize the rational, intellective aspects of the interactions that exist in an organization and to suppress the interpersonal and emotional aspects, especially those that do not seem to be relevant to achieving the task. For example, one frequently hears in organizations, "Let's keep feelings out of the discussion," or, "Look here, our task today is to achieve objective X and not to get emotional."

As the interpersonal and emotional aspects of behavior become suppressed, we may hypothesize that an organizational norm will tend to arise that coerces individuals to hide their feelings. Their interpersonal difficulties will either be suppressed or disguised and brought up as rational, technical, or intellectual problems. In short, receiving or giving feedback about interpersonal relationships will tend to be suppressed.

Under these conditions, we may hypothesize that individuals will find it difficult to develop competence in dealing with feelings and interpersonal relationships. In a world in which the expression of feelings is not permitted, one may hypothesize that the individuals will build personal and organizational defenses to help them suppress their own feelings or inhibit others in their attempts to express their feelings. If feelings are suppressed, the tendency will be for the individual not to permit himself or others to own their feelings. For example, the individual may say about himself, "No, I didn't mean that," or, "Let me start over again. I'm confusing the facts." Equally possible, one individual may say to another, "No, you shouldn't feel that way," or, "That's not effective executive behavior," or "Let's act like mature people and keep feelings out of this."

(4) Another way to prevent individuals from violating the organizational values of rationality and from embarrassing one another is to block out, refuse to consider (consciously or unconsciously) ideas and values which, if explored, could expose suppressed feelings. Such a defensive reaction in the organization may eventually lead to a barrenness of intellectual ideas as well as values. The participants will tend to limit themselves to those ideas and values that are not threatening so they will not violate organizational norms. The individuals in the organization will tend to decrease their capacity to be open to new ideas and values. As the degree of openness decreases, the capacity to experiment will tend to decrease and the fear of taking risks will tend to increase. As the fear of taking risks increases, the probability of experimentation is decreased and the range or scope of openness is decreased, which in turn decreases risks. We have a closed circuit that could be an important cause of the loss of vitality in an organization.

To summarize, to the extent that participants are dedicated to the values implicit in the formal organization, they will tend to create a social system in which the following will tend to *decrease*.

- Receiving and giving nonevaluative feedback.
- Accepting ownership and permitting others to accept ownership of ideas, feelings, and values.
- Openness to new ideas, feelings, and values.
- Experimentation, risk-taking, and new ideas and values.

If these characteristics do decrease, we hypothesize that the members of the system will tend *not* to be aware of the interpersonal impact upon others. If individuals are in social systems in which they are unable to predict accurately their personal impact upon others, and the impact of others upon them, they may begin to feel confused. "Why are people behaving that way toward me?" "Why do they interpret me incorrectly?" Since such questions are not sanctioned, much less answered, in a rationally dominated system, the confusion tends to turn to frustration and feelings of failure regarding interpersonal relations. In an attempt to maintain their sense of esteem, the members may react by questioning the honesty and genuineness of the interpersonal behavior of their fellow workers. Simultaneously, they may place an even greater emphasis upon the rational and technical interactions in which they are probably experiencing a greater degree of success. The increased emphasis upon rationality will act to suppress the feelings even more; this, in turn, will decrease the probability that the questions of confusion and the mistrust of self and others will be explored.

As interpersonal mistrust increases, and as the capacity (individual and organizational) to cope with this mistrust decreases, the member may tend to adapt by playing it safe. The predisposition will be to say those things that cannot be misunderstood and to discuss those issues for which there exist clear organizational values and sanctions. The desire to say the right thing should be especially strong toward one's superiors; toward one's peers, with whom one is competing; and toward one's subordinates, who may bypass their superiors. As a result, conformity begins to develop within an organization. Along with conformity, the interpersonal relationships will tend to be characterized by "conditional acceptance" (to use a Rogerian concept), and the members will tend to feel accepted if they behave in accordance with certain organizational specifications. Because of the existence of mistrust, conformity, and dependence, we may hypothesize that the members' commitment to the organization will tend to be external as far as interpersonal activities are concerned. By external commitment, I mean that the source of commitment to work for any given individual lies in the power, rewards, and penalties that individual may use as influence. Internal commitment exists when the motive for a particular behavior resides within (for example, self-realization). A certain amount of internal commitment restricted to rational activities may be possible in this system if the rational, intellective aspects of the job are consonant with the individual's abilities and expressed needs.

External commitment will tend to reinforce the conformity with conditional acceptance of—and especially, dependence upon—the leader. The subordinates will tend to look for cues from the leader and will be willing to be influenced and guided by him. In fact, they may develop great skill in inducing the leader to define the problems, the range of alternatives, and so on. The subordinates will tend to operate within limits that they know to be safe. As the dependence increases, the need for the subordinates to know where they stand will also tend to increase.

Thus, interpersonal mistrust, conformity, conditional acceptance, external commitment, and dependence tend to be outputs of decreasing interpersonal competence. Each of these attitudes feeds back to reinforce itself. All, in turn, feed back upon interpersonal competence to decrease it further or to reinforce it at its existing level.

All these factors tend to act to lower the effectiveness of the decision-making process. For example, in some of the situations, executives with mathematical and engineering backgrounds were observed dealing with highly technical issues, and they developed strong emotional attachments to these issues. During discussions held to resolve technical, rational issues, the emotional involvements tended to block understanding. Since the men did not tend to deal with emotions, the inhibiting effects were never explored. Since the arguments were attempts by people to defend themselves or to attach blame to others, there was a tendency for the rationality of the arguments to be weak. This, in turn, troubled the receiver of the argument, who tended to attack obvious rational flaws immediately. The attack tended to increase the degree of threat experienced by the first person, and he became even more defensive. Similar impacts upon rational decision-making were also discovered in areas such as investment decisions, purchasing policies, quality control standards, product design, and marketing planning.

Another solution is to fill decision-making activities with subordinates' gamesmanship. Some examples are:

(1) Before you give any bad news, give good news. Especially emphasize the capacity of the department to work hard and to rebound from a failure.

(2) Play down the impact of a failure by emphasizing how close you came to achieving the target or how soon the target can be reached. If neither seems reasonable, emphasize how difficult it is to define such targets, and point out that because the state of the art is so primitive, the original commitment was not a wise one.

(3) In a meeting with the president, it is unfair to take advantage of another department that is in trouble, even if it is a natural enemy. The sporting thing to do is to say something nice about the other department and offer to help it in any way possible. (The offer is usually not made in concrete form, nor does the department in difficulty respond with the famous phrase, "What did you have in mind?")

(4) If one department is competing with other departments for scarce resources and is losing, it should polarize the issues and insist that a meeting be held with the president. If the representatives from the department lose at that meeting, they can return to their group, place the responsibility for the loss on the president, and thereby reduce the probability of being viewed as losers or, worse yet, as traitors.

These games do not go completely undetected by those on the receiving end. Although they are rarely dealt with openly (because that would violate the values of the system), the top executives tend to develop their own games. Several examples that have been found frequently are (a) the constant alteration of organizational positions and charts, and keeping the most up-to-date versions semi-confidential; (b) shifting top executives without clearly communicating the real reasons for the move; and (c) developing new departments with production goals that overlap and compete with the goals of already-existing departments.

The rationale usually given for these practices is: "If you tell them everything, all they do is worry, and we get a flood of rumors"; "The changes do not *really* affect them"; and "It will only cut in on their schedule and interrupt their productivity." The subordinates respond, in turn, by creating their own explanations, such as: "They must be changing things because they are not happy with the way things are going; the unhappiness is so strong they do not tell us."

One consequence of gamesmanship and mistrust is for the executives to place greater emphasis on the use of rationality, direction, control, rewards, and penalties. In practice, this tends to mean that they begin to check on other people's work, not only to see if it is done, but also to see how it was accomplished. They also operate through detailed questioning about issues and problems that may exist at levels lower than that of the man being questioned, but for which he is responsible; for example, they may ask a personnel vice president for the capacity of a parking lot in a plant away from the home office.

The result of such action on the part of the superior is to create defensiveness in the subordinate. The subordinate now finds himself constantly checking on all details so that he will not be caught by the superior. However, the activity of the organization is not carried forward with such behavior. The result is simply one of making the subordinate (and usually *his* subordinate) more defensive. Their response is to build up organizational defenses to protect themselves. For example, in one case where executives were managing by detail, the subordinates created the "JIC" file which stands for "just in case" some superior asks. This file was kept up to date by several lower-level managers who were full time and by countless other people who were part time. The JIC file is an organizational defense against threat experienced by individuals at various levels.

Organizational defenses may therefore be developed in an organization to protect various individuals and groups. These defenses can be used to needle people, which tends to occur when the rational methods seem to fail. But since the use of feelings is deviant behavior, and since the superiors or subordinates do not have much experience in their use, the tendency may be to have feelings overdetermined—that is, feelings that tend to be much stronger than the situation warrants. Their overdeterminedness is compounded by the fact that subordinates do not tend to be accustomed to dealing with feelings.

Executives may speak of "needling" the boys, once in a while "raising hell to keep them on their toes," and so on. If these conditions continue, it is not long before the "hot" decisions of the organization are administered by using emotions. This is commonly known in industry as management by crisis.

As management by crisis increases, the subordinate's defensive reaction to the crisis will tend to increase. One way for him to protect himself is to make certain his area of responsibility is administered competently and that no other peer executive "throws a dead cat into his yard." The subordinate's predisposition will tend to be centered toward the interests of his department. As the department-centeredness increases, the interdepartmental rivalries will tend to increase. All these decrease the organization's flexibility for change and the cooperation among departments. In turn, the top management will tend to adapt to this decrease by increasing directives, which again begins to recentralize the organization.

The external commitment, conformity, interpersonal mistrust, ineffective decision-making, management by crisis, and organizational rigidity will tend to feed back to reinforce each other and to decrease interpersonal competence. Moreover, each will feed upon all the others to reinforce itself. We would conclude that, under these conditions, the tendency will be to increase the energy required to produce the same input, or someday it may decrease the output, even though the input remains constant. When this state of affairs occurs, it may be said the organization's effectiveness has begun to deteriorate.

ORGANIZATIONAL MAINTENANCE PROCESSES ARE
SELF-SEALING AND NONLEARNING

The analysis above implies that living organizations develop multilevel self-sealing processes. For example, at the lower level, the management places subordinates in dependent, submissive relationships where they utilize few of their abilities. The employees resist or fight back. Both responses are experienced, by management, as disloyal acts or signs that more direct controls are needed, backed up by increasingly belligerent or paternalistic power on the part of the top administrators. These managerial responses, in turn, trigger off further employee resistances which, in turn, trigger more controls. At the upper levels, the closedness, lack of risk-taking, and conformity lead to intergroup rivalries and win-lose dynamics which tend to make individuals, groups, or departments even more defensive, thereby increasing the closedness, lack of risk-taking, and conformity.

A second important characteristic is that the self-sealing processes are linked to many different levels of organizational units (individual, group, intergroup, and organizational). The existence of self-sealing processes that are maintained by multilevel factors make it appear, at worst, that no change is possible or, at best, that change will be difficult and will require a large team working on many different levels.

Processes that are self-sealing and dysfunctional are kept clandestine and covert in order to protect those who created them. Typically, these processes therefore are difficult to identify. Also, they are rarely made public and therefore are rarely available to public examination and test. Tests that may be made of the self-sealing qualities are usually conducted covertly, lest the system become even more defensive.

Organizations composed of multilevel, self-sealing, nonpublicly examinable or testable processes will tend to manifest a low probability for learning and experimenting. The low level of learning will feed back to make the organization seem even more difficult to change. Under these conditions, the participants may come to have little confidence in their own organizations. The lack of credibility and sense of resignation experienced by the outsiders are now duplicated and confirmed by the insider.

These qualities of self-sealing, nonlearning, low risk-taking, and inability to unfreeze causes organizational entropy and makes organizational change for increased ineffectiveness seem impossible. One might understandably conclude that our society would be doing well to arrest the processes of organizational deterioration, not to mention reversing or eliminating them.

STRATEGIES USED TO CHANGE PUBLIC ORGANIZATIONS

(1) The most frequent strategy to change organizations is to hire consulting firms to diagnose the system and make recommendations for change. Typically, these reports make specific recommendations regarding changes in structure, in policies and practices, and perhaps in the hiring of new personnel. Incompetence at all levels, if found, is usually handled more discreetly, especially if it is found at the upper levels.

(2) The next most frequently used strategy is to combine the outside consulting firm's efforts with a group of strategically placed insiders who can see to it that the recommendations are carried out. The basic assumption of this approach is that change can be made more effective if more power and control can be brought to bear upon the target areas through the use of insiders. The insiders —usually identified as whiz-kids—are given power, armed with new management information systems, and may be fortified with the feelings that they are leading a holy war against bureaucratic dry rot.

(3) Another important but less frequently used strategy is to create a new institution that will yield power over other public institutions by having financial resources to allocate to these institutions. These organizations are usually populated with people who are both bright and committed to changing things, yet are not infected by the existing organizational entropy. An example was Community Progress, Incorporated, of New Haven. This organization was created by foundation funds and given the charge to unfreeze, change, and reintegrate the human services within the city of New Haven, Connecticut.

(4) Finally, there is the mechanism of community control, or power to the people. The assumption is that the people being served—especially if .they are a minority, poor, dispossessed—know what changes are necessary. Give them the power to bring about changes that fit their needs; after all, they are the clients the service organizations are supposed to aid.

The track record of these intervention strategies is not very encouraging. Consulting firm reports find their way into the system only to be swallowed up and finally appear in locked files, drawers, or, for the systems that do not save souvenirs, in file thirteen. The whiz-kids may produce some changes but, as in the case of New York City, they manage to produce even more hostility, aggression, and mistrust between themselves and the "oldtimers." The same seems to be the case for Community Progress, Incorporated. It was soon seen as the equivalent of an independent czar to change New Haven—a perception which was accurate—and therefore feared, mistrusted, and resisted. Finally, we now know that if power to the people works, it probably works in the early stages of change when things are so bad that any change is usually for the better. However, give the people enough time, financial and other resources, and they will transform them into evidence that they cannot design systems that are more responsive and effective than those designed by the previously powerful.

Why?

(1) The concepts of change implicit in all these strategies are similar to the concept of management that created the organizational deterioration in the first place. All of them assume that effective change depends upon giving some people more power and resources, emphasizing tighter controls through newly developed management information systems, generating win-lose dynamics between whiz-

kids and oldtimers or independent bureaucratic CPI entrepreneurs and the established managerial elite.

(2) The basic assumption of this strategy is that the present employees, if given the opportunity, cannot be trusted to change the system. The second assumption is that the first one, even if valid, need not be tested publicly with those concerned. Both assumptions miss the possibility that insiders will be willing to become committed to change if they see some new hope for change (e.g., money, new tools, new management, new policies) and if their assistance is requested. For the oldtimers, to see someone else accomplish change that they had convinced themselves was not possible may be interpreted by them as public confirmation of their incompetence and withdrawal. Being made to look incompetent and irresponsible is rarely taken gracefully by people, especially if they have been incompetent and irresponsible. Consequently, they may tend to find all sorts of ingenious ways to fight the innovations. This new surge of energy, in the service of organizational entropy, is seen by those managing change as evidence of the "orneriness of rigid old-line bureaucrats." The latter may now feel insult has been added to injury. They may resist harder and win, not because their cause is right but because time is on their side. If little progress is made, the electorate may seek a new administration.

(3) The range of change strategies is activated in the name of creating new systems, unfreezing old methods, enhancing the quality of life including, for those in public organizations, renewing organizations. Yet the employees experience change processes that are no different from the managerial processes they have experienced for many years. To them, this is not organizational renewal. It is the utilization of the same old designs; the difference is who has the power, information, and resources. Consequently, a credibility gap develops. The best way to make a process of change illegitimate, in the eyes of those to be changed, is to infuse it with a lack of credibility.

LESSONS TO BE LEARNED FROM PAST EXPERIENCES

Most of the intervention strategies impose new kinds of structures, relationships, policies, and practices by whomever is powerful (the poor, whiz-kids, old-line administrators) upon those who are not. Such impositions will be integrated into the living system of the organizations if they are seen, by the employees, as legitimate, desirable, and if the employees know how to implement the changes.

But many of the public employees do not tend to believe that experimenting with learning about, and public confrontation of, issues is desirable. Moreover, even if they approved of these activities, many would not be able to perform them. Thus structural change must be preceded by changes in values, psychological sets, and the development of new competences by the employees. For example, to deal with conflict openly, to confront constructively, to take risks require different values from those presently encouraged within organizations and different skills presently held by few employees at any level of any organization (Argyris, 1968).

Why do not people have these skills? Recent research suggests that people in our culture may be programmed with certain basic values consonant with those of engineering-economic theories plus the concept of education based on rationality at the exclusion of emotionality and the emphasis upon efficiency. We may acculturate children to win, not lose; to strive to control their relationships with others; to suppress emotions; and to focus on rationality. People, when they group up and become employees, will attempt to satisfice these values. It can be shown this will invariably lead to the creation of little pyramidal-type relationships even if people are free to design new structures. They may complain about the hierarchy, but they create the same type of system whenever they are free to do so (Argyris, 1969). One can see this illustrated dramatically in a T-group (a relatively unstructured group where people are given great freedom to design any set of human relationships that they prefer). Millions of people have now attended such experiences and the overwhelming majority, in a setting openly described as being centered toward human growth (and not goals such as making money or producing a product), tend to generate human systems that are consonant with the pyramidal structure. Certain experiential learning communities, called growth laboratories, that see themselves as being more far-out than T-groups and thus even further away from corporate life, can be shown to create learning environments similar to the sales meetings of marketing-oriented, profit-making corporations (Argyris, 1972a).

Schon and I have suggested (Argyris and Schon, 1974) that most people are programmed with micro-theories of human relationships that are consonant with the micro-theories of traditional scientific management and bureaucratic theory. (We have called these Model I theories.) Moreover, like organizations, Model I theories inhibit public testing of views, genuine experimenting, and learning. Consequently people tend to depend upon culturally sanctioned deceptions to keep each other unconfirmed and blind to their actual impact upon others. The result again is self-sealing, nonlearning processes.

Finally, cases have been reported where executives have agreed with the diagnosis of their relationships as being Model I; they have agreed that these were dysfunctional; they have agreed to strive toward a more open, trustful, learning set of relationships (Model II) but, as predicted, they were unable to do so (Argyris, 1970). It takes a high degree of skill to create the new world and just believing in its values is not enough.

To return to the four intervention strategies used above, we may now add another reason why they would not tend to work. In addition to being consonant with the traditional pyramidal structure, the people implementing them—consultants, whiz-kids, public officials, and the local citizens—were probably programmed to behave in ways that were dysfunctional. Community intervention strategies will tend to bog down, no matter what structural protective relationships are designed around them, because the people will tend to be programmed to create living systems that will eventually take on the same characteristics of traditional organizations. For example, an analysis of several alternative schools that were born under positive and supportive financial, political, and structural conditions eventually failed (by their own standards). The failure could be related to the fact that all the well-meaning teachers and students were programmed with Model I behavior.

HOW IMPORTANT ARE OPPORTUNITIES FOR
CHANGE AND SELF-RESPONSIBILITY

Implicit in the argument is the assumption that public employees will tend to seek opportunities for growth, change, and self-realization. The moment one speaks of values such as growth and self-responsibility, some readers immediately may respond that the existing empirical research suggests that there are many who do not seek these qualities in their work place, or, if they do, there are wide differences in the degree to which people actually seek self-actualization. Most employees, they point out, are generally satisfied with their work. But this is precisely the point made earlier. Employees are increasingly coupling job satisfaction with alienation and withdrawal and asking simply to be paid fairly. Satisfaction and low productivity can become a new moral virtue.

To respond to the fact of individual differences, those scholars interested in human growth would not only agree that there are these differences, they would indicate that they were among the first to describe these differences (Argyris, 1960, 1956, 1954; Maslow, 1954) and to present evidence that the *actual* and *active* seeking of growth was not very frequently found. Indeed, Maslow created the concept of deficiency orientation to attempt to explain this finding.

Doubt is also expressed about the assertion that empirical research indicates people prefer not to strive toward increasing self-actualization at work. The argument is *not* based on questioning that such data exist. The argument is that the empirical results may be the result of a self-sealing process (for a more detailed discussion, see Argyris, 1972a). If organizations are designed not to permit or encourage human growth, then why should not the data show that a relatively few people aspire toward self-actualization? Employees are reality-centered and, for the most part, show constructive intent. Since they see no other viable model, then they will strive to make the system work partially by reducing their level of aspiration, partially through the informal activities noted above, and partially by withdrawal toward instrumental needs (Goldthorpe et al., 1969, 1968).

How do we know that they are reducing their level of aspiration related to growth? Perhaps they never had such needs in the first place? There are several strands of research to suggest that the valid response is that they are reducing their level of aspiration. Goldthorpe et al. (1969, 1968) found that workers were instrumentally oriented and did not seek jobs with intrinsic satisfaction. However, as they pointed out repeatedly and documented in detail, the workers desired intrinsically satisfying work but found such aspirations to be unrealistic.

Herzberg (1966) has presented data that lower-level employees —indeed, many who are part of the hard-core unemployed—value work that is intrinsically satisfying. Parenthetically, we might point out that he has cited research in countries such as Poland and Russia, where his theory has been tested and not disconfirmed (Herzberg, 1966).

Next, there is the research on satisfaction. This work is best differentiated into two parts. There is the data which come primarily from questionnaire studies showing that employees are, by and large, satisfied with their work. How is this conclusion arrived at? Primarily by responses to questions such as, "Taking everything into account," or "all in all," "how satisfied are you with your present job?" Well, if one is asking an employee to make an overall evaluation, and if an employee knows the reality of the paucity of alternatives available to him, what other answer can he give but that he is satisfied? Is this not a predictable reaction to deal with the dissonance of deciding to live in a world that is not very gratifying? If so, then this alternative needs to be ruled out with empirical research before conclusions are reached about employee motivation.

Another alternative that has to be ruled out is that the meaning of what is satisfying changes through time as the employee learns to reduce his dissonance and to live with reality. It is the assumption of all the questionnaire devices that the scale offered the respondent has meaning independent of his present existential condition. Drawing in another area for an example, when one gives people, at the beginning of a T-group, a questionnaire with a dimension on the degree of trust of which they are capable, the overwhelming number of people check off 4, 5, or 6 along a scale of 1 through 7. After a successful T-group, the same people, answering the same scale, significantly *reduce* their rating of their trusting because they have lived through a learning experience that has sensitized them to themselves and to others. However, in order to have developed these new insights, the individuals can be shown, by objective measures, to have become more trusting (Argyris, 1965). Nor can this distortion be dealt with by considering it as noise and thereby eliminating it through a large sample. The distortion has been found to exist in over 75% of the subjects in a sample of 400 representing a wide range of work situations (Argyris, 1969).

A more sensitive measure of satisfaction comes from those studies that compare satisfaction scores of people at different levels of the organization. The results are clear; the more control an individual has over his work and the more challenge it offers, the greater the satisfaction. This finding seems to hold for organizational studies as well as for national samples (for summary, see Argyris, 1964; Farris and Butterfield, 1971).

But the thrust of my argument is that these statistics, even though supportive of our view, do not respond to the basic question that behavioral scientists ought to be exploring. The basic question is what kind of a world do we wish to design? If employees will adapt (within reason) to present organizations, and granting that the organizational deterioration will not get worse (which I am willing to grant for the sake of argument), then the question to be asked is whether we would wish to explore the creation of a world in which organizations are achieving their objectives and human growth is valued.

I am suggesting that social science research ought to be encouraged to explore the creation and the effectiveness of other organizational forms and administrative processes. Only as these forms are tested will we know, with a respectable degree of certainty, whether people do have suppressed needs for growth and whether organizations that permit and encourage growth can also be efficient. If the data are negative, then the youth who have chosen to fight the present forms can be helped to see that their aspirations are not realistic.

INDIVIDUAL GROWTH AND ORGANIZATIONAL DESIGN

Once one takes seriously the notion that participants in organizations should be given opportunities to increase their sense of self-acceptance, their sense of being an origin, and their sense of competence, then a new constraint follows. The actual organizational designs should be developed under the control and management of the participants who are to work with them. One should not have organizational experts designing these organizations, giving them to the top administrators for approval, and then the administrators implementing these designs. Such a process is a regression to the past.

Does that mean that participants would not wish to use organizational experts? Of course not; such experts can play a very useful role as consultants to the participants in helping the latter develop some tentative maps of what their aspirations about their organization are.

Is everyone required to be involved? Again the answer is negative, especially when the organization is a large one. Employees are as reality-centered as anyone about the difficulties of full participation. However, groups of employees can be represented in councils where the designs are developed and information continually fed back to them about the direction and intent of the progress (Argyris, 1964; Brown, 1960).

Will not this process be cumbersome and expensive? Yes and no. It will be cumbersome at the outset because so little is known about the technology for effective participation by employees in the design of their organization. It will be cumbersome because we know that effective participation requires, at least, a relatively high degree of interpersonal competence among the participants, and that kind of competence is difficult to learn. Such participants can then create groups whose problem-solving and decision-making tend to be effective (Argyris, 1972a, 1971, 1965, 1962). If we have learned anything about participation, it is that wishing for it or deeply wanting it is no guarantee of using it effectively. However, it may not be as cumbersome and rigid as is the case in an increasing number of organizations where employees act to sabotage the organization's effectiveness. In thinking about expense, it is important to realize that much of what is now included under costs for personnel, labor relations, industrial engineering, control systems, and quality control systems are all added expenses because of the fact that organizations and the jobs within them have been designed to ignore the more complex view of man.

SOME POSSIBLE GUIDEPOSTS FOR ORGANIZATIONAL STRUCTURES

What are some of the guideposts that can be identified from research that has focused on man as a complex and actualizing-seeking system?

(1) The organizational structure, technology, managerial controls, and leadership styles require changing in the direction of increasing people's opportunity for self-control (i.e., decreasing dependence and submissiveness) and for increasing the use of more of their important abilities.

(2) Changes will probably take place in incremental steps, that the present organizational forms will be kept but relegated to use under specific conditions. Thus the traditional pyramidal structure will not be discarded. It will be kept for the decisions for which it is most effective. Similarly, an administrator will not be asked to become completely participative and to cease behaving authoritarianly. The new organization will permit a whole range of participative and

authoritarian leadership styles by defining the conditions under which each may be most effective.

Under what conditions are the pyramidal structure and authoritarian leadership most effective? Four different kinds of organizational structures and leadership styles have been developed to be used in the organization of the future. The structures and leadership styles may each be described along several continua beginning with high controlling of others and authoritarian to a sharing of power and control, to the other end of the continuum where every member, regardless of rank, has equal power for a specified set of decisions. The tight, authoritarian organizational structures and leadership styles are relegated to the programmed, routine decisions where little employee internal commitment and innovation are required. The structures and styles which enlarge everyone's power, but especially those at the lower levels, are for major policy decisions about goals of the organization and the ways by which it should be run (Argyris, 1964, 1962). Thus, the traditional pyramidal structure with its concomitant authoritarian leadership style is not dismissed; it is assigned to be used for the more routine matters of the organization.

Likert (1967) has proposed that organizations aspire toward becoming system-IV organizations. Under these conditions, leadership is based on confidence and trust, subordinates feel free to confront upward, rewards support active involvement, decision-making and goal-setting are widely done throughout the organization (although well integrated through linking processes), and the control processes are subject to the direct influence and management of the people being controlled so that the systems may be used for self-guidance and coordination.

Golembiewski (1967a) has proposed a colleague model for line and staff groups which, in effect, is a model for organizing and managing much of the administrating levels of organization. His model also focuses on structural arrangements that encourage coequal cooperation, issues defined by the participants, leadership varied according to the function being performed and the abilities of the individual to perform them.

Jay W. Forrester (1965) has been interested in using management information systems in order to decentralize the organization genuinely and provide the maximum possible autonomy for people at the lower levels. He proposes to eliminate the traditional superior-subordinate and replace it with individual self-control and self-discipline made possible by a sophisticated management information system which will provide him with adequate intelligence about his activities as well as those of others with whom he is interdependent.

Schon (1971) has developed a model of how an organization can become more open to learning and innovation. His model begins with genuine decentralization, the creation of a network of discrete parts, each working on organizational problems, responsible for their immediate integration with other parts, and continually feeding to the management information which permits it to plan through an inductive approach rather than the traditional one, which is defining for the parts through central planning and unilateral communication of the top to the lower-level parts.

Kaufman, in an exploratory paper, argues that increasing representativeness is going to be necessary in public administrative agencies. Beyond the traditional strategy of appointing representatives, Kaufman (1969) suggests the concept of organizational "ombudsman" and notes, especially in the anti-poverty programs, the insistance on participation by the poor. Marris and Rein (1967) have documented the difficulties with developing such participation. Indeed, if the experiences in industry are indicative, the public agencies have thoroughly underestimated the competence required to make participation effective (Argyris, 1970).

(3) Jobs at the lower levels may be redesigned wherever possible to increase the degree of self-control, autonomy, and the use of a larger number of abilities plus the more central ones of the individual.

Reducing dependence and submissiveness and enhancing autonomy and self-control can be begun by designing work that is more automated (Blanner, 1964; Shepard, 1971). Another possibility is job enrichment where jobs are redesigned to provide greater use of more important abilities and to encourage self-control and responsibility. Preliminary experiments with such changes have been gratifying to both individuals and the organization (Herzberg, 1966; Ford, 1969).

As we have noted above, job enrichment programs can raise basic questions that may threaten all levels of management. Such questions will, in the present state of management systems, tend to be suppressed. But if they are suppressed, the employees will soon find themselves in an intolerable position because they will not be permitted to follow the requirements of their new job. If this happens, then they will tend to see the job enrichment program as a manipulative trick.

The same kind of confrontation of management tends to occur with other changes that might be possible at the lower levels, such as the Scanlon plan, optimal undermanning, the redesign of budgets, and the like (Argyris, 1964). Needless to say, the structural and process changes implied in the suggestions of Golembiewski, Forrester, Kaufman, and Likert will require long unfreezing and reeducative processes of the people with power.

We arrive again at the conclusion that organizational change best begins at the top. If the people in power have not accepted the new values and therefore neither behave according to them nor reward others who do, then organizational change or development will not tend to be effective.

THE CREDIBILITY OF THE COMMITMENT TOWARD
MORE HUMANE ORGANIZATIONS

Employees, during the past several decades, have been subjected to many fads, especially in the area of human relations. Some of these fads have been manipulative, their intention being to overcome resistance or to increase employee satisfaction in order to accept the status quo (Argyris, 1957). Any attempt therefore to ask employees to strive toward a new ethic will probably be viewed with some suspicion.

Even if the suspicion did not exist because of past questionable activities, it would exist because employees themselves have serious doubts about the practicability of the new ethic. They know, probably better than anyone else, how difficult it would be to design and administer organizations with a view to enhancing their self-actualization. In the writer's experience, the employees are conservative on this issue. They understandably do not want to see their physiological and security needs suddenly frustrated because their organization collapsed trying to increase their opportunity for self-actualization.

This means that the employees will need visible and continual evidence that top management truly means that they wish to strive toward the new ethic *and* that they have the necessary competence to do so. The best kind of evidence is the behavior of their superiors in everyday situations.

But the new ethic is not simple to develop, and it is genuinely difficult to behave congruently with its rhetoric. Indeed, as has been mentioned above, one of the failures of scholars and practitioners has been their great misconception of how long such changes take to create even in a small group at the top. The published research available suggests that there does not exist a top management group that is so competent in behaving the new ethic that they do not lose their competence under stress. The findings that, with expert help and heavy emphasis on top management education, a group was still having difficulties after five years are probably more typical of those top management groups that are trying to raise the quality of life within their organizations (Argyris, 1972a).

MANAGERIAL PAIN AND THE
PROCESSES OF CHANGING

Recently the writer developed a model to be used to plot the degree of difficulty of any given organizational change. This model, presented below, may be a useful way to set the foundations for the logic that leads us to state that the organizational processes are paramount in thinking about and designing the organizations of the future.

Conditions for Effective Change

Low   High

Deviance from existing norms

Degree of unfreezing of old that is required

New system required to be self-corrective

Degree to which others are required

Degree of personal and system discomfort

Each dimension is hypothesized to be relevant in considering how difficult change will be. Thus, the easiest change is one that requires behavior that can be plotted on the lefthand side of the continua. The most difficult change would be one that falls on the righthand side of all the continua.

Let us hypothesize that one model of the organization of the future is that it seeks to enrich work, minimize unilateral dependence (for important issues), increase openness, trust, risk-taking, and the expression of feelings by creating a working world that emphasizes conditions that produce psychological success, confirmation, and feelings of essentiality. These conditions are almost 180° out of phase with the requirements of present organizations. Moving toward this new organizational structure and process will tend (1) to mean a high degree of deviance from existing norms, (2) to require a high degree of unfreezing, (3) to require a system that is fully self-monitoring and corrective, (4) to require the participation of others, and (5) to imply a high degree of personal and system discomfort.

This is one important reason why pain, frustration, and turmoil are very high among the executives who attempt to redesign their world to enhance self-actualization, interpersonal competence, and trust, in order to increase the effectiveness of decision-making and problem-solving.

Consequently, the changes will necessarily be slow. The people will not tend to have the competences required but will also have to unfreeze their old values and views about effectiveness. No one can preprogram the speed of unfreezing and the rate of new learning.

However, the very fact that the new behavior required is incongruent with the old means that the clients will be faced with new, puzzling, ambiguous situations. The very newness and ambiguity makes it possible to provide conditions for learning.

The implication is that, for the foreseeable future, since the new directions toward which individuals and organizations will aspire are so different from the present, it is not necessary or advisable to have complete and detailed maps of the changes that should take place.

ARE WE NOT RECOMMENDING TINKERING WITH THE SYSTEM?

There are at least two ways to evaluate the changes within an organization. The first is to focus on the formal structures, especially upon the centralization of power, information, control, and the specialization of tasks. Given this perspective, the true or real changes are those that transfer the power, control, and information to all the employees, especially the lower-level employees. The idea, at least as old as Marx, is to give the control of the technology and technocracy to the lower-level employees.

The evidence is clear that such transfer has not, as correctly predicted by Weber, brought about significant changes in the work world as experienced by the employee. For example, British nationalization did not make for a more equal distribution of power; employees did not gain more freedom, and the industrial structure did not become more democratic (van de Vall, 1970: 47). The socialist and capitalist organizational designers, by and large, operate with the same impoverished concept of man (Fleron and Fleron, 1972). The structures that they create do not therefore differ significantly. Indeed, one can argue that to give the workers ownership of organizations which cause the psychological difficulties outlined above is to place them in a double bind. If they decide to hate their job, they become politically vulnerable since they may not agress against the very system they own and for which they are politically responsible.

Recently, Hough (1969) has made it clear that Russian industrial organization and administration are based upon the traditional bureaucratic model. The Russians, he continues, express a degree of mistrust for this model for the same reasons that we have included under organizational entropy. In order to reduce the danger emanating from bureaucratic dry rot, the Russians have created group decision-making at the highest levels (a practice, familiar to Americans, known as the Office of the President and, at lower levels, matrix organizations). The Russians have also set alongside the industrial bureaucracy a political bureaucracy which is supposed to correct for any excesses and unfairness on the part of the former. Hough suggests that such corrective measures have some degree of success. Unfortunately, he does not explore the issue as to why the political bureaucracy does not develop its own excesses, as our theory predicts that it would. (After all, it too is subject to the same forces leading toward entropy.) Indeed, he implies the contrary, that the political bureaucracy has its own internal rigidities (Hough, 1969: 82-100, 184-186, 252-255).

In a few socialist countries, experiments have been conducted to alter the traditional pyramidal structure (Zupinov and Tannenbaum, 1968). The results have been modest. Difficulties still exist primarily around such factors as openness, trust, risk-taking, and effective communication within the organization.

Similar experiments have been attempted in American industry. For example, the concept of project management and matrix organization is based on more equalitarian sharing of power and information. The biggest difficulty with project management is that, too often, especially under stress, the people behave in accordance with the pyramidal structure and values (Argyris, 1967).

The same may be said for the many experiments attempted recently in creating alternative schools, free universities, and community participation. Here again, the radical structural changes were unsuccessful, primarily because people could not behave effectively in accordance with the requirements of the new structures. Thus, high school students became as authoritarian as their teachers, college students as controlling as their professors, and citizen participation groups have led to as much cancelling out of the unique contribution by members as did the smoke-filled rooms controlled by political hacks. The result is that the organizations may soon become closed to learning about how to improve their effectiveness. One can sense the degree to which people are "programmed" with behavior that is congruent with the pyramidal structure when one realizes that, after decades of experiences by hundreds of thousands of people in experiential learning, and thousands of pages of articles, individuals uninitiated to T-groups still tend to begin them with attempts to create little pyramidal systems within their group.

Why these failures? Because a change is a change for human beings only when they are able to behave in the new morality upon with the new structure is based. In order to assess if a change is truly a change, and not just tinkering with the system, one has to examine what actually goes on in the organization—how people deal with each other as they are making and implementing decisions.

Overthrowing unilateral pyramidal power relationships without reeducating human beings is not anything new, nor do I believe it will lead to effective long-lived change. What *is* new is to reeducate people to develop interpersonal and problem-solving competences that will help them become masters of their own fate, architects of their own life, and managers of their own progress. These competences are at the heart of the change strategy being proposed. Participation based on the three values of generating valid information, free and informed choice, and internal commitment requires skills most people do not have regardless of the political and social structural arrangements in which they live. To illustrate how fundamentally different this view is from those that are presently available, one need only remind the reader that, according to this view, business executives, archbishops, socialist commissars, Black Panther leaders, radical activists, and Naderites are all in the same boat, in that they tend to be manifesting the same interpersonal competence skills and therefore, given the freedom to choose and create, they will choose and create human relationships of the pyramidal variety.

It is the realization that changes are not effective until the people develop ownership of them and that, given the present culture, ownership of the new organizational morality requires behavioral changes which will take a long time, that leads the writer to focus on a change strategy that begins within the system.

Some writers have implied that such a perspective ignores the environment. I do not believe so. As internal organizational deterioration increases, the search process becomes narrower in scope, the openness to new information is diminished, the amount of learning possible greatly reduced. Under these conditions, organizations that are inherently open systems make themselves closed systems. The environment then has little influence. If organizational learning and the openness to the environment are to be increased, and if by "increased" we mean people actually behave differently, then one has to begin with the people concerned. Thus, straightening out the internal environment should lead to greater attention to and effective dialogue with, the external environment. Elsewhere, the writer has suggested that the opposite strategy suggested by some theories actually leads to stagnation among the systems already stagnating and a policy of status quo (Argyris, 1962).

Another way to make the point is to say, *if* one's concern is with human dignity, growth, and realization beyond the level of economic subsistence, then altering the political system is a good example of tinkering with the system.

There are some scholars, especially among educators and young left theoreticians, who seem to believe that, if one could do away with the pyramidal structure and all that goes with it, one then would really change the system. Aside from the fact that people would still be "programmed" to create pyramidal structures in "free situations," there is another important reason why this strategy is not apt to be effective. This reason is also based on the concept of man as one seeking self-acceptance, psychological success, and feelings of competence, and essentiality. These qualities require some degree of structure. Growth does not occur in chaos or in an undifferentiated culture. Too much choice can be frightening (Blos recently suggests this is a problem with the present adolescent). Too much variety in work can be overwhelming. Peak experiences are great but, as Maslow pointed out, they need to be followed with rest and some routine. No evaluation of performance of effort can leave individuals bewildered and with a magnified interest in finding out whether or not others would confirm or disconfirm their evaluation of themselves or of their work.

Many years ago, Lippitt and White showed experimentally that the laissez faire groups experienced a lower quality of human relationships, more discontent, and less productivity than the authoritarian or democratic groups. The laissez faire groups actually restricted the space of free movement of their members because of mutual interference and ignorance. The laissez faire climate is a particularly devilish one with which to punish people because the "leaders" can objectively say they are doing little (Lippitt and White, 1947) to reduce the members' freedom.

But, one could retort, that may be true, because human beings do not have the skills to use such large amounts of freedom effectively. This does not seem to be supported in experience. For example, T-groups typically begin with a highly unstructured situation. As the members try the traditional structure and become dissatisfied, they may swing to no structure. They soon discover that basically each individual is an incomplete human being; that he gains a sense of wholeness from interdependent relationships with others; that the only way so far developed to learn from and to use others as well as to give and help others to learn is to create some degree of order and patterning which is another way to define structure. Again, the key issue is who defines, controls, and influences the structure. If the individuals (or their representatives) can control the structure, then there is a greater probability for success.

But designing and monitoring structures is a time- and energy-consuming task. It should therefore be delegated to those decisions that are important. As suggested above, the pyramidal structure may be best suited for the routine and programmed decisions. Three other structures have been proposed for the innovative, nonprogrammed decisions that require a high degree of internal commitment on the part of the members implementing the decisions.

But again and again the research is clear. Place people in structures that sanction and reward interdependence and growth, and their tendency will be to behave in ways so as to botch things up.

The medium is the message of lasting change; the medium is human behavior.

REFERENCES

ALLISON, G. T. (1971) The Essence of Decision. Boston: Little, Brown.

ARGYRIS, C. (1973) "On organizations of the future." Sage Professional Papers in Administrative and Policy Studies 03-006.

――― (1972a) The Applicability of Organizational Sociology. Cambridge, Eng.: Cambridge Univ. Press.

――― (1972b) "Some problems and new directions for industrial psychology," in M. Dunnette (ed.) Industrial and Organizational Psychology. Chicago: Rand McNally.

――― (1971) Management and Organizational Development. New York: McGraw-Hill.

――― (1970) Intervention Theory and Method. Reading, Mass.: Addison-Wesley.

――― (1969) "The incompleteness of social-psychological theory." Amer. Psychologist 24 (October): 893-908.

――― (1968) "On the effectiveness of research and development organizations." Amer. Scientist 56, 4: 344-355.

――― (1967) "Today's problems with tomorrow's organizations." J. of Management Studies 4 (February): 31-55.

――― (1965) Organization and Innovation. Homewood, Ill.: Richard D. Irwin.

――― (1964) Integrating the Individual and the Organization. New York: John Wiley.

――― (1962) Interpersonal Competence and Organizational Effectiveness. Homewood, Ill.: Richard D. Irwin.

――― (1960) Understanding Organizational Behavior. New York: Dorsey.

――― (1957) Personality and Organization. New York: Harper.

――― (1956) "Diagnosing human relations in organizations: a case study of a hospital." Yale University Labor and Management Center.

――― (1954) "Organization of a bank: a study of the nature of organizations and the fusion process." Yale University Labor and Management Center.

――― and D. SCHON (1974) Theory, Action, Practice: Education for Effectiveness. San Francisco: Jossey-Bass.

BENNIS, W. G. (1966) Changing Organizations. New York: McGraw-Hill.

――― and P. E. SLATER (1968) The Temporary Society. New York: Harper & Row.

BLAUNER, R. (1964) Alienation and Freedom. Chicago: Univ. of Chicago Press.

BROWN, W. (1960) Exploration in Management. New York: John Wiley.

COLLINS, R. C. (1971) "Feedback or organizational change." University of Oregon School of Community Service and Public Affairs Annual Conference of the American Society for Public Administration. (mimeo)

CYERT, R. M. and J. G. MARCH (1963) A Behavioral Theory of the Firm. Englewood Cliffs, N.J.: Prentice-Hall.

EWELL, J. M. (1971) "The effect of change on organizations," in Employee Relations. Cincinnati: Proctor & Gamble.

EWING, D. W. (1971) "Who wants corporate democracy?" Harvard Business Rev. 40, 5: 12-28 ff.

FARRIS, G. F. and D. A. BUTTERFIELD (1971) "Goal congruence in Brazilian organizations." MIT Sloan School of Management Working Paper 546-71, June.

FLERON, F. J. and L. J. FLERON (1972) "Administrative theory as repressive political theory: the communist experience." Presented at the National Conference on Public Administration, American Society for Public Administration, New York City, March.

FORD, R. N. (1969) "Motivation through the work staff." American Management Association.

FORRESTER, J. W. (1965) "A new corporate design." Industrial Management Rev. 7 (Fall): 5-18.

GARDNER, J. (1968) "America in the twenty-third century." New York Times (July 27).

GOLDTHORPE, J. H., D. LOCKWOOD, F. BECHHOFER, and J. PLATT (1969) The Affluent Worker in the Class Structure. Cambridge, Eng.: Cambridge Univ. Press.

--- (1968) The Affluent Worker: Industrial Attitudes and Behaviour. Cambridge, Eng.: Cambridge Univ. Press.

GOLEMBIEWSKI, R. T. (forthcoming) "The persistence of laboratory-induced changes in organization styles." Admin. Sci. Q.

--- (1969) "Organization development in public agencies." Public Admin. Rev. 29 (July): 367-377.

--- (1967a) "The laboratory approach to organization development: the schema of a method." Public Admin. Rev. 27 (September): 211-220.

--- (1967b) Organizing Men and Power: Patterns of Behavior and Line-Staff Models. Chicago: Rand McNally.

--- and S. B. CARRIGAN (1970) "Planned change in organization style based on laboratory approach." Admin. Sci. Q. 15 (March): 79-93.

HARRIS, T. G. (1972) "Some idiot raised the ante." Psychology Today (February).

HERTZ, D. B. (1971) "Has management science reached a dead end?" Innovation (October): 12-17.

HERZBERG, F. (1966) Work and the Nature of Man. New York: World.

HOUGH, J. F. (1969) The Soviet Prefects: The Local Party Organs in Industrial Decision-Making. Cambridge, Mass.: Harvard Univ. Press.

KANTER, R. M. (1970) "Communes." Psychology Today (July): 53-57 ff.

KATZ, D. and B. S. GEORGEOPOULOS (1971) "Organizations in a changing world." J. of Applied Behavioral Sci. 7, 3: 342-370.

KATZ, D. and R. L. KAHN (1966) The Social Psychology of Organizations. New York: John Wiley.

KAUFMAN, H. (1969) "Administrative decentralization and political power." Public Admin. Rev. 24, 1.

KELMAN, H. C. (1972) "The problem solving workshop in conflict resolution," in R. L. Merritt (ed.) Communication in International Politics. Urbana: Univ. of Illinois Press.

LEAVITT, H. H. and T. L. WHISLER (1958) "Management in the 1980's." Harvard Business Rev. 36, 6: 41-48.

LIKERT, R. (1967) New Patterns of Management. New York: McGraw-Hill.

LIPPITT, R. and R. K. WHITE (1947) "An experimental study of leadership and group life," pp. 315-330 in T. M. Newcomb and E. L. Hartley (eds.) Readings in Social Psychology. New York: Henry Holt.

LONG, N. E. (1971) "The city as a reservation." Public Interest (Fall): 22-38.

McGREGOR, D. (1960) The Human Side of Enterprise. New York: McGraw-Hill.

MARCH, J. G. (1971) "The technology of foolishness." Stanford University Graduate School of Education. (mimeo)

MARRIS, P. and M. REIN (1967) Dilemmas of Social Reform. New York: Atherton.

MARROW, A. J., D. G. BOWERS, and S. E. SEASHORE (1967) Management by Participation. New York: Harper & Row.

MASLOW, A. H. (1970) Motivation and Personality. New York: Harper & Row.

--- (1969) "Toward a humanistic biology." Amer. Psychologist 24, 8: 724-735.

--- (1965) Eupsychian Management. Homewood, Ill.: Richard D. Irwin.

--- (1954) Motivation and Personality. New York: Harper & Row.

MINER, J. B. (1971) "Changes in student attitudes toward bureaucratic role prescriptions during the 1960's." Admin. Sci. Q. 16, 3: 351-364.

NEWMAN, F. [ed.] (1971) "Report on higher education." U.S. Department of Health, Education and Welfare. Washington, D.C.: Government Printing Office.

SCHEIN, E. (1965) Organizational Psychology. Englewood Cliffs, N.J.: Prentice-Hall.

SCHON, D. A. (1971) Beyond the Stable State. New York: Random House.

SHEPARD, J. M. (1971) Automation and Alienation. Cambridge, Mass.: MIT Press.

SIMON, H. A. (1960) The New Science of Management Decision. New York: Harper & Row.

——— (1957) Administrative Behavior. New York: Free Press.

SMITH, D. H. (1971) "Voluntary organization activity and poverty." Urban and Social Change Rev. 5 (Fall): 2-7.

TANNENBAUM, A. S. (1968) Control in Organizations. New York: McGraw-Hill.

TULLOCK, G. (1970) Private Wants, Public Needs. New York: Basic Books.

VAN DE VALL, M. (1970) Labor Organizations. Cambridge, Eng.: Cambridge Univ. Press.

WAGNER, H. M. (1971) "The ABC's of OR." Operations Research 19: 1259-1281.

WALDO, D. (1969) "Public administration and change." J. of Comp. Admin. 1 (May): 94-113.

WHYTE, W. F. (1969) Organizational Behavior. Homewood, Ill.: Richard D. Irwin.

WILCOX, H. G. (1969) "Hierarchy, human nature, and participative panaceas." Public Admin. Rev. (January/February).

ZAND, D. E. (forthcoming) "Trust and managerial effectiveness." Admin. Sci. Q.

ZUPINOV, J. and A. S. TANNENBAUM (1968) "The distribution of control in some Yugoslav industrial organizations as perceived by members," pp. 73-89 in A. S. Tannenbaum (ed.) Control in Organizations. New York: McGraw-Hill.

13

The Possibilities of
Nonbureaucratic Organizations

WILLIS D. HAWLEY

☐ A NONBUREAUCRATIC PUBLIC ORGANIZATION is rather like a flying saucer—it seems an interesting possibility, capable of all manner of new achievement, but few people have ever seen one and those who have differ about what it is they saw. The question of whether such a phenomenon might fly and whether it would be of use to anyone in improving the delivery of public services is the concern of this paper.

The idea of debureaucratizing government is as much admired, it seems, as apple pie, older mothers, free enterprise, and Monday night football. But discussion about it is usually rather wistful and confined to relatively modest proposals for "reform." And, since there are so few nonbureaucratic organizations about, evidence as to their contributions to the good life is hard to come by. Nevertheless, despite the uncharted nature of the territory, this paper will explore (1) the potential benefits of nonbureaucratic organizations, (2) some of the difficulties involved in instituting and maintaining such structures, and (3) some ways that the problems of nonbureaucratic organizations might be minimized.

AUTHOR'S NOTE: *I'm grateful to David Rogers, Gregory Fischer, Carolyn Hawley, Lester Salamon, and James Vaupel for their contributions to this paper.*

This inquiry focuses on the probable internal dynamics of nonbureaucratic public agencies, assuming one wanted to establish one. Only minor attention is given to the organizational effects of their interaction with clients, those they depend on for support, and other such elements of their task environment.

My analysis is largely speculative and tentative. It is informed, however, by a thorough study of two closely linked nonbureaucratic organizations which—with the help of several colleagues, including a score of students—I have been conducting for more than three years. I intend to draw on these observations (the agency involved is a two-unit public school serving a central city), though the discussion is directed at a relatively broad range of public service activities.[1]

Aside from the evidence from our own study, I make extensive use of the considerable literature on small groups though almost none of this research was carried out in actual work settings that could be called nonbureaucratic. Because such material must be used inferentially, it is necessary to lean heavily on organization theory and to maintain a certain level of generality in the discussion. Nonetheless, if this analysis does not provide those concerned with public management with some reasonably practical information on the uses, care, and feeding of nonbureaucratic approaches to service delivery, it will have missed its mark.

Let me stipulate at the outset that my purpose is *not* to contrast the alleged evils of bureaucracy with the possible virtues of alternatives.[2] Nor is it to argue that nonbureaucratic organizations are (1) a panacea or (2) desirable for all types of public service activities. Rather, the perspective taken here is that organizational forms are tools whose utility depend on what one's objectives are and what it would take, in terms of technological and human contributions, to achieve them.

HOW CAN WE KNOW ONE WHEN WE SEE ONE?

Organizations may be thought of as more or less bureaucratic; a "pure" nonbureaucratic organization is not likely to be found. Nevertheless, specifying what one might look like should clarify the discussion and make it easier to test the validity of the propositions advanced. Thus, using the work of Weber and more recent research, along with my own observations, let me suggest how we might know an "ideal-type" nonbureaucratic organization if we saw one.[3]

In a nonbureaucratic organization one would expect to find the following:

(1) Key decisions would be made collegially instead of by those atop a hierarchy of authority. While hierarchy might exist for administrative purposes, such authority would be delegated by the work group. Thus, the "vertical span" of organizational hierarchy would be zero.

(2) Specialization of roles would be minimized and those played by specific individuals might vary from time to time; status divisions would not exist.

(3) While goals may be externally established, the means for achieving them are largely determined by the organization. In this sense, the organization is relatively autonomous, and the distinction between policy and administration is blurred.

(4) Interpersonal relationships are determined by social needs and technical requirements with respect to specific tasks rather than by standard operating procedures and formal positions. Continuity of position or role is not stressed.

(5) Standardization of organizational activities (the ways problems are dealt with), the extent to which established and uniform procedure and rules govern the execution of functions, is low. Formalization is low; procedure, rules, instruction, and channels of communication are not codified or otherwise well defined.

Even though organizations may be relatively nonbureaucratic in terms of the collegiality of decision-making while being more or less bureaucratic in the way work tasks are structured (Child, 1972; Pugh et al., 1968; Hall, 1972),[4] I do not propose here to operationalize an aggregate measure of "nonbureaucraticness." Throughout this paper —unless noted otherwise—when I talk about nonbureaucratic organizations, I refer to pure or ideal types—i.e., those that are non-bureaucratic on all five of the criteria just specified.

A nonbureaucratic organization might be found in many contexts. Some, like the school we studied, may operate with considerable autonomy. Even if such agencies are parts of a larger system, they can be relatively self-contained. More likely, nonbureaucratic organizations will be found as subunits of larger, more bureaucratic agencies. Research units in environmental agencies, planning sections of public works departments, or political science departments in universities could be examples of this. And, increasingly, nonbureaucratic organizations may take the form of ad hoc task forces with responsibilities that cut across more bureaucratic work units (Bennis and Slater, 1968).

THE USES OF NONBUREAUCRATIC
ORGANIZATIONS

The problems involved in improving the delivery of public services are generally of two types. The first of these is to structure human activities and machine technologies so as to achieve a desired outcome with the least necessary expenditure of resources. Let me refer to this as the search for "technological efficiency." The second general problem facing public administrators is to motivate personnel to deliver their best efforts in the pursuit of organizational objectives. The next few pages outline (1) types of conditions in which nonbureaucratic organizations enhance technological efficiency and (2) the nature of the contributions they might make to the problems of motivating workers.

DEBUREAUCRATIZATION AND TECHNOLOGICAL EFFICIENCY

The characteristics of nonbureaucratic organizations relevant to technological efficiency that differentiate them from bureaucratic ones can be summarized in a few simple propositions.[6]

(1) Nonbureaucratic organization increases opportunities for horizontal flow of information and ideas.

(2) Nonbureaucratic organization facilitates adjustment, adaptation, and experimentation in response to new or varying conditions.

(3) Nonbureaucratic organization reduces the likelihood of internal conflict based on status differentials.

Of course, the technological efficiency of these aspects of nonbureaucratic organization depend on the nature of the work that goal attainment requires. Those conditions under which debureaucratization should lead to more effective delivery of public services are those which require the application of a variety of techniques and those for which the appropriate selection and ordering of these techniques must be determined largely by the feedback received from the object or person to be changed or served. Such conditions include:

(1) high variability of raw materials (i.e., people) the composition or nature of which cannot be well controlled;

(2) multiple ways that organizational outputs are assessed and where the salience of these criteria varies over time or among those served—i.e., where the goals of an organization are to assist its clients in meeting a range of objectives which have different values for individuals or groups served and, especially, where the resources clients have (psychological, financial, physical, and so on) also vary;

(3) areas where specific tasks to be performed with the organization's raw material are nonroutine and their consequences not readily predictable;

(4) the technical possibility to perform the functions required of organizational (or subunit) goals in small groups.

A number of public services are carried out, at least in substantial part, under these conditions. In such situations, the capacity of bureaucratization—through specialization, close supervision, routinization, and planning—to compensate for the limits and variability of human behavior becomes irrelevant, if not dysfunctional. For example, relatively nonbureaucratic organization would be instrumental in at least significant parts of activities such as schools, medical care and hospitals, social rehabilitation, police work (especially at the detective level), and research and development in all types of service areas.

NONBUREAUCRATIC ORGANIZATIONS AND MOTIVATION

There are instances where nonbureaucratic organizations will enhance agency effectiveness even though their technological efficiency is relatively low. This will occur when the costs of technological inefficiency are outweighed by the contributions to productivity that a nonbureaucratic approach to work can have on workers' motivation to excel. For example, the makers of Volvo automobiles decided in 1972—after experiencing a decline in their reputation for the mechanical reliability of their product—to modify substantially the traditional assembly line in favor of work teams which, at least within their sphere of responsibility, were relatively nonbureaucratic. Company officials, after a little more than a year of the experiment, claimed that quality had improved and, to their surprise, unit costs had not noticeably increased. Frequently, however, with debureaucratization, one might expect a tradeoff between quality and quantity where the apparent technological efficiency of bureaucracy is high. But, where public services are involved, that tradeoff may increase organizational effectiveness and long-term efficiency.

In other words, the structure of an organization can affect productivity through its impact on the willingness of employees to perform to the best of their ability. The way work is structured, the character of leadership and the individual's role in decision-making (all of which make up components of nonbureaucratic organizational structure) are part of the system of incentives an individual responds to when considering how much effort to apply to the job.

The types of motives that induce people to pursue a given activity include:[8]

- fear for physical or mental well-being
- security
- material reward, wealth
- social acceptance, need for affiliation
- pride in self: self-esteem
- normative identification with group or organizational goals
- need for achievement
- self-actualization

The first three types of motives listed here cannot be activated with maximum force by most public agencies. For example, civil service provisions reduce concern for security, and well-defined, often time-constrained, salary schedules of limited range seldom tie performance directly to material rewards. Reasonably wide spans of control and difficulties of supervision (consider teachers, nurses, policemen, welfare workers, and so on), the nontransferability of benefits among jurisdictions, and hiring-from-within policies also reduce the capacity of public agencies to motivate through negative sanctions or personal ambition.

Moreover, *if* Maslow (1954) and his followers (for example,

Argyris, 1964) are right, incentives aimed at the first three or four of the motives cited have limited capacity to engage the psychological energies of workers or to encourage them to seek out responsibility and to aspire to increasingly high standards of performance.

Thus, if there are constraints on the use of some types of incentives in certain public organizations, other incentives such as social interaction, professional exchange, personal autonomy, participation in decision-making, and other aspects of work that are facilitated by nonbureaucratic organizations probably become relatively important.

While there are a number of studies which point to the motivational potential of nonbureaucratic forms to increase the effectiveness and productivity of public employees, the support provided is inferential. The relevant findings can be summarized in several related propositions.

(1) Persons in more complex and more challenging jobs are likely to experience greater job satisfaction and less frustration (Paul et al., 1969; Shepard, 1969; Lichtman, 1970; Herzberg, 1965; Pym, 1963; Porter and Lawler, 1965; Hackman and Lawler, 1971; Morse, 1953; Gurin et al., 1960; Walker and Guest, 1952; Svetkik et al., 1964).

This finding, however, as could be true for some others under discussion here, could result from the higher status of the job rather than from inherent qualities of the work itself. However, there is some evidence that even within job types of similar status, the generalizations presented here hold true (Vroom, 1962; Mann and Hoffman, 1960; Walker, 1954, 1950).

(2) The more complex and responsible the job, the more likely workers are to identify their needs as consistent with organizational goals (Wicker, 1969, 1968).

(3) The greater the opportunity for individual growth, control, and autonomy based on competence, the greater one's job satisfaction (Carpenter, 1971; Margulies, 1969; Argyris, 1964; Morse, 1953; Kirkpatrick et al., 1964; Bachman and Tannenbaum, 1968; Davis, 1966).

(4) Persons in highly bureaucratic work situations are most likely to be alienated from their work (Aiken and Hage, 1966; Pelz and Andrews, 1966; Baumgartel and Goldstein, 1961; Udy, 1965: 701).

(5) Participation in organizational decision-making that is relevant to aspects of work the individual feels is important results in greater job satisfaction (Morse and Riemer, 1956; French et al., 1960; Baum-

gartel, 1956; Pelz, 1956; and Vroom, 1960b) and, when workers are in a position to significantly control the pace and shape of the work they do or the quality of the product they produce, such participation generally leads to increased productivity (Lewin and Whyte, 1939; Coch and French, 1948; Vroom, 1960b, 1959; Lawrence and Smith, 1955).[9]

(6) Participation in the identification of organizational or group objectives and power-sharing in general results in greater commitment to the attainment of such goals (Katz and Kahn, 1966: 332) and to consensus over goals, resulting in organizational effectiveness (Smith and Ari, 1968).

(7) Opportunities for interpersonal interaction with co-workers increases job satisfaction (Sawatsky, 1941; Richards and Dobryns, 1957) and, when interaction is related to the attainment of organizational goals, also increases productivity (Bass, 1960: 51-53).

In considering these general propositions, it is important to note that not all workers function most effectively in participatory, complex, and relatively autonomous settings (Argyris, 1960). There is reason to believe that persons with higher skills and education are motivated most in such situations (Friedlander, 1966a, 1966b; Blai, 1964; Centers and Bugental, 1966), but there is evidence that workers at all levels of organization seek greater control and discretion in their work (Smith and Tannenbaum, 1968; Halter, 1965). Moreover, it seems likely that socialization, rather than cognitive development, accounts for such findings, and, if so, resocialization to new responsibilities and expectations seems feasible (Goldthorpe et al., 1968; Wilensky, 1964; Argyris, 1964).

Age and time in service may decrease one's ability to accept new levels of uncertainty and, in general, to be adaptive to changing roles and to accept increased responsibilities. There is inadequate evidence to test this bit of conventional wisdom. Whether unionized workers, pulled by loyalties to the union and resistant to blurring distinctions between management and labor, would accept and function well in nonbureaucratic organizations remains to be seen. In the groups we studied, where most members were affiliated with the union, a union decision to strike resulted in considerable internal conflict because some felt their responsibilities to the union should not get in the way of their commitment to their clients (students) and to the organization itself (the school).

It seems reasonable to infer from the preceding propositions that nonbureaucratic organization—insofar as it increases job complexity, individual responsibility, a sense of involvement in decision-making, internalization of organizational goals, and the activation of needs for self-esteem and self-actualization—is likely to increase workers' commitments to the organization and their willingness to exert their full energies and talents. The available research leaves little doubt that this will reduce absenteeism and turnover (Vroom, 1964: ch. 6). Whether it results in increased productivity or higher-quality services or goods depends, as noted earlier, on how "sensitive" the output of the organization is to variations in human contributions (Katz and Kahn, 1966: 373-375). The point of diminishing returns at which the motivational payoffs of nonbureaucratic work environments are outweighed by the relative technical inappropriateness of non-bureaucratic forms cannot be specified. But in the public sector, where production functions are highly labor-intensive and many activities involve personal services and—as argued above—where material, technical, and supervisorial mechanisms for controlling workers' performances are of *relatively* limited value, the motivational advantages I have attributed to nonbureaucratic organizations should be relatively significant.

POTENTIAL PROBLEMS IN SECURING THE EFFECTIVENESS OF NONBUREAUCRATIC ORGANIZATIONS IN GOVERNMENT

While the potential payoffs in making more extensive use of nonbureaucratic organizations in the delivery of public services appear to be substantial, one is reminded of the old saw, "If you're so smart, why aren't you rich?" In other words, there must be some reasons why nonbureaucratic organizations are so rare. Utilizing the empirical and theoretical literature, along with my own observations of the trials and tribulations of the experimental schools I have been studying, I have tried to identify a range of such reasons.

These problems of maintaining and securing the effectiveness of nonbureaucratic organizations, which I will elaborate on in the pages that follow, can be summarized by four seemingly simple propositions.

(1) The psychological stress resulting from involvement in nonbureaucratic organizations is potentially very high.

(2) Leaders in nonbureaucratic organizations have an ambiguous status and tend to be punished for *both* nonaction and initiative.

(3) The group cohesion necessary to the success of nonbureaucratic organizations is difficult to secure *and,* at the same time, its achievement can lead to major difficulties.

(4) The problems of evaluating the quality of public services generally are complicated by various characteristics of nonbureaucratic organizations.

These are not, it seems to me, particularly surprising notions, but they cast doubt on the simple idea that ridding ourselves of rigid bureaucratic organizational systems will free our civil servants to achieve needed levels of creativity and performance. In some degree, nonbureaucratic units of government face most of the organizational difficulties that confront any public agency. I have tried, however, to limit my consideration to those problems that are peculiar to or exacerbated by the conditions I have defined as nonbureaucratic.

PSYCHOLOGICAL STRESS

As discussed earlier, certain characteristics of nonbureaucratic organizations provide opportunities for psychologically rewarding social interaction and participation in decision-making while enhancing the probability that self-esteem and self-actualization needs can be satisfied through work.

At the same time, however, working in nonbureaucratic organizations can lead to anxiety and psychological stress. Primary sources of such tension are role uncertainty, insecurity about the future, personal responsibility, and excessive aspirations. These may lead, in turn, to withdrawal, routinization of tasks, role stabilization, erratic behavior, or group conflict.

Role uncertainty: People vary in their ability to tolerate ambiguity, but most of us seek to minimize uncertainty in our relationships with others and in our perception of the tasks for which we are responsible. The absence of such definition can lead to internal tension and to group conflict (Gross et al., 1958; Merton, 1957). In nonbureaucratic organizations, tasks tend to be various, diffuse, and changing, while the mechanisms for role specification are not well developed, and, thus, role uncertainty is likely to be high.

Whatever their dysfunctions, various elements of bureaucracy let people know where they stand and, usually, what is expected of them. For example, Crozier (1964) has reported how French workers "use" bureaucratic aspects of their jobs to reduce anxiety about relations with superiors and to transfer responsibility upward in the organization. Some research on teachers suggests that they prefer more authoritarian to nondirective leadership (Anderson, 1968). And, while workers seem to prefer employee-centered supervisors to authoritarian ones, they prefer the latter to laissez faire direction (Crozier and Pradier, 1961).

In the organizations we studied, teachers and support personnel were continually seeking clarification of what they and others were expected to do. It was decided that everyone needed an individual job description. But, because the descriptions were not distributed throughout the units and because they soon required revision as roles evolved, they did little to reduce anxiety. One unit ultimately set up a big bulletin board that all could see on which was listed each person's more or less permanent responsibilities. This proved to be helpful, but limited.

Job security: Bureaucratic organizations increase one's sense of security in relation to others in that they define the roles, but their influence on one's feelings of security is not limited to that. It is easier in bureaucratic systems to specify the tasks expected of the worker, and bureaucracies often provide safeguards against arbitrary disciplinary actions, including dismissal. Indeed, Kohn (1971) has argued that persons working in bureaucratic roles may be more constrained in the scope of their behavior, but, because they are less anxious about their possible loss of position, they may be more willing to undertake initiatives and to be more satisfied with their jobs. He does not differentiate bureaucratic from nonbureaucratic organizations in the way I have here, but his findings do draw our attention to the importance of a sense of job security to job satisfaction and to self-direction. This notion is quite consistent with Maslow's idea that people respond simultaneously to various needs and that the need for security must be met, at least in strong measure, before most other needs take on their full potential for motivation. Of course, until one receives reasonable security, this uncertainty can motivate, though it is likely to direct efforts away from innovation and risk (Berliner, 1967; Thompson, 1967).

There is no reason why nonbureaucratic organizations—or organizations with nonbureaucratic components—cannot provide protection against arbitrary sanctions. But, even if such assurances can be given, individuals in ill-defined positions vested with broad responsibility for goals whose attainment seems to rest heavily on the intelligent exercise of discretion may feel more anxious about their ability to respond to charges of poor performance and more vulnerable to arbitrary actions by authorities external to the group.

Responsibility: Nonbureaucratic organizations provide the individual with considerable autonomy, but also with considerable responsibility for the attainment of organizational goals. If those working in bureaucracies do not achieve their objectives, they can always assign such failure to "the system" or to "constraints on my discretion." For example, teachers in the schools we studied indicated that, in other schools in which they had taught, they commonly explained the apparent "failure" and absenteeism of students in terms of the students' prior experience, other teachers, the rigidities of the curriculum, and so on. However, in the new nonbureaucratic setting, where teachers play a central role in determining the character of the school and what goes on in the classroom, and where group cohesion is high, the "failure" of students to meet the teachers' expectations—which were actually higher in the new school—could only be assigned to students or to themselves. Most often teachers felt the failure personally.

As I noted earlier, the technological efficiency of nonbureaucratic organizations is greater when knowledge of cause and effect with respect to organizational operations and objectives is low. And the willingness to act, particularly to take new initiatives, is related to one's sense of competence (Argyris, 1964; Vroom, 1964). Thus, broadening the scope and depth of one's responsibilities, even if desired by the worker, could result in a conservative approach to work and a sense of impotence. The more serious the individual believes the consequences of possible failure to be, the more likely it is that he will avoid coming to grips with the problems he faces. This avoidance of responsibility can take many forms, including:

(1) efforts to reduce autonomy—such as centralizing authority, establishing standard operating procedures, and insisting on stronger leadership, and

(2) focusing attention on those aspects of the jobs where success is most readily measured (Thompson, 1967: 120-122), such as teachers securing classroom discipline and police cracking down on drug users rather than drug pushers. This, in turn, may trivialize quality control and encourage routinization.

In many situations where nonbureaucratic organizations are technologically efficient, the role of the public employee is, in large part, a therapeutic one. In such cases—practical nurses, welfare workers, police, parole officers, teachers of the "disadvantaged," and the like—the anxiety and tensions seen in the client are kept at a distance. Such strategies for defending against involvement include:

- denial of one's feelings,
- routinization and ritualization of tasks,
- narrowing of responsibility and specialization,
- transference of initiatives to others,
- denial of personal effectiveness (Menzies, 1960; Lipsky, 1971).

These defenses have consequences that go beyond the direct relations with clients. They de facto restructure the organization. The resultant restructuring is, of course, a movement toward more bureaucratic forms which can, in turn, result in feelings of guilt and personal inefficacy. One sees this in the self-reproach of young, middle-class teachers and welfare workers when they find they cannot change things as easily as they had thought.

The "problem" of high aspirations: The opportunity to set one's goals, especially when first presented, often results in the setting of very ambitious targets, and the exhilaration that comes from being given significant responsibility may result in hopes for rapid and significant problem-solving. Such high aspirations are likely to be disappointed, and a sense of failure will invariably follow.

As the start-up stages of any project, one can explain unmet expectations by the newness of the effort; "It takes a while to get things running smoothly." And at the early point of any project, enthusiasm and hope may mitigate disappointment.

But, as time goes on, inexperience fails to give refuge and, ironically, even though progress is made, the group may become more convinced of its inadequacies.[10] Thus, in all new organizations, but especially in those which are nonbureacratic—where success is difficult to measure and individual responsibility is heavy—things may seem to be getting worse even though they are getting better. Perhaps this is one reason why efforts to establish nonbureaucratic structures are often short-lived.

Summary: Many of the problems noted above are not, of course, wholly unique to nonbureaucratic organizations but, for reasons I have suggested, they are complicated in such settings. At the same time, nonbureaucratic organizations are better able to provide opportunities for social support and individual achievement that can induce greater self-confidence and reduce stress.

While this may seem contradictory, it is more a paradox. If the potential of nonbureaucratic organization for improving the delivery of some urban services is to be realized, some ways need be found to reduce the kinds of psychological stress I have argued is likely in such settings. Let me very briefly—and tentatively—outline some things that might be done to deal with the sources of the difficulties. The underlying objective of these thoughts is the mitigation of uncertainty while continuing to demand high standards of performance and avoid bureaucratization.

(1) As much as possible, workers should be evaluated in terms of the products they produce or the services they deliver rather than the procedures they employ and the processes they perform. This can reduce work uncertainty and can militate against standarization and routinization of the work tasks while providing the type of information workers can use to know if they are being successful. Such feedback will encourage change and provide a basis for judging one's performance in the context of goal attainment. Moreover, such evaluation is some protection against arbitrary actions of the group or supervision from outside the group.

(2) Reports on group performance in comparative context—i.e., what agencies with similar responsibilities and clientele are doing—can help the organization keep its achievement, and especially its aspirations, in perspective.

(3) Evaluation should be developmental rather than judgmental. That is, if performance is assessed in terms of improved capacity to contribute to organizational goal attainment rather than in terms of static criteria, the threat of failure to any given task may be reduced and the

propensity to seek refuge in narrower definitions of role may be undermined.

(4) As the Chinese and Israelis have shown us, ideological consensus can be a substitute for bureaucratic structure as an effective mechanism for control that permits flexibility and adaptation in work processes. This may restrict goal adjustment, however, as I will discuss later.

(5) The problem of role uncertainty may be dealt with by continual specification and respecification of *responsibilities*. Role identities that focus on particular ongoing activities would seem to *induce* bureaucratization.

(6) Task-related communication within the work group is essential. Seeing and discussing how others are dealing with the problems they face is prerequisite to establishing the norm of mutual support and to the formulation of realistic aspirations.

(7) In-service training programs must be centered on the individual needs of workers *as they define them*. Individual needs can, of course, be dealt with in group situations, but decisions by leaders on what people need to learn and the assignment of workers to designated programs will have little consequence for reducing anxiety and psychological stress.

(8) Fundamental to the "management" of uncertainty is clarity about organizational and individual goals. Such clarification is importantly related to the maintenance of group cohesion, a point I will return to below.

Before turning from this focus on the psychological stress that may be felt by members of nonbureaucratic public agencies or work groups, let me reemphasize the points made earlier with respect to the motivational potential of such organizations. Those factors not only motivate, they probably encourage what Argyris (1964) calls "psychological success"—a sense of well-being grounded in one's work-related experiences.

An important part of the sense of self-esteem and self-confidence a person might derive from work, especially work in the public service, is the status given the position the person holds by those to be served—both the immediate client and the larger public.

In the nonbureaucratic public schools we studied, many teachers, especially those with experience in conventional, more bureaucratic schools, identified parental responsiveness to and respect for their professional status as an important source of job satisfaction. It appears that increasing the role of the teachers in school policy-making and broadening the scope of their expertise caused parents to attribute to them authority and deference usually reserved for the principal. In conventional forms of school organization, students and parents alike usually recognize that the principal is formally in charge and is the person to whom important questions must be referred.

It seems likely that the use of nonbureaucratic organization in other policy arenas, assuming the agency can be effective, will generally increase the status (but not necessarily the affection) that public employees are given by those they serve—directly and indirectly.

PROBLEMS OF LEADERSHIP

Nonbureaucratic organizations are not without the need for leadership. The organizational development and maintenance activities which often are not well performed by the group include:

(1) the identification of weakness or incompleteness in organizational design;

(2) spanning the boundaries between the work group and its environment or "superior" levels of the organization;

(3) the interpretation of needs for
    (a) information sharing and
    (b) changing patterns of behavior;

(4) mediation in cases of interpersonal conflict;

(5) identification of the strengths and weaknesses of individuals in the organization; and

(6) the solicitation of participation from the membership as a whole.[11]

It follows from our earlier discussion of the nature of non-bureaucratic organizations that leadership authority in these settings is tacitly or formally *delegated,* and leaders take on a style best described as "facilitation" rather than "direction" or "supervision." Leaders are heavily dependent on the group for whatever influence they have and may enjoy little differential status or rewards. Before I discuss some special difficulties this poses for the exercise of the leadership activities cited above, let me briefly note why facilitative group-dependent leadership is functional.

From the earliest studies of leadership (Lewin and Whyte, 1939), the bulk of the research has shown democratic, relatively permissive styles to have the greatest impact on productivity, especially where work patterns cannot be structured mechanically (Deutsch, 1968).

As Katz and Kahn (1966: 335) point out, "There is evidence . . . that the broad sharing of leadership functions contributes to organizational effectiveness under almost all circumstances." While I find it reasonable to presume that a person who is selected by the group he or she is to lead will be more effective than a person imposed on it, there is less research on this point. There is some evidence that leaders who emerge from problem-solving groups are more effective than those externally designated and that "natural" leadership is suppressed by outside assignments of authority (Borg, 1957). External control, of which leadership designation is a form, may lead to a sense of threat in an organization and to defensive reactions—including routinization (Udy, 1965: 690-691).

The absence of models: One simple reason why facilitative leadership is so difficult is that it seems unclear what it is. Almost all our concepts of "leader" involve the assertion of will, the demonstration of courage, the invocation of sanctions, or the setting of examples. So leaders are to be strong, to define relationships, and to maintain personal distance. The military is an obvious prototype ("follow me" has been the credo of the infantry officer), but the literature of public administration, especially in education and law enforcement, is replete with the images of leadership just noted. And football fans seem to find it so much easier to see why Vince Lombardi was a successful coach than to attribute the winning ways of the Dallas Cowboys to Tom Landry.[1][2]

Not only do we define leadership in heroic terms, but we prefer to think that there should be only one leader per group or organization. But, as Bales and others have shown, various leadership functions can be, and often are, performed by different group members. Bales found, for example, that in groups without formally designated leaders, persons who emerged as "task leaders" (i.e., concerned with getting the job done, solving the problem, and so on) often differed from those who emerged as "socioemotional leaders" (i.e., concerned with group harmony, individual support, and the like; Parsons et al.,

1953: 150-151). Etzioni (1965) has theorized that this complex pattern of leadership is essential and inevitable.

In our study, one nonbureaucratic unit developed multiple leaders with specialized functions, while in the other almost all leadership activities were centered in the person chosen as formal unit head. The more complex pattern of leadership resulted in some redundance and some minor coordination problems but (1) the group's esprit was significantly greater and (2) the group was more task-oriented than its counterpart. In the single leadership situation, many of the group's tensions were vented on the leader who, in turn, experienced considerable psychological stress and tended to withdraw from responsibility.

Dependency and the love-hate relationship with leaders: Earlier it was noted how bureaucratic structures are often scapegoated and used to avoid responsibility and to gain a sense of role clarity. So it is with leaders. Quite aside from the need to get things done, as outlined above, work groups seem to need leaders for social and psychological reasons. When frustrations and uncertainty set in, groups invariably look for someone to show them the way. "The way," of course, is a retreat to bureaucratization.[13] It results in dependence, loss of initiative, and a reduction in the self-confidence of group members (Bevan et al., 1958).

Under conditions such as these, leaders in nonbureaucratic organizations are damned if they take on a directive and assertive role and damned if they do not. While the successful assertive leader may be appreciated in the short run, the group—once the crisis is over—will resent the leader's intrusion on its prerogatives and his or her apparent rejection of the nonbureaucratic model. If the leader adopts the nondirective role, he or she does not remove from the group the burdens that it wishes to escape (Berkowitz, 1953). In one of the subunits, and in the overall organization we observed, we witnessed a recurrent game of "crucify-the-leader," wherein the leader is vested, murdered, buried, and—not long after—resurrected, only to be murdered again.[14] An obvious alternative to the game is the designation of a new leader. But this game is not often preferred because the consequences are uncertain and, more important, it is not easy to find someone to take the job.

Finding a willing (and worthy) candidate: In bureaucratic organizations, formal leaders seek the job for many reasons, including the discretion or power they will have, the increased material rewards, and the status. Moreover, the status schism that is a concomitant of their increased responsibility provides some protection from psychological and social assault. But few of these incentives and protections obtain for leaders in nonbureaucratic organizations.

Research on the role of leaders and their satisfactions in Israeli kibbutzim is instructive.[15] In some, though by no means all, kibbutzim, leaders seem to experience (1) more criticism from fellow workers and subordinates than do leaders in more hierarchical organizations, (2) significant dependence for action on group consensus with resultant feelings of powerlessness or frustration, and (3) reliance on persuasion rather than on formal sanctions. Observers of kibbutzim differ on how much status and reward the leaders enjoy. But even where higher intrinsic rewards of the job are reported by managers than by workers, managers seem less satisfied with their jobs and more anxious to leave the organization than do their "subordinates" (Whyte, 1973). This finding is *inconsistent* with almost all studies of the relative job satisfaction of supervisors and workers in bureaucratic organizations.

The more nonbureautic organizations differentiate and reward the leader's role, the more likely it is that the leader will tolerate the frustrations and strain that go with the job. On the other hand, the greater the rewards of the job, the greater the probabilitiy of a schism, the greater the probabilities of competition and conflict within the work group, and, in general, the more its nonbureaucratic properties are undermined. In the kibbutz, and also in the nonhierarchical work groups in China, this problem is mitigated to some extent by communal ideologies that attribute great importance to worker roles. But it is not clear that such ideological development is possible in competitive societies that do not generally support the myths that sustain communalism. In other words, the norms of nonbureaucratic organizations are importantly shaped by the environments in which the groups function. Thus, nonbureaucratic units which operate in larger bureaucratic units or those which interact with environments dominated by values that sustain bureaucracy will, at best, be partial communities facing considerable difficulty in establishing and maintaining norms that support such notions as open communication, shared authority, the importance of "lower skilled" jobs, and the like.

In the nonbureaucratic organizations we observed, the role of leader was generally eschewed. In one unit, the formal leader was chosen for her capacity to attend to details, and she sought the job in part because it served as a substitute—in terms of status—for the professional credentials possessed by most other members. In the other group, interest in the leadership emerged only when a crisis developed and the group appeared willing to surrender significant discretion to the unit head. In general, the attitude seemed to be: "who needs it?" One of the payoffs for participants in nonbureaucratic organization is that they share in the power to set organizational goals and to determine policies and practices. And since the psychic costs of such participation are great as they are, why ask for more when the rewards are uncertain at best? One can, of course, believe that the fate that befell other leaders "can't happen to me." And to be selected by one's colleagues is no small honor. So there will be those who will accept, if not seek, the job. Nevertheless, for the reasons noted, the problem of inducing the most able members of the group to accept leadership is more substantial in nonbureaucratic organizations than in bureaucratic workgroups.

Where leadership roles are not valued by the group, one might expect that those who will seek the position have a low need for affiliation with the group. The consequence of this may be a loss of group cohesion, strained communication, and intragroup conflict (Golembiewski, 1965: 117).

Dealing with leadership problems: There are no surefire ways to resolve the difficulties I have just discussed. Let me briefly note a few steps one might take to cope with some of these problems.

(1) The selection of the formal leader must be the responsibility of the members of the nonbureaucratic organization. Some protection of the leader can be secured at low cost to the nonbureaucratic character of the group by attaching honorific value to the post.

(2) The responsibilities for leadership should be broadly shared. One major function of the formal leader is to give support and recognition to emergent leaders willing and able to share authority.

(3) It is useful to be clear about who is responsible for what while, at the same time, allowing that roles will evolve. Focusing on the functions to be performed rather than on those who must perform them may help in maintaining flexibility. As uncertainty increases, role stabilization becomes more likely. Role specification *on an ad hoc and recurrent basis* may reduce the urge to bureaucratize.

(4) While it is difficult to recruit leaders, it may be more difficult, in

terms of personal hostility and group conflict to remove them. Thus, the terms of leaders should be limited (the more the differences in the priorities and values of group members, the shorter the term) and procedures for their replacement clearly specified. It is important that the norm of short tenure be established, but it is undesirable to have to remove a leader that everyone wants. Perhaps a voting provision allowing reelection only by near unanimity would meet these last two conflicting objectives. The importance of selection and removal procedures is considerable, since the authority of the formal leader in nonbureaucratic organizations rest almost entirely on the legitimacy he or she derives from the group and the congruence of her or his actions with group norms. As Verba (1961: 231) has noted, the pressure to adhere to authority is greatest in small groups where authority is democratically derived.

(5) In a later section, I will discuss the problems of accountability and evaluation in nonbureaucratic organizations. At this point it is worth noting that a major source of conflict between supervisor and subordinates in bureaucratic organizations is the former's role in judging employee performance and shaping career chances. Thus, in traditional schools, principals and teachers seldom discuss common problems except at the principal's initiative and only then in a forced and tension-filled context (Sarason, 1971: ch. 9). An obvious solution to this problem is that evaluation for continuation and career advancement must be performed by someone outside the organization. This does not mean that responsibilities for evaluation, and for professional or career development should be abandoned by nonbureaucratic organizations. The point is that this important activity, if one wants to maintain the internal openness and cohesion of the group, can only be performed in environments which minimize the threat it poses for loss of security.

(6) As noted earlier, the seemingly inexorable tendency of followers to treat formal leaders differently and to keep them at some social distance is often reenforced, even encouraged, by leaders. This problem can be dealt with through such measures as those discussed in suggestions two, four, and five, above. But the ideology of the group is also important, and norms which support openness, the right and obligation of each member to observe and comment on the work of others, the distribution of leadership tasks to more than one member of the group, and the desirability of power-sharing on an ad hoc basis, need to be developed and protected by nonbureaucratic organizations.

I have already commented on the interrelationships among the norms that become operative within the group and the organizational, social, and cultural contexts in which the group must function. When these conditions are inconsistent, ideological consensus is difficult to secure. The first step in doing so is group self-consciousness about the significance of its internal norms and the objectives they promote (intentionally or not).

GROUP COHESION

It may be obvious by now that the effectiveness of nonbureaucratic organizations is importantly dependent on group cohesion. Cohesion, which may be defined as a sense of identity with and commitment to other group members, has a number of significant consequences. The members of cohesive groups are more likely to communicate and interact frequently and openly (Deutsch, 1968: 271). This results in higher morale (Udy, 1965: 704) and lower absenteeism and turnover (Katz and Kahn, 1966: 378-379). As Robert Kahn (1972: 189) concludes in a recent review article on the meaning of work: "Workers prefer jobs that permit [peer] interaction, are more likely to quit jobs that prevent peer interaction, and cite congenial peer relations as among the major characteristics of good jobs."

Cohesive groups are more likely to have high consensus on goals to be achieved and, perhaps, on how to attain them (Deutsch, 1968: 271; Golembiewski, 1965: 94-98). High consensus results in more orderly decision-making and greater understanding directed toward other group members when conflicts arise (Guetzkow and Gyr, 1954).

As one would expect, cohesiveness reduces internal competition and increases cooperation. Experimental studies show that cooperative groups may be more goal-oriented when faced with problem-solving tasks. They are also more coordinated, accepting of members' ideas, attentive to members' problems, and more likely to derive utility from discussions (Deutsch, 1949).

Group cohesion can increase productivity. As the Hawthorne studies showed and other studies have borne out since (Raven, 1968), group cohesion and morale may be more important to productivity than the physical setting or the technological efficiency of formal work processes. However—as I will discuss later—cohesion should be seen as facilitating the objectives of the group and not necessarily those of any larger entity of which it is a part. As Likert (1961: 30) observes:

Work groups which have high peer-group loyalty and common goals appear to be effective in achieving their goals. If their goals are the achievement of high productivity and low waste, these are the goals they will accomplish. If, on the other hand, the character of their supervisor [or any other factor] causes them to reject the objectives of the organization and set goals at variance with these objectives, the goals they establish can have strikingly adverse effects upon productivity.[16]

But, in nonbureaucratic organizations, workers are more likely to be involved in organizational goal-setting, thus increasing the likelihood of congruence between group and organizational goals (in some cases, the group is the organization). It follows that, when cohesiveness is related to consensus on organizational goals, the need for bureaucratic structures to control and pattern behavior will be low (Etzioni, 1957).

Whether group cohesion has positive consequences for organizational effectiveness depends not only on the goals of the group but, as noted earlier, on whether the task of the group is such that performance is related to group effort (Raven, 1968). A great many tasks, however, are improved by group effort, since such effort may take the form of horizontal communication and mutual support in addition to direct cooperation in work processes themselves.

Furthermore, the relationship between group cohesion and success is reciprocal. As Deutsch (1968: 271) notes, "Group cohesiveness not only increases intragroup communication and group success, but group success and intragroup communication increases group cohesiveness."

Given the importance of group cohesion in nonbureaucratic organizations, there are two general questions that require consideration: (1) What are the key problems in attaining cohesion? and (2) Once cohesion is secured, what are its costs?

PROBLEMS OF ATTAINING GROUP COHESION

Normative consensus and operationalizing organizational goals: As noted above, the attainment of group consensus about what the organization should do reduces the need for bureaucratic controls and serves a number of important functions for the maintenance of an organization's nonbureaucratic character. But precisely because nonbureaucratic organizations encourage participation by members and downgrade the desirability of a generalized status hierarchy, the problem of agreeing on organizational goals is greater than in bureaucratic systems. In the latter, few people expect to actually set policy, and they are not led to believe, except at the leadership level, that they have such rights or capacities. With respect to organizational goal setting, most workers in bureaucratic settings probably have a relatively broad "zone of indifference" (Barnard, 1938).

The problem of gaining consensus on goals and then on operational meaning is greater in public than in private organizations. One of the reasons many people choose to work in government is that they hope to contribute to the improvement of their community or society. And, if they did not have such values at the outset, most public organizations, especially those that work directly with people and are more likely to be less bureaucratic, attempt to socialize their employees accordingly. Thus, organizational goals in such public organizations are likely to incorporate numerous social and political values about which there is general disagreement or confusion.

Since an individual's responsibilities in nonbureaucratic organizations are not highly specialized, the possibilities of fragmenting and particularizing the definition of goals to specific roles or subunits is reduced in comparison to more bureaucratic situations. The problems of identifying and gaining consensus on organizational objectives increase as the specificity of the objectives gets greater. After two weeks of discussion, the groups we observed initially identified some eighty "important" goals. With further discussion, these were "narrowed" to fifty-six behaviorally defined objectives. Efforts to refine the goals further resulted in some conflict. After two years of experience, the groups were able to eliminate those objectives which seemed out of reach or too specific to occupy the attention of the membership as a whole. But the point is that the very process of formulating goals is a potential source of group conflict, especially in the early stages of organizational development. Because of these difficulties, the members of nonbureaucratic organizations may resist or overlook the need to adjust and reformulate goals.

The organization as a partial community: The problems of gaining consensus on goals are related to the need to achieve widespread commitment to them. While agreement may be reached about the character of the goals and even about their priority, the intensity of feeling about them is likely to vary considerably. Such intensity must be accommodated, but it is very difficult to assess.

The variations in the intensity with which people subscribe to various objectives is an obvious source of conflict within the group. Whether individuals are seen as contributing to the organization may be related to the differential value people attach to the goals toward which they strive. This problem of attaining congruence in both priorities and relative salience of objectives can be traced in part of the fact that the organization plays a different role in the total lives of group members. This, of course, is a difficulty in all organizations, but since nonbureaucratic settings provide more opportunities for participation, some members likely will feel much more involvement in and commitment to the organization. Those who are deeply involved and committed may (1) resent the fact that for others the organization seems less important and (2) feel that they are taking more than their share of the group's responsibility.

Another way of looking at all this is that organizations, as I noted earlier, are partial communities often quite open to the influences of their environments. Group members not only differ in terms of how group-centered their lives are, but the demands placed on them by their nongroup lives are likely to vary in substance and intensity.

These difficulties are further complicated by the fact that those who are most committed to the organization are most likely to accept greater responsibility and thus to assume leadership roles. Since conflict resolution in nonbureaucratic organizations is dependent on mutual adjustment, when leaders are the most resistant to compromise, the possibilities for ameliorating interpersonal differences are not good.

Equality versus equity: In nonbureaucratic organizations, the value of equality is very important to the maintenance of collegial relationships. It is not that everyone must have equal levels of expertise. Rather, each member must be viewed as equally impor-tant—perhaps for many different reasons—to the functioning of the group. But, of course, people do differ in their centrality to the effectiveness of the group and thus some organizational members are likely to feel a sense of "inequity." Inequity occurs when a person feels that his or her contribution to the organization's outputs is going underrewarded and the contributions of others, especially those giving less, are overrewarded (Adams, 1965).

In nonbureaucratic organizations, the problem of avoiding a sense of inequity among workers is especially significant since role differentiation is seldom clear. This ambiguity can lead to concern over who is and who is not carrying his fair share of responsibilities. Difficulties in this respect would seem to be greatest when:

(1) the product is difficult to measure;

(2) people cannot see each other in process, and communication is otherwise impeded;

(3) wide differences exist in the training, ability, and social status of the group members (Worschel, 1961); and

(4) external pressure for increased output is high (Yuchtman, 1972: 592-593).

This last point deserves further mention. When an organization which is uncertain about its competence to improve performance is being pressed by clients (assuming it is vulnerable to client influence), when it faces competition, or when it is a subunit facing demands from the larger organization, members may experience considerable anxiety. As I noted earlier, in the face of threats of this kind, there may be a tendency to "rationalize," ritualize, and routinize activities, seek out "strong" leadership, and provide differential rewards to those whose contributions to outputs are most visible. Short-run improvements in output may be the result; the development of hierarchy and a breakdown in communal spirit and group cohesion are also likely.

POTENTIAL COSTS OF COHESION

As essential as cohesion is to the viability of nonbureaucratic public organizations, under certain circumstances efforts to attain it—and its actual attainment—can be dysfunctional.

Conflict avoidance through goal dilution and displacement: If the cohesion of a nonbureaucratic public organization is important to its members, as it must be if the organization is to be effective, individuals may go out of their way to avoid conflict. While conflict avoidance in some measure is essential to the effectiveness of work-groups, it can have its costs.

Earlier I pointed out that disputes over goals is a likely source of group conflict. The potential for internal friction can be reduced if (1) goals are kept vague, (2) priorities are avoided, (3) goals reflect, in effect, the lowest common denominator, and (4) differences in objectives are dismissed or redefined as differences in preferred tactics. I have discussed each of these first three points in preceding pages, and I will not further belabor their consequences. In the nonbureaucratic groups we observed, the propensity of group members to avoid conflict over objectives by stressing tactical differences was reflected in such statements as "we really agree on what must be done, we just differ on how to do it" or "we're really saying the same thing." This strategy of conflict avoidance results in a stress on differences in what individuals are doing which, if they are not to lead to stress, must be justified by emphasizing the value of individuality. It is difficult to argue with a virtue like individuality, but all manner of sins can be committed in its name. The elevation of individuality to an organizational objective is a marvelous example of what Robert Merton has described as "goal displacement," in which ends become means.

Of course, not all organizational conflict avoidance centers around the specification of goals or tactics, but I have emphasized this here because goal-setting is a group responsibility in nonbureaucratic organizations. Throughout the paper, I have pointed to other strategies of dealing with uncertainty—such as the transference of responsibility to leaders and the routinization of tasks. Since uncertainty may lead to frustration and conflict, we may think of such strategies as mechanisms for conflict avoidance. Other symptoms of conflict avoidance are deferred agendas and an unwillingness to invoke sanctions against those who are ineffective. I will return to the last of these subsequently.

Role conflict and cohesion: In positions where jobs are multi-dimensional (e.g., those in nonbureaucratic organizations), the potential for role conflict is greatest. Even though communication about responsibilities is a useful way to deal with tension, the avoidance of such interaction also may be seen as a way to reduce the possibilities of conflict (Kahn et al., 1964: 26-34). Thus, where cohesion is highly valued *and* the potential for role conflict is high, group members may cope by isolating themselves in ways that make goal-related cohesion quite difficult to secure. It is ironic that such conflict avoidance is most likely to be practiced toward those with whom interaction is very important—i.e., those with similar responsibilities.

Conformity and "groupthink": Where group affiliation is highly valued and cohesion is high, members of nonbureaucratic organizations will be concerned about clarifying and adhering to group norms. As Deutsch (1968: 273) points out:

> One of the functions of communication within a group is to establish uniform views about reality [i.e., to validate opinions, beliefs, abilities, and emotions *in terms of a social consensus*], so as to provide members with some confidence in their beliefs and to enable them to coordinate their behavior for effective action.

Once this "social reality" is established, it may be zealously protected, and those within the group who challenge it may be isolated or otherwise sanctioned. Thus, for most people, the small groups of which they are a part can exert greater pressures to conform than other forms of organizational control (Festinger et al.,1950; Homans, 1950). Asch (1965) has shown, for example, that even in experimental situations—where individual identity with the group is often marginal—peer pressures can cause people to adopt positions they know to be contrary to fact. And, just as individuals who are unsure about their capacity to solve problems tend to fall back on bureaucratic structures or "strong" leadership, people who feel threatened are likely to look to their peers for direction. Thus, under conditions of uncertainty, the tendency of individuals to conform to group judgment is strengthened (Coleman et al., 1958; Crutchfield, 1955).

Recently, Irving Janis (1972) has drawn attention to what he calls "groupthink" in decision-making settings that are relatively non-bureaucratic. Not all groups engage in it, but groupthink is a way of collective problem-solving that occurs, particularly in stress situations, "when concurrence-seeking becomes so dominant in a cohesive group that it tends to override realistic appraisal of alternative courses of action." While Janis focuses on high-level political decision-making, some of the processes he identifies that (1) suppress critical thinking and (2) increase the probability of judgmental error can be seen—perhaps in slightly different form—in the types of work-groups we have been discussing.

Victims of groupthink tend:

- to develop elaborate rationalizations for past actions that distort analyses and permit the dismissal of dissonant information;

- to punish dissident group members in the name of harmony;
- to ascribe certain preferences to leaders or to the group and to restrict considerations accordingly. Often, this decision may be individual and unstated. Thus, many different ideas are self-censored;
- to constrain the scope of analyses by assigning to others the right to make certain decisions or to consider certain options;
- to develop stereotypes of competitive groups or of those to be served in order to simplify decision-making;
- to create an illusion that the members of the group are unanimous;
- to develop a collective sense of inherent morality or indespensability.

Elsewhere in this chapter, I have attributed many of these problem-creating behaviors and attitudes to other sources and citing them again may be redundant. But it should also help in making the point, if it is not already obvious, that particular problems of sustaining nonbureaucratic organizations may stem from many sources at the same time. And the impact of such sources may be more than a sum of the parts.

Reviewing these several problems, that can be traced at least in part to the conforming consequences of cohesion, may also draw attention to the critical importance of the flow of task-related information to the viability of nonbureaucratic organizations, a theme I will return to below.

Change and conflict: Nonbureaucratic organizations depend for control on the commitment of the organization's members to either (1) the goals of the organization of (2) professional or "scientific norms." They tend, in Etzioni's (1961) terms, to be normative organizations. It may be that such work groups are particularly vulnerable to internal conflict, especially when they are faced with external threats that seem to require adaptation (Brager, 1969). In the first place, those who are most committed to organizational goals tend to be most resistant to change (unless change itself is assigned significant intrinsic value) and to resist those who would "compromise" goals or reallocate priorities. This problem is seemingly greatest where goals and means are not kept clearly distinct.

Change represents, in most cases, a potential threat to the social fabric of groups that will be involved. It is to be expected, therefore, that the more cohesive the group—unless, perhaps, cohesion is significantly based on a belief in the intrinsic value of change—the more resistant it will be to change (Trumbo, 1961).

ATTAINING FUNCTIONAL COHESION

How is it possible to achieve group cohesion while avoiding the problems I have attributed to the search for it? The query has its Catch-22 qualities, but let me hazard a few tentative ideas.

(1) Group norms that center around the desirability of certain processes or ways of doing things can be subverted by emphasizing such values as experimentation, open participation in group decision-making, and "being different." Such values can, of course, lead to a "cult of individuality," but the probabilities and the dysfunctions of such a development can be minimized if individual members are evaluated in terms of their capacity, in both group and individual contexts, to contribute to the attainment of the organization's goals.

(2) The propensity to avoid goal specification and readjustment might be reduced if (1) leaders seek to identify differences in the objectives of individual group members and to raise questions about their compatibility, and (2) there is continual feedback of both subjective and objective information about the capacity of *both* individuals and the group to meet their stated objectives. One relevant norm that can be developed in a group is the desirability of self-assessment and recurrent evaluation by peers. Another is the inherent value of individual inputs, and this norm is particularly important when there is status incongruence in the group. Finally, as Blau (1955) suggests, it is possible that organizations can develop ideological commitments to seek out and achieve new goals. This can lead to the view that change is good in itself but if emphasis is placed on the consequences of the change, rather than rewarding changes in process themselves, this problem may be controllable.

(3) A third general way to promote cohesion while minimizing its dysfunctions is to emphasize a "task-orientation." The discussion of abstract goals is a potential source of frustration and conflict. So, rather than proceed from general objectives to task prescriptions, it may be more helpful to specify problems to be solved, alternative courses of action, and the justification of individual preferences. This may serve to (1) facilitate reasoned resistance to pressures to conform and (2) reduce goal-related conflict. McDavid (1959) found that individuals in group environments that were task-oriented rather than person-oriented, while generally susceptible to group influence, were more likely to adapt to manipulation of task difficulty and were more likely to compromise with than yield to contrary group judgments. Similar findings are reported by Thibaut and Strickland

(1956), but neither of these studies tell us how task orientation can be achieved and sustained in cohesive groups. But, almost certainly, a focus on tasks must be linked to organizational goals if it is *not* to be devisive and lead to an emphasis on process rather than product.

It is important to deal with differences in individual style by avoiding debate over their inherent virtues and by asking—with as much empirical inquiry as possible—what individuals are trying to accomplish and how well they are achieving it. The "wisdom" of alternative ways of achieving organizational objectives, after all, depends first on the probabilities that different choices can be implemented. Such a focus may also help in avoiding the tendency of workers to confuse tactics and objectives. The open-classroom debate in education is a good example of how easy it is for techniques to become goals with inherent value. So, too, the notion of classroom discipline.

(4) I have emphasized the importance of objective assessments of performance in maintaining the dynamic quality of nonbureaucratic organizations. But, depending on the nature of rewards, individual measures of achievement can lead to competition that discourages productive interaction necessary to group success (Blau, 1955). One apparent answer to promoting cohesion while encouraging adaptive capacity is to focus *rewards* for performance on group achievement.

(5) There is no one way that all leaders should behave in nonbureaucratic organizations and no specific set of norms that is equally efficacious in all such settings. The leadership style and group norms that ensure productivity depend on a number of factors including the personalities and compatibility of group members (Golembiewski, 1965: 98). Thus, even more than in most organizations, the selection of people to staff nonbureaucratic agencies is a critical activity, and the selection criteria should go well beyond professional or technical expertise.

But as important as selection standards is the process. There is some evidence that cohesion, job satisfaction, and effectiveness can be enhanced if workers play a strong, probably determinate, role in the selection of their cohorts (Golembiewski, 1965: 122-123; Van Zelst, 1952). It seems appropriate, therefore, to vest the group itself with responsibility for choosing among potential new recruits and to allow, whenever feasible, work groups within the organization to evolve "naturally"—i.e., in response to the nature of the problem and the preferences of fellow workers. These tasks can best be performed when individuals have had a chance to observe potential recruits and present colleagues in work settings.

Ironically, there may be a predisposition—especially among professionals—to select new members on diversity *rather than* compatibility criteria. In the groups we observed, recruits who were "different," even iconoclastic, were valued over less colorful candidates with demonstrated records of achievement. Afterwards, such appointments were invariably bemoaned by the majority of group members, and the "different" members tended to be kept at the fringes of group decisions and activities.

Diversity and compatibility need not be mutually exclusive objectives. One can have deviant ideas but be willing to compromise and to work cooperatively on a broad range of matters. The important step in reconciling the desire to recruit compatible deviants—an objective that could be quite functional in non-bureaucratic settings—is securing the recruits' prior understanding and acceptance of the organization's objectives that the members consider most critical. And small groups, no less than large organizations, need to be self-conscious about the ways they socialize members to organizational goals and expected patterns of behavior.

The problems of recruiting to facilitate the development of cooperative working relationships may be increased where constraints on hiring—such as priorities placed on the recruitment of nonwhites, women, persons of a given height, and so on—are introduced. Such considerations, while they may advance other (and perhaps more important) goals, decrease, if not eliminate, criteria dealing with the compatibility of style and objectives.

(6) If cohesion is not to lead to resistance to change and unresponsiveness, the organization must be confronted with a certain amount of tension from its environment.[17] But tension which leads to feelings of insecurity can increase cohesion-seeking, retrenchment and withdrawal, and can induce greater conformity. As Deutsch (1968: 272) observes, the results of studies dealing with the effects of tension are not definitive. "The safest generalization seems to be that mild stress often improves group performance and increases cohesiveness while severe stress often has the opposite effects."

What is needed is what March and Simon (1958: 154) call "optimum stress." The problem, of course, is to predict the point of diminishing returns. Among the factors that might determine how much stress can creatively be dealt with are: (1) the turbulence of the environment, (2) the self-confidence and cohesiveness of the group, (3) commitment to organizational goals by group members, (4) the nature of demands, (5) organizational resources (including worker skills), and (6) the personality of the group members. Admittedly, this suggestion is vague, but being conscious of the fact that the relationship between tension and small group effectiveness has curvilinear properties may be an important first step in resolving the problem. For example, such an awareness would encourage skepticism that professionalism can effectively be countered by extensive politicization.

Organizations with nonbureaucratic subunits may induce tension by assigning similar responsibilities to two or more groups where one or more of the following conditions exist:

(1) the relative performance of the groups can lead to differential benefits to the groups;

(2) the territorial jurisdiction of the groups is ill-defined;

(3) the potential substantive scope of each group is elastic.

Probably the source of tension which is most likely to lead to change in nonbureaucratic settings is feedback to the individual worker or to the group about relative success and achieving individual or group goals.

Presumably, this discussion of the uses of tension raises the issue of how bureaucratic organizations can be evaluated and held accountable. I will turn to some thoughts on that process in the next section.

(7) A final general approach to improving cohesion without inducing rigidity and unresponsiveness is to facilitate reasonably uninhibited participation and the free flow of communication among members of the organization. This is, of course, no mean achievement. Indeed, the development of an effective pattern of communications is essential to dealing with many of the problems I have discussed. Thus, I intend to devote much of the conclusion of this paper to this matter, where I will come back to the impact of interaction on group cohesion.

EVALUATION

These days, evaluation and accountability are "the" issues among students of public administration. Everyone agrees that much must be done at the same time that they disagree about what to do. The problems of evaluating nonbureaucratic organizations seem particularly difficult.

Evaluation can be thought to have two major purposes: (1) to allow individuals and groups to improve their own performance and (2) to provide information to administrative superiors, policy makers, or clients that can be used to control, or to impose change on, the organization. Let me refer to the first of these purposes as internal, the second as external.

Both internal and external evaluations are difficult in nonbureaucratic organizations because two of the reasons such structures are useful are that:

(1) organizational goals are often multiple and diffuse, and

(2) knowledge about the best way to achieve organizational goals is not extensive or definitive.

And, in agencies where achieving change in human behavior or condition is the organizational objective, there is often considerable difference of opinion about how to measure such changes.

Evaluation used solely for internal purposes may be resisted, depending on how it is carried out, because it threatens the cohesion and social compatibility of the group. When sanctions are attached to "bad" evaluations, group members unsatisfied with their own performance will tend to avoid seeking help, co-workers may be embarrassed to give unsolicited help, and tough decisions about advancement and promotion are avoided. On this last point, one reason why senior faculty and some administrators in universities support faculty tenure provisions is that the prospect of living the rest of one's professional life with a less than fully competent colleague forces some faculty to be tougher in their pretenure evaluation of their "peers" than they might otherwise be. Once tenure is awarded, systematic evaluation is usually put aside. One cost of a stringent tenure system is, on the other hand, a status schism, with cohesion generally higher among senior (tenured) than junior faculty and some tension and relatively poor communication between the two groups, especially (and ironically) when junior faculty identify with the organization. Similarly, teachers and other workers resist merit pay notions, even when they would determine who was given awards, because such decision-making has a high conflict potential.

Evaluation is difficult when its primary purpose is external control because the threats such assessment poses may cause the organization to take major steps to seal itself off from its environment. Thus, ideologies such as tenure and academic freedom among teachers are developed and maintained in the arena of lofty principles. Professionalism among medical personnel, lawyers, and others is utilized to justify the notion that "only we are competent to judge our peers," (Downs, 1966: ch. 19). And efforts are made to sustain the myth that activities such as deployment of policy and teaching and rehabilitation practices are not political (Bendix, 1968). Of course, such practices are found in all types of public organizations, but the relative cohesion of nonbureaucratic organizations and the nature of the activities in which they are likely to be engaged may increase the probability such buffering tactics will be employed in these settings.

TOWARD A SOLUTION OF EVALUATION PROBLEMS

(1) Internal evaluation of individuals seems best carried out if rewards or sanctions—e.g., promotion, demotion, or job security—are not directly associated with it. Such rewards or penalties are best administered by persons who are not members of the immediate organization or subunit.

This does not mean that nonbureaucratic organizations should not engage in evaluation but that the internal purposes of such activity should be to develop the capacity of individuals and the group to meet organizational goals. Can evaluation in the absence of formal sanctions induce change or otherwise motivate? The importance of social acceptance by peers and the desire most people seem to have for self-esteem should provide the appropriate leverage if the objectives involved are actually valued by the group or the individual. As I noted in earlier sections of this paper, the characteristics of nonbureaucratic agencies are likely to encourage commitment to organizational goals. In any case, evaluation efforts should be individualized or at least tied to readily identified subgroups. Such a strategy would include the identification of individual or team objectives and the specific measures and types of evidence group members agree are appropriate ways to know whether objectives have been achieved.

(2) For purposes here, let me describe external evaluation as having political and administrative components. Political evaluation is aimed at holding *the organization* accountable for attaining

specified objectives. Administrative evaluation, in addition to this function, is concerned with the control, advancement, or termination of individuals.

Political evaluation is usually the function of legislative or citizen bodies. It should be concerned solely with group performance—that is, the activity of the nonbureaucratic organization as a whole. The agents of political evaluation may, of course, collect their own data and deal directly with clients. Some of the information so collected may be relevant to assessing individual performance. Administrators external to the nonbureaucratic work group should share in the responsibility for monitoring group behavior.

(3) A focus on the group, coupled with reluctance to punish short-run individual failure, may encourage the group itself to be concerned with contributing to goal-related effectiveness of its members. Moreover, there is evidence that individuals draw satisfaction from the success of the group which, in turn, encourages cohesion and collaboration (Shaw, 1960). And, emphasizing group performance in evaluation processes may provide a base from which intergroup competition can be induced. Administrative responsibility for personal evaluation in effect screens the individual from environmental threat. There are two reasons why this is important: (1) to encourage the individual to interact with clients and to develop task-related commitments outside the work group, and (2) to facilitate evaluation of individuals over time so as to encourage personal growth and permit a time perspective that can reward innovation and long-term, rather than short-run, impact.

(4) As noted earlier, the logic of nonbureaucratic organizations will be undermined by evaluation that focuses on process rather than product objectives. Product objectives are those derived from organizational goals—as contrasted with various means that might be seen as advancing such goals.

(5) If evaluation is to have developmental as well as judgmental purpose, members of the organization must be involved not only in goal-setting, but in the specification of performance criteria. Such criteria should, ideally, be subject to objective verification and should capture the range of organizational goals as well as their relative priority.[18]

These several thoughts only scratch the surface of the difficulties of achieving an effective approach to evaluating nonbureaucratic organizations. In general, I have tried to stress the importance of maintaining high levels of feedback to the group and to individuals, in terms they agree on as appropriate, with respect to goals they have agreed are important. I have also implied that the evaluation of individuals should be as nonthreatening to personal security as possible, and that reward and sanctions, when the latter are absolutely necessary, should be the responsibility of persons external to the nonbureaucratic organization.

CONCLUSION

All organizations must find the ways of linking their various parts, including personnel. In general, this can be achieved in two major ways:

(1) by established rules, routines, blueprints or schedules, and

(2) by linkages through feedback or mutual adjustment (Hage et al., 1971: 860-861).

The first type of coordination is characteristic of bureaucratic organizations and the second of nonbureaucratic organizations.

As Tannenbaum and Bachman (1968: 23) note, "The relative success of participative approaches to organizational decisionmaking depends not on reducing control but of achieving a method of control that is more effective than that of other [more bureaucratic] systems." Thus, if Type 1 coordination is to be avoided, Type 2 must be facilitated. One way this can be done is the "objective" reporting of data on job performance. Feedback, and the controls on behavior it introduces may also be achieved by pressures or information from clients and others external to the organization. But, in nonbureaucratic organizations, the most significant source of feedback and mutual adjustment is interpersonal communication.

The structuring of task-related personal interaction so as to achieve technological efficiency is necessary, but not sufficient, to assure the effectiveness of public organizations. As noted, organizations must also find ways of assuring that workers adhere to behavior patterns that are efficient and that they exert their full energy in the pursuit of organization objectives. Like the problem of control that is called coordination, the problem of motivation in nonbureaucratic organizations depends heavily on the quality and the free flow of task-relevant information, including that which workers find socially and psychologically rewarding in itself (e.g., support, encouragement, friendship, and the like). Horizontal interaction promotes cohesion and group morale (Udy, 1965: 704) and consensus on organizational goals (Tannenbaum and Bachman, 1968). As Newcomb (1959) points out, "The actual consequences of communication, as well as the intended ones, are consensus producing."

In short, interpersonal communication is of central importance in shaping the effectiveness of nonbureaucratic organizations. Thus, it seems appropriate to close this essay by drawing attention to the factors that impede or facilitate effective communication. Of course, earlier pages have already discussed a number of such factors, and I will simply outline those here. But there are other communication problems and possible remedies which I have not yet dealt with that require more extensive consideration.

PROBLEMS IN SECURING EFFECTIVE COMMUNICATION

In earlier parts of this paper I noted at least implicitly, that the free flow of task-related communication in nonbureaucratic organizations was impeded by:

- social distance between the members of the organization
- significant differences in the perceived value of individuals to the group's success
- routinization and hierarchy
- authoritarian leadership styles
- variations in the level of commitment to organizational goals
- the absence of consensus about goals and processes
- lack of clarity about organizational goals
- role conflict
- conformity and groupthink

Three other aspects of nonbureaucratic organizations affect the quality and quantity of information that is shared: (1) the absence of formal communications channels usually required by differentiation and specialization; (2) the propensity for communication to be "imbalanced"; and (3) the problem of securing optimum size.

The paradox of specialization: Students of social integration, since Weber and Durkheim, have noted that specialization of roles in social systems leads to greater interdependence and to increased interaction.

While the division of labor may lead to alienation, incomplete understanding of organizational goals, and other factors that impede communication (Raven, 1968: 290; Hage et al., 1971), the structural arrangements that follow from specialization draw attention to mutual dependence on the need for intraorganizational communication. Thus, the nature of communication between subunits and individuals can be specified, formalized, and monitored. And, somewhat paradoxically, diversification and specialization of roles may facilitate communication because people do not share common work experiences.

In collegial organizations, relative homogeneity may reduce the perception that it is necessary to seek new information from group members and, at the same time, seem to obviate the need for formal procedure for information-sharing (Brewer, 1971). Moreover, specialization means that communication is based on varieties in expertise or tasks. So, not only is dependency relatively clear, but the search for information does not imply that one does not know one's business.

The technological efficiency of nonbureaucratic work-groups is greatest where tasks are complex and knowledge about how problems can be solved is uncertain. In such situations, it is important that persons with shared experiences communicate. But, where roles overlap and responsibilities may be ill-defined, the search for information may seem to imply ignorance of one's job. And, since time pressures are great and the amount of information needed very great, it may be difficult to justify dealing intensively with only a small portion of one's responsibilities. It may seem better to "do the best I can under the circumstances." Furthermore, individuals may assume that interaction among persons of the same status is likely to have little payoff. What one needs in uncertainty is the advice of experts. This problem probably diminishes as (1) the professional expertise of individuals increases, (2) successful interactions multiply (that is, as information-sharing increases the incentives for search increase and the threat to ego diminishes), and (3) goals are clarified and responsibilities narrowed. If these generalizations are correct, one might expect, for example, that high school science teachers are more likely to share information than are elementary school teachers or that gynecologists are more likely to communicate with each other than are general practitioners. This means, however, that communication patterns themselves become partial and the sharing of *needed* information is high when the clients' problems are intense and specialized, but low when they (1) appear minor or common and (2) cut across specialties or roles.

In the schools we studied, aside from interaction associated with structured organizational decision-making, communication was greatest among teachers concerned with similar subjects and, when focused on particular students, only "failures" or "troublemakers" were commonly discussed. Ironically, teachers talked less about the successes they had had than about their general problems. Sustained discussion of seemingly intractable problems can be tough on group morale. Moreover, discussions about how to improve the school experience for most students focused not on the needs of individual students but on general approaches to teaching.

Of course, time limitations force information-sharing away from particularized interaction, but many times teachers did not know they had particular students in common and, even when they did, they were sometimes reluctant to admit they were not "having success" with students whose learning or behavior problems were not unusual.

Imbalance: One type of communication imbalance derives from the fact that the information received in urban service delivery agencies is invariably concerned with unsolved problems rather than clear successes. Thus, the incentives for seeking feedback may be low and the need for self-esteem and group pride may encourage the screening of negative information. While the degree of bureaucratization may be unrelated to the magnitude of this type of communication imbalance, the capacity of nonbureaucratic organizations to buffer their members from dissonant information is relatively low. And, because the activities they are best suited to perform are probably the most difficult to measure, refuge in assurance that one is succeeding may be harder to find in nonbureaucratic organizations. Moreover, if workers in nonbureaucratic organizations identify more with the goals of the groups they participate in, such imbalance may have particularly unhappy consequences for self-esteem and group morale.

But my primary concern is with a second type of communication imbalance—substantial inequalities in the participation of members in the group's interaction patterns. Internal imbalance of communications is a serious problem in nonbureaucratic organizations because formal channels of communication are ill-defined.

On the one hand, as groups accord their members differing degrees of status, this becomes a source of influence toward which communication shifts and from which initiatives are expected (Sherif and Sherif, 1968: 278-279). Similarly, those who are seen as most effective in groups will receive a disproportionate share of communication, and the information they receive reinforces their effectiveness (Kelly, 1951).

Patterns of communication have a tendency to become fixed in nonbureaucratic organizations, thus reducing their flexibility. Moreover, full membership participation, a major strength of nonbureaucratic organizations—both in terms of technological efficiency and motivation—is undermined when imbalance and overload become dominant characteristics of the communication pattern. In other words, when some group members decisively dominate the interactions of a group, the total amount of communication probably declines.

Of course, equality of participation is never possible, nor is its relevance to organizational effectiveness always great. But opportunities for participation in group decision-making and interpersonal interaction invariably increase morale and when complexity, the need for flexibility, or alternative solutions characterize a group's tasks, less-restricted networks seem more efficient (Raven, 1968: 291).

One's access to information and the opportunity to communicate are essential to the power one holds in an organization (Mechanic, 1962). Thus, the only way to maintain open and flexible patterns of interaction and to counter the inherent tendency in all groups to restrict communication is to maximize the full participation of members. We have come full circle.

Size and communication: If free and extensive communication is essential to nonbureaucratic organizations, then the use of such groups must be constrained to tasks that can be performed by a relatively small number of people. As Deutsch (1968: 267) observes:

The number of potential interpersonal relations increases geometrically as size increases. . . . Since there appears to be a numerical limit to the capacity to establish close associations with others, a smaller proportion of possible linkages will be formed as size increases [also see Reynolds, 1971].

From the literature on small groups one may derive several propositions concerning the impact of group size on organizational effectiveness. For example, as the size of the group increases:

(1) the need for coordination increases (Deutsch, 1968) and the less satisfactory communications are as a source of organizational control (Smith and Brown, 1970);

(2) inequality of participation among members increases (Stephan and Mishler, 1952) and personality rather than the quality of contributions becomes more important in determining the relative impact of individuals on group decisions (Deutsch, 1968);

(3) members' satisfaction declines (Deutsch, 1968) in part because the possibility of securing group action, especially action fully consistent with one's preferences, declines (Berkowitz, 1953);

(4) factions are more likely to develop, with resultant possibilities for intragroup conflict;

(5) job-related tensions increase (Kahn et al., 1964). Kahn concludes, "It . . . seems plausible that large organizations differ from small ones in that small organizations provide more opportunity for self actualization and large organizations more security and social gratification."

On the other hand, larger groups (1) provide more perspectives on problems, (2) reduce the probability that nonconforming ideas will be suppressed, and (3) depending on the complexity of the task and the talents of group members, may increase informational and technological resources available to the group. And, of course, clients of public organizations may have needs that can best be served by specialized programs and facilities that cannot be provided by small groups.

The size of nonbureaucratic organizations would seem to be conditional on several factors including, among other things: the diversity of the clientele, the nature of the resources needed to perform organizational tasks and the logistical problems in delivering them. I will not try to deal with this question here. The purpose of this brief discussion is to draw attention to different aspects of the importance of size in the workings of nonbureaucratic organizations. One implication of this commentary is that finding effective ways of linking various subunits of organizations is essential to the widespread use of nonbureaucratic organizations in the public sector. Such linkages are important not only to promote short-run technological efficiency but to facilitate information-sharing and the diffusion of innovation within large organizations employing nonbureaucratic units.

FACILITATING COMMUNICATION IN
NONBUREAUCRATIC ORGANIZATIONS

Despite its importance, communication that promotes goal attainment seldom comes about "naturally." In preceding pages, I suggested that such interaction in nonbureaucratic organizations was facilitated by several mutually reinforcing conditions that go beyond the absence of those circumstances just cited that impede such interaction. These include:

(1) group cohesion *coupled with* consensus on organizational goals;

(2) norms of openness, equality of viewpoints, and the right to participate;

(3) physical proximity;

(4) leadership concern for involvement and support of group members in decision-making;

(5) recurrent formalized reexamination of objectives and roles;

(6) an emphasis on product rather than process objectives;

(7) utilizing evaluation in ways that encourage professional development and do not threaten career security;

(8) holding the group accountable for achieving objectives it has participated in setting;

(9) sharing of power and responsibility for decision-making.

Let me briefly note some additional ways that communication on nonbureaucratic organizations might be facilitated.

First, I have already noted the importance of certain values that *allow* communication. Norms *encouraging* participation in group decision-making also should be fostered, as should the notion that "different" ideas are valuable because they are different. Hollander (1960), for example, found that, in some organizations, individuals could build up "idiosyncrasy credits," which facilitated deviant actions once basic loyalty to the group and its norms are established.

Second, efforts should be made to clarify the mutual dependence of specific group members in terms of (1) shared responsibilities for particular clients or (2) the interrelationships between the more general knowledge and experience they have.

Third, regular times should be set aside not only for group decision-making but for informal communication. Such discussions should involve only two or three members and steps should be taken to rotate the members of such parlays.

Fourth, leaders must discourage the notion that they have the answer, however attractive it might be to be thought of as the source of wisdom.

Fifth, as Janis (1972) suggests, one or more group members can be assigned the role of "devil's advocate." However, unless (a) the leader (as well as other group members) treats this role with respect or (b) it is exercised with restraint, the ideas occasioned by it may be dismissed as a predictable and inconsequential annoyance.

Sixth, role-playing might be utilized to encourage consideration of alternative perspectives on given problem—including the views of clients, supporters, and competitors.

Finally, some problems may be assigned to more than one subgroup for consideration and recommendations. This last strategy should also be used sparingly and only on the most serious problems, because it is time-consuming and potentially conflict-producing.

OTHER ISSUES FOR FUTURE CONCERN

Limits on space and my powers of analysis mean that this survey of the difficulties involved in sustaining nonbureaucratic organizations is far from exhaustive. Let me briefly take note of several matters that deserve more consideration.

First, the difficulty of integrating the various subunits of organizations with one or more nonbureaucratic parts is greater than the problems of integration in wholly bureaucratic agencies. While much of the strength of nonbureaucratic organizations comes from their autonomy, cohesion, and openness to innovation and environmental responsiveness, these factors also complicate linkage problems. The amount of integration between subunits that is necessary to achieve overall organizational objectives obviously differs, and the conditions under which demands for integration vary need specification.[19] Further, the formulation of ways of achieving such integration, without destroying the logic of nonbureaucratic organization, requires much imagination.

A second, somewhat related, aspect of nonbureaucratic subunits which seems worth noting for future consideration is the possibility that their potential for encouraging innovation may invite bureaucratization of their role by parent organizations. Udy (1965: 699) notes that the greater the autonomy of organizational subunits, the more likely it is that they will seek to expand their scope of activity. This would seem particularly true where ways of achieving goals are not well defined. Such expansion may be justified by innovation. If parent organizations wish to avoid frustrating subunits and decreasing their motivation to contribute to the organization, they must respond permissively to expansion bids. But such positive response to nonbureaucratic units may, in effect, place substantial strain on organizational resources, especially when the new activity is not revenue-producing. Under what circumstances, if any, does this theoretically reasonable situation occur and how can the rate of innovation of this type be controlled?

Third, how are the possibilities for nonbureaucratic agencies affected by differences in the extraorganizational environment in which they must perform? There is a growing body of literature on the importance of environmental conditions for the internal dynamics of organizations. While most of this work emphasizes nongovernmental organizations, much of it can be applied to the problems I have been discussing.[20] Of direct relevance, for example, is the recent finding by Duncan (1972) that uncertainty among workers is due more to the rate of change in the organizational environment than to its complexity.

Fourth, at different stages of an organization's development, the salience of various problems to the possible contributions of nonbureaucratic organizations to public service delivery will vary. It is clear in the organizations we studied that a process of group learning was taking place both with respect to personal interaction among group members and strategies for working with clients (students and parents). This developmental process would be important to map and understand so that those implementing nonbureaucratic organizations would have a better idea of what to worry about at different stages of growth.[21] It is also apparent that the learning process is very uneven and does not inexorably move toward greater success. For example, one of the organizations we observed seemed to experience more difficulty in the second year of operation than it did in the first and might not have survived without the assistance of its sister unit, which served as a model and a source of professional and emotional support.

Finally, it is intuitively sensible to believe that some people will function better in nonbureaucratic organizations than will others. It would be helpful in selecting group members and in adjusting organizational processes to employees, to know with some reliability the characteristics of those who might feel more comfortable in less bureaucratic settings. A beginning, but general, list of such personal traits might include: (1) a nonauthoritarian personality, (2) low need for social approval, (3) high ego strength, (4) high tolerance for ambiguity, and (5) high competence—or at least good training—in the skills required for the role to be played.

PROSPECTS

I noted at the outset that there were not many nonbureaucratic organizations now involved in the delivery of public services, and it is fair to ask whether this paper has any relevance to the real world of urban management. I have tried to suggest the types of concerns that might best be dealt with—at least at some point in the problem-solving process—by less bureaucratic agencies or subunits. Whatever argument might be made for the utility of relatively nonbureaucratic organizations now, there are at least two general reasons why one might expect the case to grow stronger in the future.

First, as affluence and educational levels have increased, expectations have changed on the part of both the public and civil servants. Citizens expect greater scope and quality of services (consider the demands being made on schools). Among public employees, one sees greater demands for personal autonomy, responsibility, and participation. This is true for professionals and many other workers, especially those whose jobs, for different reasons, seem to be relatively secure.

Second, as the demands for new and improved services grow, so does our appreciation of the complexity of the problems to be solved. This understanding should result in a recognition that new organizational forms are needed. Charles Lindblom (1965: 307) has observed that more sophisticated administrative systems favor "partisan mutual adjustment"—a process of decision-making not unlike that I have attributed to nonbureaucratic organizations—over centralization and hierarchy.

> As complexity increases, the case for partisan mutual adjustment, rather than centrality, becomes stronger. And some problems . . . are complex beyond any significant possibility for central coordination, a conclusion hardly hinted at in the literature of public administration.[22]

The bulk of this paper has focused on the problems involved in establishing and maintaining effective nonbureaucratic organizations with the hope of facilitating, not discouraging, this process. I trust I have given sufficient emphasis to the notion that debureaucratization is an appropriate partial strategy for improving the delivery of some public services but less appropriate (or inappropriate) for dealing with others. While it is no panacea, nonbureaucratic organization is a substantially underutilized strategy for improving the quality of public services.

NOTES

1. For an extensive description of the school and an analysis of its effectiveness, see Hawley et al. (1973).

2. The criticisms of bureaucracy are legion, but the various dimensions of these arguments are well stated in the writing of Chris Argyris (1964), Rinehard Bendix (1968), Sheldon Wolin (1960: ch. 10); and a collection of recent essays edited by Frank Marini (1971).

3. The dimensions of organizational design identified seem most similar to those suggested by Litwak and Meyer (1965). Also influential in this analysis is the work of Pugh et al. (1968); and Perrow (1967).

4. Of course, in some types of organizations, there may be a high correlation between scores on several dimensions of our measure of "nonbureaucracy." Such are the findings of Punch (1969), with respect to public schools. Perrow (1967: 208) notes the difficulties of operationalizing measures of organizational structure and comments on the relative success of a number of scholars in so doing.

5. Those familiar with the important work of James D. Thompson (1967) will recognize that what I refer to as technological efficiency he calls technical rationality. I note this because my thinking is much influenced by Thompson, although I have found that his term is confusing to those who do not know his writing.

6. These propositions are derived from a considerable body of analysis including that of J. D. Thompson (1967: 56-61), Pelz (1970: 144-148), V. A. Thompson (1965), Udy (1965: 700), Katz and Kahn (1966: 214), Havelock (1969: ch. 6); Bennis and Slater (1968: chs. 1, 3, 5); Smith (1970); Sarason (1971); Smith and Ari (1968); Morse (1970).

7. The specification of these conditons draws heavily on Thompson (1967) and Perrow (1967). In addition to Thompson's and Perrow's analyses, which I have used inferentially, the work that seems most central to the specification of these conditions is that of Eugene Litwack (1968). But Litwack is concerned with the distinction between appropriate roles for bureaucratic organizations *as opposed to* primary groups, and bases his distinctions very heavily on the importance of expertise to the problem-solving situation. In my view, nonbureaucratic organizations may or may not be primary groups and may be highly expert (e.g., a group of research scientists seeking a cure for cancer).

8. There is no consensus of which I am aware about how one might categorize human needs relevant to work (compare Cartwright, 1965). This list is meant to be representative rather than definitive. Some students are much more specific than I have been (Murray, 1938); others more general (Maslow, 1954). Also see Katz and Kahn (1966: 341 ff.).

9. Smith and Tannenbaum's (1968: 79) research suggests that the desire for participation in decision-making is not the same as desire for control over the organization. While the workers they studied wanted more control over their own jobs, they did not want to significantly reduce the power of those at "upper" levels of the organization. Indeed, there is reason to believe that productivity can decline in circumstances where workers achieve more power than supervisors since conflict is generated and goal incongruence occurs when this condition exists in formally hierarchical structures (Smith and Ari, 1968).

10. For example, one study of juries, which represent a nice analog to the nonbureaucratic organization, found that after their deliberations, juries expressed less confidence in their ability to be fair than before the experience (Bevan et al., 1958).

11. This discussion is heavily influenced by Katz and Kahn (1966: 303-308).

12. To carry this possibly obtuse example further, the television commentators, despite the fact that Landry calls most of the plays, invariably speak of Roger Staubauch's Cowboys. They were always Lombardi's Packers.

13. The tendency to vest one's fate in the hands of an apparently strong leader is not limited, of course, to work groups, and the consequence is not always bureaucracy in the Weberian sense. Soemerdjan (1957), for example, points out that, in times of external threat, revolutionary organizations are predisposed to respond to charismatic leadership. In times of crisis, charisma rather than bureaucracy may be the refuge of nonbureaucratic groups, but, in either case, the technological and motivational advantages of nonbureaucratic organization will be reduced substantially.

14. Earl Braxton brought his cycle to my attention in this way.

15. This discussion closely follows that of Yuchtman (1972). See also the numerous sources on the kibbutz cited there.

16. On this point, also see Golembiewski (1965: 103, 107). As he points out, when cohesion results from hostility toward the organization, supervision, or competitive groups, high satisfaction with work can be associated with low performance.

17. For a general discussion of the importance of such tension in securing innovation, see Hirschman and Lindblom (1962).

18. The dysfunctional character of singular criteria measures of performance, as well as some of the problems of securing composite measures, are discussed usefully by Ridgway (1956).

19. For example, the negative consequences of low levels of integration for organizational goal attainment are likely to be greatest when the power of the organization to make demands on or to compete with other groups depends on its capacity to speak with one voice.

20. For important work on this question, see Emery and Trist (1965); Warren (1967);

Terreberry (1968); and especially Thompson (1967). Only Warren, however, deals primarily with public organizations.

21. Reviews of theoretical studies which would be of help on this matter include the work of Slater (1966), Cangelosi and Dill (1965), and Sherif and Sherif (1968).

22. Lindblom's thesis has received some, but still inadequate, attention since he wrote. See Frank Marini's (1971) collection of essays by advocates of "the new public administration."

REFERENCES

ADAMS, S. J. (1965) "Inequity in social exchange," pp. 277-299 in L. Berkowitz (ed.) Advances in Experimental Social Psychology. New York: Academic Press.

AIKEN, M. and J. HAGE (1966) "Organizational alienation: a comparative analysis." Amer. Soc. Rev. 31, 4: 497-507.

ANDERSON, J. G. (1968) Bureaucracy in Education. Baltimore: Johns Hopkins Press.

ARGYRIS, C. (1968) "Organizations: effectiveness," pp. 311-318 of Volume 2 of International Encyclopedia of the Social Sciences. New York: Macmillan and Free Press.

--- (1964) Integrating the Organization and the Individual. New York: John Wiley.

--- (1960) Understanding Organizational Behavior. Homewood, Ill.: Dorsey.

ASCH, S. K. (1965) "Effects of group pressures upon the modification and distortion of judgments," pp. 393-401 in W. Proshansky and B. Seidenberg (eds.) Basic Studies in Social Psychology. New York: Holt, Rinehart & Winston.

BACHMAN, J. G. and A. S. TANNENBAUM (1968) "The control-satisfaction relationship across varied areas of experience," pp. 241-249 in A. Tannenbaum (ed.) Control in Organizations. New York: McGraw-Hill.

BARNARD, C. (1938) The Functions of the Executive. Cambridge, Mass.: Harvard Univ. Press.

BASS, B. M. (1960) Leadership, Psychology and Organizational Behavior. New York: Harper.

BAUMGARTEL, H. (1956) "Leadership, motivations and attitudes in research laboratories." J. of Social Issues 12: 24-31.

--- and G. GOLDSTEIN (1961) "Some human consequences of technical change." Personnel Administration 24, 4: 24-31.

BENDIX, R. (1968) "Bureaucracy," pp. 206-219 in Volume 2 of International Encyclopedia of the Social Sciences. New York: Macmillan and Free Press.

BENNIS, W. and R. SLATER (1968) The Temporary Society. New York: Harper Colophon.

BERKOWITZ, L. (1953) "Sharing leadership in small decision making groups." J. of Abnormal and Social Psychology 48: 231-238.

BERLEW, D. E. and D. T. HALL (1964) "The management of tension in organizations: some preliminary findings." Industrial Management Rev. 6: 31-40.

BERLINER, J. (1967) "Russia's bureaucrats: why they're reactionary." Trans-action: 53-59.

BEVAN, W., R. S. ALBERT, P. R. HORSEAUX, P. N. MAYFIELD, and G. WRIGHT (1958) "Jury behavior as a function of the prestige of the foreman and the nature of his leadership." J. of Public Law 7: 419-449.

BLAI, B. (1964) "An occupational study of job satisfaction and need satisfaction." J. of Experimental Education 32: 383-388.

BLAU, P. (1955) The Dynamics of Bureaucracy. Chicago: Univ. of Chicago Press.

BONJEAN, C. M. and G. G. VANCE (1968) "A short-form measure of self-actualization." J. of Applied Behavioral Sci. 4, 3: 297-312.

BORG, W. R. (1957) "The behavior of emergent and designated leaders in situational tests." Sociometry 20: 95-104.

BRAGER, G. (1969) "Commitment and conflict in a normative organization." Amer. Soc. Rev. 34 (August): 492-504.

BREWER, J. (1971) "Flow of communications, expert qualifications and organizational authority structures." Amer. Soc. Rev. 36: 475-484.

CANGELOSI, V. E. and W. R. DILL (1965) "Organizational learning: observations toward a theory." Admin. Sci. Q. 10: 175-203.

CARPENTER, H. H. (1971) "Formal organizational structural factors and perceived job satisfaction of classroom teachers." Admin. Sci. Q. 16, 4: 460-466.

CARTWRIGHT, D. (1965) "Influence, leadership control," pp. 1-47 in J. March (ed.) The Handbook of Organizations. Chicago: Rand McNally.

CENTERS, R. and D. E. BUGENTAL (1966) "Intrinsic and extrinsic job motivations among different segments of the working population." J. of Applied Psychology 50, 3: 193-197.

CHESLER, M. (1963) The Teacher as Innovator, Seeker and Sharer of New Practices. Ann Arbor, Mich.: Institute for Social Research.

CHILD, J. (1972) "Organization structure and strategies of control: a replication of the Aston study." Admin. Sci. Q. 17 (June): 163-177.

COCH, L. and J.R.P. FRENCH, Jr. (1948) "Overcoming resistance to change." Human Relations 1: 512-532.

COLEMAN, J. F. et al. (1958) "Task difficulty and conformity pressures." J. of Abnormal and Social Psychology 57: 120-211.

COOPER, R. (1967) "Alienation from work." New Society (July): 1-10.

CROZIER, M. (1964) The Bureaucratic Phenomenon. Chicago: Univ. of Chicago Press.

--- and B. PRADIER (1961) "La pratique du commandement en milieu administratif." Sociological Travail 3: 40-52.

CRUTCHFIELD, R. S. (1955) "Conformity and character." Amer. Psychologist 10: 191-198.

DAVIS, L. E. (1966) "The design of jobs." Industrial Relations 6 (October): 21-45.

--- and E. S. VALFER (1966) "Studies in supervisory job designs." Human Relations 19, 4: 339-352.

--- (1965) "Intervening responses to changes in supervisory job design." Occupational Psychology 39: 171-189.

DEUTSCH, M. (1968) "Groups: group behavior," pp. 265-276 in Volume 6 of International Encyclopedia of the Social Sciences. New York: Macmillan and Free Press.

--- (1949) "An experimental study of the effects of cooperation and competition upon group process." Human Relations 2: 199-231.

DOWNS, A. (1966) Inside Bureaucracy. Boston: Little, Brown.

DOYLE, W. J. (1971) "Effects of achieved status of leaders on the productivity of groups." Admin. Sci. Q. 16: 40-50.

DUNCAN, R. B. (1972) "Characteristics of organizational environments." Admin. Sci. Q. 17: 313-327.

EMERY, F. E. and E. L. TRIST (1965) "The causal texture of organizational environments." Human Relations 18: 21-32.

EMERY, F. E., E. THORSRUD, and K. LANGE (1966) Field Experiments at Christiana Spigerwerk. London: Tavistock.

ETZIONI, A. (1965) "Dual leadership in complex organizations." Amer. Soc. Rev. 30 (October): 688-698.

--- (1964) Modern Organizations. Englewood Cliffs, N.J.: Prentice-Hall.

--- (1961) A Comparative Analysis of Complex Organizations. New York: Free Press.

——— (1957) "Solidaric work groups in collective settlements." Human Organization 16: 2-6.

EXLINE, R. V. and R. C. ZILLER (1959) "Status congruency and interpersonal conflict in decision-making groups." Human Relations 12: 147-162.

FESTINGER, L. et al. (1950) Social Pressures in Informal Groups. New York: Harper.

FRENCH, J.P.R., Jr. et al. (1960) "An experiment on participation in a Norwegian factory." Human Relations 13: 3-19.

FRIEDLANDER, F. (1966a) "Importance of work versus nonwork among socially and occupationally stratified groups." J. of Applied Psychology 50, 6: 437-441.

——— (1966b) "Motivations to work and organizational performance." J. of Applied Psychology 50: 143-152.

GOLDTHORPE, J. H., D. L. LOCKWOOD, F. BECHOFER, and J. PLATT (1968) The Affluent Worker: Industrial Attitudes and Behaviour. Cambridge, Eng.: Cambridge Univ. Press.

GOLEMBIEWSKI, R. T. (1965) "Small groups and large organizations," pp. 87-141 in J. G. March (ed.) The Handbook of Organizations. Chicago: Rand McNally.

GROSS, N. et al. (1958) Explorations in Role Analysis. New York: John Wiley.

GUETZKOW, H. and J. GYR (1954) "An analysis of conflict in decisionmaking groups." Human Relations 7: 367-382.

GURIN, G. et al. (1960) Americans View Their Mental Health. New York: Basic Books.

HACKMAN, R. and E. E. LAWLER III (1971) "Employee reactions to job characteristics." J. of Applied Psychology 55, 3: 259-286.

HAGE, J. and M. AIKEN (1969) "Routine technology, social structure, and organizational goals." Admin. Sci. Q. 14 (September): 366-378.

——— and C. B. MARRETT (1971) "Organizational structure and communications." Amer. Soc. Rev. 36 (October): 860-871.

HALL, R. H. (1972) Organizations: Structure and Process. Englewood Cliffs, N.J.: Prentice-Hall.

HALPIN, A. (1966) Theory and Research in Administration. New York: Macmillan.

HALTER, H. (1965) "Attitudes towards employee participation in company decision-making processes." Human Relations 18, 4: 297-319.

HAVELOCK, R. G. (1969) Planning for Innovation Through the Dissemination and Utilization of Knowledge. Ann Arbor, Mich.: Institute for Social Research.

HAWLEY, W. D., J. McCONAHAY et al. (1973) What If There Was a School Where They Tried All the Good Ideas. New Haven, Conn.: Yale University Center for the Study of Education.

HERRICK, N. Q. (1972) Where Have All the Robots Gone? New York: Free Press.

HERZBERG, F. (1966) Work and the Nature of Man. Cleveland: World.

——— (1965) "Job attitudes in the Soviet Union." Personnel Psychology 18: 245-252.

HIRSCHMAN, A. O. and C. E. LINDBLOM (1962) "Economic development, research and development, policy making: some converging views." Behavioral Sci. 7: 211-222.

HOLLANDER, E. P. (1960) "Competence and conformity in the acceptance of influence." J. of Abnormal and Social Psychology 61: 365-369.

HOMANS, G. C. (1968) "Groups: the study of groups," pp. 259-265 in Volume 6 of International Encyclopedia of the Social Sciences. New York: Macmillan and Free Press.

——— (1950) The Human Group. New York: Harcourt, Brace & World.

HOUSE, R. J. et al. (1971) "Relation of leader consideration and initiating structure to R and D subordinates' satisfaction." Admin. Sci. Q. 16: 19-30.

JANIS, I. (1972) Victims of Groupthink. Boston: Houghton Mifflin.

JONES, S. C. and V. H. VROOM (1964) "Division of labor and performance under cooperative and competitive conditions." J. of Abnormal and Social Psychology 68: 313-320.

KAHN, R. L. (1972) "The meaning of work: interpretations and proposals for measurement," pp. 159-205 in A. Campbell and P. E. Converse (eds.) The Human Meaning of Social Change. New York: Russell Sage.

––– D. M. WOLFE, R. P. QUINN, J. D. SNOCK, and R. ROSENTHAL (1964) Organizational Stress: Studies in Role Conflict and Ambiguity. New York: John Wiley.

KATZ, D. and R. L. KAHN (1966) The Social Psychology of Organizations. New York: John Wiley.

KELLY, H. H. (1951) "Communications in experimentally created hierarchies." Human Relations 4: 39-56.

KIRKPATRICK, F. et al. (1964) The Image of the Federal Service. Washington, D.C.: Brookings Institution.

KOHN, M. (1971) "Bureaucratic man: a portrait and an interpretation." Amer. Soc. Rev. 36: 461-474.

LAWRENCE, L. C. and P. C. SMITH (1955) "Group decision and employee participation." J. of Applied Psychology 39: 334-337.

LEWIN, K. and R. WHYTE (1939) "Patterns of aggressive behavior in experimentally created social climates." J. of Social Psychology 10: 271-299.

LICHTMAN, C. M. (1970) "Some intrapersonal response correlates of organizational rank." J. of Applied Psychology 54, 1: 77-80.

LIKERT, R. (1961) New Patterns of Management. New York: McGraw-Hill.

LINDBLOM, C. (1965) The Intelligence of Democracy. New York: Free Press.

LIPSKY, M. (1971) "Street level bureaucracy and the analysis of urban reform." Urban Affairs Q. 6: 391-410.

LITWACK, E. (1968) "Technological innovation and theoretical functions of primary groups and bureaucratic structures." Amer. J. of Sociology 73 (January): 478-481.

––– and H. J. MEYER (1965) "Administrative styles and community linkages of public schools: some theoretical considerations," pp. 49-98 in A. J. Reiss, Jr. (ed.) Schools in a Changing Society. New York: Free Press.

McDAVID, J., Jr. (1959) "Personality and situational determinants of conformity." J. of Abnormal and Social Psychology 58: 241-246.

MANN, F. C. and L. R. HOFFMAN (1960) Automation and the Worker. New York: Holt.

MARCH, J. and H. SIMON (1958) Organizations. New York: John Wiley.

MARGULIES, N. (1969) "Organizational culture and psychological growth." J. of Applied Behavioral Sci. 5, 4: 491-508.

MARINI, F. [ed.] (1971) Toward a New Public Administration. San Francisco: Chandler.

MASLOW, A. H. (1954) Motivation and Personality. New York: Harper.

MECHANIC, D. (1962) "Sources of power of lower participants in complex organizations." Admin. Sci. Q. 7: 349-364.

MENZIES, I.E.P. (1960) "A case-study in the functioning of social systems as a defense against anxiety." Human Relations 13: 95-121.

MERTON, R. K. (1958) "The role set: problems in sociological theory." British J. of Sociology 8: 106-120.

––– (1957) Social Theory and Social Structure. New York: Free Press.

MORSE, J. J. (1970) "Organizational characteristics and individual motivation," pp. 84-100 in J. Lorsh and P. Lawrence (eds.) Studies in Organizational Design. Homewood, Ill.: Irwin-Dorsey.

MORSE, N. (1953) Satisfactions in the White Collar Job. Ann Arbor: University of Michigan Survey Research Center.

––– and E. REIMER (1956) "The experimental change of a major organizational variable." J. of Abnormal and Social Psychology 52 (January): 120-129.

MURRAY, H. A. (1938) Explorations in Personality. New York: Oxford Univ. Press.

NEWCOMB, T. (1959) "The study of consensus," pp. 277-292 in R. K. Merton et al. (eds.) Sociology Today: Problems and Prospects. New York: Basic Books.

PARSONS, T., R. F. BALES, and E. SHILS (1953) Working Papers in the Theory of Action. New York: Free Press.

PATCHEN, M. (1970) Participation, Achievement and Involvement on the Job. Englewood Cliffs, N.J.: Prentice-Hall.

PAUL, W. T., Jr., K. B. ROBERTSON, and F. HERZBERG (1969) "Job enrichment pays off." Harvard Business Rev. 47, 2: 61-78.

PELZ, D. C. (1970) "The innovating organization: conditions for innovation," pp. 144-148 in W. Bennis (ed.) American Bureaucracy. Chicago: Aldine.

——— (1956) "Some social factors related to performance in a research organization." Admin. Sci. Q. 1: 310-325.

——— and F. M. ANDREWS (1966) "Autonomy, coordination, and stimulation in relation to scientific achievement." Behavioral Sci. 11 (March): 89-97.

PEPINSKY, P. N. et al. (1958) "Attempts to lead, group productivity, and morale under conditions of acceptance and rejection." J. of Abnormal and Social Psychology 57: 47-54.

PERROW, C. (1967) "A framework for the comparative analysis of organizations." Amer. Soc. Rev. 32: 194-208.

PORTER, L. W. and E. LAWLER III (1965) "Properties of organizational structure in relation to job attitudes and job behavior." Psych. Bull. 64: 23-31.

PUGH, D. S., G. R. HININGS, and C. TURNER (1968) "Dimensions of organizational structure." Admin. Sci. Q. 13 (March): 65-105.

PUNCH, K. F. (1969) "Bureaucratic structure in schools: towards redefinition and measurement." Educ. Admin. Q. 5: 43-57.

PYM, D. (1963) "A study of frustration and aggression among factory and office workers." Occupational Psychology 37, 3: 165-179.

RAVEN, B. H. (1968) "Groups: group performance," pp. 288-293 in Volume 6 of International Encyclopedia of the Social Sciences. New York: Macmillan and Free Press.

REYNOLDS, P. D. (1971) "Comment on the distribution of participation on group discussion as related to group size." Amer. Soc. Rev. 3: 704-707.

RICHARDS, C. B. and H. F. DOBRYNS (1957) "Topography and culture: the case of the changing cage." Human Organization 16 (Spring): 16-20.

RIDGWAY, V. F. (1956) "Dysfunctional consequences of performance measurements." Admin. Sci. Q. 1: 240-247.

SARASON, S. (1971) The Culture of the School and the Problem of Change. Boston: Allyn & Bacon.

SAWATSKY, J. C. (1941) "Psychological factors in industrial organization affecting employee stability." Canadian J. of Psychology 5: 29-38.

SHAW, D. M. (1960) "Size of share in task and motivation in work groups." Sociometry 23: 203-208.

SHAW, M. E. (1959) "Acceptance of authority, group structure and the effectiveness of small groups." J. of Personality 27: 196-210.

——— (1954) "Some effects of problem complexity upon problem solution efficiency in different communication nets." J. of Experimental Psychology 48: 211-217.

SHEPARD, J. M. (1969) "Functional specialization and work attitudes." Industrial Relations 8, 2: 185-194.

SHERIF, M. and C. W. SHERIF (1968) "Groups: group formation," pp. 276-283 in Volume 6 of International Encyclopedia of the Social Sciences. New York: Macmillan and Free Press.

——— (1963) Groups in Harmony and Tension: An Integration of Studies of Intergroup Relations. New York: Harper.

SLATER, P. (1966) The Microcosm: Structural, Psychological and Religious Evolution in Groups. New York: John Wiley.

SMITH, C. G. (1970) "Consultation and decision processes in a research and development laboratory." Admin. Sci. Q. 15.

––– and O. ARI (1968) "Organizational control structure and member consensus," pp. 145-163 in A. Tannenbaum (ed.) Control in Organizations. New York: McGraw-Hill.

SMITH, C. G. and M. E. BROWN (1968) "Communication structure and control in a voluntary organization," pp. 129-144 in A. Tannenbaum (ed.) Control in Organizations. New York: McGraw-Hill.

SMITH, C. G. and A. TANNENBAUM (1968) "Organizational control structure: a comparative analysis," pp. 129-144 in A. Tannenbaum (ed.) Control in Organizations. New York: McGraw-Hill.

––– (1963) "Organizational control structure: a comparative analysis." Human Relations 16: 299-316.

SOEMERDJAN, S. (1957) "Bureaucratic organizations in time of revolution." Admin. Sci. Q. 2: 182-199.

SORCHER, M. (1967) "Motivating the hourly employee." General Electric Personnel and Industrial Relations Services: 1-24.

STEPHAN, F. F. and E. G. MISHLER (1952) "The distribution of participation in small groups: an exponential approximation." Amer. Soc. Rev. 17: 598-608.

SVETLEK, B. et al. (1964) "Relationships between job difficulty, employee's attitude toward his job, and supervisory ratings of the employee effectiveness." J. of Applied Psychology 48: 320-324.

TANNENBAUM, A. and J. G. BACHMAN (1968) "Attitude uniformity and role in a voluntary organization," pp. 229-238 in A. Tannenbaum (ed.) Control in Organizations. New York: McGraw-Hill.

TERREBERRY, S. (1968) "The evolution of organizational environments." Admin. Sci. Q. 12: 377-396.

THIBAUT, J. W. and L. H. STRICKLAND (1956) "Psychological set and social conformity." J. of Prospectives 25: 115-129.

THOMAS, E. J. (1957) "Effects of facilitative role interdependence on group functioning." Human Relations 10: 347-366.

THOMPSON, J. D. (1967) Organizations in Action. New York: McGraw-Hill.

THOMPSON, V. A. (1965) "Bureaucracy and innovation." Admin. Sci. Q. 10: 1-20.

TRIST, E. L. and K. W. BAMWORTH (1951) "Some social and psychological consequences of the Longwall method of coal-getting." Human Relations 4: 3-38.

TRUMBO, D. A. (1961) "Individual and group correlates of attitudes toward work-related change." J. of Applied Psychology 45: 338-344.

UDY, S. H., Jr. (1965) "The comparative analysis of organizations," pp. 678-709 in J. G. March (ed.) The Handbook of Organizations. Chicago: Rand McNally.

VAN ZELST, R. H. (1952) "Validation of a sociometric grouping procedure." J. of Abnormal and Social Psychology 47: 299-301.

VERBA, S. (1961) Small Groups and Political Behavior: A Study of Leadership. Princeton: Princeton Univ. Press.

VROOM, V. H. (1964) Work and Motivation. New York: John Wiley.

––– (1962) "Ego-involvement, job-satisfaction, and job performance." Personnel Psychology 15: 159-177.

––– (1960a) "The effects of attitudes on the perception of organizational goals." Human Relations 13: 229-240.

––– (1960b) Some Personality Determinants of the Effects of Participation. Englewood Cliffs, N.J.: Prentice-Hall.

––– (1959) "Some personality determinants of the effects of participation." J. of Abnormal and Social Psychology 59: 322-327.

WAFFORD, J. G. (1971) "Managerial behavior, situational factors, productivity, and morale." Admin. Sci. Q. 16: 10-17.

WALKER, C. R. (1954) "Work methods, working conditions and morale," in A. Kornhauser et al. (eds.) Industrial Conflict. New York: McGraw-Hill.

——— (1950) "The problem of the repetitive job." Harvard Business Rev. 28: 54-58.

——— and R. H. GUEST (1952) The Man on the Assembly Line. Cambridge, Mass.: Harvard Univ. Press.

WARREN, D. I. (1968) "Power, visibility and conformity in formal organizations." Amer. Soc. Rev. 33 (December): 951-970.

WARREN, R. L. (1967) "The interorganizational field as a focus for investigation." Admin. Sci. Q. 12: 369-419.

WASHBURN, C. (1957) "Teacher in the authority system." J. of Educ. Sociology 30: 390-394.

WEISS, R. S. and D. RIESMAN (1961) "Social problems and disorganization in the world of work," pp. in R. K. Merton and R. A. Nisbet (eds.) Contemporary Social Problems. New York: Harcourt, Brace & World.

WHYTE, M. K. (1973) "Bureaucracy and modernization in China: the Maoist critique." Amer. Soc. Rev. 38: 149-163.

WICKER, A. W. (1969) "Size of church membership and members' support of church behavior settings." J. of Personality and Social Psychology 13, 3: 278-288.

——— (1968) "Undermanning, performances, and students' subjective experiences in behavior settings of large and small high schools." J. of Personality and Social Psychology 10: 255-261.

WILENSKY, H. (1964) "Varieties of work experience," pp. 125-154 in H. Borow (ed.) Man in a World of Work. Boston: Houghton Mifflin.

WOLIN, S. (1960) Politics and Vision. Boston: Little, Brown.

WORSCHEL, P. (1961) "Status restoration and the reduction of hostility." J. of Abnormal and Social Psychology 63: 443-445.

YUCHTMAN, E. (1972) "Reward distribution and work-role attractiveness in the kibbutzim: reflections on equity theory." Amer. Soc. Rev. 37 (October): 581-595.

Part V

LESSONS FROM THE
PRIVATE SECTOR

14

Business and the Urban Crisis:
The Case of the Economic Development
Council of New York City

DAVID ROGERS

□ BUSINESS HAS BEEN INVOLVED IN EFFORTS to revitalize cities since at least the turn of the century. It was quite active, for example, in the Progressive Era, trying to help make city government more efficient and "businesslike." Indeed, partnerships between business and city hall have emerged in countless small and larger towns, reflecting a traditional American belief that business intervention with "nonpolitical" solutions was a way of saving the cities. This is an approach not at all limited to local government, as indicated, for example, by the borrowing of business executives to streamline defense procurement, the formation of the Hoover and Asch Commissions, and the hiring of such management impressarios as McNamara, even in Democratic administrations.

BUSINESS AND CITIES

At the city level, after several decades of relative inaction, there was a swing back in the 1960s, as business leaders, government officials, and even many liberals urged greater involvement of business in the urban crisis (see Birch, 1967; Chamberlain, 1970).[1] The ghetto riots, a growing disenchantment in some circles with government as the main agency for solving urban problems, and an increasing hope in the problem-solving skills of business were among the reasons for this development.

Indeed, a new "reprivatization" strategy has emerged as a way of improving the performance of government. Peter Drucker (1968: pt. 3), one of its foremost advocates, argues that we should substitute the dynamism and rationality of business for the inefficiency of government, through subcontracting out public sector programs. The Committee for Economic Development's (1971) recommendation that government-business partnerships be established to deliver public services through nonprofit corporations is another version of this strategy.

Most of the new social programs of business started in the late 1960s. Thus, the summer riots of 1967 led to the formation of the National Urban Coalition, with its numerous local branches, and of the National Alliance of Business's JOBS (Job Opportunities in the Business Sector) Program. Meanwhile, the Department of Labor, the Economic Development Administration, and other federal agencies placed great emphasis on private sector involvement in such programs as the Concentrated Employment Program and Special Impact (see Cohn, 1971; Ruttenberg and Gutchess, 1970).

Moreover, virtually every major corporation and business association developed urban-oriented programs during this period. Departments of urban and environmental affairs became commonplace, with corporations like AT&T and First National City Bank of New York, for example, embarking on major efforts at technical assistance to beleaguered urban school systems. The insurance industry, to take another example, began its two billion dollar program of loans for low-income housing in inner cities.

It is clear, then, that business has become at least somewhat committed to trying to reverse the decay of our cities, both for economic self-interest reasons—e.g., business leaders in inner cities are concerned with having an adequate labor supply—and as an expression of social conscience (for one of Ford's public pronouncements, see Chamberlain, 1970: 27). There has, in addition, been strong outside pressure on business to do something. Former Mayor Jerome Kavanaugh of Detroit gave an impassioned speech (see Birch, 1967: 24) at a Harvard Business School conference on business and the city in 1966 in which he ended with a plea to business leaders: "If not you, who? If not now, when?" And Peter Drucker (1969: 75) has pointed out that traditional "liberals" and even some "radicals" now urge business to take over many governmental functions.

It takes more than just urging, however, to keep business involved, and the experience of the past few years, viewed nationally, has not been hopeful. Business seems to have cut back a lot in its contributions to such organizations as the Urban Coalition and the National Alliance of Business by 1970 (Cohn, 1971). This cutback corresponded, and not just coincidentally, with a recession, de-creasing ghetto unrest, a rising white backlash, and a conservative trend nationally. It remains to be seen, then, how deeply involved business may become in urban problems throughout the nation in the future.

One group for whom this cannot be said, and whose action programs constitute an important effort of American business to deal with urban problems, is the Economic Development Council (EDC) of New York City, an association of over 185 large employers. Formed in 1965 to change city land-use patterns to retain and attract industry and stem the loss of male jobs, EDC has since moved on to a much broader urban development strategy that merits close exami-nation for possible use elsewhere. Since 1969, it has maintained a series of task forces of executives and professionals (e.g., lawyers) on loan and of some in retirement from the city's largest employers, attempting to transfer the management expertise of business to several municipal agencies. The goals of the program are to improve agency productivity and, where possible, to effect organizational change. EDC is now involved in four high schools, two junior highs, the school headquarters' bureaucracy, the city's superagency for welfare, manpower training, and related social service programs, the city criminal courts, and the criminal and civil branches of the state supreme court. In 1969, it began a program in the criminal courts and two high schools, and I have focused on these efforts, along with its school headquarters project, because they have been in operation long enough to merit intensive examination.

Top EDC officials contend, and with some justification, that a big part of the urban crisis is the inefficiency, low productivity, and generally bad management of many municipal agencies that, even within their present budgetary constraints, are not delivering services of nearly the quantity and quality that they might. Without condemning the civil servants as the main culprits, the council maintains that business executives trained in management techniques—planning, goal-setting, organizing, controlling, performance appraisal, work flow analysis, and the like—have a lot to offer the public sector to make it more productive, and in some instances to promote organizational change as well.

CONCEPT AND METHOD

The basic question of the program and of my assessment is: Are management skills developed in the private sector in fact transferable to the public sector? Top officials in the EDC obviously believe they are and have spent the last five years developing programs based on that assumption. Some government officials believe that they are not, arguing that conditions in government are so different. By doing a comparative study of EDC programs in several city agencies and the courts, I was able to ask the question about transferability in a more refined way. Instead of posing it as an all-or-nothing issue, I could explore what types of interventions (attempted transfers) had what impacts in what types of agencies.

Before an analysis of the program, it is important to underscore its national significance. It is the first and certainly one of the most developed efforts in recent times at the exercise of corporate social responsibility through a management intervention strategy. As such, it is an important experiment in social and organizational change. Specifically, that effort brings to bear the management and professional expertise of the private sector to the public sector, which has come under much criticism for being poorly managed; and it merits intensive study to identify the critical factors that shaped its successes and failures and that might facilitate the transfer (replication) of its productive elements to other agencies in New York and elsewhere. Indeed, one of the main implications of this program is to indicate either the futility of such private sector involvement in trying to reform municipal agencies or, as may be more likely, to specify the conditions under which it can be done well. The analysis that follows deals with these considerations.

Ideally, an assessment of a program like this would involve an evaluation study that attempted to relate in causal fashion particular program inputs to changes in agency productivity and performance.

Given the program's newness, its evolving character and great complexity, a more informative method is an analytical case study of the *process* of developing and implementing this strategy, so as to learn what is going on, how it works, and why (for extensive recent discussions of different types of evaluations, see Weiss, 1972; Caro, 1971). This involves various techniques of qualitative analysis—for example, using critical incidents to illustrate more general patterns; selecting cases of potentially greater and lesser EDC impact for description; and suggesting typologies of innovations and intervention strategies for further analysis. These techniques allow the researcher and policy makers to see a program in a holistic way and to identify the range of variables that may affect its operations. That is essentially what I have been doing,[2] relying heavily on direct observation, interviews, and a review of documents, including agency records of costs and performance.

ASSESSMENT OF EFFECTS

Several kinds of indicators are relevant in ascertaining the effects of EDC's programs. First, one can look at the extent of change in agency structure and procedures that seem related to EDC's presence. The number of such innovations, the proportion that were institutionalized in contrast to those that were dropped, and their scope of impact within the agency can all serve as indicators. The assumption is that such organizational changes will contribute to improved agency performance, even though their effects may not be discernible in the time period of the study.

In addition, where data are available and adequate, not often the case in the public sector, one can look at the extent of change in agency costs and productivity during the time of the program. Undue emphasis on a cost-effectiveness analysis, however, seems premature and misplaced for a program like this, for reasons given above.

ORIENTING CONCEPTS AND PERSPECTIVES

A critical part of the description and analysis of this program is the tracking of what happened to the EDC-initiated ideas for innovations, to find out why they got the receptions they did within the agencies. This was done in terms of three sets of variables:

(1) *characteristics of the target agency;*

(2) *characteristics of the interventions,* including both the type of innovations suggested and how they were introduced and implemented; and

(3) *the structure and operating codes of EDC itself,* especially its relations with its membership and its recruiting of corporate executives on loan (a review of various approaches to changing organizations appears in Hornstein et al., 1971).

EDC: ITS HISTORY, STRUCTURE, AND OUTLOOK

EDC represents the largest employers in New York City, including banks, insurance companies, utilities, department stores, and corporate headquarters. The top staff has defined its role, however, as representing broader interests than those of business alone.[3] The urban management program described in this paper illustrates that role. Reflecting business's concern that competent people fill entry-level jobs, it also aims to improve the management of key city agencies so that they serve other client groups as well as business.

This urban management program came about in the following way: EDC began by working on the *demand* side of the labor market, trying to attract and retain industrial jobs for the city. It soon became apparent, however, that the city's inadequate space for industry as well as its limited labor pool would limit EDC's success in that effort. Admittedly, the reasons for industry and corporate moveouts are often very complicated, but the need for manpower is critical. Furthermore, even if the industrial exodus cannot be stopped, the city has more and more white-collar jobs in both the public and private sectors[4] that its minority populations should be trained to fill. They constitute an increasing part of the city's labor pool and make up the majority of its welfare recipients, both unemployed and underemployed.

The council then began to seek out strategies to deal with the *supply* side of the labor market, especially with upgrading the occupational skills, work habits, and aspirations of minority groups. Its rationale was that it wanted to help the poor and the minorities become an employable population. This would, in turn, serve to meet the economy's manpower needs, limit the city's welfare burden, place less of a drain on city services, and perhaps encourage economic expansion.

The particular strategy that EDC evolved was to provide management assistance to the human resources agencies directly involved in developing work attitudes and skills. The school system and the Human Resources Administration (HRA) constituted natural targets in that regard. EDC's involvement in the criminal courts, on the other hand, resulted from its concern with the markedly increasing crime in the city as this affected its economic climate and public safety. Several of its members and top staff talked about reforming the entire criminal justice system and soon realized that that was too monumental a task. They finally settled on the criminal courts as a critical part of that system, whose reform alone might have positive spillover effects on other parts—the police, the District Attorneys' offices, probation, and correctional institutions.

After some dramatic early successes in the courts, EDC was then asked to help improve the management and operations of the city's Human Resources Administration, the Board of Education headquarters, and the state Supreme Court. Top officials of these agencies asked for EDC's help, but it should be added that EDC made known its availability and interest. It did not sit idly by and wait to be asked. By and large, top officials in the agencies were pleased with EDC's work, especially with the open-mindedness with which it approached them and its disinclination to impose programs or structures on them in any unilateral way.

EDC's approach was to set up separate task forces for each of its intervention efforts in particular agencies and staff them with executives on loan and with some in retirement. They were almost all provided on a voluntary basis, without any expense at all to the city or the agencies.[5] An important determinant of EDC's effectiveness in the agencies was how successful it had been in recruiting executives from its member companies, in orienting them for the task, and in keeping them motivated and productive while they were on loan, usually for a one-year period. At times, it has been a problem to get the companies to provide executives in adequate numbers when they were needed, and with the required skills, as I will indicate below. Most member companies of EDC were much more willing to give or provide services in their area of expertise than to lend executives.

EDUCATION

One of the first and most complicated agencies that EDC worked in was the school system. It has been through a series of civic convulsions in recent years, culminating in the conflict over community control in 1968. Its performance, as measured by any of the traditional criteria—reading scores, dropout rates, absenteeism, pupil suspensions—was poor and deteriorating steadily. And it had resisted previous efforts at reform and seemed likely to do so in the future (Rogers, 1968; Berube and Gittell, 1969).

Some time in the summer of 1968, the EDC staff and leadership began looking for an appropriate strategy to deal with this problem. By that time, a number of EDC's member companies felt they were hurting badly enough from the lack of employable graduates to want to take some action.

EDC had no specific expertise, however, in public education, and it certainly had no understanding of how the school system worked. It brought in a newly formed organization, the Institute for Educational Development (IED), to advise it on these matters. IED, a nonprofit educational research and development corporation, was well suited to that task. It was concerned mainly with the development and diffusion of innovations in urban schools; it had a particular interest in establishing closer linkages of business with schools; and it was close to the community control controversy, having worked on establishing criteria for evaluating the proposals that various groups had developed.

HIGH SCHOOL PARTNERSHIPS

Initially, EDC did not know whether to begin in particular schools and districts or at school headquarters. Some of its top officials wanted to take on the entire system, but they soon took IED's advice to concentrate on the high schools. To ensure that these interventions would be effective, IED did an informal inventory of the main "partnerships" throughout the nation in which business had entered into such a collaborative relationship with a high school (Institute for Educational Development, 1969a).

Perhaps the most important contribution of IED was to urge on EDC a participative strategy for intervention. The first thing that

EDC did before involving itself with several high schools on a continuing basis was to initiate a "needs analysis" at the schools under IED's leadership, which involved getting all the main participants in the school together in a group problem-solving situation. This included students, teachers, supervisory staff, parents, interested community organizations, and business; and the purpose was to open communication, develop credibility and trust among all parties, and encourage them to generate strategies for improving the school. The participants were not presented with any established agenda, but instead were asked to put forth what they saw as the main problems in the schools. They established goals, ranked them in order of importance, formulated specific programs to work toward these goals, and developed procedures for their implementation. As in most such cases, the process rarely went in that smooth and orderly a way. Nevertheless, a process was begun in these schools that was indispensible in the later development and implementation of programs.

EDC did several other things, at IED's urging, that further helped establish its legitimacy in the schools. In addition to indicating that it would not impose itself or any preconceived programs on the schools, EDC assumed a low profile, taking little credit for its involvement. There was no publicity-seeking by EDC and its participating companies and no attempt to take credit. Little emphasis was placed on the immediate manpower concerns of these companies, lest it appear that the entire project was created to serve the interests of business. Furthermore, the full minutes of all the group sessions were circulated throughout the school, to maximize the visibility of the activities. Much good will and trust developed as a result, and business had thus established its credibility in these schools.

Since 1970, EDC has been involved in four high schools, two in Manhattan and one each in the Bronx and Brooklyn. All are schools with a predominantly poor black and Puerto Rican student body and such common inner-city problems as overcrowding, underachievement, high absenteeism, dropout rates, and drug usage, and much student, teacher, and parent apathy. While they were not the worst high schools in the city, they were in serious trouble. And one of them, George Washington, in Upper Manhattan, was in complete chaos when EDC arrived in 1970.

IMPACT

It is not that easy to assess the impact of EDC's interventions in these schools. Several of the conventional types of indicators are all but irrelevant.

(a) Aggregate data on school performance—for example, attendance, dropout rates, reading and math scores, types of diplomas (academic, vocational, general), achievements of graduates (placement and performance in college, jobs), and vandalism rates—are generally not available. Neither headquarters nor the schools have ever collected them in a systematic way for very long (see New York, City of, 1971-1973). Beyond that, what data do exist are suspect in terms of their validity. Attendance data, for example, are based on counts in homerooms, but many students appear there at the start of the day, only to wander around the school and then leave. Reading scores, for another example, are taken from among students who do attend class, and this obviously excludes the many low achievers who rarely show up.

Furthermore, there are other critical, perhaps more important, variables, in addition to EDC's intervention, that affect school performance. The socioeconomic, racial, and ethnic composition of the student body, especially as this is affected by zoning decisions and neighborhood patterns over which EDC has had little control, are particularly relevant. There have been marked changes in student backgrounds in these schools just since EDC first arrived.[6]

(b) The impact of individual EDC-initiated programs—e.g., remedial and career education, anti-drug, and so on—on students and teachers is another indicator of effectiveness. EDC and IED have collected systematic data on some of these programs, and the data generally indicate improvement. Some of it may be due to a Hawthorne Effect,[7] in which case one must periodically check to see how lasting the benefits are. To give a few examples:[8] (1) Mini-schools were started in three high schools. In one (George Washington Prep), average daily attendance of those students improved from 15% while they were in the regular high school to 90%, compared with 70% among regular students. Furthermore, 60% of the students in this mini-school were passing all their subjects in its first year; compared with only 10% when they were in the regular school the year before. (2) Of the 185 students involved in an

intramural sports program at another high school, 85% had improved attendance the first year. Teachers and the principal reported a lessening of intergroup tensions at the same time, as students from diverse backgrounds played together on teams. The program existed at all four high schools, with similarly positive reports from them all. (3) An attendance program at George Washington that involved giving recognition to students with perfect attendance for two consecutive months had a very dramatic effect. Gross attendance in the school increased from 61% in January 1973 to 74% in May. Moreover, students with perfect monthly attendance increased over that same time period from 14% to 22%, while those with perfect attendance for two consecutive months increased from 7% to 15%. (4) A football team was started at that school, engendering a school spirit of the kind that had not existed there for many years. Among those athletes who were low achievers, and most of them were, grade averages increased in 1972 from 57% to 70%. Also, the number of courses they failed decreased from 3.2 per student (each semester) to 1.3 after the start of the program. Finally, semester absences per low-achiever athlete dropped from an average of about 30 to 3. (5) A new course in career education for ninth graders in another school, providing information about and exposure to different types of occupations increased students' awareness and aspirations as measured by scores on an occupational maturity test, and attendance also went up.[9]

These programs are obviously limited and incremental, and for some critics of school reform efforts by business, they may even have a "Mickey Mouse" quality. But teachers, supervisors, and students report that they often function as morale builders in the school and seem to improve student motivation and achievement.

(c) Changes in the climate of the school involve still a third type of indicator, relating to this morale-building function. At least two aspects of school climate seem relevant—first, the extent of *improved relations* among students, teachers, the administration, and the wider community (greater trust, communication); and second, the extent to which *increased innovation and experimentation* take place in the schools.

Interviews with EDC and the school officials suggest some improvement in morale and internal relationships among the various parties, resulting mainly from the participative, group problem-solving approach of EDC, in its needs analysis of these schools. But the evidence is uneven, and in some schools, it has been difficult to maintain high morale in the face of continued problems. On the positive side, EDC had an office in each of the schools, and it often became a communications center. As one EDC director recalled: "The EDC office became a place for everybody to gather. There were even teachers' union meetings there. Kids about to storm the principal's office first went to ours, and we helped them develop more constructive ways of handling their grievances. Eventually they got a big new bilingual program which was one of the things they wanted. We were liked and respected by everybody. Nobody thought we were partisan to any particular sides."[10]

The most dramatic improvement took place in George Washington High School in Manhattan, which was in a state of chaos when EDC arrived. As described in one report:

> Police were stationed at the gates and doors and throughout the halls; the street was crowded with patrol cars; barricades and riot helmets were prominently in evidence; several hundred students sat and stood on the lawn near the front entrance. Four principals had served in the school during the previous year and a temporary principal was trying to keep order until a permanent successor could be found; the parents were divided into militant factions . . . an elusive group of dissatisfied and disaffected students whiled away most of the days in the cafeteria; neighborhood merchants closed their doors when students left in the afternoon; and the threat and reality of violence against any person, young or adult, pervaded the campus [Barnes and Connolly, n.d.].

Over the next several months, EDC played a productive broker role as catalyst in getting students, teachers, supervisors, and parents together in a way that had a positive effect on the school. A committee was set up with representatives of all these groups to identify critical problems and develop alternative methods of solving them—the main ingredients of the "participative program planning process" already discussed. After being closed about 20% of the time in the spring of 1970, just before EDC arrived, the school was only closed for two half-days from mid-October through the following June and was not closed at all the following school year. EDC deserves credit for helping to improve the school climate and stimulating a few key changes in the school during that time. The cafeteria, for example, a center of dope peddling, intense student conflicts, and uniformly bad food, was reorganized and got a new director; an entirely new security system was instituted, under the leadership of a security specialist from an EDC company; and a new principal was appointed and has been there ever since. Most of these changes resulted from the work of the EDC-initiated planning committee.

Not all the schools, however, had such positive changes. Success seemed to depend heavily on the skills of the participants, especially the EDC director, the principal, and involved teachers and supervisors. It also depended on whether these participants were able to control outside forces that affected the school. For example, in the high school where EDC's office had been a communications center, morale sagged noticeably when the school was beset with major problems induced from outside, after it had so improved following EDC's arrival. Several new high schools opened in the same borough, and a series of rezonings took place that victimized this school most of all. The new schools took away most of its remaining white, middle-class students; and it was suddenly faced with many more poor black and Puerto Rican students from ghetto areas in the borough. In addition, 250 additional students were suddenly assigned at the beginning of the school year, without any prior notification.

This high student turnover, then, made it very difficult to establish common student and staff expectations as to how programs should proceed, or, indeed, to have any stable programs at all. EDC officials were aware of the problem, but felt that zoning was not one area of their expertise and that they did not want to impose themselves in such school decisions that were so complex and political. In some of the partnership schools, the educators and parent and community groups were successful in preventing or diluting a rezoning that would radically change the student body, though obviously not in this one. The result in this case was that some of the good things EDC had initiated, including improving the school's morale and spirit, were temporarily negated.[11]

Nevertheless, the developmental process that EDC began, even in that school, continued, and that relates to the second aspect of school climate referred to above—namely, the spirit of experimentation. EDC's goal in these high schools is to instill in them a built-in capacity for changing themselves or, in the jargon of organizational development, for "self-renewal." It wants to change the culture of the schools to create a climate of innovation. Specifically, this means having the capability and actually developing new curricula, instructional methods, staffing arrangements, organizational structures and procedures, and relations with employers and community groups.

Top EDC and IED officials see the partnerships as having set up modest R & D units in these high schools as laboratory sites for demonstration projects that can then be replicated in other high schools in New York and perhaps in other cities as well. Just evaluating the partnerships from the perspective of how much experimentation has taken place in these schools, the record to date is positive. As a result of EDC's intervention, the four schools contain at least a dozen innovative programs that are being further developed and refined for replication elsewhere.

The following are among the main programs that EDC has helped develop in these schools (for a summary of these programs, see EDC School Partnership Program, 1973):

(1) A *professional development center* at one school, training teachers as change agents to diffuse new curricula and instructional techniques throughout the school and have them train other teachers to be change agents as well.

(2) A *career education program* in one school that will be schoolwide in a year or two, depending on funding, and that gives students increased exposure to work, both through direct work experiences and through industry visitors to the school and materials on careers and jobs in all school subjects.

(3) *English as a second language* departments in three of the schools for students with native tongues other than English and involving most school subjects.

(4) *Remedial math and reading programs* for underachieving students, with new curricula and instructional materials brought in from industry, other school systems, or developed with EDC consultants.

(5) A *model drug education and prevention program* in one of the schools that involves counselling by peers and has since been replicated in seventeen others throughout the city.

(6) Finally, several *programs for alienated students* to improve their school attendance, motivation, and achievement, including the minischools, a "call up" program to get students back in school, intramural sports, and "Achievement Motivation" workshops with staff to help them develop a better rapport with these students.

All these innovative activities were developed at little cost to the schools, mostly through government and foundation grants and through consultants brought in by EDC and IED.

STRENGTHS

In conclusion, several observations can be made about this strategy for improving inner-city high schools through business intervention. On the positive side, the participative management approach is critical, activating a developmental process that helps the schools generate many new programs, with business assistance. Establishing an office in each school in an R & D capacity to support that process by providing resources, enlisting school headquarters support, and generating more innovative activities is also important.

Unless business or some similar outside agency comes into urban schools, neither research nor development will take place. The school staff is caught up in day-to-day problems, and there is little reward for such activity. Furthermore, few educators have the capability to innovate in any systematic way. Business, by contrast, does have that capability, having already been successful at conducting R & D activities in its own setting. The effectiveness of EDC in these four high schools indicates business's capacity to export that R & D expertise to the public sector.

Another strength of this partnership strategy is business's ability to move from short- to longer-term activities. In the early stages of such partnerships, the educators define problems in their schools as due mainly to limited resources rather than to any shortcomings in the school itself; and they keep asking the businessmen for more money, supplies, and jobs for students. Business is thus placed into what one of EDC's officials called the Santa Claus role, and, in an effort to establish credibility, it may have to play that role to some considerable extent the first year or so.

After a while, business should indicate, as EDC did, that the school and students would benefit much more if it put its limited resources into longer-term developmental activities. Though it will take time, the educators may become gradually convinced of the wisdom of this approach. If business were to take this position right away, indicating what some of the major shortcomings of the schools are, the educators would probably dismiss it as just another harsh, outside critic.

Two recent developments in EDC's program suggest some optimism for its positive impact. One is an increasing emphasis on the schools' having their own internal capability for developmental activities, so that they will not have to be perpetually dependent on an outside assist from business. This requires securing a position in the school for a full-time teacher-coordinator and for several positions to release permanent teaching staff from class periods for developmental work, as EDC has done. Meanwhile, the number of businessmen on loan in each school may then be decreased.

A second development is IED's having written a how-to-do-it manual on activating such a needs analysis and development process, so that the innovative programs begun in these four high schools can be exported elsewhere and the other schools can, in addition, develop programs of their own. The manual contains extensive documentation on how and why the process went well at George Washington High School, so that others can learn from that experience.

LIMITATIONS

It is important, however, to recognize the potential limitations of business partnerships as well. This is not to imply a rejection of the strategy, but rather an indication of the pitfalls to avoid and the ways in which it might be improved. First, it often involves incremental interventions; and one can question whether significant change can take place in this fashion, in view of the failure of such past efforts at reform as compensatory education and demonstration programs. The main reason is that the existing bureaucratic and political structure of urban school systems has generally remained intact while these programs were being implemented, and it has tended to absorb, dilute, and discredit them, just by the normally cumbersome workings of the bureaucracy. Also, the collective bargaining agreement with the teachers' union and existing by-laws and contracts with supervisors constitute tremendous constraints on new programs. If the existing headquarters structures of urban school systems can be changed and greater flexibility negotiated in union contracts, these partnerships will have a more receptive climate in which to develop. EDC is now actively engaged in a major project to reorganize school headquarters, as I will discuss below, and this may result in significant changes that will help the high school partnerships. The existing system, however, is still quite unresponsive to the need for innovation.

A particular limitation of this strategy is that one may end up not having enough control over critical variables that affect school conditions, thereby hampering efforts to generate and implement innovative programs. The example cited above on how new zoning practices so changed the student body in one school that morale deteriorated and planning became difficult illustrates the point. This suggests the importance of business assisting local parent groups and educators to become more organized to try to resist such zoning decisions, as well as intervening at higher levels to help provide more support and political power for the school.

A further problem is the difficulty of institutionalizing effective programs. Business loanees typically only serve for a year, and their departure may mean the end of an innovative program, unless the necessary expertise to keep it going is transmitted to educators inside the system or to somebody recruited from outside. To cite one example: an EDC loanee had introduced simulation games as an instructional technique in several social studies classes of one school. Teachers were reportedly enthusiastic about it and had adopted it for use, with his assistance. After he left, however, its popularity waned, and a year later, it was barely used at all.[12] Teacher turnover further exacerbated the problem. Without continuity of personnel, it is difficult to maintain innovative programs over time.

A key to the success of the partnership strategy is how effectively the innovative programs can be replicated elsewhere. Selecting an appropriate mechanism for doing that is obviously critical. EDC has decided to work within the system on this. Unfortunately, most urban school system bureaucracies, and certainly New York's, have little if any capability to replicate new programs, even when some of their staff have an interest in doing so. The vast majority of New York's school headquarters staff may fit this description; and they are, if anything, probably more skilled than their counterparts in other big cities. To rely on this approach to replication, then, does not seem hopeful.

It makes little sense, though, to completely bypass top school officials, since they have so much power to block or dilute new programs and would likely exercise it in that instance. A more productive approach would be to work through the headquarters staff while trying at the same time to retrain them, to bring in new, more competent people, and to reorganize the bureaucracy so that the delivery system is more effective. EDC is working on these strategies in New York. Moreover, key headquarters officials invited teachers from the EDC partnership schools to develop the curriculum for their innovative programs over the summer of 1973, an unusual step in view of the top-down elitist style that used to prevail.[13] This may reflect a change from the traditional approach of developing curriculum with little, if any, classroom teacher participation at all. If it indicates broader changes at headquarters, then perhaps there can be some hope.

Regardless of whether business relies on the system itself for replication or works through an outside organization, it must provide strong leadership. This means convincing headquarters, district, and school officials of the need to develop and implement the innovative programs begun in the partnership schools. A subtle balance has to be maintained between not imposing any innovations unilaterally and taking on so little initiative that the natural inertia of the system will prevent implementation. EDC's participative management strategy is probably the most effective, even though it takes a long time, but an active business role on replication is needed. A new technical assistance group recently developed by EDC and composed of business, labor, and educators (from within the New York City schools) reflects the kind of active role required.

For replication to be successful, teachers and supervisors would have to be rewarded to work on such developmental activities. On this matter, EDC has pressed for released time from the classroom, funds from school headquarters, and for new positions to be established at the schools. Perhaps new titles, other status perquisites, and some bonuses may be effective as well. Unless that is done, teachers may well feel exploited once again and not participate as actively as would be desirable.

Finally, business has to exert constant pressure on school headquarters to ensure that it plays a supportive role vis-à-vis local high schools, where new programs are being replicated. This would involve a major reorganization there, a project actually being undertaken by an EDC headquarters task force whose activities can now be described and analyzed.

THE HEADQUARTERS TASK FORCE

In 1971, after learning of EDC's reorganization of the courts, the chancellor and his board invited EDC in as a consultant organization to help improve the management of headquarters. In all the furor over decentralization, there had been little discussion about the future role of headquarters. And yet the central bureaucracy of the New York City schools was not just going to fade away, nor would it support decentralization, unless active steps were taken to make it do that. Making headquarters over into more of a service center, turning its many bureaus and other units out toward the schools and completely disbanding some of them, and sending many head-quarters staff back to the districts were among the needed reforms. Upgrading and revamping the business operations there, including payroll, audit, fiscal, personnel, purchasing, and setting up a management information system were also essential. Finally, there was a great need to clarify the functions and powers of the board and chancellor, to create new top management positions, and to bring in people with the administrative capability to fill them.

EDC took on the challenge, though for the first year or so it was engaged primarily in fact-finding and had few effective contacts with the Board of Education. Its task force did studies of particular headquarters departments and made recommendations for improve-ment. It also proposed an entire reorganization of the headquarters bureaucracy. Little if any implementation took place, however, and it was not until late 1972 that much got under way, as EDC began to push harder at that time, under new task force leadership.

The main goal of EDC in this activity was to improve the management of headquarters in the context of decentralization. This involved changing its structure, procedures, and staffing patterns; and one could assess EDC's impact by the extent to which that took place. Perhaps the most important characteristic of the bureaucracy when EDC first intervened was that central headquarters had continued to operate as though decentralization had never taken place. The bureaus, divisions, and other headquarters units still functioned in an administrative line capacity on local school operations and failed to relate to the schools in a technical assistance and service capacity. Many did not have the training or expertise to do so, and some had no interest in it. Some headquarters staff regarded the move toward decentralization as a necessary but perhaps only temporary political response to citizen protest and reportedly felt that when that, too, failed, as had all previous reform measures, then headquarters would once more assume its position as the unchallenged authority in an essentially hierarchical institution. Furthermore, some members of the central board were opposed to decentralization, and they had at least some power to shape school policy.[14] To illustrate: At least one central board member opposed inviting representatives of community school boards to participate in citywide collective bargaining with the teachers' union, though some eventually were allowed in. Later, the union's president and board members reported that they had done a very good job. In another case, the central board refused to invite these representatives to participate in discussions regarding the selection of a new chancellor and deputy chancellor.

The EDC task force worked on three critical issues to reform the New York City school bureaucracy. They included the reorganization plan to restructure positions, personnel, and departments; intensive work on improving particular business departments; and work with community school boards to improve their management.

The reorganization plan was by far the most ambitious. The main push was to separate sharply the business and administrative from the purely educational activities of the system, so that the educators could concentrate on pedagogical matters which they knew something about and stay out of administration, about which they knew little so that more competent administrators might be brought in. In addition, the plan tried to clarify the roles of the lay board and the professional staff, defining more sharply the policy authority of the board and the administrative authority of the professionals.

Several new top positions were established in this reorganization, reflecting the separation of education from administration. A strong deputy chancellor position was created, with direct responsibility for overseeing managerial and administrative functions, to allow the chancellor to devote more time to educational issues. Directors of such important business operations as personnel, audit, school buildings, and an Office of Business and Administration are all responsible to the deputy chancellor. Meanwhile, four new top positions were created to strengthen the chancellor's authority and free him to concentrate on citywide educational questions. These positions included an executive director for community school district relations, to assure closer ties between headquarters and the districts; for educational planning and support, responsible for new programs, staff training, and general educational services; and for centralized school administration that includes all school programs not yet decentralized by law—one each for the high schools and special service schools. Finally, every effort was made to limit the administrative authority of the lay board by defining its role as limited to broad policy questions.[15]

In retrospect, this was a fairly remarkable and unprecedented involvement of a business group in the workings of municipal government. Never before had any such group intervened to provide so much technical assistance or attempted to initiate and help implement such major reforms in an agency of this size (for an account of a much smaller-scale effort in New Haven by Olin's Winchester Group and Southern New England Telephone Company, see Institute for Educational Development, 1969b). And it was done at no cost to the city by executives on loan, whose salaries were paid for by their companies. Certainly, this had never been done in any big-city school system. Furthermore, EDC had supplied more than thirty executives for the effort, with expertise in particular management functions whose reform was at issue. Many of the reforms they suggested, if implemented, may result in major savings, and thereby increased funds for educational programs in the schools and classrooms.

For example, the auditing, personnel and payroll operations have been completely reorganized, with many problems having been identified and resolved (Economic Development Council of New York City, Inc., n.d.: 8-9). Faulty management of payroll resulted in new employees having to wait from six to nine months before their salary checks reflected what they should be getting; and every month several thousand others were not paid at all, while there were overpayments of about $8 million in 1971, resulting from checks being issued to teachers who had left. The headquarters task force chairman estimated operating economies achieved as of June 1973 at about $10 million a year, with the likelihood that such economies might eventually reach much higher. It is too soon to estimate how valid this projection may be, but the savings are likely to be quite substantial.

On balance, the board welcomed EDC's assistance. It had asked for the study, including a review of itself, and it recognized the value of the recommendations. Indeed, at a press conference in March 1973, at which the EDC proposed reorganization plan was unveiled after the board had passed on it, the board president acknowledged that the city school system was, in effect, "one of the largest corporations in America," but "we have been running a huge corporation like a candy store." He went on to say that "operations were often inefficient, lines of command were often unclear" (New York *Times*, 1973b).

As in all attempts at changing organizations, having creative ideas does not ensure their implementation. Throughout the project, there were long delays before the board adopted EDC's recommendations, and some were not adopted at all—for example, those relating to changes in the board's own functions, to get them out of administration.[16] The delays and inaction were due largely to the fact that the New York City school system has remained a crisis-ridden agency. The particular crises during EDC's project included a sharp conflict between the board and the chancellor over his future in the system, zoning controversies involving issues of racial balance, and many local disputes regarding the workings of community school districts. The chancellor's contract was terminated in early 1973, at a critical time, when EDC was trying to get the board to act on the reorganization plan, and that finally resolved the situation enough so that EDC could get the board to move ahead. The reorganization had actually been delayed from early 1972, when it was first announced publicly, until March 1973. Given the fate of past reform proposals, especially in this agency, that was actually not too unreasonable a schedule.

The next issue was selecting people to fill the many top positions created by the reorganization proposal. The board indicated that it wanted EDC to play a key role in screening and recommending candidates and that it would welcome EDC's judgments. And since EDC officials felt strongly that the reorganization would only be a success if competent people were put in those top management positions, they eagerly took the challenge. Throughout much of 1973, one of EDC's senior staff's main activities was to conduct an extensive nationwide search for qualified candidates. EDC's only criteria were that the candidates have the appropriate administrative and technical skills to fill these jobs.

These were not always the criteria that the board used, however, and it rejected EDC's candidates for at least three positions. In each case, it selected people inside the school system or presently employed in the city over "outsiders" EDC had recommended. Its two main criteria for selection seemed to be, first, to maintain an ethnically balanced representation, and second, to get people who would not be so independent as to challenge its active role in administration. It thus put on a façade of looking nationally for top people while apparently being much more oriented to selecting local people. Since EDC officials had spent so much time in these executive searches and took the board at its word that it wanted them to take the lead in recommending people for these positions, they were understandably dismayed when several of their choices were rejected. The fact that at least two of the key selections were revealed to them through press leaks further troubled them.

One important lesson from this experience is that, in a public bureaucracy such as this, representational criteria—in this instance, ethnic ones—inevitably affect staff appointments at all administrative levels. EDC staff did not initially take this into account, operating instead on apolitical, "objective" criteria. This is not to argue that ethnic balance should be an important criterion on such decisions, but only that it is now and that such political criteria are as relevant and rational in their own way as others, given the stakes that various ethnic interest groups have in how the school system works. EDC and its successors would do well, then, to take ethnic considerations more into account, along with technical ones, in such decisions.

On the more positive side, there has been substantial implementation of EDC's reorganization plan. New positions and departments have been set up. Lines of authority and responsibility have been altered and clarified. Virtually all headquarters personnel, both administrative and educational, have been shifted around and jobs redefined to take into account the new dual hierarchy of management and education. Also, the board did select well-qualified people for some of the top management positions, despite its differences with EDC over some appointments. There has, then, been some movement toward reform.

Perhaps equally as important as the reorganization are the individual projects of EDC executives on particular business operations at headquarters. This is an area in which business has expertise and where it can perform a useful public service. The New York City school system has a $2.5 billion annual expense budget, and yet has never undergone any kind of systematic audit. Indeed, it does not even have what could legitimately be called an auditing department. Neither does it have any management information system. Moreover, in its warehousing operation, it has a thirty-five-year inventory on some items, mainly because it does not have adequate records and keeps trying to buy at lower prices, almost regardless of need (for an analysis of the mismanagement of New York City schools, see Rogers, 1970a). The amount of administrative incompetence, inefficiency, and waste are mind-boggling. One contributing factor is that key management positions are held by former teachers and supervisors with little if any administrative ability, and yet they are not held accountable for their actions. By contrast, there are able people who enter the bureaucracy with management and technical skills, but they tend to remain outside the informal politics of the system and have no incentive to stay, since their salaries are generally lower than those of people who have come up through the ranks from the classroom. The result is mediocrity in many administrative positions (for a description and analysis of the personnel practices of the New York City schools, see Rogers, 1968: 285-295 and 1970b).

EDC started in this field by making a list of the business problems it thought were most important, on such matters as payroll, purchasing, audit, personnel, and the like. It then went over these with top professional staff and with the lay board and got started on those that they mutually agreed should have highest priority and would be most workable.

In each project, EDC had an executive on loan with expertise in that particular field. He would first do a thorough study and documentation of how the unit had been working, usually in collaboration with one or more EDC colleagues. The reports would then be made available to the chancellor and lay board, with recommendations for improvements. Major changes in structure, procedures, and personnel would sometimes be recommended, as it became clear how mismanaged these departments were. The fact that the chancellor and board were so shocked by the well-documented reports indicates how little they knew what was going on and how far out of control the bureaucracy had got.

These studies were strong and in given instances startling, and EDC was tactful in the way it went about doing them. On most reports, it followed a policy of getting input from the department it was studying to such a degree that they became almost co-authors. This provided some professional upgrading of department staff, helped build a relationship of trust that enabled EDC to gain entree and do the study in the first place, and then eased the staff's resistance to many of its recommendations. Many staff in the business departments welcomed EDC's presence as giving them management assistance that they had never had before. Depending on what happens over the next couple of years in implementation, this could conceivably be one of the most important of all EDC's activities. The need for improved management in the New York City schools is so great that EDC's efforts could have far-reaching implications, not only in New York, but nationwide.

The third EDC activity in its headquarters project was to provide management assistance to community school boards and districts under decentralization. In June 1973, EDC selected a community school board in a middle-class area of Queens that it felt was representative enough of many others for such assistance. Its staff interviewed all key interest groups in the district and prepared a report on how it could be better managed. The main recommendations paralleled those EDC made at headquarters. Nonpedagogical, administrative, and business activities were to be delegated from the community superintendent to a new deputy, to be in charge of all business-related matters. This would free the superintendent to spend more time on education issues.

EDC's goal is to see if it can replicate the programs it develops in this district or, if not, to develop different ones elsewhere. It sent out letters to all thirty-two community boards, offering to explain what it had done at headquarters and to provide management assistance, if invited. As of this writing, ten boards have asked for such assistance, and, even at this early stage, it looks as though EDC has been identified as an organization that has worked vigorously to reform headquarters and make it more supportive of local needs. Its credibility has been enhanced by the fact that it has touched base with all involved community groups in the Queens district and has indicated to local groups that it does not want to impose any preconceived plans on them but rather to help them do things for themselves. On the community side, the superintendent, his staff, and school board members have expressed strong interest in seeking out such assistance. They reacted positively to EDC's recommendations. Many lack confidence in their capacity to handle effectively the complex issues they face; and they have a lot of faith in the capacity of business to help. Such unrealistically high expectations could lead to disenchantment with business later on, if it fails to deliver as hoped, but EDC could probably temper those expectations and make them more realistic.

The ultimate test of all these intervention efforts in the headquarters project, of course, is their implementation, and the record has to be incomplete at this time. A key consideration affecting the likelihood of future success is the power base and political skills that EDC is able to develop, as it continues its efforts at reform. The fact that it is gaining credibility locally, as well as among the civic groups and headquarters staff who want change may be a positive sign in that regard. Its successes in the criminal courts and its capacity to document in considerable detail managment problems in the school bureaucracy have added to its legitimacy. Now in its third year of operation at school headquarters, it has much more influence than before and could begin to speed up the implementation of its recommendations for reform.

One dilemma that slowed EDC down in the past was the conflict between the lay board as its client—and therefore feeling it had to work collaboratively with the board—and having to go "outside" and mobilize pressure on the board to get its reform recommendations implemented. It chose not to do the latter and was in a weakened position as a result. Yet, if it had gone outside and the board had found out, it probably would not have been able to continue its work. Actually, EDC's client is as much the wider citizenry of New York as it is the board.

EDC has now embarked on a significant reform strategy in this regard, to make the school system much more efficient and responsive to citizen groups than it was in the past. Its efforts to improve management at school headquarters and the districts have indicated many inadequacies in the way the current citywide board is selected and functions. It is now working with various agencies to establish new selection procedures for the board and chancellor and to, in essence, have a whole new trustee group more representative of citywide interest groups and serving as an advisory body only to the chancellor. This trustee group would replace the present board and reflects EDC's perception, with which I agree, that a basic change in the whole top management and policy-making structure is necessary before too many significant improvements can take place in the schools.

THE COURTS

Perhaps the most successful and certainly the most dramatic of all the EDC efforts at organizational change in city government is its project in the criminal courts. If one wants to develop a model of how to change organizations by outside intervention, in this instance of how a business group might change city agencies, the EDC courts project provides an excellent example. Delineating the main elements of that case, in terms both of what EDC did and of the structural characteristics and reactions of the courts to EDC's presence, suggests a standard against which to assess other efforts of this type.

EDC began to work with the criminal courts in early 1970.[17] Though the courts are sometimes seen as one of the most independent and insulated branches of government, conditions were right for EDC to offer assistance when it did. Court officials had a great sense of urgency to do something, since they were under heavy public attack for failing to respond to the tremendous increase in crime in the 1960s. They had more than 400,000 new cases a year in New York, and they fell so far behind in handling them that many had to be dropped. Plea bargaining became common practice, just to get cases off the calendar. The jails were overcrowded with defendants awaiting trial, and many more, some with long criminal records, walked the streets free on bail for months or even years before the judges could get to their cases (see Miller, 1972: 1).

One of the appellate court judges, having heard of EDC's interest in the criminal justice system, invited a couple of its officials in to take a look around and see what might be done. They could scarcely believe what they saw, so great was the chaos and disorganization. There was little relation between the organization chart and lines of authority; there was nothing to coordinate or integrate people, activities, and positions; everybody seemed to report to everybody else; lacking a chain of command from the top, the bottom ran itself, and with such inefficiency that the judges in the busiest courthouse averaged just over three hours in court a day. What structure existed was extremely fragmented into numerous specialized parts or courtrooms. The same case might go through a dozen or more of these parts, each time with a separate judge who had to start from scratch. Record-keeping had broken down in a way that reflected this fragmented structure to such an extent that pieces of information on a particular case might be scattered in many different courthouses throughout the city. In each one, the proceedings were recorded in pen, reminiscent of nineteenth-century practice. Few if any incentives or rewards existed for improved work performance by court officials, given the poor records and the emphasis on seniority and traditional civil service examinations. And cost controls were all but nonexistent.

Perhaps one of the best single indicators of how far the deterioration of the courts had gone was the appearance of the courtroom. "It was just like Grand Central Station," reported one EDC official. "You couldn't even hear the judge speak, the place was so chaotic." "Some courthouses," said an EDC official, "were perpetually strewn with garbage with litter, inside and out, and decrepit beyond belief" (Miller, 1973: 3).

The board chairman of a large life insurance company that was one of EDC's members came up with the idea that since a lot of problems in the courts were processing ones—e.g., how to speed up the flow of paper work through the system—EDC should recruit executives who have that management expertise. He further indicated that, since his industry was also involved in extensive processing activities, there should be a direct transference of skills developed there to the courts. That turned out to be very much the case. Metropolitan Life, Equitable, and New York Life all offered the services of two to four executives per company per year, and they did an excellent job.

When EDC first moved into the courts, it did several things to ensure its acceptance. It indicated that it had no axes to grind, that it was not looking for personal credit, and that it had no intention initially of trying to impose new structures or procedures on the courts. Instead, it asked the judges and other court officials to indicate what their problems were and how EDC could help. It was only much later, after considerable groundwork had been laid through EDC's documentation of the courts' vast mismanagement and inefficiency that it made any recommendations. Over the short term, in a fashion similar to the high school partnerships, it helped the courts obtain financing for new court facilities and for space renovation studies.

Indeed, court officials exerted the same kind of pressure on EDC that the educators had. "They wanted more money and bodies in the court for day to day work," reported an EDC task force member, "and we kept trying to resist that, not because we don't want to help but because we thought we could help more in other ways."

The most important thing EDC did was complete documentation from the bottom up of how the system worked, as a basis for eventually making recommendations for change. This gave EDC as an "outsider" group with no prior knowledge of the courts considerable legitimacy with many court officials; and it made it very difficult for those judges and court administrators who resisted EDC's recommendations to be able to do so authoritatively.

There were two parts to EDC's early activity in the courts. One was some work flow and measurement-type studies of particular parts, looking at management procedures and results. The other was a fairly extensive organizational structure study to find out more why work flows and efficiency were not what they might be (for a good summary of this work, see Economic Development Council Task Force, 1970).

Unlike in the schools, there are fairly clear impact data on EDC's courts project. The case backlog, for example, was cut from 59,000 when EDC's reorganization was implemented in 1971 to 14,000 a couple of years later, the first such reduction in eighteen years. The average length of cases went down from nine to three and one-half weeks. The number of defendants in detention was cut from 4,200 to less than 1,300.[18] Furthermore, a study by a New York City accounting firm and member of EDC, covering the period from January 1971 through June 1972, indicated a saving of more than $6.7 million a year and one-time savings of $48.5 million (Peat, Marwick, Mitchell & Co., 1973).

It would be a mistake to attribute all these improvements just to EDC's recommendations. There were other projects attempting to improve court operations going on at the same time that undoubtedly contributed. But EDC, along with these other reform projects, deserves credit.

Equally as relevant as cost and performance improvements, and directly responsible for their existence, were the many structural changes that resulted at least in part from EDC's studies. Taken together, they constitute an impressive array of management innovations. First, the overspecialized and highly fragmented parts structure was simplified and converted where physically possible to all-purpose parts. This meant that the same cases were much less likely to be handled over and over again in different parts, each time with a different Legal Aid lawyer and with separate records. Second, a clear and logical organizational structure was set up, with a strong top management in the position of the administrative judge and his executive officer and with explicit lines of authority and reporting relationships all the way down the chain of command. Much closer supervision and management control existed than ever before, with the creation, in addition, of new supervisory judges for each borough. Also, there was much better communication than ever before throughout the courts, especially from the administrative judge's office.

All these administrative changes were, in turn, reflected in sharp improvements in discipline within the courts. There were, for example, more sanctions imposed on lawyers for lateness and nonappearances, resulting in more judge days worked over the year and more output per judge per day. This is to say that performance measures had been developed, and they were enforced. Lawyers who appeared late or not at all were often reprimanded by the administrative judge, who even sent letters to the presiding judge on some of these cases. Lawyers regarded this as a strong sanction and improved their performance accordingly.

A third change was the creation of internal training programs for all court security officers. There was, in addition, a new statistical staff, concerned with developing an information system so that better internal monitoring and program planning could take place as well. EDC prepared procedures manuals covering all clerical actions in the courts that are now being used.

The benefits of improved management, discipline, and internal controls in the criminal courts have extended beyond the courts to the wider criminal justice system. The activities of attorneys, corrections personnel, police, and probation officials, for example, were much more closely integrated into those of the courts in a way that resulted in fewer bottlenecks and vastly improved management. To give two of many possible examples: the execution of warrants for court appearances is now done much faster than in the past. The tracking by police of people who jump bail has also been improved because of pressure from the administrative judge. Since the police know that those warrants will result in court action, given the speedier processing of cases, they now see it as worth their while to bring in people who have jumped bail. The number of disappearing defendants has consequently decreased, and the percentage of warrants executed went from fifty to ninety.

Perhaps the most significant organizational change of all was a subsequent administrative unification in New York County and the Bronx between the Criminal Court and the Criminal Branch of the State Supreme Court. EDC officials had recommended this, based in part on their observation that so many felony cases start at the Criminal Court and then move up. "Why not appoint the same administrative judge for both courts?" suggested the EDC Supreme Court task force, and in 1973 that was done. It stopped a lot of duplicate work, constituted another step toward limiting administrative fragmentation, and generally should speed up the judicial process.

In conclusion, EDC seems to have been quite successful in its efforts to affect structural change and improve the workings of the criminal courts. Again, it would be a mistake to impute all these changes to EDC alone, since many judges and court officials, especially the Administrative Judge, as well as outside agencies, were also responsible. However, the Administrative Judge and several top judges of the Appellate Division have said on numerous public occasions that EDC's studies and recommendations have indeed had a major impact on the courts. Some of these judges attended EDC's semi-annual board meeting in December 1972 and gave unsolicited testimonials to its work. They said that it was the most significant thing of this kind that had ever happened to them while working in the courts.

EDC INTERVENTIONS IN COMPARATIVE PERSPECTIVE:
TOWARD A THEORY OF SOCIAL INTERVENTION

EDC has been involved in other agencies as well as the ones discussed above, including the Human Resources Administration, its many operating agencies, and the Civil Branch of the State Supreme Court. Though EDC has had successes in these other agencies, its greatest success, at least in the short run, has been in the criminal courts. This is not to say that its task force in the criminal courts is necessarily more skilled in affecting organizational change than the others. Rather, it is to suggest that a broad constellation of conditions existed in the criminal courts that contributed to success there, the absence of which made it much more difficult in other agencies. A comparison of these various interventions, then, should begin to indicate what conditions are necessary to produce change in municipal agencies.

Agency characteristics: Obviously, some agencies are easier to change than others, and that seemed to be the case in the EDC experience (for some propositions relating organizational characteristics to the rate of program change, see Hage and Aiken, 1970; also see Rogers, 1971: ch. 2). The criminal courts differ from the other agencies in which EDC is involved in several characteristics that are associated with greater success in the courts case.

(1) First, the *goals* of the courts are much more explicit, clear, and measurable than those in the schools and social service agencies. A main problem in the courts is a procedural, not a philosophical one—how to reduce the backlog, speed up the time it takes defendants to move through the system, decrease the number in detention, awaiting trial, and the like. From an administrative viewpoint, then, the goals of the courts are to process as many cases as rapidly and efficiently as possible, consistent with dispensing

justice. Those were essentially the goals of the EDC task force, enunciated by one EDC official as the "swift application of fair justice in the courts." The courts have other goals as well, including a concern for judicial rights and rehabilitation, but the implementation of these is often hampered by administrative problems, given the overcrowding in some jails and the case backlog. Indeed, court officials often commented on how improvements in management resulting from EDC's intervention helped secure defendant rights by limiting the likelihood that they would have to sit in the Tombs (the city prison) for months on end, awaiting trial.

(2) *The authority and power structure of the courts,* as fragmented as it was, was much less so than that of the Board of Education, the high schools, and the Human Resources Administration. One of the main reasons for this in the latter three agencies is that they are flooded with state and federally funded programs, each with its own guidelines. Collectively, they further increased the internal fragmentation in these agencies, making them literally unmanageable. Through federal monies and an increasing public interest in reforming the criminal justice system, the courts may be moving in the same direction; but they have nowhere near the number of such programs as the other agencies.

(3) In the criminal courts, *the resources and commitment to change of their top management* constitute still another condition favorable to implementing EDC's recommendations.[19] The Administrative Judge has much greater control over the courts and consequently more power to implement reform proposals than did the Chancellor of the Board of Education or the Administrator of the Human Resources Administration. Both of the last two were strongly committed to change and had in fact invited EDC in; but they had to contend with numerous vested interests inside and outside their agencies who opposed reform. In brief, the Administrative Judge had so much more power than the heads of the other two agencies that he was able to implement EDC's recommendations much more easily than they. In fact, it was Judge Ross who was so central to EDC's successes in the criminal courts. His commitment to its recommendations, his leadership, and his capacity to implement new programs were critical. As a former City Council President, State Assemblyman, and State Supreme Court Judge, he has many political contacts and skills; at the same time, the structure of his court gave him a power base that his counterparts in the other two agencies lacked. The backing of Presiding Justice Stevens in the Appellate

Division, first department was equally essential. Stevens asked EDC in, and Ross was not even appointed until after EDC's first major study was completed and approved for implementation by Stevens.

(4) Another factor was the *number and diversity of subunits within the agency,* what sociologists would call its internal complexity or degree of structural differentiation. Again, while the courts certainly suffered from overspecialization and fragmentation, these conditions were not nearly as advanced there as in the schools and the Human Resources Administration.

There are two reasons for this. The greater number of discrete state and federally funded programs in the schools and HRA, as already discussed, is one.[20] A second is the tremendous difference in size and scale between them. The criminal courts, though part of a larger system of criminal justice, had themselves an annual budget of only $12 million. By contrast, the Board of Education's annual expense budget alone is over $2 billion, and HRA's is closer to $2.5 billion. The relation of this to innovation is that the greater the internal complexity in an organization, the greater the number of informal vested interests who can function as negative veto groups to delay implementation or dilute new programs by redefining them as they are implemented.

(5) The relation of these agencies to their environments was also a factor affecting their receptivity to EDC's efforts at reform. In the case of the courts, there was sharp *criticism and pressure for change* from citizen groups of all political persuasions. A concern with law and order, with increasing crime, and with the failure of the courts to respond effectively was widespread. Given this sense of urgency, court officials were under enormous pressure to improve their agency's performance—or at least to appear to be doing so.[21] Though there was certainly much outside criticism and pressure on the schools for change, it was more sporadic, and there was not the same degree of consensus among citizen groups as to what was wrong. Moreover, the concern about crime may have been greater than that about education. As for the Human Resources Administration, there was increasing public criticism of its welfare agency for mismanagement as this related to the increasing welfare rolls; and as that criticism increased, EDC had an easier time gaining legitimacy for its efforts at improved efficiency and organizational change.

The nature of the EDC interventions: Regardless of how receptive an organization may be to change, the nature of the outside intervention is critical also. Interviews with EDC and agency officials suggest several differences that may further explain the greater success in the courts.

(1) The fact that the management problems in the courts were so amenable to *work flow analysis* meant that there was much greater transferability of business expertise to this agency than to others. EDC could thus identify the task in a way that was appropriate to its capabilities. The problem in the courts, then, was largely managerial and did not involve broad policy and political considerations to the same degree as in the other agencies. To be sure, there was strong resistance to change by some judges and court administrators, but it did not hamper EDC nearly as much as the ethnic politics of HRA and the school system or the desire of the lay board to control all administrative decisions in education.

EDC's goals in the courts, then, were very concrete—e.g., speeding up the process of justice delivery—without getting trapped into policy or ideological arguments about what is "justice" or how to solve all problems of the criminal justice system. The goals of the other task forces, of innovation and making the agency more adaptable and effective were more abstract and difficult to implement and measure.

(2) The courts task force emphasized strongly *the recruiting of management analysts with agency-relevant skills.* This was done as well in the Board of Education headquarters task force, and one of its strengths was a result of that selective recruiting effort. It was done first, however, in the criminal courts task force and paid off a lot in terms of its productivity. The fact that those insurance company executives gained the respect of many court officials, despite their knowing nothing initially about the courts, is some indication of their effectiveness. In the high school partnerships, where the goals of innovation and creating a climate of "self-renewal" were so broad, it was difficult to recruit people with specialized management skills except those in organizational development. And that specialty itself is not well codified or developed.

(3) Beyond just recruiting people with relevant skills, the leaders of the courts task force were sensitive to the importance of *letting their companies know they were doing a good job.* Frequent meetings were arranged with the board chairman of the company, court officials, the executive on loan, and top EDC officials, in which contributions of the executive on loan were lauded by court officials in the presence of their superiors. This was done on the other task forces, but not to the degree or as early as the courts project.

(4) *The internal structure and patterns of supervision in the task force* were still another critical factor. The courts task force was more closely managed and supervised and probably more effective as a result. As one informant related: "The criminal courts task force had lots of soldiers rather than captains. No high-level executives from sales were in that group. Instead, they had solid people in case processing. This was critical. The other task forces had more 'high-status' men and were consequently more democratic and maybe not as effectively supervised."[2] [2]

The backgrounds and characteristics of the executives as they affect the management of task forces are all-important. Older, retired executives who may be more individualistic in style than organization-minded, younger managers tend to be harder to supervise. High-level executives on loan are probably also more difficult to control. Another group who may pose different but equally as severe problems are many from urban affairs departments, some of whom have limited administrative competence. They are sometimes placed in urban affairs as a dumping ground, because the company has no other place for them. While they have empathy for the poor and sensitivities about how to deliver services in the public sector, they often do not have the management expertise required for many EDC development efforts. People with both qualities are needed—with management skills and an understanding of public sector problems. Those loanees EDC recruited who did have both usually did an outstanding job.

(5) A hallmark of the courts project was its extensive *bottom up documentation* to indicate how the agency worked. This gave legitimacy to the task force and limited court officials' resistance to EDC's recommendations. It was simply too difficult to contradict hard data of that sort. After a while, EDC began to know much more about how the courts functioned than many judges and court officials; and while this was threatening to some, it also established EDC as more than a casual consultant. Indeed, it indicated a commitment and seriousness that further increased court officials' respect for EDC.

By contrast, the early work on HRA and the Board of Education did not have such bottom up documentation. Much of the HRA study during its first year was an analysis of the central headquarters functions; and the Administrator indicated that he already knew about that and needed instead studies of the actual delivery of services at much lower levels. After that, the HRA task force started the other way, under a new director who had previously coordinated one of the criminal courts groups and was skilled at such documentation. In the school headquarters project, the first year was conducted with almost no documentation at all. Some studies were done, but most reports were given orally by EDC to the Board of Education. Much valuable time and executive talent were thus used in less than the most efficient ways. Over the last couple of years, that condition has been corrected and with more positive results.

(6) Finally, the courts project involved looking at the *total organization* of the criminal courts and attempting to reform them on a systemwide basis, recognizing, of course, that it could not all be done at once. The same thing was done at Board of Education headquarters and HRA.

Even from such preliminary investigation, then, some suggestive hypotheses have been generated about the conditions under which a business group like EDC can transfer management expertise from the private to the public sector. It is important as well, however, to recognize the limitations of this strategy, so that future decisions either supporting or curtailing it are made in an informed way.

LIMITATIONS OF A BUSINESS, URBAN MANAGEMENT STRATEGY

A look back at history is often a good way to gain some perspective on programs like EDC's, and it is rewarding in this instance. There is a striking similarity between the EDC programs and theories of administration underlying them and those of reformist business leaders during the Progressive Era. As several historians of that period have indicated, businessmen were very active in the movement for the elimination of patronage and spoils in municipal governemnt and greater efficiency (see Hofstadter, 1948; Haber, 1964; Weinstein, 1969). In the process, they made several assumptions about how to change government agencies and about what to change. One was that government should become *nonpartisan* and employ only people with administrative competence. Another was that government needed *strong executive rule* to override parochial local interests, especially ward politicians, and to plan effectively; and still another was the importance of *separating politics from administration.* By and large, businessmen used the corporation model in trying to reform the public sector and were convinced that centralization of authority under a single appointive manager was the most effective strategy. Both the city manager and commission movements followed from this. The emphasis, then, was on efficiency, procedures, a strong center, technical expertise, and the elimination of party and other politics from government.

POLICY VERSUS ADMINISTRATION

Business executives trying to improve municipal agencies have tended to concentrate on management efficiency considerations, the primary area of their expertise. But improving productivity and promoting institutional change in the public sector often involves serious *policy* and *political* questions—for example, selecting the agency's primary target populations, establishing program priorities, setting goals, and reconciling those that conflict. This involves political bargaining and tradeoffs and goes well beyond simple management improvements. It also involves having some substantive knowledge about the agency's activities—education, welfare, criminal justice, health, and the like. Serious question exists as to whether many business executives have much skill on these matters, since few have had experience in government. Beyond questions of skill are those of ideology—namely, the view by some businessmen that politics should be kept separate from administration. A main issue for students of public administration trying to get government agencies to work better is how to harness and, if possible, change their politics. EDC task forces and executives have begun to understand and deal more effectively with these public sector politics as a result of their New York City experience. Businessmen elsewhere would have to do likewise if they were to be successful.

THE POLITICS OF THE PUBLIC SECTOR

This raises the question as to whether public sector politics even allows for the transferability of management techniques from the private sector. All large-scale organizations have their own politics; but those of government—party, civil service, union, neighborhood, city versus state—are quite different from those of business; and they have a profound effect on management, especially on prospects for implementing new programs.

A whole different set of forces thus affects the management of public sector agencies. For example, the political career interests of elected officials and the influence of civil servants and client groups have much more impact on decisions in the public sector than any "objective" considerations of "need" or "rational management." Indeed, these concepts are only defined in the context of such interests. Any attempt to change the public sector involves working through the politics that those interests reflect. They have to be harnessed in a positive direction, since change requires the support of a large coalition of participants to override the usual vested interests of agency officials and some client groups in the status quo. Many businessmen may not be knowledgeable about such politics, lack confidence in dealing with them, prefer to ignore them, or dismiss them as immoral. Again, EDC has had to learn through direct experience how these politics get played out and has been increasingly effective in many agencies as it came to understand them more.

CORPORATE COMMITMENTS TO EDC-TYPE PROGRAMS

This raises another critical question—namely, the commitment of business to such urban management activities. Top EDC officials are genuinely dedicated to getting business deeply involved in helping make city government more effective, but it has not been possible thus far to get more than a small number of corporations committed to this program to the point where they will supply a regular cadre of competent executives on loan. Only 13 of EDC's 125 members have donated the services of more than 3 executives since EDC began this program in 1969. Yet the program's success requires that EDC have a large and steady supply of executive talent. Part of the problem has been EDC's failure to define its needs for particular types of executives more sharply, make its program visible to more of its membership, and engage in an aggressive recruiting effort.

Another part of the problem is the fact that some executives on loan often have been marginal within their companies at the time, and sending them out was a way of delaying having to make a decision on their future within the company. If they are not well qualified for this work, it reflects badly on EDC and the company, but it is very difficult for EDC to simply turn them away. It is in the weak bargaining position of needing as many people as it can get. Yet it could be more effective in recruiting good people if it got the companies to give greater recognition to these executives for this public service activity, for example, in their house organs, and it should strive to protect their career interests in their companies by indicating to top management their contributions to EDC's work. This will help improve their performance with EDC while on loan, since they will not be constantly preoccupied with what may happen to them when they return to their companies.

A critical problem for EDC that further relates to the corporate commitment question is providing training and continuity for executives on loan. The usual length of service is one year, a good part of which is spent either in phasing in or phasing out. The first several weeks are spent learning about the complex workings of the city agency, while the last couple involve some concern about what awaits them back at the company and affects their productivity while they are there. City agency officials understandably resent having to orient these executives when they first arrive, only to have them leave after such a short time and have to keep undergoing the same process every year. Moreover, there is sometimes a time lag between one executive leaving and another coming on. Corporations should be asked to guarantee a steady supply of executives so that this does not take place. They should also make available those who have already served to be on call when needed, to give training and orientation to those just arriving. A two-year term of service would help. Some initiative is required on EDC's part to make sure that these things take place, and it is now working on the problem. In thinking about how the EDC program can be packaged and replicated elsewhere, much attention should be paid to ensuring that these executive recruitment problems will be solved.

CITY AGENCIES AND LARGER SYSTEMS

The prospects of improving productivity and reforming municipal agencies depend not only on recruiting large numbers of competent executives but on changing the larger systems in which these agencies are embedded. Rational management at HRA requires changes in patterns of funding by federal and state agencies; that in the criminal courts requires changes in the police, the district attorneys, and correctional institutions; and in the school bureaucracy and high schools on establishing more linkages with other youth-serving agencies. To take HRA as an example, as long as the federal government proliferates discrete categorical programs that duplicate and conflict with one another, it is difficult to plan and coordinate services in that agency.

The same argument can be made about reforming the criminal courts. Fundamental changes are required in the entire system of criminal justice of which the courts are a part, to ensure more coordination and the provision of rehabilitative programs. One of the things that may be required is a computerized information system, used by all agencies in the system. Any attempt to change the criminal courts without linking them into this wider criminal justice system and without changing the other agencies in the system may thus have limited benefits. As indicated above, EDC has been effective in establishing better linkages throughout the system, even though it was concerned primarily with improving just the criminal courts.

The problem with urging intervention in a whole network of agencies is that it sets unrealistically high standards for a program like EDC's. Ultimately, everything is related to everything else, and one has to decide where his limited resources will have the greatest impact and make some pragmatic choices about what agencies one wants to change. Furthermore, dramatic, short-term results may be possible through intervention in just a single agency like the criminal courts, notwithstanding the fact that it is part of a larger system. A useful strategy would be to see such single-agency programs as merely a first step in a bigger intervention strategy that tackles networks of agencies whose operations are interdependent.

CONCLUSION

Business may well get increasingly involved in the future in trying to improve public sector agencies and, in some cases, in actually taking over their programs. The need for outside assistance, especially in municipal agencies where inefficiency and mismanagement are so rampant, is very great. The fact that revenue-sharing and federal decentralization are likely to increase, giving states and cities much more responsibility and autonomy than they ever had before, means that tremendous efforts will have to be made to improve their capability.

As the programs described above indicate, businessmen do have skills that are directly relevant to that task. They were obviously quite successful in the courts; and even in the high schools, the Board of Education bureaucracy, and the Human Resources Administration, where the problems are much more complex and the bureaucratic obstacles to change much greater, it has had successes as well. An assessment in depth of how it is doing in those agencies is needed, from the perspective of exploring the strengths and outer limits of that management assistance strategy. That is the purpose of the case study that the author has under way. Even at this early stage of the investigation, the potential benefits from such business involvement seem great enough to merit its further development. The continued urgency of problems in the cities and the limited success of past efforts at agency reform dictate that any strategy with promise should be pursued. This one seems to have such promise and certainly merits further assessment and development in the future.

NOTES

1. From the turn of the century through the 1960s, many family-based firms in cities were replaced by branch plants and offices of national corporations whose managers had limited community concerns. Norton Long has appropriately labelled them "bureaucratic birds of passage"; and numerous community power structure studies have documented this change.

2. This is a preliminary paper, summarizing early findings from a case study of EDC that the author has under way. The study assesses EDC's urban management programs in several city agencies for purposes of developing some generalizations about how to intervene effectively in such agencies. It is intended both to advance social science knowledge and to assist business and government officials in the further development of such social change efforts.

3. The discussion to follow is based on extensive interviews with top EDC and IED officials.

4. In 1970, EDC published its study of employment in clerical jobs in New York, indicating that perhaps as many as 30,000 such jobs had not been filled, due largely to a shortage of applicants.

5. Federal funds have recently become available to pay the salaries of a few members of EDC's Supreme Court task force. They are the only executives on loan whose companies have not paid their salaries while they were with EDC.

6. The one in the Bronx dropped from 37% white in 1969 to 5% in 1973. During the same time period, whites declined in the Brooklyn school from 19% to about 8%, while those in George Washington in Manhattan declined from 28% to 10%. Most of these white

students were from middle-class homes and were replaced by blacks and Puerto Ricans from ghetto areas. These data were complied by EDC and local school officials from ethnic census data taken on each school by school headquarters.

7. By Hawthorne Effect, I mean the tendency for people involved in a demonstration program to be highly committed to its success, contributing to very effective performance initially, only to decline after that, as the novelty and need to prove the program's effectiveness may wear off.

8. The directors of these programs kept records on attendance and pupil achievement from which the data were drawn. Since the school and EDC were quite concerned about getting some assessment of these programs, they were careful in gathering the data. The fact that small numbers of students were involved as well meant that the data were far more reliable than those on the entire school.

9. There were no control groups with whom to compare these students in innovative programs, and the data must be interpreted in that light. They are suggestive, however, of some positive changes in student performance that may be related to their new school experiences.

10. Though one would expect such positive statements from EDC people to promote the program, the perceptions of many teachers and supervisors about the partnerships in their schools were also favorable. All four principals, for example, unequivocally endorsed EDC's efforts; and several gave public testimonials about them. The text quote comes from an interview conducted in May 1973.

11. This information was obtained from interviews with EDC and school officials.

12. This information was obtained from interviews with EDC and school officials.

13. This information was obtained from interviews with EDC teachers and school headquarters officials.

14. This information was obtained from interviews with school officials and EDC.

15. This information was obtained from interviews with EDC officials (see also New York *Times*, 1973a).

16. This information was obtained from interviews with EDC and school officials.

17. Interviews with EDC and court officials formed the basis for much of the analysis to follow. An early account of EDC's impact on the courts is provided by the 1971 *Annual Report of the Criminal Court of the City of New York*, Mr. Justice David Ross, Administrative Judge.

18. This information was obtained from interviews with EDC officials.

19. See Argyris's chapter in this book for a further discussion of this.

20. See Pressman's chapter in this book for an analysis of how the proliferation of federal programs has hampered the management of city agencies.

21. One recent development making management reform much more difficult in the state supreme court was former Governor Rockefeller's order for more stringent punishment in narcotics-related crimes. This has loaded the court with many more cases than it could initially handle very efficiently.

22. What constitutes the most appropriate task force structure may well depend on the nature of the work they are doing. If it involves the transferring of known techniques and solutions, then a style of close supervision and a lot of structure seem appropriate. If, on the other hand, the task force is searching for new ways of managing public sector agencies where the transfer of business skills is less clear and where considerable experimentation is required, then a more collegial and less bureaucratic approach seems desirable.

REFERENCES

BARNES D. and R. CONNOLLY (n.d.) "Report outlining the needs analysis and participative planning strategy." Institute for Educational Development.

BERUBE, M. R. and M. GITTELL [eds.] (1969) Confrontation at Ocean Hill-Brownsville. New York: Praeger.

BIRCH, D. L. (1967) The Businessman and the City. Cambridge, Mass.: Harvard University Graduate School of Business Administration.

CARO, F. G. [ed.] (1971) Readings in Evaluation Research. New York: Russell Sage.

CHAMBERLAIN, N. [ed.] (1970) Business and the Cities. New York: Basic Books.

COHN, J. (1971) The Conscience of the Corporations. Baltimore: Johns Hopkins Press.

Committee for Economic Development (1971) Social Responsibilities of Business Corporations: A Statement on National Policy by the Research and Policy Committee of CED. New York.

DRUCKER, P. [ed.] (1969) Preparing Tomorrow's Business Leaders Today. Englewood Cliffs, N.J.: Prentice-Hall.

——— (1968) The Age of Discontinuity. New York: Harper & Row.

Economic Development Council of New York City, Inc. (n.d.) "Responsibility means accountability." Seventh Annual Meeting Report, New York.

Economic Development Council Task Force (1970) "Organization study of the New York City Criminal Court." Harold A. Finley, Coordinator, October 28.

EDC School Partnership Program (1973) An Evaluation and Recommendations for Future Actions.

HABER, S. (1964) Efficiency and Uplift: Scientific Management in the Progressive Era, 1890-1920. Chicago: Univ. of Chicago Press.

HAGE, J. and M. AIKEN (1970) Social Change in Complex Organizations. New York: Random House.

HOFSTADTER, R. (1948) The American Political Tradition and the Men Who Made It. New York: Alfred A. Knopf.

HORNSTEIN, H. et al. (1971) Social Intervention. New York: Free Press.

Institute for Educational Development (1969a) "Partnerships." Industry and Education Study 2, New Haven.

——— (1969b) "Business methods in reorganizing administration of an urban school system." Industry and Education Study 1, New Haven.

MILLER, J. N. (1972) "New York group produces instant court reforms." National Civic Rev. 61 (March).

New York, City of (1971-1973) "High school profiles." Board of Education.

New York Times (1973a) March 11.

——— (1973b) March 13.

Peat, Marwick, Mitchell & Co. (1973) "Cost saving analysis of administrative improvements in the New York City Criminal Court." Economic Development Council, March.

ROGERS, D. (1971) The Management of Big Cities. Beverly Hills: Sage Pubns.

——— (1970a) "The failure of inner-city schools: a crisis of management and service delivery." Educ. Technology (October): 27-33.

——— (1970b) "The New York City school system: a classic of bureaucratic pathology," pp. 131-136 in A. T. Rubenstein (ed.) Schools Against Children. New York: Monthly Review Press.

——— (1968) 110 Livingston Street. New York: Random House.

RUTTENBERG, S. H. and J. GUTCHESS (1970) Manpower Challenge of the 1970s. Baltimore: Johns Hopkins Press.

WEINSTEIN, J. (1969) The Corporate Ideal in the Liberal State, 1900-1918. Boston: Beacon.

WEISS, C. H. [ed.] (1972) Evaluating Action Programs. Boston: Allyn & Bacon.

15

Municipal Monopolies Versus Competition in Delivering Urban Services

E. S. SAVAS

☐ OUR CITIES ARE NOT WORKING WELL. Sanitation, safety, transportation, housing, education—even electricity—all seem to be failing. The taxpayer complains about waste, inefficiency, and mismanagement, and blames his public servants.

Yes, many of our urban services *are* inefficient and often ineffective and unreliable, primarily because they are poorly designed for the job at hand. Like any poorly designed, complicated apparatus, they often seem to be out of order as one part or another breaks down. Our municipal systems must be redesigned so that they function better and have fewer critical parts.

Our industrialized society, which is itself very complex and interrelated, provides a guide. Just think of all the different people—from farmer to supermarket clerk—whose efforts must mesh in order for a slice of bread to reach your table. Any one participant could break the chain, including the man whose guild card authorizes him (and only him) to pump gas into the baker's delivery van. Nevertheless, we manage to get our daily bread after all. That's because there are many sources of flour and numerous individual bakeries: no one has an effective monopoly. Furthermore, products

[473]

can be stockpiled, and so there is always fertilizer, wheat, flour, and even bread and frozen rolls, stored at various points in the system.

The city, however, is uniquely vulnerable to service shutdowns —and it does not have the option of moving to the South, starting a branch in Hong Kong, or going out of business. After all, a principal function of government is to provide, or at least regulate, those services that are deemed to be monopolies; and so the city furnishes public sanitation, police, and fire services, while the state government regulates the private power and telephone companies. These are all monopolies of the most vulnerable sort, for their services—unlike flour—cannot be stockpiled or imported.

Therein lies a key problem of municipal management: monopolies, whether public or private, tend toward inefficiency. Since most city agencies are monopolies, their staffs are automatically in a position to exercise that monopoly power for their own parochial advantage —and efficiency is rarely seen as an advantage. In short, we have unwittingly built a system in which the public is at the mercy of its servants; many municipal agencies are malfunctioning monopolies which no longer serve the public interest, but their own. However, the inefficiency of municipal services is not due to bad commissioners, mayors, managers, workers, unions, or labor leaders; it is a natural consequence of a monopoly system. The public has created the monopoly, the monopoly behaves in predictable fashion, and there are no culprits, only scapegoats.

In addition to this inefficiency, monopoly systems are inherently unreliable because of their vulnerability to strikes and slowdowns. Legislators who do not seem to understand the fundamental workings of the system continue to demand that public employees behave as though they did not possess monopoly power. The New York State legislature, for example, persists in drafting futile no-strike edicts and looks to compulsory arbitration as the latest cure. That's like King Canute asking the sea to pretend it is a pond and telling the tides they must cease and desist. The U.S. Congress did much the same thing—and achieved equally spectacular failures— in its naive dealings with monopolies such as the Postal Service and the railroads.

Employee groups favored with a monopoly can always arrange work slowdowns and carefully contrived absenteeism to achieve the effect of a strike, while getting around no-strike laws and avoiding prosecution. The government is then left with trying to prove there was a conspiracy when a tenth or a third of the work force suddenly took ill or started diligently following some obscure, trivial, but time-consuming work-safety rule.

There is much evidence to support the belief that public monopolies are inefficient (Savas, 1971). New York City alone provides several examples, including this striking statistic about its police force: during the 25 years between 1940 and 1965 the number of policemen was increased by fifty percent, from 16,000 to 24,000, but the total number of hours worked by the entire force in 1965 was less than in 1940! The increase in manpower was entirely consumed by shorter working hours, longer vacations, more holidays, more paid sick leave, and a longer lunch period. By comparison, during the same period the length of the average work week throughout the United States declined by only eight percent. Thus, except to the extent that a better-rested police force can perform its job better, the public realized no apparent benefit from the fifty percent increase in manpower.

Education. A similar phenomenon occurred in the city's education system: while the number of teachers increased by fifty percent and, in addition, one teaching aide was hired for every two teachers, the number of pupils remained relatively constant. Under these circumstances one would surmise that the pupil/teacher ratio in class had declined markedly. However, only a slight reduction in class size was achieved. Teachers simply worked fewer hours and were relieved of some of their duties. Parents can judge for themselves whether this has resulted in better preparation by the teachers and hence better education for their children.

For inefficiency in public services, one need look no further than the state-operated mass-transit bus lines in New York. Seventy percent of all mass transit rides take place during rush hours, and therefore relatively few bus drivers are needed between rush hours. Nevertheless, efficient "split shift" scheduling of manpower does not occur. Instead, drivers are paid for fourteen hours while actually working only eight hours a day; they are compensated handsomely for the hardship of taking a paid four-hour break after their four-hour morning stint, and then are paid at the time and a half overtime rate for their four-hour afternoon tour of duty. Imagine where baseball would be if hot-dog vendors at Yankee Stadium insisted on being paid for a forty-hour week, fifty-two weeks a year!

These brief examples suggest that local services which are provided through public monopolies have moved beyond the control of urban management, and no longer serve the public interest effectively and efficiently. It is important, therefore, to examine whether urban services can be provided through the competitive marketplace and whether there are any advantages to the public in doing so.

PUBLIC MONOPOLY VERSUS PRIVATE
COMPETITION: A CASE STUDY

It is sometimes possible to find situations where fair comparisons can be made between public and private services. Often these are minor services, and the findings attract little attention. However, circumstances in New York City made it possible to examine a highly visible, vital service in this way, and, predictably, considerable public interest was aroused.

The mayor's Office of Administration studied refuse collection in New York by comparing the performance of the city's agency, the Department of Sanitation, with that of the competitive private carting industry (Savas et al., 1970). Data from the private sector were available because the industry is regulated by the Department of Consumer Affairs. The latter receives copies of all service contracts, as well as audited financial statements of each carting firm, and sets an upper limit on rates that can be charged by the private cartmen.

First, let us examine the private carting industry in New York. Almost all of its business is from commercial customers ("trade waste"). Stores, shops, restaurants, office buildings, institutions, manufacturing establishments, and the like comprise the bulk of their work. There are several classes of carting operations. Class 1 firms are of particular interest: they collect and transport putrescible waste (garbage) and rubbish. Class 2 firms are licensed to collect and transport nonputrescible wastes. Class 3 firms can collect and transport only waste which originates in the business operations of the vehicle owner.

The Department of Sanitation serves only residential dwellings and tax-exempt facilities (although occasionally employees are arrested for illegally picking up commercial refuse and taking bribes for doing so; New York Times, 1971a, 1971b). The costs of the agency are paid out of general tax revenues, not by user charges.

The relative magnitude of public and private refuse collection activities is shown in Table 1. It is obvious that the private sector is quite large; in fact, it collects a majority (fifty-five percent) of the total solid waste produced in New York City, although, as pointed out above, the residential segment of the marketplace is the near-exclusive domain of the municipal agency. (That domain is not completely municipal. Some residential cooperatives choose to buy superior private services rather than accept free but inferior public service.)

A fair and proper comparison can be made between the refuse collection activities of the Department of Sanitation and the Class 1 cartmen. They use similar kinds of vehicles and pick up similar kinds of waste along regular routes. Although the Department of Sanitation performs other functions as well—for example, street cleaning and snow removal—it is relatively easy to isolate the collection costs in their annual budget.

Table 2 presents an item-by-item comparison of collection costs, and reveals the astonishing fact that it costs the city more than twice as much as the private sector to collect a ton of garbage—$39.71 compared to $17.28! This finding, bitterly attacked by the union involved, and half-heartedly disputed by the department, was subsequently corroborated by an independent study (Citizens Budget Commission, Inc., 1972a), which updated the figures and showed that departmental costs had risen two years later to a level of $41.11 per ton, compared to a range of $17.41 to $21.50 per ton for the Class 1 cartmen. Other supporting figures were reported by enterprising journalists (Kazan, 1971; Cordtz, 1971; Phalon, 1971), as great public interest was generated in this subject (New York Times,

TABLE 1
COMPARISON OF PUBLIC AND PRIVATE REFUSE
COLLECTION ACTIVITIES IN NEW YORK CITY[a]

Collecting Agency	Millions of Tons Collected	Employees	Vehicles	Firms	Locations Serviced
Dept. of Sanitation	3.4	12,000	1,850	1	705,000
Class 1 firms	2.1	2,200	985	450	125,000
Class 2 firms	.29	n.a.	850	600	n.a.
Class 3 firms	.15	n.a.	1,503	n.a.	n.a.
Others[b]	1.9	n.a.	n.a.	n.a.	n.a.

a. For the period July 1, 1968 to June 30, 1969.
b. Includes construction and demolition waste.
n.a. = not available.

TABLE 2
COST COMPARISON OF PUBLIC AND PRIVATE
REFUSE COLLECTION IN NEW YORK CITY[a]

Expense Category	Cost per Ton Collected	
	Public[b]	Private[c]
1. Direct labor	$12.93	$ 7.81
2. Indirect labor	4.59	1.92
3. Paid leave	4.55	.85
4. Pension and fringe benefits	9.09	d
5. Administration and overhead	2.43	2.66
6. Gas, oil, grease, and other supplies	.38	.79
7. Vehicle and plant maintenance	2.81	1.41
8. Public liability	.17	.40
9. Depreciation or debt service	2.76	1.44
Total[e]	$39.71	$17.28

a. For the period July 1, 1968 to June 30, 1969.
b. From "Annual Progress Report and Statistical Reviews of the Department of Sanitation," except for items 5, 8 and 9, which are from the Bureau of the Budget and the Office of the Comptroller.
c. From the Department of Consumer Affairs, based on maximum charges.
d. Included in items 1, 2, and 3.
e. Not included are costs of licenses, taxes, and garage space, which are zero for the city agency. Disposal costs and profits are also not included.

1971c, 1971d, 1971e, 1971f; Wall Street Journal, 1971; New York Post, 1972a, 1971).

It might be argued that the comparison is improper, because the customers—and hence, the collections—differ. Savas (et al., 1970) estimates, in fact, that the amount of refuse at a typical private pickup site is about twice as much as that at a sanitation stop, and so somewhat higher productivity is to be expected of the cartman (Savas et al., 1970). However, offsetting this are three countervailing factors which act against the private operator:

- He goes onto the premises to pick up refuse, whereas the sanitation worker picks up from curbside.
- His route includes a relatively large amount of unproductive driving time between his widely scattered customers, compared to the sanitation truck which has a densely packed route.
- He must gear his collection schedule to the specific needs of his customers, unlike the Department of Sanitation.

For those unpersuaded by this reasoning, it should suffice to look at one more compelling piece of evidence. In Douglaston, a part of New York with spacious, one-family homes, the Department of

Sanitation provides *twice*-a-week service, from *curbside,* and at a total annual cost of $*207* per dwelling. Four miles away, just across the city line, is the community of Bellerose—which is much like the Douglaston area. It contracts with a New York City private carting firm for *three*-times-a-week service, from the *back* of the house, all at a cost per dwelling of only $*72* per year—about *one-third* the cost of the municipal agency!

Statistics from other cities corroborate the finding that competition produces lower costs for refuse collection. In San Francisco, it is reported that the average family pays only $30 per year for service provided competitively by private firms. Brisk competition in Boston leads to a cost of $18 per ton for residential collections by private cartmen under contract to the city (Phalon, 1971), a figure almost identical to the New York figure for private firms. Furthermore, Bostonians rate the service much higher than any other municipal service.

Labor costs account for the large observed differences between public and private collections in New York. To begin with, sanitation uses three men per truck, whereas the private firms use two. Second, wages and fringe benefits in the city agency are about twenty-five percent higher than in the private firms. Finally, management and supervision seem far superior in the private firms, and this is probably manifested in more tons collected per hour. Supporting evidence for this view of close attention to the business can be found by looking at the figures for vehicle down-time: thirty-five percent for the department in 1970 compared to only five percent for the private cartmen. The explanation is obvious: if a man owns only one or two trucks, as most cartmen do, and if his family's livelihood depends on them, he cannot afford to have them out of commission more than a day or so a month; he makes sure they stay in working order.

The three factors responsible for the extravagant costs of the public service—overmanning, overpaying, and underworking—are a consequence of the essentially monopolistic position of the city agency with respect to residential refuse collection. The remedy for this condition, competition, is self-evident, but it must be administered with care. Sudden abolition of the Department of Sanitation is neither wise, humane, nor politically feasible. (In fact, many feel that there is *no* change which will meet the test of political feasibility.) A sensible strategy would be gradually to reduce the department's refuse-collection service, in scope and in size, while the private cartage industry is given the opportunity to expand and provide some of the service now supplied by the municipal agency.

Two major strategies can accomplish this (Savas et al., 1970). One is to "load shed"—that is, to stop servicing certain classes of customers. Those customers can then purchase private services either entirely at their own expense or with the aid of a subsidy. The other strategy is to "contract out"—that is, for the city to purchase private services directly by competitive bidding.

"LOAD SHEDDING"

Candidates for load shedding include tax-exempt properties which are used for commercial activities, and institutions such as private schools and hospitals. (Under the present anomalous situation, organizations that *do not* pay taxes *do* receive "free" refuse collection service while those that *do* pay taxes *do not* get it.)

Another customer class suitable for load shedding consists of one- and two-family dwellings. Such units can readily arrange for private collections, as in many suburban communities. In New York, as cited above, the municipal cost for servicing such homes is three times the private cost. Alternatively, multiple dwellings can be dropped as a class and left to make their own arrangements for private collections as is done in Chicago and Philadelphia; the property owners include these costs in the rent. Yet another way to reduce the size of the municipal monopoly is to provide no municipal refuse-collection service to buildings constructed after a certain future date. Properties subjected to load shedding would receive appropriate reductions in their real-estate taxes.

As another approach to load shedding, the level of service could be reduced. That is, the municipal agency might cut back to once-a-week service and, if the property owner wanted more, he could purchase supplementary service from the private sector.

An incidental benefit of allocating costs in this way is that a closer regulatory link is established when a producer of waste is forced to pay more directly for its removal. This has profound advantages at a time when society is interested in reducing the amount of waste burdening the environment.

"CONTRACTING OUT"

As with load shedding, there are several distinct and nonexclusive approaches that a city can use to contract with private firms for refuse collection. The most straightforward arrangement is to divide the city into areas and to auction off individual areas to the lowest bidder. Each area should be of a size that will attract maximum competition; practically speaking, this means that the area should be small enough to be serviced by a one-truck operator. Ideally, enough time should elapse between auctions of individual areas to permit many firms to evaluate each prospect. Not all areas need be contracted out; those which promise the greatest potential savings should be first on the list. In addition, contracting out on this geographic basis may be appealing for those neighborhoods in which private services have many customers and the municipal service relatively few, as in business and industrial areas.

Contracting for services can also be considered for certain customer classes. For example, one can drop municipal service to public schools and colleges, public housing projects, jails, public hospitals, fire houses, museums and libraries, municipal office buildings, and so on. Each of these facilities would then contract with the private sector for service. Even though the city would pay the full cost, net savings would be realized due to the high marginal cost of direct refuse collection by the municipality.

Geography and customer class represent two guiding principles for differentiating the service and contracting out identifiable segments. A third principle is the kind of service. For instance, bulk refuse—furniture, refrigerators, mattresses—is usually collected separately and on a different schedule from normal household refuse. The same is true of leaves and plant clippings in neighborhoods with trees and lawns. Specialized collections of this sort could be considered candidates for private handling on a contract basis.

To summarize, this case documents a startling disparity between the costs of refuse collection provided by a municipal monopoly and that provided competitively by the private sector. However, it must be stressed that the conclusion calls for *competition* rather than a *monopoly;* it does *not* call for *private* services in perference to *public* services. There are many locations in the country where private collection service is a monopoly, or a collusive oligopoly at best, and the situation is hardly in the public interest. In *those* cases, competition can best be spurred by *government* entry into the business. The delivery of public services will be most efficient if those services are offered in a truly competitive market situation, whether the provider is a public agency or a private firm. Quality or cost or both will suffer if the provider, public or private, has, in effect, a monopoly.

A subsequent analysis by Young (1972: 73) also concludes that it is both possible and desirable to harness the forces of competition to provide efficient collection of refuse. However, selfish interests in government agencies and in industry and misguided public regulatory bodies may instead bring about more regulation, more exclusive franchising, more industry protection, and ultimately less competition and more public and private monopolies.

THE NATURE OF PUBLIC GOODS

"Public" is one of the most overtaxed and ambiguous words in the lexicon of social policy. It is used to refer to providers, regulators, and recipients of a wide array of goods and services. Consider a private firm, which is owned by numerous individual members of the public, whose shares are traded publicly, and which is therefore often termed "publicly owned." One can thus speak of a "publicly owned"—but actually private—enterprise such as an investor-owned telephone company or hotel chain which supplies private services (communications, lodging) to the public with (in the case of the former), or without (in the case of the latter) extensive regulation by a public agency. Some clarification and precision is obviously in order. For the present purpose, it is sufficient to focus on the concept of individual and collective goods (where the generic term "goods" means both goods and services). As will be seen, "individual" and "collective" are more accurate descriptors than their usual surrogates, "private" and "public."

Individual goods. Individual goods are characterized by their exclusion property. That is, when an individual good is consumed, it is not available for consumption by someone else. This is obviously true of food, clothing, and shelter, for example, and also for haircuts, tickets to the ball game, and restaurant tables. Although others may be able to acquire identical goods at a different time, or similar goods at the same time—e.g., adjacent seats at the stadium or the restaurant—the consumption of a unit of goods by one individual renders that unit completely unavailable to anyone else.

Collective goods. By way of contrast, collective goods are characterized by their *sharing* property.[1] When a collective good is "consumed," it is still available in undiminished quantity and quality to be shared by others. For example, more than one person can watch the same television program. Other examples of collective goods include the walls around medieval cities, armies, priests, and public officials.

This simple dichotomy is not really so simple, of course. Aqueducts and power lines may be considered collective goods, but the water and electricity drawn from them are individual goods —except for the water used to fight a large fire whose extinguishing will benefit many. A police department is a collective good, but any finite force on patrol can rapidly be "consumed" by individual calls for service. Central Park in New York is a collective good but its quality as a good varies greatly depending on the number of people sharing it.

Some collective goods can be thought of as micro-collective and others as macro-collective. The former are shared by a small, identifiable portion of the populace, such as a neighborhood or a group interested in horses and bridle paths. The latter cannot be so localized in terms of their beneficiaries.

The point of this disquisition is the relation between the nature of the goods and the means of paying for them. Individual goods are usually paid for by the individual consumers. In somewhat similar fashion, micro-collective goods can be charged to the consuming group on a collective basis; for example, a local watchman might be hired by a cooperative block association, and tennis enthusiasts might pursue their recreational fancy by forming a club which is financed by members' dues.

Macro-collective goods, often called public goods, are properly paid for by the public at large, for their benefits cannot be charged to individual consumers or small collective groups. However, from this reasonable arrangement, it is easy to leap to the unwarranted implication that public goods paid for by the public through payments to the public tax collector must be provided *to* the public *by* a public agency *through* public employees. There is no logical reason for the mode of payment to bear any relation to the ultimate mode of delivery of collective goods.

THE COMPETITIVE PROVISION OF
PUBLIC GOODS

When one looks at municipal public goods in terms of their individual or collective nature, it becomes possible to think of ways of fostering competition in their provision.

INDIVIDUAL GOODS

With respect to supplying individual goods, competition can be encouraged rather readily as such goods are natural candidates for a market mechanism. In fact, many individual goods are already subject to user charges, even though they are supplied by local government agencies: many transportation, culture, recreation, entertainment, and health services are typically provided on a fee basis. The individual user chooses to partake of a particular good, is aware that he is consuming it at a particular time and place, and is used to paying for it.

One obvious way, therefore, to enhance competition is to load shed—that is, to stop providing such individual goods via public agencies and to allow the private sector to supply them, under competitive conditions, directly to the users. Many similar goods are already provided exclusively by the private sector, and their extension is straightforward. For example, the field of recreation is one where government plays a relatively modest role anyway. Privately owned and publicly available gardens, golf courses, tennis courts, gymnasiums, swimming pools, riding facilities, and amusement parks abound. One can readily conceive of a local government operating only general-purpose municipal parks and playgrounds, while relegating entirely to the private sector all athletic and recreational activities which require special-purpose facilities. If necessary, government can reduce the cost of market entry, and hence enhance competition, by providing space for the facilities or even paying for their construction, and then contracting out the operation of the facility through competitive bidding. A variant of this approach can be found in New York City, where the Parks Department has turned over its tennis courts to private firms for winter operation as a franchise. The firms enclose the tennis courts within inflatable structures and sell time to players on an hourly or seasonal basis. The city agency operates the courts during the summer and charges user fees.

Much of the transportation service in urban areas is provided through the private sector, by personal automobiles, parking lots, taxis, group-ride services, and private bus lines. Government-operated bus lines and parking lots almost invariably charge user fees and therefore, in principle, are candidates for load shedding. Indeed, private bus lines charging the same fare as municipal bus lines in the same area are often able to make a profit, while the government lines require subsidies. Even in congested Manhattan, the introduction of express buses serving certain neighborhoods during the rush hours at a relatively high fare brought profits to the private firm responsible for this innovative service.

Where the user charges come close to the cost of providing individual goods, there is little reason to have a municipal agency provide these goods; it is likely that load shedding to the competitive, commercial marketplace can be carried out effectively and with savings to the users. The problem arises where the user charges are substantially less than the cost of providing the service in question. Society deems that some goods, although individual ones, *ought* to be consumed and that the public at large will thereby benefit; therefore, a subsidy is in order. Museums and cultural institutions are cases in point, and so are municipal bus operations. A ride on a bus represents an individual good, and a fare is usually levied as a user charge, but for many bus services this is not sufficient to defray the cost of the ride. In fact, many private bus lines have demanded and obtained operating subsidies from the municipalities they serve. It appears that, more often than not, these lines were in a monopoly position, immune to incentives for more efficient performance, but the basic problem appears to be that the economic incentives favor private automobiles over mass transit. While automobile travel is subsidized through highway construction and oil-depletion allowances, and as long as the external social costs of automobiles (pollution, congestion, space) are not internalized, the need will exist for substantial compensating subsidies for mass transit. Given such subsidies, whether in the form of initial capital investment for equipment and garages, as operating subsidies, or as exclusive bus lanes, it should become possible to create a situation where several private organizations can be induced to compete in providing bus service. Even where only a single route is involved, organizations serving nearby areas constitute potential competitors who could vie for the opportunity to assume that one route.

There seems to be no shortage of entrepreneurs willing to enter the field of urban transportation, as witnessed by the appearance of unauthorized "gypsy" cabs in such widely disparate municipalities as Moscow, Belgrade, New York, and Hong Kong. The gypsy cab industry in New York grew in response to the market opportunity presented by the authorized, oligopolistic, protected-by-regulation taxis, which provided poor service to ghetto areas. The gypsy cabs give better service at a lower price to those areas, and have quite naturally extended their operations throughout the city.

In Hong Kong, privately owned minibuses appeared illegally but proved their value when they reduced the city's vulnerability to a strike by regular bus drivers. The minibuses were legalized, and although they charge one and one-half to two and one-half times the standard bus fare, they account for twenty-two percent of all passenger trips (Meier, 1972). In comparison to the municipal buses, they offer guaranteed seats, convenient pickup, less noise, and greater cleanliness. Economies of scale are insignificant in this business, and therefore two-man partnerships of owner-drivers, offering two-shift coverage, are the norm. Under these circumstances, entry into the market is relatively easy and a fully competitive condition has been maintained.

Vouchers systems. Not all municipally supplied individual goods are subject to user charges. Education is a prime example of such a good. However, because it can be purchased for full cost from the private sector or at a (privately) subsidized rate from parochial institutions, the government-issued product is also a candidate for load shedding. In this case, the voucher system commends itself as a mechanism for realizing the benefits of competition. Under this system, a family receives a voucher good for one year's worth of grade school education, for example, and can use the voucher to enroll the child in any certified private or public school. The school subsequently converts the voucher to cash by turning it in to the public agency that issued it.

One might argue that a completely competitive system of education could lead to such diversity in curriculum and achievement that students transferring or being promoted to other schools would be badly served. There is no reason for anguish; higher education in the United States is a good example of a competitive system, with both private and public colleges in many different state systems, yet these problems do not arise in any serious form. Regulation of schools would continue to be exercised by government boards, which would prescribe general standards for curriculum and reading achievement, for example, while leaving the pedagogical details to the individual schools. An important function of the state would be to inspect, measure, and report on the performance of the different schools.

Vouchers can be used to introduce competition into the provision of any individual good which society decides should be supplied to each individual or family. For example, refuse collection could also be handled by vouchers. Every property owner would be issued a voucher annually, let us say, and he would have a choice of using it to purchase refuse collection services from any organization that offers them, including his municipality's sanitation department. Competition for his voucher would be based on the quality and quantity of service—convenience, cleanliness, quietness and frequency of service, and the amount picked up.

The use of vouchers satisfies an important principle: it subsidizes the *consumer* rather than the *supplier* of a service. Inasmuch as the whole intent of a program that encourages competition is to improve the quality and reduce the cost of service to the citizen, it makes good sense to bestow upon the consumer/taxpayer the direct power to evaluate and choose his suppliers. Subsidizing the producer, a common government practice, is an inferior policy in this respect. When government subsidizes the consumer, one of its roles is to help its citizens become enlightened and discriminating consumers of municipal services; it can do so by evaluating and publicizing the efficiency and effectiveness of those services.

The ultimate in subsidizing the consumer rather than the producer is simply to give the consumer—through a family assistance plan or a negative income tax, for example—the money to purchase individual goods in the marketplace that would be created by load shedding. However, this approach would not satisfy the societal goal that everyone consume certain goods—by force, if necessary. Some families would undoubtedly forego expenditures on their children's education, while others would throw their garbage into the street or in front of someone else's property. Vouchers have the virtue of directing purchases in accordance with the community's social values. However, it should be recognized that there are ways to get around such constraints. Thus, a black market has developed in food stamps, which are a form of voucher; authorized recipients have been known to sell them at a discount for a more flexible buying medium—namely, cash.

Something analogous to the voucher system is emerging in health services and is having the effect of stimulating competition with some of the health agencies of local government. The growth in coverage of medical insurance plans and the introduction of Medicare and Medicaid have expanded the options available to those needing medical care. In particular, the traditional clientele of county and municipal hospitals—the low-income group—is no longer restricted to those facilities. Their enrollment cards in these programs serve, in effect, as vouchers entitling them to purchase their medical care from other suppliers. The effect of this newly acquired freedom of choice in New York City has been to reduce the demand for the municipal product. (The quasi-governmental agency that runs the municipal hospitals in New York appears to have reacted in classic bureaucratic fashion: it holds on to patients who should have been discharged [Citizens Budget Commission, Inc., 1972b], thereby inflating the bed-occupancy rate, and is—commendably—looking for new activities to occupy its staff and new kinds of patients—e.g., alcoholics—to populate its wards.) The net effect of all this is equivalent to a strategy of load shedding the hitherto municipally provided individual good of medical treatment.

COLLECTIVE GOODS

As was demonstrated in the preceding section, load shedding is the basic strategy for creating alternatives and stimulating competition in the process of supplying individual goods; however, the strategy can be applied as well to collective goods. This is best done with micro-collective goods, as defined above. The example given there of private watchmen is an instructive one. In many big-city neighborhoods, local residents, dissatisfied with the level of public safety, have augmented the local police force with their own direct resources. These residents are, in effect, levying a special tax on themselves and purchasing private guard services or volunteering for guard duty for their block or building—and thereby buying protection at a lower cost than the municipal monopoly could provide it. This phenomenon represents a variation of load shedding; the municipality does not withdraw policy services from the area, it merely fails or refuses to provide more. In such cases, when the local citizenry wants more of a certain kind of service, or more frequent service, or a higher quality of service, they can resort to the marketplace and acquire it at their collective expense (or provide equivalent services in kind).

The process illustrates a general approach that may be applied to induce competition in the supply of many collective goods: load shedding, followed by the formation of a micro-collective, which then provides the service either through a "do it yourself" policy or by purchasing it.

Another illustration of this process can be found in New York City, which operated a number of large food markets where shippers and distributors came together. By helping to form cooperative associations of the users to run the markets, and also by directly contracting out to the private sector the management and mainte-nance of the markets, the city was able to increase its net rental income from the markets by almost one million dollars annually, thus cutting its overall costs in half (Patton, 1973).

The broader lesson from this successful effort is that consumer groups to purchase micro-collective goods can be sought, or organized, on a geographical basis—a building, a housing develop-ment, or a neighborhood—or on an interest-group basis, such as the aforementioned tennis enthusiasts or merchant groups.

Local merchant associations and chambers of commerce in New York and elsewhere have purchased special street lights and litter baskets for their streets, as well as trees and benches. This action, also, is a response to implicit load shedding by the municipality and tends to encourage competition with the conventional suppliers (and municipal purchasing agents) of such street furnishings.

Besides load shedding, contracting out is the other basic strategy for creating competition in the supply of collective goods. Unlike load shedding, subsidized or not, contracting out involves a contract between the municipality and a private firm which will provide a specified service at a price to be paid directly by the municipality to the contractor. Under the right conditions, potential contractors will compete vigorously for this business.

A wide variety of municipal services can be provided by the private sector under contract. Refuse collection was previously discussed in detail, but even as fundamental a life-saving service as fire fighting is done on a contract basis by private firms in Denmark (Philip, 1954: 60), and the municipality of Scottsdale, Arizona, does the same. (Volunteer fire companies are legion throughout the United States, and, by analogy with voluntary safety patrols, can be viewed as competitors to nearby municipal fire departments, for they provide a visible alternative.)

Emergency ambulance service is another vital function which can be provided under contract; this is done in parts of Los Angeles and New York, for example, and in many rural areas of the United States. Voluntary groups are also quite common, but there is some question about their overall performance level, and similar questions have been raised about volunteer fire services.

Police services can also be provided under contract; the New York City Housing Authority is experimenting with private guards in some of its housing projects instead of using its regular patrolmen at twice the cost. Private guard services have long been used to maintain security in municipal hospitals, welfare offices, and construction sites. A specialized private detective agency in Ohio hires out skilled, professional narcotics agents to small-town police departments (Newsweek, 1973).

Another service that lends itself to performance by contract with the private sector is street cleaning. Norwalk, California, received five bids ranging from $44,000 to $56,000 to perform the work that the government department performed at a cost of $62,000 (Warren, 1966: 227). A number of municipalities in California therefore have their streets swept in this way, and, while Wall Street, the bastion of capitalism, is cleaned by a government bureaucracy, socialist Belgrade contracts out to a free-enterprise cooperative to clean its streets!

Snow removal can also be done under contract. Montreal's Department of Roads divides the city into fifty-two sections, each with about twelve linear miles of street. In a comprehensive document which should serve as a model for other cities and services (Montreal, City of, 1966), the department issues detailed performance specifications which include the provision that plowing of streets and sidewalks should be completed within eight hours after the end of the snow fall. Brisk bidding takes place as private firms compete for the business, and they post sizable performance bonds as assurance when they submit their bids. The fact that the municipal agency clears the snow from five of the fifty-two sections means that it retains (and exercises) the in-house ability to intercede effectively if a contractor fails to perform according to specifications. The private firms compete with each other not only at the bid auction but also on the job, as they attempt to demonstrate their ability in order to be assured of an invitation to the bidding next year.

These five factors—precise performance specifications, competitive behavior by suppliers of the service, the posting of performance bonds, systematic evaluation of the supplier's performance by the purchaser, and the credible threat of intervention and competition by the municipal agency—are essential elements for successful contracting out.

City streets have to be cleared not only of litter and snow, but also of abandoned cars. In New York, this troublesome function was contracted out at a profit to the city, as auto wreckers submitted negative bids and paid for the privilege of picking up such autos in specific parts of the city. Only in Manhattan, because of congestion that reduces their productivity, did the firms submit positive bids, requiring the city to pay them for each car picked up. An effective procedure was developed to identify legally abandoned vehicles and to inform the contractors of their location.

While on the subject of hauling away automobiles, a look at tow-away programs is instructive. In New York, police employees operate the tow trucks which remove illegally parked cars from the central business district, but in Los Angeles the job is performed by the private sector, with various authorized tow truck operators acting as agents of the city.

Street maintenance and construction is another area which lends itself to contracting out. A detailed study by New York City's Office of Administration, following its earlier study of private refuse collection, cited above, disclosed that in 1971 resurfacing of city streets with two inches of asphalt overlay cost $57.17 per ton of asphalt in direct labor, materials, supervision, and overhead—not counting equipment depreciation, pension costs, and indirect over-head—while similar work (of better quality) by private paving contractors in the New York metropolitan region cost $18.00 to $20.00, only a third as much. (The national average for federally aided construction was $8.92 per ton.) A series of newspaper articles provided support for these analytical findings with field observations in colorful journalistic terms (New York Post, 1972b, 1972c, 1972d). A private contractor was quoted as saying that he could resurface twenty-six miles a day, compared to the two miles completed by a city crew.

To quote from the city administrator's report,

There are inherent reasons why city government cannot function as efficiently as a private contractor. . . . It is clear that municipal enterprises function under handicaps. Labor productivity is influenced by civil service rules and a union-management situation entirely different from that in private industry. A municipal worker costs more per unit of work. Other handicaps result from the prevailing attitude that watchdog systems are needed to prevent municipal officials from stealing public funds.

The point is basically that the rules of the game handicap productivity in a municipal enterprise. A good manager will be able to do better than a poor manager, but it will be nearly impossible for him to do so as well as he could in private industry, playing under a different set of rules.

One way of strengthening the position of management in the municipal enterprise is to provide the comparison and competition of an alternative system, private contracting. Another yardstick for comparison is the practices of other jurisdictions. No complete survey was made, but it appears that all resurfacing in the area, other than in New York City, is performed by private contractors. This includes the Port of New York Authority, Triboro Bridge and Tunnel Authority, the States of New York and New Jersey, and various counties, towns, and cities.

The main arguments against outside contracting are (1) that a change from present practices will cause difficulties and (2) that the contractors will rig their bids and not compete with each other, thereby raising prices. The first argument is undoubtedly true. The second is not serious, since even in the worst case the outside contractor will have to work more cheaply than city forces to get any work. Coping with collusive bidding or restrictive practices is not inherently worse for resurfacing than for reconstruction, which is currently contracted out. . . . The city should retain the capability for resurfacing, if it is needed.

Whereas street resurfacing in New York City is a good candidate for performance by competitive contracting, repair of "potholes" in that city's streets is not. The reason for this is that data on the location and frequency of pothole occurrence is virtually non-existent. This makes it impossible for someone to bid intelligently on a specification which requires the contractor "to repair all potholes that may occur" in a certain street segment. Furthermore, the absence of good information on the lifetime of a resurfaced street segment makes it impossible for a contractor to respond sensibly to a broader specification "to maintain a (given) street segment free of potholes," for the contractor is unable to weigh the tradeoffs of pothole repair versus street resurfacing. Finally, the technical difficulty (although not impossibility) of identifying individual potholes makes it senseless to let a contract on a "per pothole" basis.

Another municipal function that can be, and often is, contracted out is the maintenance of street lights and traffic lights, and the painting of traffic markings. Municipal construction can also be performed by the private sector on a contract basis. The director of municipal construction in New York suggested, for instance, that private enterprise be permitted to build municipal office buildings and day care centers, as critics pointed out that municipal construction costs in New York are ten percent greater than private ones (New York Daily News, 1972: 23).

The social services offer another broad category of opportunities for contracting out. Under a variety of programs, local governments contract with private agencies (usually nonprofit ones) to operate day care centers, head start programs, job training institutes, mental health clinics, social service centers, regional centers for the mentally retarded, and so on.

In a different realm, but still susceptible to contracting out via the private sector, are school lunch programs. Institutional catering services are well equipped to handle this function.

A very different type of service is the provision of electric power to a city. Some cities operate their own power plants, while others are served by a local utility, functioning as a "natural monopoly." There is a way, however, to introduce competition even into this tidy picture. If the city owns the distribution network, as does Detroit, it is, theoretically at least, in a position to purchase power for its residents from any supplier. As the interconnection of power grids proceeds apace, it is becoming feasible to start thinking of buying power contractually and competitively from any one or more of several different potential providers, including distant ones. This is not a realistic possibility as long as power supplies are barely adequate to satisfy local demands, but the picture could change if energy-conservation programs come into effect, and if power is fully priced as the scarce good it really is.

Incidentally, until the 1950s, New York had its own power stations which produced the electricity needed for running the subways. These served as a useful yardstick for comparing the performance of the local public utility.

The foregoing examples of competition in the provision of public goods through contracting out to the private sector generally refer to visible services provided directly to the citizen/consumer. There is another whole class of local government activities—staff or back-room services—not generally visible to the public or in the public

consciousness as collective goods which nevertheless offer opportunities for efficiency through competitive contracting out. All sorts
of consulting services and engineering work can be provided by the
private sector, as well as data processing system design, programming,
computer operations, and key punching. A New York study showed
that city key-punch operators were only half as productive as
private-sector key-punching services.

Fabrication and painting of traffic and parking signs can also be
done under contract, as can the repair of municipal vehicles such as
ambulances, garbage trucks, and police cars. A close examination of
productivity in New York's Sanitation Department garages showed
that mechanics in the private sector, working to conservative "Blue
Book" standards, were twice to five times as productive as the
department's mechanics. This finding was used by departmental
management to reform the shops and thereby to reduce the
disparity.

With respect to ambulance repairs, an audit by the Office of the
Comptroller (n.d.) showed that repairs on five ambulances which
cost the city garage $773 in labor alone could have been done by
commercial garages at a total cost of $527 for labor *and* materials.

Even the most prosaic of municipal paper-processing activities can
be a candidate for contracting out to the private sector. Referring
again to the New York experience, substantial savings are achieved
by the tax collector when he contracts annually with local banks,
after competitive bidding, to have them receive, open, tabulate,
deposit, and account for individual income-tax returns.

Public housing authorities can contract with the private sector for
custodial services, painting, exterminating, and other routine maintenance work in their developments, and parks departments contract
with nurseries and horticultural services for plantings, and with
engineering construction firms to repair structures and facilities.

When analyzing the potential cost savings of contracted services,
one must not neglect to include the overhead costs of preparing,
negotiating, processing, and supervising contracts, and the cost of
any delay that may be caused by the contracting procedure itself. As
an illustration of the latter point, the New York Parks Department
has observed that vandalized facilities are targets for additional
damage, and their prompt repair by more expensive municipal
workers is more cost-effective than delayed repairs by less expensive
contracted work.

Up to this point in the discussion, competition from only the private sector has been implied. This need not be the case at all. The benefits of competition may also be reaped when it is possible to organize and structure interagency competition within government or even between governments.

Parking enforcement illustrates this point. In cities which have parking enforcement agents in a department other than the police department, and if both departments can and do issue summonses for the same kinds of violations, then systematic comparisons of their relative productivity and cost-effectiveness can be used to improve the performance of both agencies (Savas and Berenyi, 1971).

Unfortunately, politicians campaigning on an economy platform, aided and abetted by students of public administration and management consultants, have devoted their energies to reducing and eliminating competitive behavior among government units, on the erroneous assumption that such competition was invariably, by definition, a wasteful duplication of effort.

A refreshing and thought-provoking example of deliberate intergovernmental competition comes from Ljubljana, in Yugoslavia—of all places. The city fathers there required the services of city planners in connection with a particular project and solicited formal bids for the work *not only from the city planning agency of Ljubljana, but also from the city planning agency of Zagreb,* a rival city! According to an American observer there (Birch, 1973), he had never seen city employees anywhere work as hard as Ljubljana's city planners, who were desperately trying to avoid the humiliation of losing their own city's work to their professional rivals.

THE LAKEWOOD PLAN

The notion of structured interagency and intergovernmental competition has been refined, formalized, and carried out to its logical conclusion in the Lakewood Plan. Lakewood is a city of 60,000 in Los Angeles County, and, as the originator, has given its name to the plan which—in that county alone—is followed in full by some twenty-five other cities and to some extent by more than seventy cities. For a detailed discussion of the political origins and evolution of the Lakewood Plan, the reader is referred to Warren (1966), but for the purpose here it is sufficient merely to summarize the main features that he describes.

Six months after Lakewood was incorporated, in 1954, its city government had only ten employees (city administrator, finance officer, city attorney, city clerk, and several administrative assistants and secretaries), but nevertheless provided the full range of municipal services to its residents. It had done so by entering into a series of specific service contracts with Los Angeles County to provide some services and by joining or retaining membership in several special districts which provided the remaining services it needed.

During the next fifteen years, this notion of contracting out to other governments for municipal services was embraced by twenty-eight newly incorporated cities. By 1963, there were 1,437 separate contracts—covering fifty-five different services—in effect between seventy-four "purchasing" cities and the "producers": sixteen departments of Los Angeles County and five special service districts. The five districts provided fire protection, library services, lighting and lighting maintenance, park and recreation services, and sewer maintenance; costs were covered by district taxes levied on member cities.

Some of the services offered by the special districts were also available from Los Angeles County. The full list of services marketed by the county was the following: library services, sewage maintenance, park maintenance, recreation services, street-light maintenance, assessment and collection of taxes, emergency ambulance service, hospitalization of city prisoners, personnel staff services such as recruitment, examination, and certification; city prosecution, building inspection, engineering staff services, subdivision final map check, subdivision engineering, school fire safety officers, weed abatement, state health-law enforcement, city health-ordinance enforcement, mobile-home and trailer-park inspection, milk inspection, rodent control, mental-health services, tree trimming, animal pound services, election services, street construction and maintenance, bridge maintenance, street signing, street sweeping, traffic-signal maintenance, traffic striping and marking, law enforcement, traffic-law enforcement, motorcycle patrol, jail services, disaster-law enforcement, business license issuance and enforcement, school safety, and crossing guards. Treasury and audit services were also offered at one time, but the county decided that there was a potential conflict of interest.

Contract agreements generally have a five-year life, but either party may terminate a contract on July 1 with two months' notice. Prices are set to cover full costs and are of four types: taxes for the special districts; statute-set fees for such services as tax assessment and collection; individual user charges for such things as animal licenses; and, for the majority of services, production costs. For the last, labor is charged on a productive work-hour basis, with fringe benefits added, plus departmental overhead costs. Equipment costs include depreciation, and material costs include warehousing and handling costs. Also added in is a general county overhead charge. There is some flexibility in negotiating costs between the seller and the buyer. The threat to terminate the contract can lead to greater efficiency or to the elimination of certain unwanted cost factors. For example, Norwalk insisted on and obtained patrol services by one-man police cars (and even got the county to establish the policy that Los Angeles County policemen serving in Norwalk be Norwalk residents).

The ability of Los Angeles County to maintain a market depends on its capacity to provide more favorable cost/benefit ratios than communities can obtain with other options, and they do have other options. A city can:

- organize its own service,
- purchase service from another jurisdiction,
- enter into a joint management arrangement with another jurisdiction,
- participate in a self-governing special district, or
- buy services from one or more private firms.

There is pressure on the county to perform, for a loss of customers leads to a loss of jobs, a loss of influence, and a loss of efficiency in those services which exhibit economies of scale. Even though some cities lack the tax base to consider all the alternatives listed above, the fact that some cities can do so is sufficient to expose the county to competitive forces.

Under these circumstances, there is a danger that the county will behave in a way to reduce competition and protect its market. Indeed (with no evidence of official sanction or encouragement), county employees have exerted pressure on municipalities to retain their services and to prevent the incorporation of areas which might form their own fire departments. Laws have been passed to restrict the ability of cities to utilize alternative delivery systems. For example, after a year of service, a city can no longer withdraw from a special district unless such action is approved by public referendum.

The county has been suspected of subsidizing the contract cities, at the expense of its unincorporated areas. However, there are both internal and external constraints on such potential behavior. As an internal constraint, the calculation of production costs, and hence the recommendation of prices, is carried out by the County Auditor/Controller, hence reducing the possibility of price discrimination and manipulation at the department level. The external constraint, of course, is the threat of incorporation by areas which feel they are being shortchanged, and, on the other side, complaints by noncontract cities that the county is violating the state's full-pricing law, which threatens their autonomy and their employees by creating pressures for them to contract out also.

One effect of such regulatory, counterbalancing pressures is that the county maintains excellent information on its service costs and performance levels. This has been good for county officials and for the cities. The former have improved their production efficiency and their internal communication and coordination; the latter use this information for their own cost-benefit studies, supplementing the market information they obtain from private bids and from each other. (The formation of the California Contract Cities Association further enhanced this flow of information.)

In short, Lakewood Plan cities (contract cities) can be viewed as consumer cooperatives that purchase municipal services. Their governments express community demands, but need not produce the goods themselves. This arrangement means that the demand unit can be a small community with relatively homogeneous preferences, the production unit for each service can be of any economic size and unrelated to the size of units which produce other services, and the best market price can be obtained. Cities have not had to make uneconomic investments in buildings, equipment, and personnel, but have been able to obtain the economics of scale of larger units, where they exist, without the loss of authority to negotiate "custom-tailored" services from private firms or from other jurisdictions at the most favorable price. The market-like aspects of the Lakewood Plan create pressures for efficiency and a responsiveness to communities that did not exist as long as the county maintained an effective monopoly.

The results of the market interaction in the Lakewood Plan demonstrate that the benefits of competition in the private sector may also be realized in the public sector. "This quasi-market pattern appears to provide a basis for structuring political fractionation in such a way that basic service standards are maintained, differing preferences in the public sector can be satisfied, and higher levels of efficiency and responsiveness induced in a large-scale producer" (Warren, 1966: 260).

The largest contract city in Los Angeles County has a population of about 90,000. How can the benefits of this approach be applied to large cities? The answer may lie in the growing political pressure for decentralization and neighborhood government in such cities. If lower-tier district government units were created and had autonomous authority over certain services, it should be possible to create a competitive environment not unlike that described above. Existing citywide service departments would sell their services, in competition with the private sector and with local districts.

Such a system will work under the following conditions:

(1) independent suppliers of services exist or can be created;

(2) at least some of the local districts have the resources, including managerial ability and political power, to change suppliers, and do so as the situation warrants;

(3) the suppliers place a value on gaining or retaining such customers, and post performance bonds or otherwise demonstrate tangible evidence of such value;

(4) local districts are able to prepare useful performance specifications and to evaluate suppliers' performances;

(5) local districts provide enough information to permit intelligent bidding.

These conditions are becoming more prevalent. And therefore the time is approaching for us to restructure and redesign our city governments in order to inhibit monopolistic behavior and to stimulate and institutionalize competition.

NOTE

1. The author is indebted to Professor Roland Artle of the University of California (Berkeley) for his stimulating insights on this subject.

REFERENCES

BIRCH, D. (1973) Private communication from Harvard Business School.

Citizens Budget Commission, Inc. (1972a) Reducing Refuse Collection Costs in New York City. New York.

――― (1972b) The New York City Health and Hospitals Corporation. New York.

CORDTZ, D. (1971) "How come it costs so much to run the city?" New York Magazine (November 22).

KAZAN, N. (1971) "Can free enterprise speed up our garbage collection?" New York Magazine (July 12): 41-43.

MEIER, R. L. (1972) "Evolution of the minibus system," chapter 9 in M. D. Mesarovic and A. Reisman (eds.) Systems Approach and the City. New York: American Elsevier.

Montreal, City of (1966) Specifications for Snow Clearing. Montreal: Department of Roads.

New York Daily News (1972) "Let private pros build: city aide." (May 10): 23.

New York Post (1972a) "Garbage dispute ripens." (January 20).

――― (1972b) "A 2-hour work day: road crews keep city and drivers in a rut." (May 8): 3.

――― (1972c) "Pothole city: slow men working." (May 9): 3.

――― (1972d) "Asphalt jungle, says contractor." (May 10): 3.

――― (1971) "City hall split by call for private sanitmen." (May 19).

New York Times (1971a) "Sanitation worker arrested in inquiry." (May 25).

――― (1971b) "City charges some sanitationmen charge for collecting trash." (May 26).

――― (1971c) "City may use private refuse haulers." (April 6): 1.

――― (1971d) "Suppressing urban realities." Editorial (April 7).

――― (1971e) "City budgetary default." Editorial (April 30).

――― (1971f) "Slimming the city budget." Editorial (May 4).

Newsweek (1973) "Rent-a-narc." (August 27): 25.

Office of the Comptroller (n.d.) State of New York Report NYC-38-73.

PATTON, D. K. (1973) Press release from the office of the Economic Development Administrator of the city of New York, June 21.

PHALON, R. (1971) "Private garbage hauling is being studied by city." New York Times (June 6).

PHILIP, K. (1954) Intergovernmental Fiscal Relations. Copenhagen: Institute of Economics and History.

ROBBINS, I. D. (1971) "Cutting city government costs." Wall Street Journal (May 19).

SAVAS, E. S. (1971) "Municipal monopoly." Harper's Magazine (December): 55-60.

――― and J. BERENYI (1971) "Systems analysis of parking regulations, violations, and enforcement activities." Office of the Mayor, City of New York, January.

SAVAS, E. S. et al. (1970) "Refuse collection: department of sanitation vs. private carting." Office of the Mayor, City of New York, November.

WARREN, R. O. (1966) Government in Metropolitan Regions. Davis: University of California Institute of Governmental Affairs.

YOUNG, D. (1972) How Shall We Collect the Garbage? Washington, D.C.: Urban Institute.

16

Increasing the Role of the Private Sector in Providing Public Services

LYLE C. FITCH

☐ THIS CHAPTER DEALS WITH ISSUES regarding greater dependence on the private sector to perform services which otherwise might
be performed in-house by government agencies.

The discussion concentrates on two main areas.

First is the production of what I call public interest services—those
goods and services which government undertakes, for whatever
reason, to supply or to have supplied to its constituents. I am most
concerned with tax-financed services which, if they are to be
privately supplied, involve contracts with government. In this
context, the contracting process and its improvement, and particularly the use of incentives to stimulate good performance are all
important.

Second is the use or stimulation of the private sector to promote
economic development in ghettos, thereby relieving governments of
responsibilities and activities which they would otherwise have to
assume. While this is a bit outside the usual definition of "privatization," it raises many of the same issues and deserves treatment in a
discussion of the private sector's public role.

The paper of necessity omits a number of other important related
issues. One is the general subject of pricing versus financing particular

services by public prices or charges instead of general taxation. Another is the voluntary contributions which businessmen and business firms can make toward the solution of problems confronting government, ranging from broad studies of public policy issues (the Committee for Economic Development) to lending management experts to assist government in improving internal management (the New York Economic Development Council).

INTRODUCTION

APPREHENSION ABOUT LOCAL AND STATE GOVERNMENT

Interest in privatization has been stimulated by three factors, all of which have been magnified over the last decade.

One is the growing cost of state and local government services, in turn caused by rapid increases in the amount of personal and other services purchased, and by the rate of cost inflation in the public sector (greater than that of any other major economic sector save private construction).

The second factor is the apparent inability of urban governments to cope with the increasing demands thrust upon them, as manifested by growing congestion, pollution, crime rates, delinquency, and dirty streets, worsening housing for low-income groups, declining pupil performance in ghetto schools, deteriorating public transportation, and declines in other indexes of civic quality. These conditions in turn reflected in part (a) the increasing magnitude of the problems faced by local governments, particularly governments of large cities inundated by the new wave of immigration from rural to urban areas; and (b) the political and organizational inflexibility of many local governments, and their inability to adapt to changing economic and demographic conditions and demands for service.[1]

The third factor concerns the widely held belief that public bureaucracies are inherently less efficient, in the economic sense, than are private firms dominated by the profit motive. In addition to the ideology which regards private enterprise as the natural order of things and the public sector as a necessary evil, several reasons are advanced for thinking that the public sector is inherently less efficient than the private.

(1) Bureaucracies are subject to the instinct to survive and expand which characterizes most organizations, including private firms; but private firms are limited by the demand for their products, which must compete in the market with the products of many other firms. With tax-financed government services, the taxpayer-consumer is in theory the ultimate choice maker, but "rational" taxpayer control is inhibited (a) by lack of identity between taxes paid and services rendered and the fact that taxpayers and service consumers in many cases are different individuals, and (b) the fact that most taxpayer-voters have no very good idea of the types, amounts, and qualities of public services they receive for their money.

The organizational impulse for survival and expansion is therefore subject to the discipline of consumer demand in the private sector (for an analysis of this point, see Niskanen, 1971).

(2) Whereas the "outputs" of private firms are clearly identified as the goods and services which they sell in the market, many of the outputs of public agencies whose "market" is the political process, are more or less nebulous. Moreover, they consist not only of the services delivered to consumers, in the conventional sense, but also of the benefits involved in providing inputs—the jobs, contracts, and organizational objectives (such as survival and expansion) described above. The beneficiaries of government production are therefore not only the direct consumers but also the government officials, decision makers, and employees, and private firms which purvey inputs to governments. All these interests exert pressure on the political process to maintain and expand the particular services.[2]

(3) The aims and objectives of government bureaucracies and the standards of service to which they should aspire tend to be defined by the professional bureaucrats and their constituencies, rather than by the demand of consumers in the market.

With functions which involve rendering services to clients, a common complaint against the public bureaucracies is their tendency to be governed by bureaucratic routine and the preferences and conveniences of the bureaucrats rather than the needs of the clients. Thus, public services are frequently made available only in places and times inconvenient to clients—health and welfare offices conventionally are open only during the daylight hours when the important part of their clientele is also likely to be working.[3]

THE POLITICAL PROCESS VERSUS
THE MARKET PROCESS IN
ALLOCATING RESOURCES

Public operating agencies resist the types of control dictated by considerations of economics and public administration. Requests of program budget agencies for definition and justification of the purposes and mission of operating agencies are likely to be regarded as irrelevant. In many cases, the real justification for maintenance or expansion of an operation, or the initiation of a new project, is to be found, not so much in public benefits as in the fact that jobs will be provided.[4]

As previously noted, "outputs" of the political process are of larger scope than the "outputs" of the production-for-market process, and in the political process the distinction between outputs and inputs tends to be blurred—inputs (defined in the economic sense) tend to become ends as well as means. The political process acts to bring public, quasi public, and private interests together for the purpose of securing larger allocations of resources for particular purposes than would be dictated by consumer preferences operating in the market.

This point is often obscured in discussions of relative advantages of the private versus the public sector. The real issues concern not who is responsible for production but rather the market process versus the political process of allocating resources. Greater efficiency (in the economic sense) may be obtained by introducing neo-market techniques such as financing public services by user charges rather than general taxes, thus allowing consumers to exercise choices directly.[5]

Those are primarily allocation questions, however. Here we are concerned primarily with *production* efficiency rather than *allocation* efficiency.[6]

ADVANTAGES AND DISADVANTAGES OF
PRODUCING PUBLIC SERVICES "IN-HOUSE"
VERSUS CONTRACTING OUT

The following discussion examines first the possible advantages and disadvantages of having public interest services performed in-house (directly by the employees of the jurisdiction responsible for furnishing the service), or by contracting out to other governmental jurisdictions or private organizations, profit or nonprofit.

ECONOMIES OF SCALE

Economies of scale imply the existence of elements of overhead whose costs per unit can be reduced by spreading production over a larger number of units than the jurisdiction requires. Examples of situations where there may be economies of scale are a specialized computer which can be employed only part-time by the work of a single government jurisdiction or a highly skilled engineer or other specialist who, if he worked full-time for one jurisdiction, would have to spend part of his time doing work that could be performed by a less-skilled lower-salaried person.

An example of the entire production process which combines several scale economies is a high school desiring to offer a variety of courses; the pupil-population must be large enough to provide adequate enrollments for each of the courses if the per pupil costs are to be kept within acceptable limits. Here scale economies are associated with a function which needs an administrative organization large enough to bring together closely related activities and facilities in a way to achieve the potential advantages of cooperation and avoid excessive duplication of effort. Other functional examples include hospitals and other health services.

Another set of economies has to do with geographic scale—for example, air pollution control in a metropolitan area or flood control in a river valley. Still another set is associated with both functional and geographic scale—for example, intraurban transportation in a metropolitan area.

The foregoing examples indicate that there are two different bases for scale economies, one having to do with the use of specialized resources (men, machines, and so on) and the other having to do with the size of jurisdiction required for effective planning and policy-making. In principle, most activities having to do with specialized resources can achieve scale economies by contracting out. The supplier achieves his economies in turn by serving a number of clients, public or private.[7]

Scale economies concerned with basic policy-making, however, in a democratic society cannot be easily contracted out, particularly insofar as they concern the proportion of a community's resources which it desires to spend for particular kinds of public services, and the quality of services required.

Still another dimension of scale is political influence. Large agencies can bargain more advantageously than can small agencies with other organizations, including other government jurisdictions and higher levels of government, labor unions, and suppliers. This element of scale economies cannot be contracted for, except indirectly by the small organization contracting with the larger, which in turn can bring its own size to bear in obtaining costs and other concessions from suppliers. (Los Angeles County presumably can bargain more effectively with the federal government and the state of California than could individual smaller jurisdictions which it services under Lakewood-type contracts.)

Similarly, a small contracting jurisdiction may be at some disadvantage in dealing with a larger public or private organization, which is likely to be in a monopolistic position. The contracting-out jurisdiction ordinarily may be able to discontinue the contract and provide its own service, but on penalty of sacrificing economies of scale. Where the services required are only occasional, like specialized planning, engineering, or legal services, or the occasional use of specialized machinery, there may be a number of outside purveyors from whom to choose, giving the jurisdiction an opportunity to take advantage of competition. Where the scale economies require a large organization with investments in capital equipment or maintenance of a large, specialized force—in other words, where the supplier is a monopoly—alternatives are less readily found.

FREEDOM FROM CONVENTIONAL BUREAUCRATIC CONTROLS

Many (not all) general governments operate under restrictive budgetary, personnel, and other controls designed to avoid the abuses of spoils systems and otherwise preserve the proprieties, to avoid excessive spending and other "waste" by operating agencies, and ostensibly to provide governments with competent personnel.

Personnel systems are probably the most invidious types of control. They tend to freeze employees into positions and to prevent management from disciplining uncooperative employees; only cases of extreme recalcitrance or misbehavior will justify actual discharge. They prevent lateral transfers or entry into the service at any but the lowest grades, thus effectively blocking competition for jobs and the chances of recruiting ambitious people. Civil service examinations are designed to reduce arguments by overemphasizing "objectivity," and knowledge items having small relevance to such job requirements as qualities of leadership and integrity and demonstrated past performance—qualities which cannot be easily measured and hence tend to be underemphasized (Savas and Ginsburg, 1973).

Such dysfunctional qualities of civil service systems commonly reflect employee pressure, which tends to emphasize continuity and seniority over competence as qualifications for higher-level positions, and by employee unions which emphasize the traditional union goals of more pay, less work, and job security.

Budgetary controls are another hindrance to administrative efficiency. Most municipal budgeting, and much budgeting at other levels, is primarily for controlling spending, rather than for the purpose of making rational allocations of resources (Wildavsky and Meltsner, 1970: 311).

As ordinarily administered, budgets afford no incentives for efficiency. Heads of operating agencies (the bureau chiefs) commonly attempt to maximize their budgets; this is how they display their power, influence, and political acumen.

The lack of incentive for managerial aggressiveness and initiative in effecting economies and finding better ways to deliver service is magnified by the common budgetary practice of recapturing all savings, however realized, by operating departments. The tendency is furthered by the one-year appropriation practice under which appropriations lapse if not spent or obligated by the end of the fiscal year.

The year's-end scramble to obligate unspent funds may not be an entirely bad thing, however, if it gives agencies some incentive for economies over the course of the fiscal year. More general approaches have been advocated, whereby operating agencies could spend at their own discretion a substantial part of any savings realized by administrative economies with respect to any particular service.

These are only broad tendencies in present systems. Two contrary forces should be noted. First, there is a long tradition of obeisance to principles of public administration which stress effectiveness and efficiency, along with economy. Most administrators and politicians pay at least lip service to the tradition. Professional administrators and auditors increasingly take responsibility for furthering it.

Second, program planning and budgeting has given more canny program administrators another tool, and hence the incentive, for supporting budget requests with PPBS-type evidence of program effectiveness.

The increasing use of state and local governments of professional managers, either as chief executives responsible to municipal and county legislatures or as assistants to elected chief executives, is an indication of public interest in effective management. Although there is no index of the quality of management in cities, the city manager cities probably rate higher on the whole than do the nonmanager cities. They also are more likely to examine the options available to get various services performed, including that of contracting out.[8] Professional managers are also more able to modify, get around, or otherwise deal with the types of bureaucratic controls commonly found in local governments.

Possibilities of escape from unions and collective bargaining. Some see contracting out as an escape hatch from not only debilitating civil service regulations but also the more immodest demands of public employee unions and their political muscles. While circumstances differ in various communities, however, it is a safe generalization that politically powerful employee unions can muster the power to prevent competition from employees of private firms. The Bacon-Davis Act, along with state and local legislation requiring the payment of union scales for work done on public contracts and union pressures on governmental jurisdictions to buy only from union shops testify to this strength. Moreover, once a public employee union has gotten a firm hold on a governmental function, any attempt to escape by resorting to private contracts will be considered union-busting and dealt with accordingly.[9]

TAKING ADVANTAGE OF COMPETITION AND
THE PROFIT MOTIVE

In a market system dominated by private enterprise, the chief guarantor of product quality, the chief incentive to efficient operations, and the chief force operating to hold prices reasonably close to production costs are competition, coupled with the profit motive. One of the main objections to the way in which government bureaucracies operate lies in their tendency to disregard and place their own convenience over the needs and wishes of their clientele, which is attributed in turn to absence of any counterpart of the profit motive.[10]

The economic advantages presumed to attach to competition are the most commonly advanced argument for hiring private, competitive firms to produce as much as possible of the economy's public interest services. Before buying this viewpoint, however, it is well to be clear on several points.

(1) It is not always necessary to compete in order to enjoy the fruits of competition. Government bureaus may compete with each other (though such competition has usually been decried by public administration experts as "duplication of effort") or with private firms.

(2) Competition appears to work best when purchasers have a variety of alternatives from which to choose and are aware of the intrinsic differences among products. Consumer information is deemed so necessary an element of the proper functioning of the market that many governments undertake to see that producers provide it, frequently over the producers' objections. A large part of the popular advertising of different brands of goods seeks to conceal real differences among different brands, withhold relevant information from consumers, and accentuate irrelevant or slight differences (for a famous statement of this tendency, see Chamberlin, 1962).

(3) Individual consumers have much trouble with complicated technical services which can be performed only by specialists and which they, the consumers, cannot understand. Survey after survey of television and automobile repair shops indicate that the majority are, in varying degrees, dishonest. Such situations raise a demand for public control by licensing producers or inspecting their products. As for a wide range of other products, the acidulous reports of the various consumers' organizations, governmental and private, and of self-appointed public defenders such as Ralph Nader remind us that

> competition by itself does not afford adequate consumer protection. In defense of competition, it must be said that protection of consumers is generally easier with competition than without it.

Finally, it should be noted that economies of scale and the advantages of competition are ususally not available in the same package, since economies of scale require large-scale operations, which in turn usually imply a degree of monopoly.

REDUCING GOVERNMENT ADMINISTRATIVE RESPONSIBILITIES

Another potential advantage of contracting out has to do with the decreasing administrative burden on general government. The advantage can be particularly great when the service concerned has to be performed by numerous small units.

A case in point is the use of private towing services to remove illegally parked or disabled cars from public streets, which takes at least a certain amount of police attention away from more important duties, the more so because police forces seem to be incapable of handling towing functions efficiently. Another possibility is in the field of low-income housing (where private firms frequently perform badly and gain ill repute). Turnkey construction for public housing authorities, as an alternative to closely supervised public contracts, was reported to have had considerable success and to have reduced construction costs by an average of some fifteen percent.[11]

Another administrative problem has to do with the propensity of public agencies, reinforced by civil service and related institutions, to expand or at least perpetuate themselves. Thus, agencies or sub-agencies set up to accomplish a particular purpose frequently find ways of continuing long after the need for them has been satisfied. Attempts to eliminate dead wood commonly elicit much cater-wauling and threats of political reprisals, and necessitate heavy expenditure of political capital. Such hazards can often be circumvented by contracts with nongovernmental agencies, both profit and nonprofit.[12]

DISADVANTAGES OF CONTRACTING OUT

The foregoing discussion concentrated on the potential advantages of contracting with, or otherwise arranging for, nongovernment agencies to provide public interest services. It was noted that most of the advantages have various off-setting drawbacks and are limited as to the conditions under which they may be realized. This is not to dismiss the notion of using the nongovernmental sector, but only to emphasize that the advantages must be pursued with caution. In addition to the drawbacks previously mentioned, there are some general problems which must be weighed in each case.

THE PROBLEM OF QUALITY CONTROL

Controlling the quality (as opposed to the cost) of public interest services is a problem whether they are produced by governmental or nongovernmental agencies. For goods and services which can be identified, weighed, and measured, or tested as to performance and use, tests concentrate on how well the product meets specifications. The problems are primarily administrative—who inspects the product and how well, and what the motivations are for doing or not doing the job.

Where the product is not easily measurable as to quality *or* quantity, the apparatus of control involves product inspections, investigations of complaints (as of faulty service), and monitoring of production processes.[13] Internal controls are subject to various intramural pressures—for example, hostility of administrative agencies toward auditors. External controls are vulnerable to friendly relationships between representatives of contracting agencies and contractors' representatives,[14] political pressure, and outright bribery.

In this respect, contracting out has no clear advantages over government in-house account production; in fact, the difficulties of quality control in many cases may be greater with private contractors. Competition, as observed earlier, is not an adequate means of assuring quality, as demonstrated by the widespread abuses in the appliance repair industries. Moreover, once a large contract involving substantial amounts of work is signed, competition ceases, and other control processes take over. Malperforming contractors, to be sure, are subject to being dropped from consideration for future contracts, depending on their political clout, but the possibility of such retaliation may be poor consolation for bad performance on an important project.

One way of reducing inspection problems is to use only firms which have demonstrated their competence and reliability in past performances. Use of this expedient, however, may be complicated by the requirements of competitive bidding, and the necessity of giving serious consideration to low bidders unless they are conspicuously unqualified. Also, the continuous use of one or a few firms lends itself to the creation of friendly relationships, which may make difficult the exercise of appropriate controls.

In general, the more nebulous and unmeasurable is the end product (police protection being a good example) the more difficult is the problem of inspection and control. Does this consideration militate against contracting out? Not necessarily, I think. Much depends on whether the contracting jurisdiction is dominantly an economic or political model.

CONTROLS IN ECONOMIC MODEL VERSUS POLITICAL MODEL COMMUNITIES

The economic model presumes a hard-nosed contract manager capable of setting up specifications for the public interest service concerned and insisting that the contractor follow specifications. Even police services could be contracted for, if certain other barriers (such as the impropriety of vesting employees of private firms with powers to arrest and subdue other citizens) did not intervene.[15] In the political model, where provision of inputs (jobs and contracts) are considered as valid, if not strictly "legitimate," prizes of governmental control, the general level of citizen satisfaction (as registered by squawk indexes, among other devices) becomes a more important servomechanism in the control process, and services can probably be better controlled through political channels than by monitoring contracts with private firms.

Government-produced services may be more costly, but bureaucracies may be more responsive to the demands of politically potent citizens. To the degree that the bureaucracies are subject to political sanctions, they may be more responsive to the demands of politically potent groups; other groups—notably, the minorities—are likely to be ignored until they can muster the power to make themselves bothersome. Nonetheless, the quality of services delivered is seldom an issue of urban politics, except around election time.

In the political community, contractors are expected to make political contributions in order to be eligible for contracts. Contributions may take the form of outright bribes and graft, but such open breaches of the proprieties are a favorite hunting ground of politically ambitious public prosecutors (even those belonging to the party in power), and the more popular form is the campaign contribution—outright grants, subscriptions to fund-raising dinners, and so on. Such potlatch may be expected to take its toll by raising the costs of contract services and loosening the assiduousness of inspections, though the more cautious political operators will insist that work be at least passable, and only the more venal will tolerate such practices as mixing concrete with salt water (a trick reported to have been recently revived in Florida).

The political community, therefore, has more to gain from contracting out than does the economic community, if it could get the same results; in reality, the factors which militate against efficient production in the public sector also militate against getting highest-quality results from contracts.

This is not to say that the values of the political community are utterly deplorable. To take one instance, public service jobs have long been one of the few available ladders up the economic and social scale available to the immigrant groups of large cities; this was true in the past of the Irish, the Jews, and the Italians, and it is beginning to be true of the Blacks and Spanish-Americans in today's larger cities. And while the services may cost more and be technically less efficient than if performed by civil servants who pass more rigorous examinations, the total situation, including the opportunities of participation, may be more acceptable to the minority groups than better services provided by "white" workers,[16] if this were in fact the option.

In this milieu, contracting out may be seen by middle-class groups anxious for better municipal services as possible means of escape from the highly political civil service. It can serve another function as well—that of providing minority firms with opportunities not available if they had to compete in the private sector. Such demands are reinforced by federal, state, and local civil rights legislation enacted over the past decade. Other firms doing business with governments are likely to be confronted with demands for "equitable" representation, or even overrepresentation, of minority group workers, qualified or not.

Another relevant model, which is an offshoot of the political model, might be called the entrenched bureaucratic, in which the bureaucracies themselves wield large political power. The monopolistic powers derived from civil service regulations are in many cases enforced by powerful public employee unions. Such bureaucracies are frequently permeated by petty and not so petty corruption, which is most difficult to uproot because the members, the honest along with the dishonest, close ranks to protect their own from prosecution (see, for example, Pilleggi and Pearl, 1973). In such systems, inspectors and other bureaucrats often form alliances with contractors and others whom they are supposed to be watching. Or, in another version of the game, they exact petty tribute from suppliers, building contractors, and others who do business with the city. Since the entrenched bureaucracies tend also to be impervious to demands for greater effort and better services, particularly if such might require changes in established ways of doing things, they again generate desperate longing for other service delivery systems. The political muscle of the bureaucracies, backed by the public service unions, usually enables them to resist changes, such as contracting out, which would attenuate their own service monopolies. They are also prone to resist (1) incentive systems (the teachers organizations for years have fought differential pay for superior performance or for harder or more disagreeable assignments, such as posts in ghetto schools); (2) the very notion of productivity; and (3) changes in work routines which would disturb ongoing rackets.

To summarize the main point of the foregoing discussion, it is that contracting out public interest services to the nongovernmental sector is unlikely to provide significant means of escaping lowering production costs or improving service quality in either the political or the entrenched bureaucracy models. Many of the forces which inhibit efficiency by forestalling efficient performance (in the technical economics-engineering sense) by the bureaucracies operate in like manner on private contractors.

PROBLEM INDUSTRIES

Construction industry. The fact that the private sector, in competition, does not necessarily produce outstanding economies or slow the rate of inflation is demonstrated by this black sheep. The industry has achieved a rate of inflation equal to the cost inflation in state and local government and has not been notable either for technological innovation or for high productivity.

There are several reasons for this state of affairs. First is the nature of the building trade unions which dominate the construction industry, and their immoderate penchant for high wage increases and featherbeds. Second, many of the trades have held down the labor supply by restricting union entry—notably, of minority groups. Third, the construction industry, particularly the home building industry, has been characterized by small contractors; few large corporations, with enough economic and political influence to employ techniques of mass production and cost reduction, have emerged. Fourth, unions, contractors, and suppliers have collaborated to impede technological improvements, by maintaining archaic building codes, restrictions on new materials and prefabricated products, and the like. These practices have been furthered by the exemption of unions from the application of antitrust laws. Fifth, the unions have been politically powerful enough to maintain rules requiring the use of union labor and payment of union scales to public employees employed on publicly financed projects.

The transit industry. Private ownership and management have largely gone down the drain. Reasons include: (1) a long period of declining volume owing mainly to increasing motor vehicle competition, (2) union insistence on uneconomic pay scales and work practices regardless of their effect on the volume of business, (3) stodgy management, due partly to the difficulty of attracting good managers into a declining industry; and (4) a record of shady practices and looting the assets of established firms by unscrupulous operators (for example, the New Haven Railroad, the Washington Mass Transit System, and many others).

The common pattern has been for private companies to be taken over by municipalities or public authorities, as private firms have pulled out of the business. In some cases, the effect has been beneficial, as public agencies have consolidated operations of numerous private firms (Dade County, Florida, and Allegheny County, Pennsylvania, are examples). In some cases, the move to public ownership was made to relieve transit systems of tax obligations, and to facilitate public subsidies; in some of these instances, private firms were retained to manage the transit systems after they had passed into public ownership. There is no evidence, as far as I know, to indicate that private managements have been any more or less successful than public-employee managers, and given the record of weak management in the transit industry generally, there is little reason to think that a breed of superior transit managers can be recruited by private firms.

GRAFT

Contracts are one of the most common and lucrative sources of corruption in government. The abuse has been only diminished, not eliminated, by public bidding and other formalities designed to improve the integrity of the process. Private contractors doing business with the government are still one of the principal sources of campaign funds, and of support for shady politicians.

The widespread abuses which attended the Federal Interstate Highway program remind us that the public trough is still a tempting alternative to bank robbery. A House of Representatives subcommittee reported in 1963 that various investigations by different agencies had uncovered "fraud or carelessness involved in right-of-way acquisition in 24 states, shoddy or deliberately dishonest construction practices in 21 states, payola accepted by highways department employees in some states" (cited in Cook, 1966: 116).

The problems of overreliance on the private sector to perform public-interest services efficiently and competently are highlighted by a report on the federal low- and moderate-income home ownership program which found that about twenty-four percent of the new homes and thirty-nine percent of the rehabilitated homes under the program had significant defects which were charged to HUD's failure to adequately inspect homes before they were sold. It was charged also that HUD countenanced block-busting and real estate speculative activities under FHA home mortgage insurance programs, and that these programs were utilized in Detroit, Boston, New York, Philadelphia, Washington, Chicago, Newark, and St. Louis to sell deficient homes to moderate-income persons.

Subsequent indictments of FHA officials and private real estate and mortgage company officials pointed out the existence of fraud and bribery, as well as laxity, in FHA operations. There have also been instances of poor sites and shoddy construction in both new and rehabilitated housing developments in a number of metropolitan areas. There apparently was a great emphasis on production, to meet production targets, and insufficient emphasis on quality control [U.S. Congress, 1974].

The widespread irregularities in defense contracting programs have been regularly publicized, leaving one prominent critic, Gordon Rule, to comment that U.S. industry is incapable of producing a quality product on time at a reasonable cost, and to charge that industry is "smug and perhaps rightly so" because it knows no punishment will ever be imposed on it (Cook, 1966).

Evidence of the temptation of the contracting process is to be found also in the number of prominent New Jersey politicians in jail or under indictment for accepting illegal gratuities from contractors and otherwise perverting the contracting process. And, of course, this is the case of former Vice President Agnew.[17]

PUBLIC SUSPICION

Revelations of corruption in government contracting have generated a public suspicion which at least matches and often overrides public acceptance of private enterprise and competition as the normal way of life. For example, several years ago New York City Mayor Abraham Beame, then comptroller, deemed it politically profitable to charge the Lindsay Administration with having greatly increased the amount of contracting out done by city government agencies, and with using the contracting out device as a means of paying high salaries (presumably to individuals whom the administration wanted to favor). This illustrates another problem of contracting with private firms which involves paying high salaries to obtain superior expertise. Such contracts tend to be viewed with suspicion by the public at large and by media reporters, whose own modest pay scales probably fan their zeal to report such conditions.

In the attempt to improve their integrity, or at least the appearance thereof, government contracting agencies have over-reacted by imposing a wide variety of regulations, standards, and other requirements, many of which have the effect of putting a greater strain on honest firms than on dishonest firms, which can often find some way of beating the regulations, if only by buying the cooperation of government contracting officers.

OVERZEALOUS AUDITORS

Government contracts with consulting firms for expert services frequently raise another difficulty. In the bidding and negotiation phase, the consulting firms frequently overcommit the services of their top experts (which government agencies think they are buying), planning to perform much of the work by people of less exalted stature. Firms will optimisitcally or deliberately underestimate costs, hoping to recoup later by obtaining contract extension, thereby cutting corners. Many of the reports produced by consulting firms are more distinguished by the elegance of their covers and formats than by the quality of their intellectual work.

On the other hand, government contractors, particularly at the federal level, must contend with government auditors who in many cases have no hesitation in disallowing what the contractors regard as legitimate expenses, rewriting contracts to conform with their own interpretations of regulations, and otherwise making life difficult.

The high cost of obtaining government contracts, the limitations on salaries and other costs frequently imposed by government regulation, and the problems raised by zealous auditors make government contracting for the typical small firm, and for many larger firms, a chancy business. The risks impel many firms to limit the amount of government business they will seek, and some now go after government contracts only because of ancillary advantages (such as access to information not otherwise available).

Of course, giant firms such as Lockheed—and smaller firms whose output is crucial to certain government programs—have received special assistance to keep them in business, and such occasional bones doubtless encourage the rest of the pack, particularly the hungrier ones.

ALTERNATIVES TO CONTRACTING OUT

THE PUBLIC CORPORATION

The public corporation or utility is a common answer to the political, administrative, or geographic incapacities of local jurisdictions. The federal government has stimulated the creation of many single-purpose local agencies, such as housing and redevelopment authorities, out of lack of confidence in the ability of general-purpose governments to implement programs with reasonable speed and probity. Other agencies have been set up to escape financial restrictions laid on local governments by state constitutions or statutes, and still others, as we have seen, to escape the civil service and other administrative restrictions. Most such agencies are enterprise-type, supported mainly by user charges; some, including school districts and special districts created to operate on a metropolitan or regional scale, have limited taxing powers or are supported from tax levies.

One of the primary purposes of the large port authorities has been that of assembling large amounts of funds for capital-intensive facilities—roads, bridges, ports, airports, and so on. The usual financial instrument is the revenue bond backed by projected revenues from the venture, revenues from already existing facilities, and occasionally by the full faith and credit of the general government.

By and large, such entities confine themselves to functions specifically enumerated in their charters, but some, with wider grants of power, have taken on a wider range of functions. Thus, the San Diego Port Authority has taken over many functions relating to urban redevelopment and housing. The explanation for such expansion lies in progressive, imaginative leadership and a history of demonstrated administrative competence. Neither of these things can be legislated.

Public corporations, particularly of the enterprise type, ordinarily can pay higher top salaries than can general governments financed by taxes, and hence can better compete for good managers. Even so, good management appears to depend on a combination of happy circumstances (the right man being in the right position at the right time), as well as on the diligence of boards of directors (and the authorities who appoint directors) in seeking out and thereafter supporting competent people. The management records of the public authorities and corporations have been mixed; some have been regarded as outstanding, while others have fallen into disrepute and scandal.

In the absence of competitive firms or of other "yardsticks," it is frequently difficult to determine whether a public authority, or any monopolistic enterprise, is being efficiently managed by the tests of industrial engineering and conventional administration.

The Tennessee Valley Authority and the National Aeronautics and Space Administration are two examples of very large organizations with reputations for innovational, capable management. Both from the outset were highly self-conscious about management, defining their goals and devising ways of attaining them efficiently and expeditiously.

That the public corporation form, and the greater flexibility afforded thereby, is not necessarily sufficient to improve public-interest service is shown by the recently established U.S. Postal Service. The transition from a more conventional bureaucratic agency to a public corporation was marked by an accelerated deterioration of service brought about in part by reductions in personnel for the purpose of reducing costs and making the system more nearly self-supporting. What will happen in the long run and whether the claimed large-scale efforts at modernization and mechanization will improve services remain to be seen. One suspects, however, that permitting more competition by private agencies would have beneficial effects on the system overall and might considerably enhance the quality and variety of services available.[18]

Reputations for good management may depend upon credit ratings rather than upon outstanding efficiency in operations or upon qualities of imagination and willingness to innovate. The authorities generally have the reputation of being more solicitous about their bondholders than about undertaking public service ventures that might incur financial losses. Or they may use their financial resources for ventures chosen by authority management for its own reasons, rather than ventures chosen by elected officials.[19] The freedom from general government bureaucratic control, in short, is likely to involve also the propensity to interpret public needs in light of authority management's own interests and preferences.

The private utility corporation is only a short step along the spectrum from the public corporation. It typically provides a public service which could be and in some cases is provided by public corporations or even by general government administrative agencies. The public-interest character of such service justifies both their regulation and, for services in which competition among companies would be wasteful or inconvenient (for example, telephone services), a grant of monopolistic privileges to the designated firm.

The chief differences between public corporations and regulated private utilities lie in the following:

(1) The method of selecting boards of directors. Public corporation directors may be chosen by elected officials (governors, mayors, legislators, and the like) and are supposed to represent the public interest. Private corporation directors nominally represent the stockholders but are in fact likely to be selected by bank and other commercial interests (political interests often are more or less surreptitiously represented, also); once established, boards tend to be self-perpetuating in that members select new members for ratification by stockholders.

(2) Financing. Public corporation financing is primarily through bonds or other, usually tax exempt, obligations of indebtedness. Private utilities are financed by taxable bonds and stock, the latter representing so-called equity investment. Since regulated companies are assured a "fair rate of return" on invested capital, the advantage lies in leveraging the earnings to provide relatively high rates of return on common stock. The risk to stockholders (and bondholders) lies primarily in market fluctuations rather than in losses of the enterprise itself.

The chief advantage of a regulated private utility over a public corporation would therefore seem to lie in the possibility of attracting more capable management through various compensation devices (stock options and the like) available to private corporations. Whether or not such devices actually attract better managers or produce more efficient operations is an open question. (*Efficiency* here connotes industrial engineering and human relations techniques, as opposed to financial manipulation.)

On the other hand, public service offers certain rewards in the form of prestige and satisfaction which are thought to appeal to an increasingly large group, notably many able younger people.

Comparative studies of public and private enterprise performance tend to be inconclusive, partly because of the difficulty of measuring performance for purposes of comparison. However, a study of communications enterprise by the Communications Workers of America (a not completely disinterested source) indicates that privately controlled and operated companies provide better service, maintain higher wages and better working conditions, and impose charges no higher than those of public companies (see Koontz and Gable, 1956: 106).

Few would argue that the Port of New York Authority has not been better managed than Consolidated Edison. The Penn Central debacle and the repeated plundering of the New Haven before it was acquired by the Penn Central have not to my knowledge been matched by any public enterprise scandals. And if the Postal Service, first as a government agency and then as a public corporation, is unable to cope, telephone service provided by private utilities has likewise been deficient. One can only conclude that good management and efficient performance depend upon factors apart from the question of whether the corporation is public or regulated private.

Moving along the spectrum to private nonregulated, or less-regulated, firms which take government contracts, there may be little advantage in the degree of freedom achieved from bureaucratic administrative controls, what with the plethora of red tape, officious auditors, and so forth.

Whatever advantage exists lies in substituting competition for regulation.

SUBSTITUTING PURCHASING POWER FOR
DIRECT GOVERNMENT SERVICES

Food stamps, rent supplements, and education warrants have in common the notion of supplying beneficiaries (usually low-income beneficiaries) with purchasing power limited to a specific type of service (food, housing, or education) and thereafter letting him (the consumer) make up his own mind. In some cases—for example, education warrants—the consumer can choose between services offered by a public agency or private firms. In other cases, as with food stamps, he is limited to private firms.

In addition to providing consumers with a wider variety of choice, this approach has the theoretical advantage of involving consumers and private firms in more or less normal market relationships. The degree of normality depends upon the administrative arrangements for using the stamps, warrants or other money substitutes. One common complaint against food stamps is that the users are made to feel conspicuous by being required to use special checkout counters or submitting to other special requirements. The rent supplement programs have been handicapped by the variety of restrictions on the type and quality of housing eligible for rent supplements, which restrictions had the effect of substantially reducing consumer freedom of choice.

As to the use of education warrants for the purpose of giving parents wider choices over the schools attended by their offspring, thereby introducing competition among schools, it would be surprising if these had much impact, save on the disadvantaged children of urban ghettos who need educational improvement the most, where it would only work to facilitate school segregation. Middle- and upper-class parents would be even less inclined than at present to send their children to schools patronized by less-capable and less-motivated children. In that case, the greatest of the numerous problems of contemporary education appears to lie in narrowing the achievement gaps between children of poor minority families and other children. While there are cases of individual schools or individual programs which appear to have had considerable success, few programs have yet appeared which can be replicated on any considerable scale. Many such programs appear to depend heavily on the special skills or leadership qualities of gifted teachers or administrators.

The use of money substitutes in connection with special categories of goods, as an alternative to providing direct services, is one aspect of a larger question: If particular families are deemed to require more purchasing power than they have hitherto been getting, why not simply give them more money to be used as they choose? While this course would satisfy the economic criterion of affording maximum choice to consumers, it raises several objections. First, society presumably has some responsibility for setting priorities in the use of the public funds which go to supplement the incomes of needy families. Second, the preferences of taxpayers who supply the funds should be given some weight in choices of how the funds are to be used.

FEDERAL CONTRACTING FOR
PUBLIC SERVICES

Private concerns already provide many of the inputs utilized by governments in the process of producing and delivering public-interest goods to final consumers; they also directly produce and deliver many final products. The total procurement by the federal government as of fiscal 1970 was estimated at $57 billion; I have found no comparable figure for state and local government procurement purchases, but the volume in some areas—notably, road and other construction—is large.

The federal government's needs for high technology have given rise to the government-oriented corporation, particularly in the more research-and-development-oriented industries—those concerned with aeronautics, missiles, electronics, scientific instruments, and the other aspects of high technology. The lively government demand in the space and defense industries promoted a corresponding growth of corporate capacity in these areas and promoted the growth of the subtribe of corporations specializing in government work.[20]

As the defense and space business declined, so did the industries that had depended thereon, not without great social losses to employees and communities, and financial disappointment to over-optimistic stockholders, but with much less fuss than would have attended government agency cutbacks of any such magnitude.

It was notable, however, that the government-oriented corporations produced by the defense and space age exhibited relatively few of the benefits usually associated with competition and the profit motive. Rather, they acted as organizational mechanisms for handling the kinds of projects for which they were largely designed. In the process, they gained little experience in mass production for the consumer market, in cost control, and in marketing. Attempts to enter the civilian economy—and in particular to branch out for the purpose of meeting new social and civilian technological needs—have largely been unavailing. Corporations have variously run Job Corps camps, experimented with new learning techniques, undertaken systems research on transportation, crime, and other problems, and ventured into housing and new town development, with little notable success. Most such ventures have subsequently been dropped.

It should be observed, however, that the change of emphasis in government programs under the Nixon Administration and the decline of funds available for many of the earlier experimental programs were responsible for the decline in the innovative ventures of government-oriented, as well as more conventional, corporations. In the absence of any established market for innovational projects, private funds could not be secured for highly speculative ventures (particularly after the stock market debacle of the late 1960s), and the federal government was the only possible source of funds. Unfortunately, the record of social experimentation during the 1960s was at least as strewn with failures and abortive experiments as was the field of military hardware, and social experimentation became a prime target of the hard-nosed business executives brought in by the Nixon Administration.[21]

The increase in the volume of purchases from the nongovernmental sector in postwar years has led to substantial improvements in the management of contracts, the nature of working relationships between government agencies and contractors, the control of product quality, and other aspects of efficient interaction. NASA has given much attention to the techniques of organizing and managing the efforts of hundreds of private contractors in conjunction with its own large organization. The fact that it achieved a better reputation than have the defense agencies is doubtless due in part to the fact that its top direction from the beginning vigorously supported and was skilled in the arts of good management. Moreover, being more vulnerable to political public criticism than the "military-industrial complex," it could less afford management mistakes.

Notwithstanding the development of greater sophistication in procurement, the procurement policies of many of the federal agencies, as shown by the recent report of the Commission on Government Procurement (1973), still leave wide room for improvement.

ORGANIZATIONAL EXPERIMENTATION IN THE 1960s

Any analysis of contracting out must take account of the greatest experiments in this direction of recent times—those generated by the New Frontier and Great Society anti-poverty programs of the 1960s. These encouraged the creation of new organizations—quasi-governmental, private nonprofit, and private for-profit—in the form of community action agencies, concentrated employment programs, neighborhood service centers, model cities programs, organizations to provide community services, training, and so forth, and in many cases outside the framework of the existing administrative and political establishments. In political terms, Democratic national administrations offered jobs and services to build up party loyalty, bypassing in the process the states and recalcitrant large-city administrations. Blacks and Spanish-Americans gained control of citywide coordinating agencies much as the Irish, Italians, Jews, and other ethnics had earlier gained control of other municipal functions and power centers.

It is now widely recognized that the programs in the mass did little to improve life in the ghettos, aside from the direct impact of the funds they distributed, and there is still debate about the feasibility of "maximum feasible participation" by the poor. Daniel Patrick Moynihan charged that programs only led to "maximum feasible misunderstanding," diverse and contradictory goals, and general muddleheadedness and confusion. Others claim that clarification has been fatal to the real purposes of the programs—to politicize the ghettos and the poor, confirm them in the ways of democracy (meaning voting for national Democratic candidates), and to put money into the pockets of active participants.

CONTRACTING FOR MUNICIPAL SERVICES

Municipal governments contract for many public services, though there is no information as to the proportion of services provided "in-house," the proportion provided by nongovernmental agencies (including private firms), or the proportion which might be so provided. The latest detailed data come from a 1963 survey by the International City Management Association (1964). Although the survey is ten years old, nothing has happened since that time which would greatly change the basic picture except for the variety of social programs spawned by the federal anti-poverty and related programs of the 1960s.

The ICMA report found that nearly seventy-five percent of the 1,007 reporting cities contracted for one or more municipal services. By size category, the number reporting the use of contract services was largest for the 50,000-100,000 group (eighty-three percent) and smallest in the under-10,000 group (sixty-eight percent).

The major services contracted for, in order of frequency, were:

- Street lighting: Designing, installing and maintaining street lights: forty-five percent of reporting cities. Ninety-three percent of the contracts were with private utility companies.

- Garbage and sanitation services: Forty-five percent of the cities contracted, mainly with private operators. Twenty percent reported contracting with other government jurisidictions, mostly for industrial waste services rather than for collection and disposal of garbage from private residences. Subsequent surveys of refuse-related services indicate that the proportion of contracting out is rising.

- Health services: With contracts ranging from exchange of information to contracts for specific services. Most contracts were with other government jurisdictions; only ten percent were with private agencies, and most of the latter were for emergency ambulance service.

- Animal pounds and related services: Many of the contracts were with the ASPCA.

- Tax collection and assessment: Including preparation of assessment rolls and collection and distribution of taxes. Twenty-five percent of the cities reported such contracts; ninety percent were with other government jurisdictions, usually counties.

- Water supply: Forty percent of the contracts were with private companies; most of the remainder were with nearby cities for joint use of water facilities.

- Law enforcement: Mainly contracts with other jurisdictions. The most frequent type of contract with private firms concerned provision of radio communications facilities.

- Street maintenance and cleaning: Only ten percent recorded contracts, mainly with private operators.

- Protection services: Seven percent of the cities contracted, mainly with other government jurisdictions. (The city of Scottsdale, Arizona, is still the outstanding case of provision of fire protection services by a private corporation.)

- Building and safety code enforcement: Four percent contracted, mostly for building inspection services, and mostly with other jurisdictions.

The study is concerned largely with items ordinarily paid for by taxes and from expense budgets. The proportion of capital construction (construction of roads, buildings, and other public facilities) contracted out is almost certainly much larger.

Being confined to city governments, the ICMA study did not cover adequately (or cover at all) some of the fields of most interest to this report, including public assistance and related social programs, health and hospitals programs, education, housing, transportation and manpower development, and other anti-poverty programs. Public assistance and health services are largely the function of state or county governments, and education and public transportation are largely the responsibility of special districts or public corporations, and many anti-poverty projects were contracted for directly by the federal government.

The findings of the 1963 ICMA study were essentially corroborated by a 1973 Urban Institute Study, sponsored by the National Commission on Productivity. Although it commented on contracting out experiments in public health and the use of contract management services in education, the Urban Institute report did not cover social services, health services, housing, and other matters of interest here.

The principal reason for contracting out was cost savings through use of more specialized equipment. Cities are more prone to contract for highly specialized services or services needed only occasionally, particularly those which require heavy equipment and specialized manpower—here scale economies become dominant considerations.

Offsetting factors are contractors' profit margins and tax costs. In addition to cost disadvantages, however, cities reported that private contractors suffer other operational disadvantages. One is the difficulty suffered by any outsider in coordinating activities with those of insiders who "know their way around." In any organization, good personnel relationships are a paramount requirement for getting things done. Second, private contract operations require thorough quality control and inspection, particularly because the profit motive creates an incentive to cut corners. Municipal officials are prone to skimp on quality controls, first because they are costly, second because they require skilled and incorruptible personnel, and third because of the hesitation to interfere with private contractors' operations (the contractor is presumed to be the expert and is being paid for the service).

Both the ICMA and Urban Institute studies paid little attention to the political factors in the production of municipal services, the conflicts between the public bureaucracies and the private contractors for new jobs and other perquisites involved, and the means used by competing bidders to gain their objectives. Without an understanding of these factors and of the climates in which they operate, any study of contracting out and the potentialities thereof is incomplete and likely to be misleading. Also, little attention now is given to the use of policy research studies—planning studies, engineering feasibility studies, evaluations, and so on. The reasons for contracting out such work frequently are more numerous than appear on the surface. They include the desire to pass the buck for difficult decisions to "outside experts," in which case the main function of the contracting firm may be merely to support the conclusions already reached by insiders, or to delay action or the necessity for decision by the familiar technique of making another study.

INNOVATION (INCLUDING TECHNOLOGICAL)
IN MUNICIPAL SERVICES

Local-level public services have lagged, it is alleged, as to innovation generally—managerial, organizational, social programming, and planning, as well as technological. The result has been not only failure to reduce costs of ongoing services and the aforementioned great inflation in municipal service costs, but inability to cope with new problems and increasing demands for services. In this connection, it is notable that there has been little organized research and development at the municipal level—certainly, nothing remotely comparable to the large sums spent by the federal government for R & D in various fields—and that the surveys of contracting out have found few instances of contracts for research and development or contracts for the introduction of innovations. Private firms, moreover, have paid little attention to public-interest services (aside from areas such as utilities, construction, and so on, in which they have been long entrenched) and have devoted few resources to product development for that market.

The picture, in fact, is somewhat mixed. Products which have been developed for a more general market have found ready application in municipal services, an outstanding example being the use of computers. Police forces have adapted technologies developed primarily for the military, along with technologies for rapid retrieval and communication of information, though police officials have been much more interested in the adoption of new hardware than in the development of human relations and other techniques which are largely more important for what they do.[22]

Public health appears to be another field where officials have kept reasonably abreast of technological developments, and, in some notable cases, technologies have been developed in public health research units (isoniazid, the anti-tuberculosis drug, was developed in the New York City health research laboratory).

Fire departments, on the other hand, have been notably slow to advance technologically, and prevailing techniques of firefighting are little advanced over those of a generation ago.

A number of reasons have been advanced for the innovational backwardness of the public bureaucracies. One, which has its genesis in Max Weber's turn-of-the-century analysis of bureaucracy, and expressed by Bruce Smith's comment that "all bureaucratic systems seem to suffer from a built-in entropy, a tendency to run down in efficiency. Goal displacement or the deflection of initial objectives into such other aims as organizational maintenance, is a familiar phenomenon" (Smith and Hague, 1970: 72). The more charitable view would be that the specialists of the municipal bureaucracies have professional interest in improving their operations, but that they resist innovations imposed from the outside, and, being characteristically short of funds and pressed by the day-to-day exigencies, have little time for inventing and installing basic improvements.

As for the possibly innovative role of the private sector, which the federal government has used to such good effect, the story goes somewhat as follows. Public administrators by and large have not been effective in thinking up techniques of improvement, or even in defining and communicating basic problems which lend themselves to hardward and software solutions.[23] Private firms, for their part, are inhibited by lack of knowledge of municipal department operations and, consequently, of the kinds of improvements which might make public services more effective. Also, they are put off by the uncertainty of the market and (given the sloth-like pace of bureaucratic decision-making) the difficulties of selling new products, which difficulties are enhanced by the political connections of established suppliers.[24]

In this situation, municipal officials have been stymied by the lack on display counters of products for solving many of their insistent problems, as well as by the lack of time and tools even for identifying and defining problems. One of the results of this failure is frequently manifested in the hiring of consultants, who are frequently called in to solve problems which have been misconceived or improperly defined.

There have recently been several important breakthroughs in this circle of frustration. One was the School Construction Systems Development project conducted by Educational Facilities Laboratories, Inc. (a nonprofit corporation established by the Ford Foundation) under a grant to Stanford University for the purpose of improving the quality of school building systems. The project, which had a substantial immediate success and produced results which were widely beneficial elsewhere, brought together the following elements.

(1) Market aggregation. A number of school districts in Northern California, with plans to build several dozen schools, agreed to

purchase school building components developed by the project, assuming the success of the initial demonstration.

(2) A thoroughgoing analysis of the physical requirements of school building plants, in recognition of the fact that conventional school architecture was inadequate for the needs of modern education.

(3) Drafting specifications, based on the surveys of educational needs, for five component systems: basic shell, lighting, partitioning, air conditioning, and flooring.

(4) Sending the specifications to interested manufacturers, with invitations to develop and submit bids on the newly developed systems. To assure responsible performance by the various manufacturing firms concerned, it was specified that each award winner would act as his own subcontractor with respect to handling his own product in building the demonstration schools.

(5) Otherwise redefining the roles of contractors and, particularly, unions who would be engaged in the assembling of the newly developed components. This involved negotiating agreements with the California Central Trade Union Council, from which agreements were obtained involving all trades except plumbing.

The competition attracted several competitors for each of the systems.[25] As a result of the experiment, thirteen schools costing $25 million were built by school districts in the original consortium. In addition to incorporating the improved new systems, the schools were moderately less expensive than conventional schools.

More important, however, were the ramifications of the project. The firms which competed for, but did not win, the original awards nevertheless had new innovational products to sell. It is estimated that such products are now used in twenty-five to thirty-five percent of all schools built in the country. For example, in Florida, more than sixty schools have applied the general approach and used these newly developed products, at a net estimated saving of about six percent of costs of conventional construction.[26]

The principal elements—defining needs, writing specifications, and getting competitive bids on the specifications—of course are not new, having been highly developed by NASA. The technique, however, has seldom been utilized for the highly fragmented municipal services market, either in developing new products or providing ongoing services. The potentiality was in part the stimulus for the creation of Public Technology Inc.

PUBLIC TECHNOLOGY INC.

Public Technology Inc. was originally launched under the auspices of the International City Management Association, with the assistance of a grant from the Ford Foundation. Its broad purpose is to stimulate the development, adaptation, and use of "technology," broadly conceived, to improve the production and delivery of public, particularly municipal, services and to facilitate the selection of urban administrative problems.

In the course of initial surveys, the PTI confirmed several points about the innovative deficiencies of municipal administrators. First, they lack knowledge of technologies already available, and how to find, adapt, and use them. Second, they lack experience, and hence are cautious about, contracting with private firms or nongovernmental agencies except in already-established areas. Third, they lack facilities and knowledge for doing more technical kinds of operations analysis and more sophisticated decision analysis. Fourth, they have not been aggressive in stimulating innovation, although many are interested in innovation and receptive to at least proven new techniques.

PTI is proceeding along several lines, including the following:

(1) It is working with methods of market aggregation, following the technique of getting commitments or expressions of interest from enough governmental units to provide incentive for private firms to invest in developing needed technologies. A major device is the users' committees, with representatives of municipal administration agencies (such as fire departments), managers' offices, members of professional organizations, and other specialists to discuss mutual problems and formulate technological needs. An example is the need for a device which will enable firemen at the end of the hose directly to control the pressure and volume of flow from the hose at a given time. The Grumman Corporation has invested around $150,000 thus far in developing a control-from-nozzle device.

(2) Compiling technical operating manuals for specific government operations. The principal project thus far is a manual for motor fleet management and replacement policies.

(3) Provision of operations analyses of specific problems, such as optimizing the location of firehouses and routes for refuse collection. Such projects presumably will serve as the basis for later "how-to" manuals.

(4) Sponsoring an innovation extension service, with "extension agents" placed in a number of cities for the purpose of stimulating interest in technological possibilities and assistance in the application of various technologies to local problems.

PTI has therefore made a promising start in generating the kinds of self-analysis needed for the identification of technologically related urban needs, and interesting private firms in developing products (which might include hardware, software, and possibly even services) to meet the needs. This is somewhat different from contracting with nongovernmental firms to provide specific categories of public services financed by general taxation or granting monopoly privileges or nonexclusive franchises to operate in the public enterprise field.

POSSIBILITIES OF MORE USE OF THE PRIVATE SECTOR IN PROGRAMS AFFECTING THE POOR

I will consider three types of program, classified according to beneficiaries:

- Programs benefiting persons able to work—those not incapacitated by age or by physical or mental infirmities.[27]
- Programs primarily benefiting those unable to work by reason of physical or mental incapacities.
- Programs benefiting the poor generally.

PROGRAMS CONCERNED WITH IMPROVING EMPLOYMENT AND CAREER PROSPECTS

Until the 1960s, "manpower development" was long dominated by the state employment services and vocational educational training programs.

JOBS

The most notable attempt to mobilize the private sector was the JOBS program, which attacked the problem of providing the jobs for the "hard core"—those at the end of the employment queue. The National Alliance of Businessmen was formed to support the effort. Federal funds were made available to firms which would contract to employ specific numbers of such candidates, and subsidies could be had for such facilities as new plants in ghetto areas. In the first year of the program, several hundred thousand pledges were made and some 250,000 ostensibly hard core were actually employed. Some corporations (e.g., Chrysler) entered into federal contracts; others (e.g., General Motors) preferred to keep clear of commitments to the federal government.

It developed later that the apparent success of the JOBS program was largely due to the economic boom and the labor shortage which prevailed during its initial years. The program quickly collapsed when unemployment rose during the recession of 1969. Moreover, many skeptics questioned whether the JOBS pledges, including those supported by contracts with the federal government, were actually being filled by people from the end of the employment queue—the harder of the hard core. Since the definition of hard core was loose, it could be met by almost any unskilled, or even moderately skilled, unemployed person. During the labor shortage, most of those actually hired probably would have been hired in any event, though the JOBS program may have stimulated some employers who would not otherwise have done so to recruit in the ghettos.

There are four characteristics of the present-day economy which tend to perpetuate the poverty of the able-to-work poor.

(1) The chronic shortage of jobs, which has been disproportionately concentrated in large-city ghettos. Ghetto unemployment in turn reflects to some extent the outmovement to the suburbs and nonurban areas of manufacturing and other goods-handling jobs which sustained former generations of urban immigrants.

(2) Many workers remain poor because they are stuck in low-pay, dead-end jobs. These in turn reflect to some extent employer personnel practices which can be changed, given the will to do so. Of this, more later.

(3) Entrance to jobs is blocked by widespread use of credentialization, including the requirement for high school diplomas, professional certificates for nonprofessional jobs, and so on, which is more or less irrelevant to the requirements of the task to be performed and serves to keep many people out of jobs they could otherwise perform.

(4) Restrictive practices of labor unions (the construction unions are especially notorious) keep people out of high-paying jobs.

In addition, there are several characteristics of those at the end of the labor supply queue which militate against their employment:

- Lack of training does inhibit the employment of many, problems of credentialization and overqualification notwithstanding.
- Some simply have poor work habits or lack motivation for working continuously.
- Many people have psychological hangups respecting finding and holding new jobs—fears of the unknown, inability to adjust to working rules, and so on.
- Many have personal problems—trouble with the law, with creditors, or family problems which get in the way of their showing up regularly for work.
- Many individuals labor under combinations of such problems with the addition of others such as poor health, which may result from bad health habits or physical defects.

THE ROLE OF PRIVATE FIRMS IN
MANPOWER PROGRAMS

The manpower programs saw considerable participation by private firms, ranging from the creation of the imposing National Alliance of Businessmen (headed by Henry Ford II) to small, privately run training programs. The principal for-profit possibilities were the following:

Contracts to operate specific types of training programs. The fact that so much of the emphasis of the manpower programs was on remedying the deficiencies left by the failure of the public school systems led to the search for new educational approaches, some having to do with basic skills and some with vocationally oriented training. Some of the training was provided under contract with already existing educational institutions, including vocational schools. Other training centers were set up by nonprofit community institutions. But of particular interest here is the fact that opportunities for contracts led to the creation of numerous for-profit firms; some of these were organized and run by blacks or

Spanish-Americans, others by whites—mainly, whites with pretensions to some expertise in the manpower field. A few of the firms compiled successful records, at least in organizing good programs and carrying through contracts; some, like New York's Regional Industrial Training Association, Inc. (organized with the encouragement of the city's Manpower Career and Development Agency) misused funds and failed abjectly.

Operation of Job Corps camps. As previously mentioned, a number of large corporations (Xerox, Litton, and others) contracted to run the out-of-slum residential training centers with varying degrees of success. One element of success was the attitude of the contracting corporation and whether it regarded the task as primarily a public responsibility or only as another way of making money.

In any event, the cost per resident year of the Job Corps camps was substantial—in the magnitude of $8,000—which led critical congressmen to make unfavorable comparisons with the somewhat lower cost of sending students to major private universities. Another criticism was that training students away from the slums did not necessarily equip them for working in the urban environment to which they would return.

Development of new educational technology. Some corporations (Time, Inc., Corn Products, and others) sought to capitalize on the notion that educational programs could be improved by greater use of "learning" machines—audio visual equipment, talking typewriters, programmed instruction with computers, and the like. Financing for these ventures came largely from the appropriations provided under the Education Act of 1965. For whatever reasons, the new technologies failed to produce overnight revolution, and hence the market for the new hardware and software being peddled by hopeful corporate educational divisions never fulfilled early hopes.[28] Federal funding dried up, and with it, the spurt of interest of most of the corporations.

This experience supports the common supposition that private firms require bright prospects of early returns before they will commit large amounts of energy and effort to socially oriented ventures. Unprofitable ventures are antithetical to the basic rationale of private enterprise, and even if such are undertaken for public relations or other reasons, their direction is likely to be assigned to has-beens or second-raters.

THE CONTINUING CHALLENGE OF MANPOWER DEVELOPMENT

Even though interest in manpower development has faded in recent years, it still remains a subject of prime importance, particularly in the large cities with their concentrations of unskilled blacks, Spanish-Americans, and other immigrant minorities. The challenge lies in equipping people to undertake the jobs available in central cities—the white-collar jobs of the predominant office industries, the many service jobs necessary to maintain the cities' technology, government jobs, or jobs having to do with public-interest services which have been one of the fastest-growing industries, and the numerous less-exalted service jobs having to do with building maintenance, retailing, food handling, hotel keeping, and the like.

The greatest contribution the private sector can make toward relieving the burden on municipal governments (and, less directly, on state and federal governments) imposed by the high concentration of unskilled and disgruntled people in central cities is to accelerate the process of getting them into the economic mainstream by providing them with jobs and opportunities for advancement.

Despite all the attention, stimulated particularly by the JOBS program, to the peculiar training-employment problems of the urban disadvantaged, few employers and personnel departments have the knowledge to cope successfully with the problems. Not infrequently, top corporation management is converted to the need for new approaches, but personnel managers and departments, who are likely to be judged—or think they will be judged—by their success in recruiting and maintaining a viable labor force, are reluctant to break out of their accustomed ruts. Two major problems are (1) an overemphasis on credentialization (requirements of high school diplomas for work not requiring high-school-level education, for example), and (2) the tendency of personnel departments to keep unskilled recruits in the lowest-level jobs and to pay little attention to problems of building ladders of advancement. Recruitment for supervisory and other higher-level positions is typically from outside the firm rather than from lower-echelon personnel. Such practices contribute to the "dead-end job" problem and also to high rates of turnover in firms.

Improving opportunities within firms. Attacking this problem requires two major approaches. The first has to do with more systematic planning for advancement and promotion within the firm itself. This requires identifying promising candidates for promotion to higher-level positions, and taking measures to see that the personnel are trained and otherwise qualified to move upward.

The second approach involves restructuring the table of organization to provide more higher-level opportunities. This involves working out ways of giving employees more responsibility as they are prepared to assume it. While this approach may seem highly theoretical, there is enough accumulated experience to show that it can actually be done in many organizations.

Still other approaches involve use of accumulating knowledge respecting group dynamics and such techniques as forming production teams who work under the general direction of group-selected leaders and those whose pay is based on the output of the group as a whole.[29]

THE PROBLEM OF FEMALE HEADS OF FAMILIES

Since a large proportion of poor families is headed by females, and since the category of aid for families with dependent children has been the chief component of the rapid increase in welfare expenditures over the last several years, the problem of finding employment for mothers who are heads of families, and making it easier for them to hold paying jobs, has received much attention. Day care has been the standard answer.

Day care, it was thought, would serve several useful purposes: (1) making possible the advantages presumed to attach to the work status, including an appreciation of the problems of "earning a living," and inculcating middle-class attitudes thereby, and (2) placing young children in "learning environments" which would better prepare them for later education and work experiences than would the care of their mothers.

In the early rounds of enthusiasm for "workfare," there was considerable backing in Congress and the federal agencies for extensive federal support of day care, which appeared to offer an opportunity for nongovernment enterprise in the form of nonprofit groups organized under the aegis of community action agencies, other groups, or private firms. The enthusiasm waned when the high cost of day care began to be appreciated, and the competition of other demands on the federal budget drained away the support for welfare reform generally. Several other flaws developed in the original, logical basis for providing day care in order to allow mothers to work who would otherwise require AFDC support:

(1) A basic obstacle for employment of female family heads is the lack of jobs available to ghetto residents and the high rate of unemployment among ghetto females.

(2) Good day care turned out to be very expensive, costing as much as or more than welfare support for unemployed mothers; poor day care is inferior to the care most mothers will provide for their own children.

(3) Aside from the lack of job opportunities, American culture seems to be missing some ingredient which makes day care possible and successful in such other cultures as the Russian and Israeli.

One question is whether day care can be provided by private for-profit firms on any considerable scale. Experience thus far has not been encouraging. When it appeared that the federal government would provide substantial financing for day care, several organizations to franchise day care blossomed—most in the hope of making a fast buck, it would seem—and thereafter collapsed.

PROMOTING GHETTO ECONOMIC DEVELOPMENT

The ghettos, it has been observed, resemble less-developed countries in many ways, including the lack of indigenous entrepreneurship and business leadership, the scarcity of savings available for investment, and the lack of a market owing to low incomes.[30] One approach to this set of problems is "black capitalism"—i.e., "minority-group" capitalism. Most objectives of black capitalism, if realized, would ease the burden on government of providing special services to ghetto residents, particularly those necessitated by unemployment, delinquency, and diversion of leadership talent into crime, dope, and other illegitimate activities.[31]

Black capitalism, as an objective, implies simply that blacks (or other impoverished minorities) should own and operate more businesses than they do. It should be distinguished from community ownership and control, which is quite a different concept—that business enterprises established in the ghetto should be owned and controlled by the ghetto community.[32]

The arguments for black capitalism are summarized by Hetherington (1971: 19) as follows:

> Any program for economic development of the ghetto must include measures aimed at promoting entrepreneurship among ghetto residents to the greatest possible degree [and] to make a meaningful and symbolic social and economic contribution to the solution to urban poverty, black capitalism must be developed far beyond the meager experience of "mom and pop" establishments known in the ghetto.

The central concept of community participation and control, on the other hand, is that ghetto business, production, and productivity can flourish only by involving and changing the attitudes of the larger community (including its ingrained hostility toward business enterprises).

Black capitalism, therefore, looks to such programs as management training and technical assistance for black (read minority group) entrepreneurs, abundant and cheaper credit, insurance coverage (denied to black concerns by many insurance companies), and access to markets (for example, patronage of ghetto manufacturers by national distribution firms, subcontracts with larger manufacturers, and so on).

A number of federal programs have ostensibly offered financial assistance to community development corporations and other ghetto-based and ghetto-located enterprises. In general, such programs have been heavily impeded by local political opposition, often flowing up through national channels, cautious and incompetent program administrators, and pettifogging lawyers intent on nailing down all legal contingencies instead of getting programs into operation (see Hetzel, 1971; Faux, 1971).

The fact that governmental gestures toward economic development in city ghettos have been so ambivalent and feeble, however, is small consolation for the withdrawal of federal support in recent months and the abolition or attempted abolition of such vehicles as the Model Cities Program, the Office of Economic Opportunity, and the Economic Development Administration. Improvement of federal programs, not their abolition, continues to be the great need.

The existing political-governmental structure having failed badly in promoting ghetto economic development despite a plethora of financing and other assistance programs, there is need for increased efforts by private firms (banks, insurance companies, and management companies in addition to production and service firms). There has in fact already been a considerable effort and participation through such programs as

- locating plants in ghetto areas;
- assisting local community development corporations to start new enterprises;
- providing high-risk loans, in most cases insured by Small Business Administration or other federal programs;
- providing management assistance for minority enterprises;
- easing the terms of franchises in order to facilitate participation by ghetto enterprises;
- providing markets for the products of ghetto firms (for a description of various types of effort, see National Industrial Conference Board, Inc., 1971).

Most of the possibilities that come to mind have been exploited, though not systematically and often not vigorously and continuously. One need is for renewed federal interest and a coherent and sustained federal program which would systematically involve existing private firms to facilitate ghetto-sponsored enterprise on at least as advantageous terms as those offered other enterprise. The most radical aspect of this suggestion is the demand for coherence and continued commitment of the federal government, in contrast to the half-baked measures, designed more for publicity than for serious implementation, which have characterized so much of the manpower and economic development efforts of the past.

The climate for such programs appears to be improving, as indicated by the fact that many of the more far-out radicals of the 1960s are turning from revolution and violent protest to trying to enlarge the black political power base.

OTHER PROGRAMS FOR THE POOR

This section considers programs for the nonworking, aged poor and for the poor in general. They include general assistance (welfare), health care, social services, and services to meet other special needs, notably including housing, transportation, and recreation services, along with special programs having to do with health, delinquency —including treatment for drug addiction and alcoholism, and various anti-delinquency programs for young people.

Private agencies, both profit and nonprofit, already offer a variety of services for the poor financed wholly or partly by public funds. They include medical and hospital care, nursing home care, and various kinds of social services, housing for welfare clients, and day care and other services of custodial institutions. Some such private institutions operate solely with government funds; many others depend as well upon private contributions, or, as is the case with many nursing homes, partly on private fees as well as on government support for indigent patients.

There is little published information respecting the extent to which private organizations are involved in providing this category of public interest services, the financial arrangements under which they operate, or the problems of regulating and controlling them. One can, however, make several generalizations. First, the institutions through which services are provided in different communities vary widely, depending on how the service has evolved. Second, the quality of service provided by private organizations ordinarily will depend (a) upon the precision with which obligations of the contractor are defined in the contract, and (b) the quality of inspection and control. The poor and infirm are particularly vulnerable to exploitation by for-profit institutions. Competition may be relatively ineffective where the clients have no options, lack knowledge of available options, or have no power to exercise options. Without persistent inspection and control, conditions in such institutions as nursing homes and day care centers can be miserable.

Third, there is no reason to think that private firms, just because they are private, have any superior prowess in invention and innovation in social fields in which they have had no expertise or previous experience. Expertise may, of course, be recruited from government agencies or universities. University and ex-government specialists not infrequently form their own for-profit consulting firms, and these may be useful resources, offering the various advantages associated with nongovernment agencies. Specialists may be found as well in nonprofit agencies, frequently marked with a greater spirit of zeal and dedication.

As for large corporations—such as the aerospace corporations—which have specialized in innovative technology, there is little to show that the art of innovation is easily transferred from technological to social-governmental problems, a point demonstrated by the failure of most large corporate ventures into solving urban problems.

What could a private firm do with the assignment of administering public assistance funds, providing a roster of services to the aged, or rehabilitating alcoholics, delinquents, or ex-convicts? The following section considers at length some special problems of public assistance administration.

PUBLIC ASSISTANCE ADMINISTRATION

Could welfare administration be improved, the rolls be cut, and funds be more effectively channeled if welfare administration were handled by private agencies? Administration of public assistance programs, particularly in the large Northern cities, has come under increasing fire in recent years, due partly to the enormous growth in public assistance rolls. There are no conclusive explanations of the reasons for the increase, which continued during the period of high prosperity of the 1960s (the first boom period in which welfare rolls had risen), but they seemed to include the increasing aggressiveness and political power of welfare recipients themselves and the emergence of welfare rights organizations, the increasing availability of legal services to protect welfare rights, and the fact that welfare administrative agencies were increasingly staffed with blacks and Spanish-Americans, who identified more readily with their clients than did previous generations of welfare administrators.

Nonetheless, it is difficult, at least for the author, to make a convincing case for the use of private agencies, particularly for-profit firms. To be sure, private charities preceded professionally administered public welfare and compiled a good record; many private organizations still exist, and some operate publicly funded programs. Also, the bad old political machines more or less systematically took care of the urgent needs of the poor, in return for their votes. One of the objections to the modern system of administration is that it deprived welfare clients, by forcing them to deal with faceless bureaucracies, of the sense of personal interest and protection formerly afforded by local political operators.

Whatever validity this point has would be magnified if welfare administration were removed even further from the political process by vesting it with private, nonpolitically responsible firms.

Public assistance, who gets it, and the conditions under which it is granted, involve one of the most delicate of the community's political questions, that of transferring income from those presumed to have earned it to those who have not. It has many of the same overtones as does the administration of justice. It is difficult to see how a private organization could successfully manage the amount of conflict inevitably associated with welfare programs.

Second, it would be difficult to formulate an incentive system for a private firm administering welfare, which would at once be conducive to "efficient" management and at the same time be fair to clients. Incentives for reducing welfare rolls or preventing their growth would hardly be politically tolerable. A cost plus fixed fee system provides no efficiency incentives. Perhaps a fixed price system relating to the applications process, with penalties for letting ineligible people get onto the rolls, would be a most effective incentive, but this would involve extensive audits and would be subject to many of the administrative problems that now exist.

The complexity of public-private arrangements in such social program areas as health services and general social services, plus the lack of information in these areas, prevents detailed comment here respecting opportunities for expanding the use of private organizations therein. At present, the field is a largely unexplored jungle. The experience of the 1960s makes clear, however, that many of the postulates on which were based anti-poverty, education, and other special programs were of dubious validity, and that more extensive and careful efforts are needed to gain valid information for reforming existing social programs and launching new ones. Obviously, there is a continued role for private agencies in organizing, administering, monitoring, and evaluating research and demonstration programs.

PERFORMANCE-CONTRACTING

One of the principal objectives, which thus far has not been systematically pursued even at the federal level, is that of incentives for successful performance. This, rather than privatization per se, is the central objective of a more economically motivated contracting system.

There are several kinds of incentives, but here we will concentrate on money payments for successful performance. Performance tests may be simply of the pass-fail variety, under which contractors are paid for successes and penalized for failures. Or incentives may take the form of bonuses scaled to degrees of success. There has been no systematic collection of information on the use of such incentives. A few instances have been reported in public construction contracts (such as bonuses for completing projects on schedule), and in one or two refuse collection contracts which call for small bonuses if no complaints are made about the service.

Performance-contracting thus rests on "performance accountability" as opposed to process accountability (adequacy of procedures) or fiscal accountability (observance of fiscal proprieties). Of course, there are usually primitive incentives in the latter two categories in the form of nonreimbursement in case of blatant irregularities.

Performance accountability depends in turn on the activity's lending itself to quantitative measurement—e.g., tons of refuse collected, performance of pupils on specified tests, and the like. Doubtless much more could be done to improve output measurement and win acceptance therefor than has been done thus far. Legally enforceable contracts, however, depend on the ability to distinguish clearly between program accomplishment and non-accomplishment, or measure the volume of work performed, or degrees of success, in a manner which will meet legal challenge in contract disputes.

This raises another delicate problem—that of the incentive for contractors to use biased procedures or to falsify reports of results. A case in point was the Texarkana educational performance contract, where the contractor was to be paid for each pupil in the contract training program whose test performance rose by a specified amount and penalized if the average fell after the ensuing six months. Spectacular results were reported, but then it developed that the contractor's training program had concentrated on specific preparation for answering the test questions. The Texarkana and like cases do not refute the validity of performance-contracting, of course, but only emphasize that the responsible contracting authorities cannot relax their efforts to achieve integrity in procedures and reporting.

In any case, there are circumstances in which private contractors can help keep honest the government bureaucrats who urgently desire successful results from particular programs or demonstrations.[33] Private agencies have less resort to the range of bureaucratic tricks for fudging objectives, hiding inadequate performance, and for organizational survival in the face of adversity, partly because private contracts are more likely to come under the fishy eye of government auditors and investigations by the suspicious media.[3][4]

Reluctance to admit failure is not peculiar to private agencies or to the operation of the profit motive. Alice Rivlin (1971) observes that one

> reservation about the desirability of social experimentation concerns the honesty with which experimental results will be recorded. No one likes to fail. Rightly or wrongly, the administrator of a successful experimental project will receive more acclaim and greater opportunities for advancement than the administrator of an unsuccessful project.

Such circumstances afford a strong temptation to cheat—"to choose the most favorable measuring instruments, to 'lose' the records of children who fail or patients who die, to coach participants on what to say to the evaluator or how to meet the test question" (Rivlin, 1971: 112).

Clear definition of objectives is a first requirement of performance-contracting (as it should be for all government-sponsored projects and programs). Suggestions such as turning over to private firms responsibility for such broad objectives as rehabilitation of alcoholics, training of unskilled persons, or whatever are as naive as awarding a contract for building a road without specifying its width, type of construction, construction techniques, or quality of materials.

Certainly, private organizations, either singly or in combinations, are a major resource for research and demonstration in the social program field. Though their use by the federal government has been less than uniformly successful, a great deal has been learned about contract-letting and management which over time should improve performance (see Danhof, 1968). Few state and local governments have the personnel or the institutional procedures to apply the knowledge accumulated to date; at these levels, contracting is a backward art. This deficiency is the more serious in that present trends are to place an increasing share of the responsibility for social, environmental, and economic development programs on the state and local governments.

In considering how to remedy this situation, moreover, one is stymied by the lack of knowledge of how private organizations are now used in providing some of the more important types of public interest services, including health, public assistance, and related social services. Moreover, most state and local governments have been little interested in research and demonstration and are not inclined to spend much thereon.

Then there are technical problems. Random experimentation and demonstration projects are inherently limited in what they can "prove" and in the dissemination of the knowledge acquired. Systematic experimentation on a wide scale, on the other hand, is expensive and difficult to organize and monitor.

Under the circumstances, one of the most useful subjects for research and demonstration may be how to improve the use of resources possessed by private agencies, the process of contract management in general, and the application of incentives for good performance. In other words, improvement of contracting (or other means of involving private agencies) could itself be the subject of a major research and demonstration effort.

With respect to research and demonstration projects in general, the best rule is to take expertise and organizational skills where one finds them and to put together optimal resource combinations. This frequently involves consortiums of nongovernment agencies, and on occasion collaboration between nonprofit and for-profit organizations.

On the negative side, it should be emphasized that the objectives of efficiency and holding down costs are not universally accepted values, particularly among those who value "participation," with greater involvement of minority group members and leaders, above technical efficiency. More specifically, black firms and government administrators are likely to resist various forms of accountability —program, process, and *fiscal*—on the grounds that these are "whitey's" ways of continuing domination over the black community.

It should be said, however, that such attitudes are not confined to the black (or other minority) political communities—as previously mentioned, contracts and extracurricular ways of profiting therefrom have long been major prizes of governmental control and main sources of political patronage. Such attitudes are characteristic of predominantly political, as opposed to predominantly economic, communities.[35]

Offsetting these attitudes are a growing impatience of ghetto residents with low levels of public services and a growing realization by black community leaders that services can be improved only by acceptance of more efficiency-oriented techniques of decision-making and administration.

FEASIBILITY OF INCENTIVES IN PUBLIC-INTEREST SERVICES

This section considers whether there is a place for an organized effort to promote innovation in the social program field and to recruit more private resources to work on social programs. As noted in the last section, the greater use of incentive devices, as through performance-contracting, would be, if successful, an important innovation. The concept of social program innovation is somewhat analogous to that underlying the work in technological innovation now being carried on by Public Technology, Inc.

It should be observed that conditions in the social program field are somewhat different from the situation in technology application which gave rise to PTI.

- The specialists of various disciplines who dominate the social program field and who operate in concert at all three levels of government, along with their academic counterparts, have been more active in defining needs than have been the more technologically oriented services (which have been also more localized and parochial). There is, of course, the danger that professional organizations will devote themselves to maintaining orthodoxies (as do the professional educational organizations), but this danger is offset at least partly by the lively interest of academic social scientists—witness the ferment in education during the last two decades.

- The specifications of an experiment or demonstration in the social program field are usually less easy to formulate and to implement than are building systems and hardware, and even operations analysis techniques.

- There is a hardy race of nongovernment management and program analysts, of both profit and nonprofit organizations, who specialize in finding answers to social programs' operational and policy problems. Their specialty being software rather than hardware, they are less inhibited than are the hardware manufacturers by the lack of a mass market for their innovations. On the other hand, most of their innovational ideas require demonstration, which few are prepared to finance in the absence of funding from foundations or governments.

I nonetheless conclude that there is need for a PTI analogue in the social program field, for several reasons.

(1) The degree of professional specialization in social science disciplines often results in too little attention being paid to systems effects, positive or negative, of particular professional activities. An instance is the indifference of administrators in many professional fields (excluding administration itself) to matters of administrative efficiency. (Medicine is one of the worst offenders.) A multidisciplinary effort (a concept much touted but little implemented) is essential for the task of promoting social innovation.

(2) A systematic effort is needed to improve contracting and contract management at the state-local government level, similar to but broader than the work of the Commission on Federal Procurement. To take one aspect, writing specifications for services to be performed is a highly specialized art, one done poorly by most federal government agencies and even worse, in my observation, by less-sophisticated state and local government administrators. Performance-contracting is particularly difficult—defining program objectives, projecting costs, preventing fudging to make results appear better than they actually are, the nature of the contracting process (negotiated contracts will ordinarily be better than sealed bids)—all are matters with which most state and local administrators are unprepared to cope.

(3) There is at present no organized effort to discover whether incentives for effective performance may be expanded, and where private organizations might take over more of the service production function which otherwise would be performed in-house by governments.

(4) State and local government administrators as a class are babes in the woods about contracting for services generally and consulting services in particular. A common error is to misconceive the problem at the outset—thus, a situation might be perceived as a lack of recreation facilities when it is really a problem of lack of transportation, or vice versa. There is needed, particularly for municipal government, a service of consulting on consultants which works with the client to define his problem and advise him on hiring a

specialist-consultant to resolve it. (The concept requires that Consultant 1 be entirely independent of Consultant 2.)

(5) The above-listed needs will be the greater now that the federal government is cutting back its own initiative in program development and putting greater emphasis on state-local governments (aided by more flexible federal grants). This at once affords the opportunity for state-local governments to experiment with service patterns and puts a burden on them to show that they can perform adequately.

FUNCTIONS OF A PUBLIC SERVICES CORPORATION

The emphasis of a PSC would be on innovation in social programs and private-sector involvement therein. Following are the main functions contemplated for a PSC.

(1) Working with committees of social program administrators to identify several problem areas in which private contractors might make contributions or in which new contracting techniques offering incentives for performance might be introduced.

(2) Formulating specifications and model contracts for projects identified under 1.

(3) Sponsoring studies of existing contract programs to compile information about formats for contracts and contract administration. Developing improved formats for negotiating contracts, and administration and auditing, which will protect the contracting government without needlessly hampering contractors. (In other words, seek ways of cutting red tape.)

(4) Supplying resources for, or helping to find, demonstration projects involving the use of performance contracts or contracts featuring other types of incentives.

(5) Stimulating interest in the use of private organizations and of incentive-contracting through
    (5.1) regional conferences and workshops,
    (5.2) collaboration with committees of program administrators in various fields, and
    (5.3) maintenance of a clearinghouse to compile and disseminate information on existing practices and new programs of incentive-contracting.

(6) Developing, initially on an experimental basis, a consulting service for state-local governments directed at making the best use of consultants

by working with clients to define problems, develop specifications for consultant contracts, and choosing the appropriate consultant for the particular task.

SUMMARY AND CONCLUSIONS

(1) This chapter concentrates on using nongovernmental organizations, both for-profit and nonprofit, to reduce the burden on state and local governments of providing public-interest services. Two main approaches are explored:

(1.1) developing additional opportunities for, and improving the techniques of, contracting for the production of public interest services, with emphasis on social programs,[36] and

(1.2) private action to stimulate employment and raise incomes in poverty-prone urban areas.

(2) There are a priori reasons for thinking that private firms as a class can provide public interest services more effectively and cheaply than can government agencies, but little hard evidence to support the point.

(3) Nevertheless, governments can benefit by arranging for private-sector agencies to perform public interest services by taking advantage of one or more of the following:

(3.1) advantages of competition, not only between private and public sector agencies but also between private organizations;

(3.2) economies of scale and specialization—these may be obtained by contracting with specialized private firms, or larger governments, or by combining with other governments to produce the service;

(3.3) escape from the rigidities of personnel, budgetary, and other central controls imposed by general government agencies;

(3.4) greater freedom to discontinue contractors or services which prove to be less effective;

(3.5) use of incentive devices, such as bonuses for superior performance.

(4) There are also a number of disadvantages.

(4.1) the danger of becoming trapped in a relationship with a single firm;

(4.2) the extralegal activities frequently practiced in order to get and hold contracts—political contributions, kickbacks, collusion among bidders, bribery of inspectors, and so on;

(4.3) the difficulty of drawing contracts for services that cannot easily be defined or identified in a way which permits the use of incentives (most government contracts pay little attention to efficiency incentives);

(4.4) the necessity of closely monitoring and inspecting the processes and output of contractors, and

(4.5) the difficulty of compelling a deficient contractor to perform satisfactorily.

(4.6) public suspicion, arising from past experience that where there is private contracting there will be boodling.

(5) Governments should weigh the above-listed advantages against the disadvantages, particularly the respective projected costs of providing service in-house or under private contracts. The answers may come out differently for the same service in different communities. In some cases, the best attack on high costs of municipal services may simply lie in vigorous efforts to reduce costs, as by inserting productivity clauses into labor contracts, and improved management and supervision.

(6) City governments already make substantial use of private contractors for routine services which may be performed by either government or private organizations, and specialized services. There is little quantitative information, however, about the role of private organizations in providing many of the social services, particularly those which are largely the responsibility of states, counties, and special districts.

(7) Contracting at the state-local level is a backward art with respect to such matters as defining objectives, use of incentive devices, contract administration, supervision, and control and auditing.

(8) There is both opportunity and need for greater involvement of private organizations in public interest services, particularly social services.

(9) A continuing need in the social program area is more experimentation and demonstration. Here, private organizations also can play an important role; however, financing must come mainly from government and philanthropic sources. One promising field for experimentation has to do with the use of incentive-contracting, in

which payments are related in various ways to the achievement of specified objectives, such as the quality of services performed. More work is needed in formulating objectives and specifications, preparing models for contracts, and formats for administrative supervision and control.

NOTES

1. Three main complaints concern (1) overlying and overlapping layers of local government, (2) governments too small to realize economies of scale in decision-making or operations, and (3) governments too large to be responsive to the special needs and sentiments of local neighborhoods.

2. Government employees' unions are a case in point. They exercise not only the economic power deriving from their ability to withhold services and otherwise disrupt or expedite the production process, but also their political power stemming from the number of votes they control. In a city like New York, where organized employees constitute about ten percent of the total labor force and (with their families and friends) control at least twenty percent of the vote, the concerted pressure of the public employee unions is a potent force.

3. Under prevailing practice, "most services to clients are offered on a model of a 'walk in,' a client who presents himself at a place where diagnosis and treatment can take place—a hospital, a clinic, welfare office, even a community action center. . . . Ideally the client should be preprocessed before much is done. What this means is that, ideally, the professional person or decision-maker wants information about the client before contact is made with him.

"What is more, service agencies usually are structured on what used to be called 'banker's hours'—a 9:00 a.m. to 5:00 p.m. schedule of work. Over and beyond that, the high specialization of services often means clients must move episodically to different specialists in different agencies. Such a system seems to be designed to dissuade clients from making demands for service rather than to meet client demands. The client is required to adapt to the agency personnel's schedule rather than his own schedule. . . . It must be recognized that life is more comfortable for the professional that way and these agencies must be regarded as being professional rather than client-centered. The capacity of professionals to subvert the goals of the client is enormous" (Reiss, 1970).

4. Construction unions, for example, are commonly among the strong proponents of public construction projects.

5. The applicability of user charges depends upon the type of public service rendered, and in some cases upon the socioeconomic characteristics of the beneficiaries. Public finance economists distinguish between "public goods" and "merit goods." "Public goods" have the characteristics that their benefits are widely diffused and that increased consumption by one beneficiary does not directly subtract from the consumption of another beneficiary (examples of such goods are the national defense and the administration of justice). "Merit goods" have benefits which are more direct, and most might be provided through the market process, but they are considered sufficiently important to the public welfare to justify government action to assure their provision.

Another basis for the distinction between pure public goods and merit goods is exclusivity. Pure public goods are "nonexclusive" in the sense that, once they are produced, would-be beneficiaries cannot readily be prevented from consuming them; they therefore must be financed from general taxation. Merit goods, such as public education, are financed by taxes because it is thought important to make them available to a wide range of beneficiaries including those who would not or could not buy them if they were sold at prices measured by production costs.

6. Technically, improving production efficiency implies increasing the ratio of outputs to inputs.

7. Or possibly by serving jurisdictions large enough to realize all economies of scale, in which case they could realize the economies without contracting.

8. Controversial programs such as fluoridation, however, often can be pushed more readily by an elected chief executive backed by a party organization (see Wilson, 1968).

9. Again, circumstances differ. The pay scale of New York City sanitation workers, who collect from residences, exceeds wages of private carters who collect from commercial establishments. This differential is one major factor in the large unit cost differential of collection by the city and by private carters. It has been said that the threat to contract out refuse collection was a factor in the sanitation men's union agreeing to productivity clauses in its most recent contract and for the substantial increases in productivity which are reported to have actually occurred.

10. This definition of efficiency returns to the "economic model" of government operations which defines outputs as public interest services delivered to final consumers. As previously mentioned, the "political model" defines outputs somewhat more broadly to include all the prizes (including jobs and contracts) associated with control of government. In the political model, the chief objective of government agencies is to increase their budgets rather than the spread between sales and costs of goods and services produced.

11. Under turnkey, housing contractors built on speculation under a general understanding with housing authorities that the authorities would buy the structures if they met broad specifications and standards.

12. Projects involving large amounts of capital equipment—machinery, laboratories, and the like—or large staffs of professional and skilled personnel are found almost exclusively in the larger for-profit corporations, since the amount of capital required to finance them can ordinarily be raised only by private corporations.

13. David Z. Robinson (1971) has distinguished three types of accountability in research work: (1) program accountability, having to do with the quality of the work carried on and whether it met the goals set for it; (2) process accountability—whether the procedures used to perform the research were adequate, and whether experiments were carried on as promised; (3) fiscal accountability—whether funds were expended as stated and items purchased were used for the project, and so on.

14. The armed services are frequently suspected of an unseemly tolerance of such relationships, and well-publicized incidents such as the firing of A. Ernest Fitzgerald and the attempts to silence Gordon Rule are not calculated to allay such suspicions.

15. In this culture, law enforcement by long tradition is performed by public employees, and there seem to be no good arguments for changing the tradition.

16. The question of whether "better" services would be preferred in the ghetto to more job opportunities for ghetto residents is seldom put to the test, however; services in the ghetto are likely to be miserable no matter who is responsible for providing them. With some services—notably, education—old-line civil servants have been quite incapable of adapting to the different cultures and mores of the new ethnics. In New York, where the Jews have dominated the educational system for more than a generation, the examination system was systematically manipulated to exclude the new ethnics (until recently a southern accent could disqualify a candidate for a teaching job). The frustration produced by this situation led to the controversial school decentralization plan, by which the new ethnics hoped to secure a larger share of control of the schools. To the disappointment of many of decentralization's proponents, the quality of education was not improved thereby; noted

sociologist-educator Kenneth Clark observed sadly that local school boards and administrators were more interested in power and politics than in improving the performance of the pupils.

17. A lead in a New York *Daily News* (1973) story read, "High officials in suburban Baltimore have told federal officials that it was virtually impossible to land a consultant's contract when Spiro Agnew was county executive in the 1960's without making a cash contribution to political figures here." Such allegations indicate an all-too-common aura of suspicion about the contracting process.

18. The Postal Service opposes such competition on the grounds that competing agencies would take the more profitable business and leave the Postal Service with the unprofitable. If the latter is to be subsidized, however, there is an argument for subsidizing it from general taxation rather than by above-cost charges on other postal services.

19. Thus in New York City Robert Moses used the financial resources of the Triborough Bridge and Tunnel Authority, of which he was chairman, to build the 59th Street Coliseum, which had little to do with the original purposes of the TB&TA.

20. Examples of government-oriented corporations were North American Aviation, Lockheed Aircraft, General Dynamics, McDonnell-Douglas Corporation, Grumman Aircraft Engineering, and Thiokol Chemical. They do not include such giants as General Motors, General Electric, General Foods, or Alcoa, although these have done large amounts of government contract work (see Weidenbaum, 1969).

21. Significantly, a large number came from that section of the American economy known as the military-industrial complex. While they may have had a hand in steering firms toward experimental social program ventures, they had small success in most of these ventures, probably did not ever believe in them, and, once they got control of the reins of government, abandoned them in the interest of economy.

22. The acceptance of or indifference to new technologies varies greatly among departments. Motor scooters were on the market for thirty years before the New York City police force recognized their potential for patrol work. New York City lagged far behind Chicago in developing a communications and dispatch system which would enable prompt and efficient response to calls.

23. In some cases, innovation is inhibited by lack of communication in departments. Thus, observation of the New York City fire department indicates that there is no organized process by which accumulated knowledge of rank-and-file individuals respecting operational and technical improvements is communicated up the command hierarchy to levels which might be in a position to do something about them. This organizational failure both frustrates lower-level personnel (to the consequent detriment of morale) and deprives the department of a valuable source of ideas.

24. An interesting case in point was the struggle to get the traffic department of the city of New York to replace a brand of parking meter which could be readily burglarized, vandalized, and was otherwise dysfunctional. The adoption of the one burglar-proof meter then on the market was accompanied by inordinate political commotion, allegations of bribery, and the indictment and subsequent suicide of the president of the company which manufactured the superior meter.

25. In some cases, there was considerable dissension within corporations as to whether to enter the contest; it is reported at least one corporate vice president lost his job over the issue.

26. Information from Ezra Ehrenkranz, the architect-director of the original project.

27. This raises the difficult question of determining when a person is incapacitated, particularly when the incapacity takes the form of psychological maladjustment or other attitudes which prevent one from holding a job. Some of the programs considered here are directed toward reducing such causes of inability to work.

28. One of the reasons was the opposition of many educators who feared technological unemployment.

29. In setting up methods of group compensation, it is necessary to avoid the kinds of abuses which in the past have given piece-rate and other incentive systems a bad name among workers and unions.

30. Theodore Cross (1969) reported that fewer than a dozen black-owned firms in Manhattan had more than ten employees.

31. It is charged also that ghettos have been exploited in the fact that they pay out more in taxes than the value of what they receive in benefits. This point is debatable; various studies show quite different results, depending on methodology.

32. The latter movement, which has gained considerable support, is a major departure from the existing business structure, although somewhat similar to such relatively rare community enterprises as municipal utilities, a few producer cooperatives, and a number of consumer cooperatives.

33. In an experiment with a suggested technique for rapid rehabilitation of New York City old-law tenement buildings, a private nonprofit organization was selected by HUD to be prime contractor and evaluator, with actual construction work being handled by the building contractor who had devised the rapid-rehabilitation technique. The experiment (which was improperly labelled a "demonstration") proved conclusively that the technique was not feasible, and the evaluation so reported. The HUD people immediately in charge were reluctant to accept the findings and brought pressure on the evaluator to change its report. The incident demonstrated the need for (1) an independent evaluator of professional integrity, and (2) a reasonable amount of objectivity on the part of the sponsoring government agency.

34. See note 35 for a reference to recent events relevant to the point.

35. Two stories in the New York *Times* of August 12, 1973 bear upon this conflict in values. The first on page 1 of the news section, was headlined,

Beame Accuses Drug Unit of Losing Local Control
Says Its Privately Run Programs Have Become Wasteful
Addiction Agency Calls Charges False and Scurrilous.

The second, on page 24 of the news section, began, "The Federal Government has cancelled more than $6 million in manpower training contracts held by business concerns in New York and New Jersey because, it said, the firms are failing to provide disadvantaged workers with on-the-job instruction that would enable them to get and hold semi-skilled and skilled jobs."

36. The term I use here to designate loosely those services aimed at improving the economic status and general well-being of poor communities and poverty-prone families and individuals. They include health services, day care and similar social services, housing, transportation, manpower development, and community development, as opposed to the housekeeping services performed by urban governments.

REFERENCES

CHAMBERLIN, E. H. (1962) Theory of Monopolistic Competition: Reorientation of the Theory of Value. Cambridge, Mass.: Harvard Univ. Press.

Commission on Government Procurement (1973) Report Summary. Washington, D.C.: Government Printing Office.

COOK, F. J. (1966) The Corrupted Land. New York: Macmillan.

CROSS, T. (1969) Black Capitalism. New York: Atheneum.

DANHOF, C. H. (1968) Government Contracting and Technological Change. Washington, D.C.: Brookings Institution.

FAUX, J. (1971) "Politics and the bureaucracy in community-controlled economic development." Law and Contemporary Problems (Spring).

HETHERINGTON, J.A.C. (1971) "Community participation: a critical view." Law and Contemporary Problems (Winter).

HETZEL, A. J. (1971) "Games the government plays: federal funding of minority economic development." Law and Contemporary Problems (Winter).

International City Managers Association (1964) "Contracting for municipal services." Management Information Service Report 240.

KOONTZ, H. and R. W. GABLE (1956) Public Control of Economic Enterprise. New York: McGraw-Hill.

New York Conference Board, Inc. (1971) Business and the Development of Ghetto Enterprise. New York.

New York Daily News (1973) August 11:2.

NISKANEN, W. A., Jr. (1971) Bureaucracy and Representative Government. New York: Aldine-Atherton.

PILEGGI, N. and M. PEARL (1973) "What happens when cops get caught?" New York Magazine (July 23).

REISS, A. J., Jr. (1970) "Servers and served in service," in J. P. Crecine (ed.) Financing the Metropolis: The Role of Public Policy and Urban Economics. Beverly Hills: Sage Pubns.

RIVLIN, A. M. (1971) Systematic Thinking for Social Action. Washington, D.C.: Brookings Institution.

ROBINSON, D. Z. (1971) "Government contracting for academic accountability in the American experience," in B.L.R. Smith and D. C. Hague (eds.) The Dilemma of Accountability in Modern Government. New York: St. Martin's.

SAVAS, E. S. and S. G. GINSBURG (1973) "The Civil Service: a meritless system?" Public Interest 32 (Summer): 70-85.

SMITH, B.L.R. and D. C. HAGUE [eds.] (1970) The Dilemma of Accountability in Modern Government. New York: St. Martin's.

U.S. Congress (1973) "Report to the Joint Economic Committee, Subcommittee on Priorities and Economies in Government." Ninety-Third Congress, First Session, March 5.

WEIDENBAUM, M. (1969) The Modern Public Sector. New York: Basic Books.

WILDAVSKY, A. and A. J. MELTSNER (1970) "Leave city budgeting alone," in J. P. Crecine (ed.) Financing the Metropolis: The Role of Public Policy and Urban Economics. Beverly Hills: Sage Pubns.

WILSON, J. Q. (1968) "City politics and public policy," in J. Q. Wilson (ed.) City Politics and Public Policy. New York: John Wiley.

APPENDIX

FOR-PROFIT VERSUS NONPROFIT ORGANIZATIONS

Although there is less difference between these two types of organizations than one might suppose, particularly with respect to organizational motivation and competitive instincts, they differ in several important ways, and there are great differences among individual organizations.

In general, business firms treat the government "market" like any other market—as an opportunity for profit. They will produce public interest services only where operations promise to be as profitable as other opportunities. This is not to say that business firms are exclusively concerned with maximizing profits or that they do not undertake mainly public service-related activities (such as serving on advisory committees, lending personnel to assist government agencies, and undertaking other civic functions) with no motivation save a sense of civic responsibility and public relations.

The nonprofit organizations differ widely as to purpose, motivation, and modus operandi. Here I distinguish two basic types:

(1) Those created to serve a special social purpose, frequently having to do with the interest of a particular, presumably disadvantaged, clientele, such as children, old people, unwed mothers, and the like.

(2) Those concerned with specific types of services, such as health, research, and other types of public interest services. This category covers agencies as diverse as the RAND Corporation, the Hudson Institute, and other think tanks, created primarily to serve government and financed mainly by government grants and contracts; and voluntary hospitals, the Red Cross, and other service organizations which are financed mainly by contributions or combinations of fees and contributions.

In performing public interest services, the nonprofit agency has a number of advantages over the private firm. First, its personnel are more likely to be dedicated to the social purposes of the organization than to financial reward and may therefore be more concerned with turning out a product of professional integrity. This is probably one reason why the main nonprofit think tanks have had more staying power than those set up in the 1950s and 1960s by various large corporations (TEMPO, WABCO, and a number of others now defunct) to promote a public sector market for corporate services.

Second, the nonprofit organization is professionally dedicated to "doing its thing" as long as its funds hold out, whereas the private firm quickly loses interest in activities with no prospects of profit.

Third, the nonprofits, including the universities, have done much more in the way of "basic research" and social experimentation than have the private firms, again for the simple reason that the private firms are impelled to activities offering more immediate profits.

What of the efficiency advantage presumably adhering to the profit motive? To the extent that the drive for profit leads to cutting corners and skimping on product, true efficiency may be lost. Efficiency, however, is likely to depend more on competition and professional attitudes than upon profit per se. Rate regulation assuring a fixed rate of return and cost-plus contracts are not conducive to high efficiency. The nonprofits have their own efficiency incentives: first, as a class, they are not lavishly endowed with funds and must economize; and second, for many types of work they must compete vigorously with each other and with for-profit consulting firms.

This emphasizes once again that for-profit firms, nonprofit agencies, and government bureaucracies have no overwhelming inherent advantages, one over the other, for performing many types of public services. Much depends upon the particular agency, its track record, the quality of its management, and the political milieu in which it operates.

Moreover, circumstances may change drastically over time. In the middle 1950s, the New York City Sanitation Department was able to protect business firms from overcharging and mulcting by private carters by offering such firms a competitive service at cost to the city. By 1970, city collection costs were far above private carters' charges. (The New York City Sanitation Department serves residences, while commercial establishments historically are served by private carters.)

Part VI

BARRIERS TO CHANGE

17

Public Service Unions and
Public Services

SUMNER M. ROSEN

INTRODUCTION

□ UNIONS OF PUBLIC EMPLOYEES represent a major new phase in the development of American unionism and collective bargaining. Controversy has accompanied virtually every step in the development of these unions and, as their size increases and their energies grow, their impact on government at all levels will generate further controversy. These unions now come in all shapes and sizes; they include small associations of employees in suburban towns; blue-collar workers; educated professionals like teachers, engineers, architects, social workers, nurses, doctors, and so on. Some are genteel and respectable, others loud and truculent. Policemen and firemen are organized in many large cities and smaller towns, and more are organizing each year. Union membership of government workers doubled between 1960 and 1970, while total union membership increased during this period by only fourteen percent. In 1972, the largest union of public employees, the American Federation of State, County, and Municipal Employees (AFSCME) numbered 529,000 members; AFSCME, the American Federation of

Teachers, and the American Federation of Government Employees were by far the fastest-growing unions in the 1962-1972 decade. AFSCME was eleventh in size in 1971, ninth in size in 1972 (U.S. Bureau of Labor Statistics, 1972a; U.S. Department of Labor, 1973). If the projected merger of the American Federation of Teachers and the National Education Association comes to pass, the merged organization will rank among the largest unions in the country.

These developments have occurred rapidly; events have frequently outpaced the readiness of public or legislative opinion to adapt to them. Public employee unionism thus continues to divide opinions, professional and public. Few now argue against the right of public employees to organize, though only a few years ago the view was widely held that government as sovereign could not be required to recognize or negotiate with employee organizations. Less well established is the view that strikes by public employees do not necessarily offend or infringe the sovereign rights of government. While statutes and courts vary in their treatment of such strikes, it is now abundantly clear that public employees who are sufficiently aggrieved or provoked will and do strike, regardless of what the law holds. Accordingly, attention has turned from the question of how to prevent strikes to that of how to deal appropriately and justly with the grievances and demands of public employees, and some progress has been made.

PRODUCTIVITY AND THE PUBLIC

Less readily disposed of is the issue of the impact of collective bargaining on public services. Increasingly, the public requires more and better services; increasingly, the public treasury limits, often severely, the resources available to provide these services; consequently, the pressure grows for improving the productivity of public employees, and public employee unions, as well as governments, feel this pressure. Complaints increase that rising wage and salary scales for public employees are not accompanied by commensurate improvements in services rendered, with reference to such indices of performance as school reading scores, crime rates, dirty street, delays in processing official paper, court delays, and the like. Unions are often perceived as obstructing progress, imposing excessive staffing patterns, preventing modernization of procedures, insulating employees from scrutiny and evaluation, tying management's hands, and so on. Large sections of the public are increasingly critical of government, dissatisfied with the services they receive and with the tax costs of these services. Middle-class and professional "good government" forces are articulate and effective in voicing these criticisms; others "vote with their feet" and leave the city; still others—the poor, who cannot leave—express their alienation and anger through direction action—attacking firemen, teachers, or the police, heaping garbage in the streets, and the like. The targets of these attacks are often the employees and the unions which represent them. There is increasing public discussion about how to curb the "excessive" power of public employee unions in order to cut taxes and restore the ability of governments to manage their affairs. A recent study of labor relations in New York City offers an example, more extensive perhaps than some of such complaints:

> With respect to the managerial policies of New York City govern-
> ment organized civil servants pursue three goals: to increase the
> number of civil servants, to decrease the amount of work individual
> civil servants must perform, to restrict the amount of managerial
> control or discretion over their specific work situations [Horton,
> 1972: 104].

Such views are frequently heard, but they must be seen in the perspective of the realities which are found in public service. In 1970, unions of public employees had 2.3 million members; another 1.8 million belonged to associations. The total of 4.1 million represented thirty-three percent of the 12.5 million government employees, the union totaled only nineteen percent, quite small in comparative terms. By contrast, union membership constituted sixty-five percent of employment in manufacturing and is also high in truck transportation, utilities, railroads, mining, and other key sectors of the economy; in few sectors where unions matter does the proportion fall below thirty percent. Because events in cities like New York or Philadelphia, where public employee unionism is relatively strong, capture headlines, people assume that these cities are representative of the state of public sector unionism and collective bargaining generally; in fact, they are quite exceptional for the country as a whole. Further, the contractual provisions which constitute the real muscle of unionism in the private sector occur much less frequently in contracts with public employee unionism. A comprehensive survey of contract provisions issued by the Bureau of Labor Statistics (1972b) analysed 286 agreements covering 613,490 employees; classified by numbers of employees, this survey found:

- a majority contain no-strike provisions;
- only forty percent provide a union role in disciplinary proceedings;
- only fifty-five percent provide for binding arbitration of grievances;
- only fifty-eight percent deal with promotion procedures and criteria;
- only seventeen percent provide for any form of union security, and the most frequent among these is the relatively weak agency shop.

While it is important to recognize that the headline-making events which occur in large cities may well portend developments elsewhere, strongly unionized jurisdictions are still the exception. Further, the most active situations are still very much in flux and are changing. In New York City, for example, the modern era of collective bargaining is less than a decade old; how confidently could one have outlined the shape and thrust of unionism in the private sector after ten years under the Wagner Act? At that time, unionism appeared to many as destined to transform many sectors of private industry; cries of alarm and predictions of doom were abundant. In retrospect, the business system has shown a remarkable ability not only to survive but to prosper in an era of unionism. By analogy, one can argue that predictions that, unless curbed, public employee unionism will paralyze or bankrupt the American state and local government are, at a minimum, highly premature. In some respects, as is argued below, unionism indeed offers new opportunities to government. It is too early to predict what these opportunities will mean and what the responses of the unions will be, but one can offer some informed speculations based in part on collective bargaining in the private sector, in part on close observation of recent events, and in part on emerging trends in public sector unionism.

One generalization seems justified: public sector unionism is likely to exert significant force over the long run in the direction of rationalizing public administration in both policy formulation and the management process. It is the latest in a long list of forces which, since the late nineteenth century, have formed and shaped in this way the administrative and managerial systems of state and local governments.

THE UNION CHALLENGE,

THE MANAGEMENT RESPONSE

Where they have strength, unions of public employees present important challenges to public employers. What responses can government make? What results can be expected? Governments, like the private sector before them, must learn new ways and develop new capabilities in response to this new challenge. Much of the conventional wisdom of public administration, rooted in the pre-union era, will prove unwieldy or obsolete as collective bargaining develops, matures, and stabilizes. Rules and precedents will mean less; results and the ability to manage employees well will mean more. Decisions and policies in the labor-management sphere will become more centralized and more important in day-to-day management. Collective bargaining will supplant or substantially modify the operation of civil service (for a full discussion, see Stanley, 1971).

Public administrators with managerial responsibilities will need to be trained as managers and must learn to operate under constraints and pressures which have not previously been important to them. But change need not mean catastrophe; rather, it offers opportunities to government agencies to improve their performance and their ability to meet contemporary needs, often more effectively than has so far been the case. Unionism, in other words, is a challenge to management, as it always has been; the real issue is how managers meet this challenge.

A major pressure on governments at every level is to provide adequate services at reasonable cost. Put in other words, employee productivity has become a central issue of governmental conduct at the state and local level. The phrase "a fair day's work for a fair day's pay" has moved from the private to the public sector. Because collective bargaining and the need for a more productive and responsive work force have both emerged at the same time in our history, they have become linked in public discussion. But we need to remember, first, that the issue of productivity, which is complex at best, is at least as acute where unions are weak or nonexistent as it is where they are strongest. Second, authors like Horton assume that strong unions mean weak management and therefore low productivity, but the lessons of the private sector point in a different direction. It is true that, where management is weak or incompetent, unions can acquire de facto control of the workplace and veto power over changes which appear to reduce their power or inconvenience their members. But such cases are the exception. More characteristic, as the literature on labor-management relations demonstrates, is the process by which unionization leads to changes in management which replace an older, ineffective group with managers who are better trained, stronger, more competent, and more assertive. Where strong unions face strong managers, union-generated pressures are a continuing force on management to improve its capabilities and performance. Managements which do not have to face unions are often lazy; they solve problems the easy way, often at the expense of legitimate workers' interests and in ways which lower morale and productivity. When unions are able to prevent this from happening, management must then look more deeply for the underlying causes of their problems. These lessons from private sector experience apply at least as much to the public sector.

DIFFERENCES AND SIMILIARITIES IN
THE PUBLIC SECTOR

To what extent can one rely on private sector precedents in discussing the public sector? Clearly there are important differences. Politics play a far greater role; public administrators and managers are often chosen and evaluated by criteria which are far more explicitly political than any considerations related to managerial competence or efficiency. Similarly, unions of public employees, particularly the more traditional ones like the uniformed groups, have operated far more actively in political life and in a political style than do their private-sector counterparts. In New York City during the Wagner period, for example, collective bargaining issues were resolved primarily in accordance with that style. At that time, the city made the rules and operated them to its own advantage. However large a union's majority within a unit, minority unions were permitted to operate, pitting group against group. The city retained control of the final disposition of grievances. There were no impartial procedures to settle impasses; they were resolved by the mayor, and political considerations played a major role (for further details, see Rosen and Rapoport, 1970: 630-632).

CRAFT-LIKE UNIONS

Some unions dating from that time, like the police and firemen's groups, resemble the earlier benevolent societies of craftsmen more than they do contemporary unions. Internally, they are likely to continue to be close-knit, wary and distant with regard to other organized groups, eschewing direct labor action—strikes, slowdowns, work to rules, and so on—for closed and quiet negotiations, and seeking to preserve close ties to political circles. They are like the craft unions which still prevail in important parts of the private sector. They are an important but—in relative terms—diminishing segment of public employee unionism. And the bargaining process which involves them may be as unstable as it is in, say, the newspaper industry, which faces similar problems—separate unions, fragmented bargaining, intercraft jealousies, resistance to technological change, and the like. Only merger or coordinated bargaining can deal with problems of this kind. But note that they are not unique to the public sector and should diminish in importance as separate unions give way to merged or amalgamated unions.

Members of other craft-like unions—such as the teachers—often bring some "liberal" actions to bear along with the more purely defensive activities which are their major reason for being. In the early years of the struggle for Negro rights, teachers' unions were often allied with civil rights groups and in some cases sought to secure actions through collective bargaining which would address some of the problems of inner-city schools. Partly in self-interest, partly from political conviction, reflective of their middle-class and ethnic origins, teachers' unions have fought against cutbacks in education budgets. In collective bargaining, they often raise issues —class size, professional standards, support for professional edu-cation—which are justifiable in terms of improving education as well as improving the benefits provided to members. In fact, for many questions, the two cannot be separated; if teachers seek the opportunity to share in decision-making, they may well improve the quality of education by bringing a different professional perspective to bear, if management is prepared to take it seriously.

These activities should not be exaggerated; like all unions, and particularly like all craft unions, the teachers' union uses the members' collective power primarily to protect their job rights and to advance their own interests. These efforts have in the past led some teachers' unions, particularly in big cities like New York and

Philadelphia, into activities which polarize rather than bridge racial or class barriers. These unions resisted the movement toward community control and have opposed most efforts to devise methods of accountability for educational results, even when the performance of the school system has been abysmal. In their search for allies, they have joined forces with some of the most backward elements of the labor movement, despite their own professed commitment to liberal values. But these activities, in the heat of the volatile urban climate of the mid-sixties, do not necessarily point the way to the future, nor do they accurately predict the values and goals of teachers' unions in many smaller cities and in suburban and rural areas.

Much depends on the character of the national leadership which emerges as teacher unionism grows, and much on the political setting for public education in the years ahead. Present trends indicate a national leadership which will play a conservative role, will stress the need to defend teachers against pressures for community control and accountability, and will ally itself with the dominant, conservative elements in the American labor movement. So long as the major urban areas are politically as weak, nationally, as they have been in recent years, the pressures of the minority populations for more voice in the schools and more teacher accountability will be contained, and this conservative direction validated in the eyes of the membership. But if we can restore some of the national sensitivity to urban problems which began to shape national politics in the 1960s, the leadership of the teachers' unions will need to respond, and union policies should become less parochial and defensive.

THE INDUSTRIAL UNIONS

Despite the importance of the teachers, the craft union model is far from the whole story. In big cities, at any rate, and to a considerable degree in statewide bargaining also, the newer unions of public employees increasingly resemble industrial unions more than craft unions; in New York City, the multilevel citywide unit is already dominant, despite large craft-like units among the uniformed groups and the teachers. Transit workers in New York and other cities were affiliated with the CIO and have operated as industrial unions for many years. In these cases, while different occupational groups and levels seek different bargaining goals, the differences are not of the dramatic and disruptive kind which characterizes rivalries between—e.g.—police and firemen's unions or the different newspaper crafts. The industrial form is likely to predominate, in the view of some observers (compare, for example, Zagoria, 1972: 163).

NEW BARGAINING RELATIONSHIPS

These unions have bent their energies extensively in the search for valid and reliable bargaining mechanisms, free of precisely the kind of political distortion which has been the rule. In this effort, they have been able to work together and in the process to move even the more traditional unions away from their "insider-influence" relationship with the employer toward a posture of arms-length negotiation more characteristic of the private sector. Many dramatic moments in recent years in New York have been the result of this effort, most notably the short strike by bridge tenders and other city workers in July 1971, which disrupted traffic in and around New York, on the issue of pension improvements. This was a protest against the refusal of the governor and the legislature to abide by their own rules which, in the Taylor Law of 1967, mandated collective bargaining and pledged that the employing governments would honor the agreements which bargaining produced. It was a political action against the political subversion of bona fide collective bargaining.

Older styles of organization and operation will continue to be found for some time, but there is a discernible trend which should make collective bargaining less overtly political. Some union spokesmen, particularly in the AFSCME, are committed to placing the bargaining relationship as close as possible to the private sector model: outside of the political process, stable bargaining relationships, negotiating all issues at the bargaining table, writing mutually binding agreements, and utilizing the necessary degree of economic pressure to equalize bargaining power and to ensure that the bargaining is serious and in good faith.

This is a normative goal rather than a fully realizable one. The political realities will never disappear from the public sector; they were never ever wholly absent in the private sector and in the recent past the politics of inflation and unemployment have been major presences at private sector negotiations. Yet the normative goal is important because it represents a significant change from the past and because movement in this direction is both real and sustained.

THE CHALLENGE TO MANAGEMENT

Can we expect state and local government bodies to develop a more purely managerial approach to employee relations and collective bargaining? Will not the public's stake in the bargaining results ensure a degree of political involvement which will substantially exceed the norm in even the most dramatic private sector battles? One view is that held by Horton and others, that bargaining is root and branch a political process and that the bargaining table is simply a façade which both conceals and misleads the public (compare Raskin, 1972). Such views are exaggerated and wide of the mark.

Clearly, political decisions are required to implement collective bargaining commitments; taxes may go up and services may be reduced or reorganized. But the process of determining how many people will be employed, what they will do, how much they will earn, and what their benefits will be are primarily managerial in nature. They are the responses which public managers make to the needs they face; in this respect, they are in principle no different from the managerial decisions. Increasingly, they will be seen as such; already, the impact of collective bargaining is diminishing the primacy of civil service rules and procedures in settling these issues, and this will continue (Rosen and Rapoport, 1970: 630-632; see also U.S. Department of Labor, 1972). The impact of collective bargaining is not accurately portrayed by Horton and others, who see all-powerful unions facing prostrate, helpless governments. The essence of collective bargaining is the substitution of joint decisions for unilateral decisions; it gives employees through their unions an effective voice in the decisions which directly and irreversibly affect their lives. The process is similar whether it takes place in the overtly political atmosphere of a governmental unit or in the privacy of a business office. The fact that public officials may be poorly trained or motivated to deal effectively with unions is not an objection to collective bargaining, but a problem which must be solved, just as it becomes necessary to provide school principals with new skills to meet new educational problems or to computerize public records so that order can be brought out of chaos. Collective bargaining is primarily a managerial, not a political, challenge to government. It is well to recall Stanley's (1971: 138-139, 144-145) conclusions, based on careful study in a variety of cities:

> What is happening to government achievements under union pressures? Do unions impair government's ability to translate the

will of the people into effective action? . . . Judging by specific bargaining outcomes, by interviews, by government reports, and by press reactions, the answer to such questions is 'not much.' . . . In the bargaining process unions are necessarily restrained by prevailing practice in the private sector and in other governments, by fiscal realities, by the need to do business in the future with management, by management's skill and resolution in bargaining, by impasse resolution procedures, and by management's ultimate willingness to "take a strike" and apply sanctions.

THE POLITICAL DIMENSION

Of course, public funds are raised by tax levies, while private funds are not coerced by state power. And, of course, public services are mandated as private goods or services are not, though some in the private sector are as vital as any which government provides. These are valid points, but their importance can be exaggerated. There is no law which requires administration in a unit of government to be fragmented rather than unified, amateur rather than professional, or unskilled in labor relations; nor is there any which requires a mayor or governor to accept any union demand which is forcefully or dramatically presented. Elected officials can argue the merits and develop support for their position, just as well as private managers can, perhaps more effectively since they are skilled in reaching the public and the media beyond the capabilities of most private managers. The equivalent in public bargaining of the uneconomic demand, which threatens the economic viability of the private enterprise, is the politically unacceptable demand, which threatens the official's public mandate. If that mandate is real, the evidence can be developed, often quickly and clearly, to rebut and refute union demands.

Division and fragmentation are inevitable in the public sector, if only because of the division of powers among executive, legislative, and judicial branches. In addition, different agencies within a system of government frequently have their own constituencies and processes which, even when they are informal, can nevertheless be highly effective and resistant to rationalized control. Local communities may exert important leverage over parts of the system which are nominally centrally controlled. Finally, public employee unions can wield direct political power, influencing the election process itself. Diffusion of power clearly complicates collective bargaining and makes effective management difficult to establish and sustain. But it is important to note that, just as a government may tend toward fragmentation, the same holds for unions. Based in different units of government, with different ethnic compositions, different histories, and different leadership styles, public sector unions may be suspicious of or hostile to one another even when this division undermines their collective self-interest. The pervasiveness of political pressures and the tug of political interest operates to weaken unity and promote rivalry. As indicated earlier, Mayor Wagner demonstrated great skill in exploiting these differences in the early years of unionism in New York, and other examples can be found.

The chief executive of a state, city, or town must bring to the management task a range of skills and a degree of patience which exceed those required of most private sector managers, though the latter seldom recognize that this is so. To state that a task is difficult is not, however, to agree that it is impossible. Government has faced and surmounted many challenges which taxed the imagination and appeared to defy solution. Acquiring the necessary authority and mandate to deal effectively and constructively with public employee unions constitutes the present and future challenge to public administration. One of the great incentives to success is the fact that government will not be able to solve the issue of more effective services without solving the question of how to establish and maintain effective systems of collective bargaining. Since it will not be possible to dissolve, coopt or contain these unions, they must be made a part of the process through which the issue of better services is addressed.

THE POTENTIAL CONTRIBUTION OF UNIONS

When governments move in this direction, they will learn that public sector unions are not inherently indifferent or hostile to efforts to improve services and increase employee effectiveness, nor do they necessarily resist efforts at reorganization, rationalization, or decentralization. Note that while the United Federation of Teachers bent major efforts first to prevent and then to emasculate the decentralization of New York City's schools, District Council 37 of AFSCME, representing municipal hospital workers, actively supported legislation to mandate community boards for each municipal hospital (Gotbaum, 1973), and opposed the UFT in one hotly contested local school board election where community control was at issue.[1] Both unions actively supported programs to train low-level employees in schools and hospitals, in order to improve employees' abilities to provide services and to improve the supply of skilled employees. Clearly different unions interpret their self-interest differently and may show widely different willingness to take short-run risks in the interest of long-run goals. The differences among unions, some of which were cited earlier, produce a diversity of response.

But there is no evidence that employees object to higher standards of performance or to the introduction of better methods; what they do object to is unilateral imposition of change without consultation, without the right to object, without protection against job loss, without compensation for being asked to take on new or added tasks.

ROOTS OF CONFLICT

In addition to the political dimension, a major difference between public and private sector unionism is that the basis for class conflict is absent. However muted, transformed, or ameliorated by sustained economic growth, workers and owners in the private sector have interests which are at root in opposition. The search for methods to improve worker efficiency and productivity derives from the unending search for profits. In the absence of joint exploitation of others, owners and workers must and do confront one another on this basic issue. It sets limits to even the most far-reaching schemes for collaboration and joint problem-solving, such as that exemplified by the Scanlon Plan (Lesieur, 1958). Even so, where the settings were right and the parties faced problems which required their cooperation for mutual survival, such efforts accomplished much in some areas of the private sector.

The basis for class conflict is not found in the public sector. Public employee unions, like all unions, are committed to curb management's unilateral authority, to establish joint determination of the rules which govern employment, and to achieve higher levels of economic benefits for their members. But once the legitimacy of these purposes is recognized and the conditions established to make them possible, the major cause of employee unrest—arbitrary and unilateral authority and denial of the right to bargain over benefits—will have been dealt with. These preconditions do not yet prevail anywhere in the United States, and their establishment will take a long time. But inevitably they must come to pass, just as the legitimation of private sector unionism did. At that point, public administrators will discover, perhaps to their surprise, that they and the unions representing their employees *share* certain fundamental goals. Among them are the legitimation of public services in a society which still accepts a private enterprise ideology, and the need to provide a broad fiscal base for funding public services—particularly in cities. Public sector unions dealing with competent managers in a setting which accepts their legitimacy and recognizes their rights will go far to promote these goals.

To realize the potential contribution of public sector unionism to the resolution of the controversy over public services, these developments are needed. The sections which follow discuss these in turn.

MOTIVATING MANAGEMENT

When unions articulate their demands, they are imposing new standards of planning and foresight on managers. In the early days of the Lindsay Administration, a major demand from the Municipal Labor Committee, representing the major unions, was for competent and professional leadership in the city's Office of Labor Relations. To the degree that unionism leads to the improvement of the quality of municipal management, collective bargaining will make a major contribution to the process of meeting the need for better services and more productive and effective labor force. And employees will welcome such improvement; their sense of pride and of self is often ruined and their morale destroyed by long years of work under incompetent and unmonitored leadership. Just as, in the words of the late Cyrus Ching, employers get the labor relations they deserve, governments get the level of performance they deserve. The first requirement for good services is good management; if managers know their business—as business knows—employees produce and gladly.

But public sector management is very often poorly trained, badly motivated, and incompetent; unions do not create this problem. Neither do they make effective management harder; as pointed out earlier, unionism forces a higher standard of competence and performance on managers. As for productivity and performance, the reality is far different from the stereotype. Measures to improve the quality, training and performance of managers, particularly at middle and lower levels, are long overdue and are fully compatible with a setting which includes strong unions and vigorous collective bargaining. Here the major problem appears to be instituting an effective system of incentives which would identify and reward good managers and would penalize poor ones. It is harder to do this than to design good management training programs, using the resources of both the professional management schools and the schools of public administration. As mayors, governors, county executives and other key executives increase their efforts to recruit middle managers and first-line supervisors from the ranks of the college educated, they must also begin to institute systems for assessing management performance and for rewarding superior performance.

It is a truism in the private sector that "what is counted is what counts" and public administration must learn how to adapt this approach to its own settings and systems. The present system of incentives which most directly affect middle managers is political in nature; not much will change until those incentives are replaced by others which assess competence in managing people and in improving productivity and the quality of services provided. The Police Commissioner of New York recently showed that it is possible to hold managerial subordinates accountable for performance, and other examples could be found. The task is difficult, but feasible, given the present state of the art of management, if the top is serious about tackling it. In a well-managed system, some direct incentives to workers on a group basis would be possible; similar systems occur, though rarely, in the private sector.

CAREER DEVELOPMENT

The second major element in any productivity program would be the design and deployment of a career development system, a potent stimulus to increased worker motivation and development, which has been almost totally neglected in public administration and government. Most proposed "reforms" of civil service stress increasing lateral entry and increasing the system's attractiveness to the better educated. The stated objectives have value, but involve risks which are seldom recognized. Effective recruitment of this kind without other steps lowers the morale of incumbents; the message of such programs is that government does not believe that its own employees can be trained and promoted to fill upper-level jobs. A career development system, by contrast, would deliver precisely the opposite message. Moreover, it would *reduce* resistance to the recruitment of qualified candidates from outside the system, provided that the selection-promotion process provided opportunities for those with motivation and ability to increase their skills, status, and earnings.

Many civil service administrators would argue that such opportunities already exist. Their major defect is that they place extraordinary burdens of proof on employees to demonstrate their ability to qualify for promotion, and provide little or no training or development of incumbents. Savas and Ginsburg (1973) report that, in fact, an "inverse merit system" prevails in New York—that the lower one's examination score, the more likely that one would be hired. Even in its virtually exclusive preoccupation with hiring, civil service appears to have failed. But the preoccupation is what needs to be highlighted; civil service has focused on hiring to the virtual exclusion of other aspects of employee relationships. By contrast, many private employers place far less stress on hiring qualifications, far more on developing the latent skills and abilities of employees hired, and on recognizing and rewarding them. The works of Herzberg et al. (1959; Herzberg, 1966; Ford, 1969), of Likert (1967), and of others influential throughout the private sector have had little impact on public administration.

Management in the private sector is beginning to recognize that developing employee potential is not only worthwhile but that its benefits can be quantified; investments in human resources, to improve employees' educational levels and skills—which are both necessary in today's urban labor market—are in many cases amply justified by the returns which such investments generate. To survive, such programs in the private sector must pass the test of profitability but in the public sector other, more flexible and socially responsive standards can and should be used. All the relevant tests appear to strengthen the case for moving in this direction. Improving the upward mobility opportunities of employees also speaks directly to the problem of minority group employment in government.

A system which provides career development opportunities is bound to be more attractive to potential employees and more interesting and rewarding for incumbents. Efforts to apply good management methods in a setting of this kind will inevitably be easier to make than in a setting where employees see any change as threatening, because they have no alternatives and no opportunities.

A career development system does not happen by itself; it must be planned and developed with some care, using techniques of task analysis to map promotional or career ladders. A good career system can only work if the resources are provided to equip those who want it with the training and the education needed for them to fill upper-level jobs competently; here educational institutions need to be brought into the picture, as has been done successfully in some of the paraprofessional programs of the past decade. Finally, criteria of eligibility for upward movement and standards for judging relative merit require major change. Formal civil service tests should in the future play a less dominant role; they should also be made far more responsive to real job needs and more "culture free."

This, in turn, raises the issue of civil service rules and structures in a collective bargaining era. I have argued elsewhere (Rosen, 1973), as have others, that collective bargaining increasingly renders obsolete a civil service system which was erected to protect employee rights in a nonunion era, and that bargaining can improve the access of minorities to public employment. In creating a career system, civil service can play the watchdog role, to assure that basic merit principles are preserved. But civil service need not have the exclusive or even the dominant role in deciding who is ready to move in or up. Line agencies need to develop strategies which would enable them to take some of the decision-making power back from the civil service or personnel departments which have acquired most of it. This would require them to develop their own manpower planning and development competence and to validate their ability to administer manpower programs professionally and nonpolitically; this is a tall order, but necessary. They can get help from the educational institutions, particularly those with analytic and training competence. If an agency can present a program of upward mobility which relies heavily on a professional analysis of job requirements and job links and involves professionals in the training which both prepares employees for promotion and assesses their ability—functions which could well be developed in schools of management—a basis will be laid for reducing the dominant or exclusive role of civil service in determining job requirements, eligibility criteria, and tests which decide the ranking of candidates.

This "disestablishment" of civil service functionaries can only be successful if there is political support at the top. If this support is withheld or compromised, any diminution of traditional civil service control will be seen by many as a threat to the merit principle. Employee groups are likely to be the most skeptical of all; even a proposal which is said to be in their interests will be received with caution or hostility, if only because it is unilateral.

UNION-MANAGEMENT COOPERATION

This brings us to the third necessary element of any productivity effort. In an era of increasing scope for collective bargaining and increasing unionization, a government which is serious about improving productivity must put the effort within the framework of union-management cooperation. This is a concept, developed in the private sector, which has much greater promise in the public sector of the future. The private prosperity of the 1950s and 1960s renewed the confidence of the unions in traditional collective bargaining and led them to abandon the nascent effort to develop other mechanisms for solving problems (see Rosen, 1968). But penury, not prosperity, is the characteristic state of governmental financing at the state and local levels; this provides a strong stimulus to unions to look beyond traditional collective bargaining to safeguard the long-run economic well-being of their members. Government and union alike are aware of the resistance to tax increases on the part of the public and of the growing demand for more effective utilization of manpower in the labor-intensive activities of government, where capital substitution cannot offer much. Moreover, the "new" unions, like their CIO-era predecessors, take a broad view of social responsibility rather than the exclusively protective role of the unions founded in earlier periods. Jerry Wurf, President of AFSCME and AFL-CIO vice president, has fought within the AFL-CIO for enlightened foreign and liberal domestic policies, much as Walter Reuther did. His principal affiliate leader, Victor Gotbaum, Executive Director of D.C. 37, has long been the vanguard labor spokesman in New York City on controversial issues of race, justice, urban policy and other questions.

Proposals which offer these unions the opportunity to play a socially constructive role will not be seen as contradicting or threatening further narrower bargaining and protective functions. If the legislative basis of union rights and roles is secure, the political setting reasonably stable, and the competence and good faith of the employer side credible, these unions should welcome the chance to establish joint mechanisms for solving joint problems (compare Lesieur, 1958, for description of how such cooperative arrangements operate).

In such a setting, a union-management framework could be established for the frank airing and resolution of problems of efficiency, productivity, and the organization of services. In these discussions, the myth of the managerial prerogative needs to be set aside. The focus must be on identifying and solving problems, which are admittedly of common concern and which require collaboration or worker cooperation for their resolution; the focus must not be on grievances or inequities which belong on the bargaining table; the two spheres must be separated. Problems must be raised and disposed of systematically; if employees raise questions or propose a better method for solving a problem, they are entitled to answers complete with a rationale for making a decision. Decisions should be decentralized, with individual departments or other units free to dispose of issues raised there, within some agreed limits. Problem-raising conferences need to be scheduled regularly, and management needs to provide the information necessary to deal with a problem of suggestions. To the extent possible, *at all levels decisions should be joint, not unilateral.*

These structures and procedures are not difficult to design and put in place, once the decision is made to go ahead. The critical issue is to establish a relationship which will support joint problem-solving and will not at the same time appear to either side—but particularly to the employer side—to invite any undermining of the basic collective bargaining process. Success requires security on both sides, not only at the top, but at subordinate levels. Middle managers who feel threatened by a process which both welcomes and seriously appraises employees' proposals for solving problems will resist any serious effort at union-management cooperation as compromising their toughness or independence. Each side must look to these problems and deal with them seriously and effectively.

NOTE

1. D.C. 37 is a citywide general union with a large proportion of minority members.

REFERENCES

FORD, R. N. (1969) Motivation Through the Work Itself. New York: American Management Association.

GOTBAUM, V. (1973) "Bending the facts to the thesis." New Leader (May 14). (This is a review of Horton, 1972.)

HERZBERG, F. (1966) Work and the Nature of Man. Cleveland: World.

––– B. MAUSNER, and B. SNUDERMAN (1959) The Motivation to Work. New York: John Wiley.

HORTON, R. D. (1972) Municipal Labor Relations in New York City: Lessons of the Lindsay-Wagner Years. New York: Praeger.

LESIEUR, F. G. [ed.] (1958) The Scanlon Plan. Cambridge, Mass.: MIT Press.

LIKERT, R. (1967) The Human Organization. New York: McGraw-Hill.

RASKIN, A. H. (1972) "Politics up-ends the bargaining table," pp. 122-146 in S. Zagoria (ed.) Public Workings and Public Unions. St. Louis: American Assembly.

ROSEN, S. M. (1973) " 'Merit' systems and workers," pp. 143-153 in A. Gartner et al. (eds.) Public Service Employment: An Analysis of Its History, Problems and Prospects. New York.

––– (1968) "Union-management cooperation: is there an agenda for tomorrow?" Proceedings of the Twenty-First Annual Meeting of the Industrial Relations Research Association (December): 81-89.

––– and N. RAPOPORT (1970) "City personnel: forces for change," in L. C. Fitch and A. H. Walsh (eds.) Agenda for a City: Issues Confronting New York. Beverly Hills: Sage Pubns.

SAVAS, E. S. and S. G. GINSBURG (1973) "The civil service: a meritless system?" Public Interest 32 (Summer): 75-77.

STANLEY, D. T. (1971) Managing Local Government Under Union Pressure. Washington, D.C.: Brookings Institution.

U.S. Bureau of Labor Statistics (1972a) Director of National Unions and Employee Associations 1971. Washington, D.C.: Department of Labor.

––– (1972b) "Municipal collective bargaining agreements in large cities." Bull. 1759, Washington, D.C.

U.S. Department of Labor (1973) Press release, August 22.

––– (1972) "Collective bargaining in public employment and the merit system." Washington, D.C., April.

ZAGORIA, S. (1972) "The future of collective bargaining in government," in S. Zagoria (ed.) Public Workers and Public Unions. St. Louis: American Assembly.

18

Federal Programs and Political Development in Cities

JEFFREY L. PRESSMAN

☐ FEDERAL URBAN PROGRAMS—their goals, their structure, and their impact—have constituted a central subject of recent political debate in this country. The future shape of these programs has been an issue in the conflict between the President and Congress, and the signals from Washington have been closely watched by officials at the local level who are the recipients of federal aid. Regardless of the outcome of the struggles over general revenue-sharing, special revenue-sharing, and categorical programs, it seems clear that the federal government will continue to spend substantial sums of money on urban programs and that federal officials will continue to have an interest in what cities do with that money.

In this essay, I will focus on the problems posed for federal programs by weak local leadership and by the lack of arenas in which federal and local actors can bargain effectively. (By "effective bargaining," I mean a negotiating process that will have an impact on real programs in the outside world.) Drawing on recent experiences in federal-city relations, I will suggest ways in which federal programs might be redrawn in order to deal with these problems and achieve their goals more fully.

TABLE 1
FEDERAL AID PAYMENTS IN URBAN AREAS HAVE
INCREASED SUBSTANTIALLY, 1961-72 (in millions of dollars)

Function and Program	1961 Actual	1964 Actual	1969 Actual	1972 Estimate
National defense	10	28	30	31
Agriculture and rural development:				
Donation of surplus commodities	128	231	313	294
Other	27	40	104	81
Natural resources:				
Environmental protection	24	8	79	773
Other	30	10	101	170
Commerce and transportation:				
Economic development	—	158	104	147
Highways	1,398	1,948	2,225	2,646
Airports	36	36	83	117
Urban mass transportation	—	—	122	289
Other	1	5	5	6
Community development and housing:				
Community action program	—	—	432	549
Urban renewal	106	159	786	975
Public housing	105	136	257	570
Water and sewer facilities	—	36	52	110
Model Cities	—	—	8	420
Other	2	17	75	278
Education and manpower:				
Head Start and Follow Through	—	—	256	97
Elementary and secondary	222	264	1,262	1,457
Higher education	5	14	210	113
Vocational education	28	29	179	393
Employment security administration	303	344	449	327
Manpower activities	—	64	530	1,271
Other	3	7	77	704
Health:				
Hospital construction	48	66	89	113
Regional medical program	—	—	19	63
Mental health	4	8	50	66
Maternal and child health	18	34	139	203
Comprehensive health planning and services	29	48	80	150
Health educational facilities	—	—	106	117
Medical assistance	—	140	1,731	2,074
Health manpower	—	—	28	193
Other	—	4	54	283
Income security:				
Vocational rehabilitation	37	61	247	400
Public assistance	1,170	1,450	3,022	5,581
Child nutrition, special milk and food stamps	131	168	482	1,690
Other	3	16	148	510

TABLE 1 (continued)

Function and Program	1961 Actual	1964 Actual	1969 Actual	1972 Estimate
General government:				
Law enforcement	—	—	17	464
National Capital region	25	38	85	158
Other	—	9	27	145
Other functions	—	2	—	5
General revenue sharing	—	—	—	2,813[a]
Total, aids to urban areas	3,893	5,588	14,045	26,848

a. Tentative estimated impact calculated on the basis of population includes both direct pass-through and discretionary state allocations (Source: Special Analyses, Budget of the United States Government, Fiscal Year 1972, p. 241).

PROBLEMS IN FEDERAL URBAN PROGRAMS— AND SOME ATTEMPTED SOLUTIONS

RECENT GROWTH IN FEDERAL AID

Through new legislation in the 1960s, Congress extended both the range and the level of federal involvement in cities. Major national programs were developed in new fields of activity like manpower training, and new aid was provided for established local government functions like mass transportation and sewer and water systems. In both the Economic Opportunity Act of 1964 and the model cities program of 1966, Congress broke with precedent by authorizing aid to local communities for a relatively unrestricted range of functions. And in 1968, the national government initiated a program of aid to local law enforcement—a traditional preserve of local government.

Table 1 shows the sharp rise in federal programs of aid to cities in the years 1961-1972, from approximately $3.9 billion to over $26 billion.

FRUSTRATION, CONFLICT AND RESPONSE

Although the growth in federal expenditures has been impressive, there have often been gaps between federal goals and program outcomes at the local level. There is a small but growing literature on the reasons why high hopes in Washington are frequently frustrated during the implementation process (see, e.g., Derthick, 1972; J. T. Murphy, 1971; Pressman and Wildavsky, 1973).

Federal policy makers have not been the only ones to complain about the results of urban programs; local officials have expressed dissatisfaction as well. Like foreign aid, urban aid efforts have generated considerable friction between donor and recipient organizations. Federal officials have usually diagnosed this friction as stemming from confusion and frustration caused by the fragmentation of federal programs and the lack of communication between federal and city actors. Proceeding from this diagnosis, federal efforts to reduce tensions in the federal system have taken the following forms:

(1) Communication. To overcome confusion and ignorance about substantive programs and about each other's plans, the federal government has initiated a number of policies designed to increase information available to intergovernmental actors and to facilitate communication between them. In 1959, Congress established a continuing agency—the Advisory Commission on Intergovernmental Relations—for study, information, and guidance in the field of intergovernmental relations. And in 1961, President Kennedy established Federal Executive Boards in ten of the nation's largest cities, in order to improve communication among federal agencies.

During the Johnson Administration, there was a flurry of activity in the fostering of communication. Vice President Humphrey was designated as the administration's liaison with mayors, and the President issued a memorandum calling on all federal officials to take steps to ensure closer cooperation among federal and local officials (for a discussion of various attempts at the fostering of intergovernmental communication, see Leach, 1970). President Nixon, during his first month in office in 1969, created an Office of Intergovernmental Relations under the direct supervision of then Vice President Agnew. The new office, the President declared, would "seek to strengthen existing channels of communication and to create new channels among all levels of government" (Leach, 1970: 179). There has appeared to be strong and continuing support for communication.

(2) Comprehensive Planning. Once federal and local officials start communicating with each other, what form—according to federal pronouncements—should that communication take? A prime vehicle of intergovernmental conversation is the writing (on the part of the local recipient) and evaluation (on the part of the federal agency) of

a comprehensive plan, which has become a standard part of the application for urban funding. The creation of such a plan—featuring the identification of total community needs and available re-sources—has been required for participation in the poverty, model cities, and Economic Development Administration programs, among many others.

Comprehensive planning is designed to encourage local com-munities to inform themselves about various funding sources and to match those potential resources with the problems they might help solve. In this way, it is hoped that ignorance and confusion about programs can be diminished. But even if local leaders were able to inform themselves completely about available programs and the uses to which they might be put, the leaders would still find that those programs are generated by numerous, fragmented agencies which often work at cross-purposes to each other. It is difficult to see how comprehensive planning can make this fragmentation disappear. Something more is needed.

(3) Coordination. The antidote to fragmentation, expressed in numerous congressional and presidential policy directives, is coordi-nation. For example, in the statute which gave life to the poverty program, Congress, in the Economic Opportunity Act of 1964, Section 611, authorized the director of that program to "assist the President in coordinating the antipoverty efforts of all federal agencies." In the legislation creating the Department of Housing and Urban Development, Congress, in Section 3(b) in the Department of Housing and Urban Development Act of 1965, provided that "the Secretary shall . . . exercise leadership at the direction of the Presi-dent in coordinating federal activities affecting housing and urban development" (for further discussion of attempts to coordinate federal urban programs, see Sundquist with Davis, 1969). And President Nixon has shown his concern for coordination in the urban field by creating an Urban Affairs Council to "coordinate programs and provide a forum for the discussion of interdepartmental problems that cut across jurisdictions" (Leach, 1970: 180).

Although these attempts at "coordination" are designed to cure the fragmentation of the federal organizational effort in cities, it is obvious that a new phenomenon has been created—the proliferation of coordinators. With one official responsible for coordinating anti-poverty programs, another charged with coordinating urban programs, and still others (the Secretaries of Labor and HEW) directed to coordinate manpower programs, it is obvious that numerous programs fall simultaneously into the jurisdictions of *each* of these officials (see Sundquist with Davis, 1969: 13 ff.).

To complicate matters still further, various federal agencies in the urban field have stimulated the formation and development of counterpart "coordinating structures" at the local level. Thus, OEO has had its local partner in Community Action Agencies; HUD has worked through City Demonstration Agencies; and Labor has had both the Cooperative Area Manpower Planning System and Concentrated Employment Program. Each of these local agencies has been charged with "coordinating" a wide variety of urban programs—over most of which it has had little effective power.

WHY THESE SOLUTIONS ARE INADEQUATE

Federal policies of (1) increasing communication, (2) encouraging comprehensive planning, and (3) creating centers of coordination have all been put forward as means of lessening conflict in the intergovernmental system and of furthering the achievement of federal aims in cities. The above methods of dealing with federal-local friction are based on an assumption that the fundamental problems involved are those of inadequate information and confusion about organizational jurisdiction. Technical methods have been put forward to deal with what have been thought to be technical problems.

But these solutions and the diagnoses that lie behind them do not go to the heart of the problems involved in federal-local relations. For federal-local tension is not merely the product of misunderstanding and lack of communication. Federal and city policy makers often disagree with each other (and among themselves) about the nature of urban problems and what to do about them. There have been sharp differences of opinion between federal and local officials with regard to program goals, procedures, funding levels, and the proper recipients of federal grants.

If actors and organizations have conflicting policy preferences, then the technical methods of communication, planning, and coordination are unlikely to resolve the differences among them. More discussion and gathering of information might only result in pointing up differences between the organizations. As for coordination, this much-used term is often proposed as a cure for fragmentation, but it does not offer much guidance to one who wishes to make or to understand policy.[1] If organizations disagree about objectives, then coordination may mean that one wins and the other loses. Alternatively, bargaining between them may result in a solution which is somewhere between the opposing preferences. In fortunate circumstances, an integrative solution can make both parties better off than they were. But in no case is coordination among conflicting parties a bloodless and technical process. Certainly, the creation of multiple coordinators and coordinating boards has not eliminated conflict between federal and local agencies.

Even when basic disagreements between federal and local actors are recognized, the present structure of intergovernmental arenas (federal-city liaisons, Federal Executive Boards, intergovernmental relations offices) does not allow for effective resolution of those disagreements. This is because decisions made in these "communications-facilitating" bodies are rarely binding on governmental policy. Federal-local communications channels have often stimulated talk, but they have proved frustrating to local leaders who are more interested in concrete commitments (decisions as to which group will receive federal money; approval or disapproval of projects; appropriation and delivery of funds). As for comprehensive planning exercises, they have frequently been treated by both federal and local officials as an isolated part of the application process—a sort of college board examination for cities—having little if any effect on local decisions regarding budgeting and personnel (for an example of this attitude toward comprehensive plans, Sundquist with Davis, 1969: 193).

A basic problem in the communication-planning-coordination strategy lies in its emphasis upon building commitments to initial goals, rather than on creating a framework for collaborative day-to-day actions. This separation of policy formulation from implementation is fraught with danger for the ultimate achievement of program objectives. For even if overall objectives are agreed upon by federal and city officials, that agreement can rapidly dissipate into disagreements on the "details" of implementation—funding procedures, project site locations, the extent of participation by competing local groups, and so forth.

REVENUE-SHARING

Revenue-sharing—the provision of money by the federal government to cities with few or no strings attached—might appear to deal
more directly with some of the basic sources of friction and
discontent in federal-city relations. Local officials would be relieved
of many of the restrictive administrative guidelines they so often
denounce. Furthermore, the power asymmetry and irritation produced by constant local dependence on federal officials' decisions
would be eased. With an infusion of relatively unencumbered
financial resources, city officials would have less need to beg.

But there are serious political problems connected with proposals
for revenue-sharing. First is the problem of identifying the proper
recipient in a local system characterized by fragmented governmental
authority. There have been continuing battles among state, county,
and municipal governments over the distribution of both general
revenue-sharing and (proposed) special revenue-sharing funds. With
respect to the latter category—which would earmark funds for a
broadly defined function such as "community development"—
competition seems likely between general-purpose local governments
and special units (such as redevelopment agencies and housing
authorities) for control over money and programs.

Another political problem is that, although revenue-sharing may
reduce the power asymmetry between federal and city governments,
it may *increase* the resource imbalance among groups within the city.
If a goal of federal urban policy is to redistribute resources in cities
to people and neighborhoods who presently lack them, categorical
grants can have the advantage of earmarking funds for use in certain
poorer areas. (For example, the Elementary and Secondary Education Act's Title I program for compensatory education only
provides funds for schools which meet certain poverty criteria.)
Unrestricted revenue-sharing might be divided up by powerful
interests in a community, with weaker political actors getting short
shrift.

Finally, merely increasing the flow of revenue to cities does not
deal with the quality of political leadership and political institutions
at the local level, factors which can have an effect on the
implementation of federal programs. Proposals for distributing funds
on the basis of population offer no incentives for local governments
to expand the scope and effectiveness of their programs. And even
those revenue-sharing proposals which attempt to reward local

government capacity[2] end up defining that largely in terms of planning capability or administrative efficiency.

In his State of the Union Address on January 22, 1971, President Nixon declared that revenue-sharing would have a dramatic effect on local governments: "If we put more power in more places, we can make government more creative in more places. For that way we multiply the number of people with the ability to make things happen—and we can open the way to a new burst of creative energy throughout America." In fact, it is not at all clear that increased financial resources can by themselves create strong and responsive government at the local level. It has been observed that there is considerable variability in local leaders' uses of outside resources (see R. D. Murphy, 1971). If local governmental leaders have made a practice of avoiding conflict and underutilizing their own resources, why should the new revenue make them suddenly adventurous? It is more likely that such leaders would use the money either for local tax relief or for meeting the increasingly aggressive demands of city employees' unions.

Some observers of city politics have recently noted the extent to which the achievement of federal urban program goals is dependent upon the existence of strong local political leaders who can offer support for the programs (see, e.g., Derthick, 1972; J. T. Murphy, 1971). More than revenue-sharing is needed to facilitate the emergence of such urban leaders. As a panacea for the problems of federal urban programs, revenue-sharing falls far short.

THE OAKLAND EXPERIENCE

The problems and tensions of federal-local relations, noted above and also observed by various students of urban policy, were evident in a long-term study of Oakland, California, in which I was a participant.[3] In Oakland, local and federal officials argued over program funding levels, federal guidelines, and which groups or institutions should receive federal money in the first place. Local government officials were particularly angered by certain federal programs—the poverty program, model cities, Neighborhood Development Program, Concentrated Employment Program—which provided resources for black community leaders who were openly hostile to the City Council.

The federal response to intergovernmental difficulties in Oakland followed a predictable pattern. Under the auspices of the Federal Executive Board in the area, federal and local task forces were set up to communicate with each other and to formulate priorities. But the task forces did not have authority to make binding decisions, and both federal and local officials soon grew weary of the priority-setting exercise.

A more ambitious exercise in federal-local goal-setting occurred when the federal Economic Development Administration decided in 1966 to launch a $23 million public works and job creation program in Oakland. Federal representatives, city officials, and local busi-nessmen all committed themselves to the goal of providing jobs for the hard-core unemployed. The EDA would provide the money for building public works, and the local recipients would undertake to contact and hire the unemployed. Although the outlook for the project in 1966 appeared to be hopeful, the "details" of imple-mentation—land fill, contracting procedure, construction disputes, disagreements over organizational jurisdiction—combined to delay and nearly end the project. Federal officials had assumed that, if the various participants could agree on goals, then implementation would be no problem. But the separation of policy formulation from implementation, and the lack of attention paid to mechanisms for ensuring continued collaboration, proved disastrous. Furthermore, strong local leadership, on which the federal government had counted for support, did not exist in the city. Thus, what was designed as a "demonstration program" proved to be a textbook illustration of difficulties encountered in administering federal urban programs (for an analysis of the EDA Oakland Project, see Pressman and Wildavsky, 1973; for further discussion of Oakland city politics, see Meltsner, 1971; May, 1971; Pressman, 1972).

CONCLUSION

We have observed that federal programs can be vitiated by the absence of strong political bodies at the local level, and by the lack of effective arenas where federal and local officials can bargain with each other. Obviously, technical solutions—planning, "coordinating," and so on—are not adequate to deal with what are essentially political problems: the creation of viable sites for the exchange of resources and the mobilization of public support for programs. In other words, the capacities of political systems themselves (both federal and local) must be increased. Social scientists have dealt with these issues in the study of political development. But "political development" has been discussed by these scholars in a number of different ways, and we must be more specific in our analysis and our advice.

AID STRATEGY AND POLITICAL DEVELOPMENT

DEFINITION AND ALTERNATIVE APPROACHES

Although many meanings of political development have been suggested, students of the subject usually include two elements in their definitions: (1) the ability of a wide range of people to participate in politics and make demands on government, and (2) the ability of government to satisfy those demands (see, e.g., Pye, 1965; Almond, 1965; Ilchman and Uphoff, 1969: 48). I will define political development as "the capacity of a political system both to articulate people's needs and to respond to them effectively." There must be a balance between the demand for political resources and the supply of them, but this balance cannot be achieved by the choking off of effective demand.

Scholars have differed on the conditions which they identify as the prime correlates or determinants of political development (for a helpful summary of diverse views on the subject, see Packenham, 1964). One common view is the *economic* approach, which treats political development as primarily a function of a level of economic development sufficient to serve the material needs of the people and to enhance a reasonable harmony between economic aspirations and satisfactions (see Packenham, 1964: 110-113; see Lipset, 1959: ch. 2).[4] If a man's material needs are satisfied, argue the adherents of this approach, he can spend the time needed to participate in politics—and he may have financial resources with which to participate. On the supply side of the political development definition, economic development can provide a government with the wherewithal to meet increased demands.

Another group of writers has taken the *administrative* approach, arguing that political development is primarily a function of the administrative capacity to maintain law and order efficiently and to perform governmental output functions rationally and neutrally (see Packenham, 1964: 113-115). The case for political development in terms of advanced administration was argued persuasively by Max Weber (see Gerth and Mills, 1946: ch. 8).

Still another approach is provided by those who stress widespread social *mobilization* or *participation* in public affairs as the primary facilitator of political development (see Packenham, 1964: 115-117). Karl Deutsch has stated that modernization—increases in literacy, urbanization, exposure to mass media, and the like—has expanded the "politically relevant strata of the population," has multiplied the demands for government services, and has thus stimulated increased governmental capabilities (Deutsch, 1961).

AID DONORS AND POLITICAL DEVELOPMENT

How do aid donors view the effects of their assistance on the political development of the recipient? After interviewing fifty-four foreign aid officials, Robert A. Packenham (1966: 229) concluded:

> Our examination shows that while the declared purpose of American foreign aid is to help create "a community of free nations cooperating on matters of mutual concern, basing their political systems on consent and progressing in economic welfare and social justice," the doctrines of AID and aid administrators in other agencies indicate little explicit attention to political development.

The author found that the approach to political development relied upon most often was the economic one, and even then it was only implicit (Packenham, 1966: 211). In a memorable quotation, a Chief Planning Officer in one of the aid regions told Packenham: "You know, one thing I've never been clear about is what our fundamental policy is on the question of whether we're trying to promote democracies or not" (Packenham, 1966: 211-213).

Although many federal donors of urban aid tend to view their assistance as financially beneficial and politically neutral, my interviews with federal officials[5] and my observations of their activities in Oakland did uncover some examples of donors' concern with development of the recipient polity. They were not merely technical people, ignoring politics. Some Economic Development Administration officials espoused the economic approach to political development; if minorities had jobs and money, these officials argued, they could begin to play a larger role in the city's political system. (EDA people tended to combine this economic approach with an endorsement of group pluralism. If the blacks, the mayor, and businessmen were all given some resources, then they would come together to talk about specific programs, instead of fighting each other with empty slogans.)

Other federal officials in Oakland followed the administrative approach to political development. Department of Labor representatives persisted in encouraging the mayor to develop an efficient manpower planning unit, on the grounds that increased political authority would eventually flow to those with demonstrated administrative capacity. And a third group of donors favored the participation approach; OEO representatives and HUD sponsors of model cities talked about the crucial need to mobilize poor people to press their demands in the public arena.

THE IMPORTANCE OF POLITICAL INSTITUTIONS:
LOCAL AND FEDERAL

Although certain federal officials have given attention to the political development effects of their programs, it is not clear that the economic, administrative, and participation approaches are sufficient to go to the heart of the problem. In Oakland, even if people in the city received more money, the lack of effective groups and parties (see Pressman, 1972: 513-514) would make it difficult for the citizens to use their new financial resources in the political arena. Even if the city government's administration were made more efficient, local political leaders would not automatically move to expand the city's jurisdiction and increase its responsiveness to diverse citizen demands. And even if more poverty groups were mobilized, the amorphous nature of Oakland politics would make the transition from the federal program arena to the city's electoral arena a difficult one.

Thus, the weakness of political institutions and the absence of political leadership have combined to limit both citizens' effective demands on government and government's willingness and ability to respond. The virtual nonexistence of political groups and parties have deprived leaders of support for programs and information about citizens' preferences, and the elected leaders themselves have shown a marked tendency to avoid conflict and to contract their job jurisdiction rather than expand it. This pattern of local political behavior has meant that federal officials have lacked effective partners on the local level who could help them implement programs.

One desirable policy aim, therefore, might be the strengthening of local political institutions. This side of political development has been less thoroughly treated by social scientists. As Packenham (1966: 202) has written: "More than anything else in recent years, political development has meant participation and social mobilization. The need to harness, control, organize, and put to work this process of social mobilization has been less well perceived than has the need to get it started." One student of political development who *has* defined the process in terms of harnessing, channeling, and organizing political action is Samuel P. Huntington (1965). Huntington (1965: 393) defines political development as "the institutionalization of political organizations and procedures." Institutionalization is "the process by which organizations and procedures acquire value and stability" (Huntington, 1965: 394). Thus, for Huntington, political development takes place as societies learn "the art of associating together," a phrase he takes from De Tocqueville (Huntington, 1965: 386). Without strong political institutions, says the author, society lacks the means of defining and acting upon its common interests. Huntington's treatment of political development as the strengthening of political institutions seems well directed to the Oakland experience, in which the weakness of such institutions has been a prime reason for citizens' inability to press effective demands and the political system's inability to respond.

A particularly important institutional weakness in Oakland has been that of the city government, which has lacked both the jurisdiction and the will to take action in the policy areas of redevelopment, housing, education, manpower, and poverty programs. Ilchman and Uphoff (1969: 86) have noted the existence of a "vicious circle of weak authority" in developing countries: Governments lack control over other institutions and resources; the demand for participation in government is therefore low; and given a low demand for governmental resources, the government has trouble in getting *its* way with other actors. The authors conclude: "One thing is clear. The more skillfully the statesman uses the authority he has, the greater [nongovernmental actors'] demand will be for influence and a share in authority." Thus, it can be argued that more aggressive action on the part of government will result in more demands being made in the governmental arena—intensified attempts to win electoral office and to influence the behavior of elected officials. Because competitive electoral activity can provide a way for citizens to influence the allocation of public resources, and also offer incentives for public actors to build constituencies and join in framing common programs, it is worth searching for policies that make electoral activity more likely. If federal policies can help to increase the worth and the scope of local government, this might encourage electoral attempts to take over that government.

An increase in the power of local government does not, of course, equal political development in the local system. After all, definitions of political development have usually included the capacity of a political system to articulate and respond to a wide range of citizen demands. And a frequently expressed concern of federal urban policy makers (at least until recently) has been that the responsiveness of local government to the demands of citizens—particularly poor people and minorities—should be increased. Thus, federal policy might well be directed in ways which would encourage local governments to be both stronger and more open.

Strengthening elected officials and making local government more responsive to the needs of poor people are not mutually exclusive policies. As Duane Lockard (1970: 9) has stated, the dispersal of local power to nonelected bureaucracies has made it extremely difficult for poor and black people to influence policy outcomes.

> The dispersal of power to the housing bureaucracy, urban renewal authorities, the police department, and welfare agencies have done much to bring to climax the conditions of the large city. If black people are the victims of these scattered powerful bodies, and they assuredly are in city after city, and there is little means to assert contol over these agencies (also true), then it follows that one factor making for the explosiveness of the present situation is dispersal of power.

The development of strong and open local institutions would make a city's political system better able to articulate and respond to its citizens' demands, as well as providing potential sources of support for federal programs. If federal policy is to be concerned with local political development, then it must go beyond economic, administrative, and participation approaches to build and strengthen political institutions themselves.

Thus far, the discussion has focused on political weakness and development on the local level. But it would be inaccurate for federal officials to assume that local inadequacies constituted the only obstacles to the successful completion of federal programs. Indeed, as mayors often complain, the federal aid system itself is characterized by a lack of effective arenas in which the exchange of political resources can take place. Just as there is a need at the local level to develop the capacity to respond to citizens' needs, there is a need in the federal system for institutions and processes which will respond to the concerns of local officials.

TOWARD FUTURE POLICY

FEDERAL AND LOCAL ASSESSMENTS

Because a local political system can have a strong effect on the outcomes of federal programs, it behooves federal officials to assess early the capability of local political leaders and institutions. If local political officials lack the power to influence other institutions and people, then they will be of little value as "partners" in federal programs.

The strength of local institutions varies from city to city, and federal policy ought to take that into account. In the mid-1960s, the federal Economic Development Administration expected local officials in both Chicago and Oakland to "lean on" local recipients of EDA funds in order to convince them to hire unemployed minorities. In Chicago, strong mayoral leadership produced compliance on the part of the recipient; in Oakland, there was no political leader or group who could produce the same result.

Federal agencies ought not to expect that all local governments and leaders are potentially helpful vehicles for their purposes. In cities where such is not the case, measures might be taken to increase institutional capacity.

For their part, local officials should try to assess the likelihood of a given federal program's actual arrival and completion. Before entering into planning exercises and telling local citizens about potential federal funding, city leaders should try to get commitments both on money and on federal assistance in the implementation stage. The cycle of bright promises, raised expectations, and ultimate disappointment has been repeated too often.

BUILDING INSTITUTIONS

There are a number of ways in which future policy could be designed to increase the strength and responsiveness of both local governments and the federal system itself. First, the federal government could help city governments to increase their financial resources by providing increased credits toward federal income tax for the local taxes a citizen had paid. This would not be a policy of providing funds to cities on the basis of population criteria, with the hope that the money would then stimulate local government to increase the scope of its activities. Rather, the tax-credit system would provide incentives for local governments to raise their own revenue. Local revenue increases would then be tied to local effort, with the federal government providing an assist.

Second, federal agencies could tie funding to local actions which would expand the authority, capacity, and openness of local government. During 1970 and 1971, for example, the HUD area office with responsibility for Oakland took a number of steps designed to encourage the city government to expand its authority and increase meaningful citizen participation. Funds for HUD projects were held up in late 1970 because a broadly based citizens' committee was not judged to be participating sufficiently in redevelopment decisions (Oakland Tribune, 1970). Furthermore, the HUD area director strongly urged the Oakland City Council to involve itself more deeply in the redevelopment process by requiring the council to review in detail and then pass upon Redevelopment Agency projects. And HUD returned plans for a City Center project, on the grounds that city government leaders had not had adequate opportunity to examine them first. (It is no coincidence that the HUD area director who made these decisions was formerly the administrative assistant to the mayor of Oakland. He had observed the performance of Oakland's elected officials at close range.) Another HUD attempt to expand city government's authority came in 1971, when the HUD area director held up $3 million in housing funds, citing the Oakland Housing Authority's administrative difficulties and its failure to provide social service programs for tenants. The HUD official urged the City Council to take on more authority in the housing field and offered federal assistance for city hall efforts to exercise greater control over the Housing Authority (Oakland Tribune, 1971). Thus, a federal agency was directly encouraging Oakland's elected leaders to abandon their habit of conflict avoidance and to act in a more aggressive and more public manner.

Besides making city government stronger, federal policy can also be directed toward making that government more open to the policy demands of a wide range of citizens—especially the poor, who are most in need of public social services. In the past, participants in Oakland's federally funded poverty programs have found it difficult to make an effective demand on the city's governmental resources. Rather than merely encouraging participation for its own sake, federal policy ought to devise ways of ensuring that participation will mean something in terms of ongoing programs and the allocation of resources. Policies can be designed which will encourage the building of links between poverty organizations and the city government. For example, in Oakland's model cities program, city government representatives and West Oakland community representatives each have a veto over program decisions. Before a project is approved, both groups might signify their assent. This "double green light" approach means that participation can have a real effect on programs.

Third, local and federal officials can enter into longer-range agreements which can both strengthen local elected officials and create effective sites for bargaining in the federal process. In late 1970, the HUD regional office in Chicago entered into a negotiated agreement with the city of Gary, Indiana. Under the terms of the agreement, HUD pledged to fund a list of projects desired by the city. In return, the city committed itself to achieve federally set objectives in a number of areas: equal opportunity employment; building of low-cost housing; modernization of codes; and improvement of city capacity. HUD made a particular effort to bolster the position of city government and of the mayor. Formerly, HUD programs had their own diverse client agencies. But under this arrangement, the city government was treated as the client and the mayor was designated its spokesman. On the HUD side, one person represented the HUD regional administrator and had decision-making power for all agency program areas (Model Cities Service Center Bulletin, 1971).

This kind of arrangement is tailored to avoid the constant battling over guidelines which so often irritates both donors and recipients. Unlike unrestricted revenue-sharing, it does tie federal funding to specific city efforts in certain fields, but the standards are set as the result of federal-city negotiation. In this situation, discussions between federal and city representatives are not merely attempts at fostering communication and understanding; the negotiations result in real commitments on programs and resources. Furthermore, federal cooperation is not limited to the goal-setting stage; the federal agency pledges itself to help the city on a continuing basis in the process of implementation. Thus, the gulf between policy formulation and implementation may begin to be closed. (Following the Gary model, HUD initiated a program of "annual arrangements" with a growing number of cities in the early 1970s. By March 1973, the federal agency had completed annual arrangements with 79 cities; National Journal, 1973.)

Of course, there are many potential roadblocks to the success of such arrangements. Federal and city officials may not agree during the negotiations on a mutually acceptable package, and autonomous local agencies may fight the mayor's attempt to increase his authority. Furthermore, what happens if one or the other side fails to live up to its promises? The difficulties of transmitting and following through on commitments have been amply illustrated by past experiences in federal-city relations, and they also apply here. Any judgment on the outcome of the annual arrangements will have to be deferred until we know how these obstacles have been dealt with or avoided, but I mention that strategy here as a way in which federal policy might be designed to strengthen local political leaders and to build a continuing framework for collaboration between federal and local actors.

PAYMENT FOR PERFORMANCE, NOT FOR PLANNING

If federal donors desire to stimulate the development of strong and responsive political institutions at the local level, they ought not merely to share revenue with cities and then hope that the additional money will make city governments more responsive to citizens' needs. Rather, they should explore ways of rewarding cities for specific achievements: raising local taxes to finance expanded services, instituting an equal employment policy for city jobs, or building a required number of low-cost housing units. Revenue-sharing should be based on a formula which rewards those city governments which are making a concrete effort to increase their capability and expand their services.

Too often, federal programs require local planning rather than evidence of local performance. EDA has required preparation of an Overall Economic Development Program before funds can be received; HUD has insisted upon preparation of a "workable program." Cities have been careful to produce the required plans, but there is no guarantee that the plans will be related to decisions involving the allocation of city resources.

Instead of treating planning as an isolated exercise, with implementation either forgotten or thrown in at the end, it would be more useful to consider the planning process as including the totality of governmental decisions: in budgets, in personnel choices, in the daily operation of departments. The separation of planning from implementation, which is encouraged by requirements for "comprehensive" and quickly drawn plans, means that implementation is forgotten and the plan itself becomes the finished product. To be effective, planning must be linked to the operations of government and must provide for implementation. If federal rewards are tied to performance, then local officials have an incentive to carry out such plans that will help them improve performance; the distinction between planning and implementation is broken down. Federal assistance ought not to be given simply to help cities improve their "planning capacity"; rather, money should be available for policy analysts and staff assistants who can help city officials actually carry out projects. (In order to influence federal action, local officials might jointly demand that federal planning funds be made available for needed staff work and be related to projects that will actually materialize.)

SUMMARY

Because intergovernmental problems in the administration of federal urban programs involve more than confusion and misunderstanding, such problems are not likely to be ended by joint committees created to facilitate "communication," "planning," or "coordination." Revenue-sharing, which would appear to deal with one cause of intergovernmental friction by reducing the resource imbalance between federal and local governments, may increase other resource imbalances between have and have-not groups within the city.

Furthermore, it is far from evident that the mere infusion of financial resources would cause city governments to expand the scope of their activities and attempt to meet more citizen demands. A weak local political system, and the absence of leadership in city hall, can deprive federal officials of a local force needed to aid program implementation. And within the federal system itself, there is a shortage of arenas in which donor-recipient bargaining over political resources can take place and have a concrete effect. Therefore, it seems reasonable to urge that future programs should concern themselves with both local and federal political development. But political development—defined here as "the capacity of a political system both to articulate people's needs and to respond to them effectively"—has been treated in different ways by social scientists. Some students of the subject have followed an economic approach, while others have focused on administrative or partici- pation paths to political development. And a number of federal officials have followed one or another of these approaches in defining the goals of their programs.

But more money, increased administrative efficiency, and height- ened levels of participation may not by themselves lead local governments or intergovernmental systems to be stronger and more open. Therefore, federal policy ought to be framed with the specific intention of increasing the capacity and responsiveness of political institutions themselves. And this might be done in ways that reduce the need for constant planning exercises and administrative haggling between levels of government, which are present sources of dis- content among both donors and recipients. By building the frame- work for commitments to continuing joint action, instead of frustrating goal-setting, both federal and local officials can gain.

NOTES

1. Caiden and Wildavsky (forthcoming) point out that coordination may have various contradictory meanings: efficiency, reliability, coercion, and consent. As a guide for action, they point out, the injunction to "coordinate" is useless.

2. Such as the proposal put forth by Senator Humphrey and Representative Reuss in 1971.

3. The Oakland Project was a student-faculty research enterprise at the University of California (Berkeley), headed by Aaron Wildavsky and funded by the Urban Institute. Members of the project worked in city agencies and conducted research on city politics and policies. From 1969 to 1971, I worked in the mayor's office, specializing in federal-city relations.

4. Lipset links economic development to democracy.

5. During the summer of 1970, I interviewed thirty federal officials who had had some familiarity with Oakland programs. And in 1972, I interviewed twelve more federal agency representatives, concentrating on their views of their local counterparts.

REFERENCES

ALMOND, G. A. (1965) "A developmental approach to political systems." World Politics 17 (January): 183-214.

CAIDEN, N. and A. B. WILDAVSKY (forthcoming) A Constant Quantity of Tears: Planning and Budgeting in Poor Countries. New York: John Wiley.

DERTHICK, M. (1972) New Towns in Town: Why a Federal Program Failed. Washington, D.C.: Urban Institute.

DEUTSCH, K. W. (1961) "Social mobilization and political development." Amer. Pol. Sci. Rev. 55 (September): 493-514.

GERTH, H. H. and C. W. MILLS [eds.] (1946) From Max Weber. New York: Oxford Univ. Press.

HUNTINGTON, S. P. (1965) "Political development and political decay." World Politics 17 (April): 386-430.

ILCHMAN, W. F. and N. T. UPHOFF (1969) The Political Economy of Change. Berkeley: Univ. of California Press.

LEACH, R. H. (1970) American Federalism. New York: W. W. Norton.

LIPSET, S. M. (1959) Political Man. Garden City, N.Y.: Doubleday.

LOCKARD, D. (1970) "Value, theory, and research in state and local politics." Presented at the annual meeting of the American Political Science Association.

MAY, J. V. (1971) The Politics of City Revenue. Berkeley: Univ. of California Press.

MELTSNER, A. J. (1971) The Politics of City Revenue. Berkeley: Univ. of California Press.

Model Cities Service Center Bulletin (1971) "The Gary arrangement." Volume 2 (February/March): 4-6.

MURPHY, J. T. (1971) "Title I of ESEA: the politics of implementing federal education reform." Harvard Educ. Rev. 41.

MURPHY, R. D. (1971) Political Entrepreneurs and Urban Policy: The Formative Years of New Haven's Model Anti-Poverty Project. Lexington, Mass.: Heath Lexington.

National Journal (1973) "New federalism: annual arrangements." Volume 5 (March 3): 301 ff.

Oakland Tribune (1971) "U.S. prods council on housing." (June 18).

——— (1970) "OCCUR approved by council." (October 30).

PACKENHAM, R. A. (1966) "Political development doctrines in the American foreign aid program." World Politics 18 (January).

——— (1964) "Approaches to the study of political development." World Politics 17 (October): 108-120.

PRESSMAN, J. L. (1972) "Preconditions of mayoral leadership." Amer. Pol. Sci. Rev. 66 (June): 511-524.

——— and A. B. WILDAVSKY (1973) Implementation: How Great Expectations in Washington Are Dashed in Oakland; Or, Why It's Amazing that Federal Programs Work at All, this Being a Saga of the Economic Development Administration as Told by Two Sympathetic Observers Who Seek To Build Morals on a Foundation of Ruined Hopes. Berkeley: Univ. of California Press.

PYE, L. W. (1965) "The meaning of political development." Annals of Amer. Academy of Pol. and Social Sci. 358 (March): 4-13.

SUNDQUIST, J. L. with D. W. DAVIS (1969) Making Federalism Work. Washington, D.C.: Brookings Institution.

19

Constraints on Urban Leadership, or Why Cities Cannot Be Creatively Governed

PETER A. LUPSHA

☐ A RECENT ARTICLE IN THE *American Political Science Review* (Pressman, 1972) on the conditions of mayoral leadership opens with this quote from the Kerner Commission. "Now as never before the American city has need for the personal qualities of strong democratic leadership." This is nice rhetoric, and a fitting plea for one of the grand old myths of our political system. Without doubt, most Americans believe it. They believe they elect their politicians, particularly the local ones, to provide democratic leadership and popular responsiveness for the people of their city. But, while Bob Salisbury can write, and Charles Adrian (1972) can cite that the mayor "presides over the 'new convergence,' and, if the coalition is to succeed, he must lead it," the realities of urban leadership are often far different from our democratic myth and rhetoric.

T. S. Eliot once wrote that between the idea and the reality falls the shadow, a fact many urban politicians learn only after they have sweated blood and won an office they now have to earn. Democratic myth and political science as we "do it" is of little use to the candidate turned elected official and "urban governor." His reality, a learning situation for us all, is often that elected urban officials cannot govern even if they want to.

Nondecisions, despite the insight and good will of Bachrach and Baratz (1970), are usually not the willful choice of spiteful men protecting vested interests and the status quo. The reality of the urban political "game" is that there are usually few options, and those that do exist offer such a bad tradeoff between low benefits and high negative reinforcements that they are, or appear, rarely worth the risk.

Leadership, when it does occur, is more often a product of serendipity and "Fortuna" than of any ongoing process or structure of urban leadership or office which facilitates the making of innovative or responsive anticipatory decisions. Indeed, in retrospect, one realizes that the popular emphasis (during the halcyon city days of the sixties) on personalities as mayors—Lee of New Haven, White of Boston, Cavanaugh of Detroit, Stokes of Cleveland, Hatcher of Gary, Maier of Milwaukee, Lindsay of New York, and so on—was an ill-fitting cover over the reality we all knew but were afraid to admit: that urban leadership required something other than office. It required, as Pressman (1972) notes, a pyramiding of resources —power, staff, prestige, publicity, money, and the like—which are often hard-to-find commodities in the urban marketplace, particularly if the incumbent lacks the sex appeal and luck to make the most of his limited hand. A careful reading of Alan Talbot's (1967) study of Richard Lee's mayorship of New Haven, for example, can only lead to the conclusion that Lee was personally canny and possessed a large measure of luck. He had the intelligence to pick bright staff and the "Fortuna" to be first in line for urban renewal. This he combined with his Black Irish wit to charm his Kennedy clansmen into creating the illusion of a Model City (Powledge, 1970). The reality for Dick Lee and the people of New Haven, as for most American city dwellers, was often far different from redevelopment's press releases or the media's image via programmed tours. The reality is that most mayors, elected officials, or appointed chief executives cannot lead, for everything is stacked against their potential for leadership.

DEVELOPMENT OF PROPOSITIONS

The purpose of this paper is to attempt to develop a set of testable propositions on the limits of urban leadership. Here leadership is defined as the ability to make decisions and pass policy which innovate, anticipate, or change the issues, ideas, needs or demands of the polity's environment. Thus, it is distinguished from managership or system maintenance in which the tendency is to look backward and make decisions in light of the precedent of past action. Such a definition of leadership, I believe, is what Bachrach and Baratz (1970) are seeking, what Dahl (1961) admires in Mayor Lee, and what students of cities, from riot commissions to "new left" critics, demand from big-city mayors. The point of this essay, however, is that such leadership is impossible, given the reality of our economic, social, and political environment.

It is my belief that we have erred in our myth-building when we expect mayors, managers, and other elected city officials to lead, for the constraints and ratio of reinforcement are such that leadership is nearly impossible, and management and incremental reaction become the only possibilities. To expect elected or appointed city executives to lead and be responsive to the popular need and will is to provide expectations where there is little capability of performance. While leadership and popular responsiveness are linked in all the propositions in this study, they are two very distinct and often polar concepts. Eulau's (Wahlke et al., 1962) notion of trustee versus delegate encompasses the key distinction. But more important on the urban scene, the demands for innovative, dynamic leadership (i.e., change politics) often confront head on the typical demands passing over a mayor's desk. These are, namely, demands for jobs, a light fixed, a pothole filled, a noisy neighbor punished, teenagers removed from corners, and animal defecation removed from sidewalks. Citizens want these responses, city administrators have their list of demands, and the major social issues which press and pundits headline cry for solution. My point is that each requires a different style of leadership and a different set of leadership tactics and cues. In this paper, my stress is on popular responsiveness as defined in terms of the major social issues needing solution, and why there simply is not enough time in a day or energy in a human to cope with these issues, given the other daily demands. Change politics, to be responsive to major social demands, requires a different urban governance game than that available under the present social, economic, and political rules.

The common view for citizens and newly elected officials alike is to think that once one has office, one has the power to get things done. Most urban elected officials soon learn that office-holding is not sufficient to guarantee that one can do something. Citizens, however, rarely become acquainted with this reality, and so live with the expectation that they can select men who will lead. An elected leader, however, when taking office, often finds that the reins of power are but a rope of sand and that, in order to simply keep things running, one must make deals and compromises that one had expected to reserve for new priorities.

The person active in politics or observing the decision-making process quickly realizes that one of the key elements in successful policy formation rests with agenda-setting. He who sets the agenda sets the parameters of discussion and the limits of policy initiation. In order to be able to innovate or to attempt to initiate decisions which might anticipate issues or which would at least lengthen precrisis lead time, one must control agenda-setting. But this kind of control is usually out of the hands of newly elected officials and often falls into experienced ones only after several terms in office. We shall shortly see where agenda-setting lies, but for now it is important to note that it is frequently beyond the grasp of those elected officials who are believed to be responsible for policy formation.

Reaction and reactive politics are the essence of urban decision-making today and a key aspect of maintaining the urban crisis. But, given the structure, process, and daily operation of urban government, crisis and crisis reaction often provide an urban chief executive with one of his few opportunities for going beyond the system and attempting to achieve change. As Kevin White, mayor of Boston, once noted, "I hate these constant crises, but without them would we get anything done?"

The reasons for reactive rather than anticipatory policy formation at the local level are manifold and will be developed in the propositions that follow. One must note, however, that, in the maze that comprises the urban governmental system, no chief executive acts alone. A hedge of management surrounds his every turn. Routine actions are generally given positive reinforcement by middle-level managers for whom routine is the preferred operating position. Change politics or innovation, however, is usually heartily decried. Herbert Kaufman (1972) notes this middle management aspect of organization and the ways in which the administrative process can be used to constrain leadership and promote non-decision-making. Typical of the tools available to urban bureaucracy are: (a) controlling information, (b) taking agenda-setting initiative,

(c) protecting existing practice, and (d) selective administrative enforcement. Through these means and others, the "middle managers" in an urban system can choose to overload executive leadership with routine, thus constraining the time and opportunity structure available for new directions, or can provide strong negative reinforcement for innovation actions and policy risk-taking which threatens routine. At the same time, such middle management provides constant positive reinforcement and psychic inducement for the management of routine. Thus, it is not surprising that, over time, the bright elected executive) like a "bright" Skinner pigeon—learns to avoid the "shocks" of risk-taking in favor of the rewards of nondecision and develops a predisposition for policy avoidance.

- Proposition one: That local elected and chief executive leadership is constrained in its scope of leadership and popular responsiveness by department heads and agency officials whose claimed expertise and positional longevity provide them with "veto" potential as well as a strong position in agenda-setting.

- Proposition two: That local elected and chief executive leadership is constrained in its scope of leadership and popular responsiveness by middle-level bureaucrats in departments and agencies who—often protected by Civil Service tenure—can act to "veto" policy by controlling information and access and by nonimplementation or enforcement of policy.

- Proposition three: That local elected and chief executive leadership is constrained in its scope of leadership and popular responsiveness by middle-level bureaucrats in departments and agencies who may selectively overenforce certain policies as a tactical weapon to "create" harassment and pressure on the chief executive from other constituencies and thus maintain the agency's "veto" control over policy outputs.

Though the actions or inaction of the operational managers of the city can wear an executive down and predispose him to risk avoidance, the two critical limitations on policy innovation and anticipatory decision-making in the typical American city are the budgetary cycle and the electoral cycle. It is to these cycles and their interaction that we must give our attention.

When a citizen votes his candidate for mayor or city councilman into office or sees a new city manager appointed, he—and often the new role incumbents themselves—expect that something is going to happen. A new mayor or council slate, while often selected for their personality attributes and group contacts, normally also possesses a platform of policy proposals or agenda they would like to see enacted. Similarly, a new city manager is not only appointed for his background and past administrative performance, but also with an eye to how his policy world view fits the city's conception of its goals and needs. For better or worse, it is expected that things are going to change. But, more often than not, they do not. In order to understand why they do not, it is necessary to recognize and examine the lead and lag time essential to the process of urban governance.

New officials rarely enter a brand new setting; rather, they come to an ongoing system and must initiate their tenure on the legacy of their predecessors. In particular, they must live with their predecessors' budget and the budgetary cycle. Some cities, particularly the larger metropolises, operate with a fiscal calendar year (July 1-June 30) and use zero-base budgeting, some type of performance budgeting, or any one of a number of variants of Planned Programmed Budgeting Systems (PPBS) which allow for greater executive oversight—and thus leadership—in monitoring the performance of departments and agencies.

> ● Proposition four: That local elected and chief executive leadership is
> constrained in its scope of leadership and popular responsiveness by its
> ability, or lack of it, in monitoring, controlling, or having oversight over
> the ongoing performance of city departments and agencies.

Most cities, however, do not have monitoring or performance controls on operations. Indeed, most cities still operate with line-item budgets where there is no internal way to measure performance, and many cities still use a standard calendar year (January 1-December 31) in budgetary planning. The difficulty with the standard calendar year in budgetary planning is that it normally requires anticipatory borrowing to fund policies and programs because revenue will not be forthcoming until later in the year. Thus, decision makers are constrained by this added uncertainty and are more likely to be predisposed to accepting less innovative incremental funding—"five percent over last year for all departments, and we will fiddle for whatever else is needed."

> ● Proposition five: That local elected and chief executive leadership is
> constrained in its scope of leadership and popular responsiveness by

simple structural constraints like the use of a calendar year budget which increases uncertainty and narrows the scope of the decisional agenda.

- Proposition 5A: That local elected and chief executive leadership is constrained in its scope of leadership and popular responsiveness by simple structural constraints like the use of line-item budgets which focus attention on past allocations and narrow the scope of the decisional agenda.

Moving beyond the impact of these simple structural constraints, a more critical and telling constraint on leadership is the lead-time involved in the budgetary cycle. Before a city council arrives at the point where it votes the next year's budget up or down and passes it on to the mayor for his final approval or disapproval, a large number of intermediary steps and months have passed. Since these steps and time-lags are often-overlooked aspects of the decision-making process, it is worth examining them in some detail.

If a city is operating on a fiscal year budget cycle, city council hearings, committee reports, and final budgetary approval are normally planned for late April or early May. In terms of steps in the budgetary process, however, this generally means that departments and administrative agencies have to get started on planning their funding requests in October and November of the previous year. These requests are then submitted to the city's chief executive officer in late December or early January.

Depending on those serendipities of leadership (personality, elite membership and friendship, personal energy, federal government connections, personal policy goals) and the structural supports possessed by the city's chief executive (staff, tax base, partisan support, veto power, appointment and removal power, and so on), much or little time will be spent by the city's chief executive officer in meetings with department and agency officials on their budget requests. If the mayor or manager feels he has the trust of agency heads or the political clout or statutory veto power to bring them into line, he may attempt to exert personal leadership. But, as the department heads bring their requests to the mayor, they have initial control over budgetary agenda-setting. While the mayor may be able to shape departmental needs to his leadership desires, the tradeoffs are likely to be on the margin, for the department heads have past budgetary allocations, "earmarked" allocations, replacement needs, and agenda-setting initiative weighing on their side of the decisional equation. The mayor can "plea bargain" for his position, but, given constraint recognition on his part and the realization that there are a myriad of other departments to meet with, the individual budgetary compromises are likely to favor routine over policy innovation.

- Proposition six: That local elected and chief executive leadership is constrained in its scope of leadership and popular responsiveness, in the early stages of the budgetary cycle, by (a) having to earmark large portions of city resources to the ongoing "housekeeping" functions of city departments, (b) by being placed in a position of having to respond to an agenda initially set by department heads. Because the mayor is placed in a responding rather than initiating position, the scope of executive leadership is constrained, for the mayor must engage in "one-on-one" tradeoffs, informally, before his overall place in the budgetary cycle has been defined.

If the structural and environmental setting is such that the city's chief executive officer can exert some policy leadership over the departmental budgetary agenda-setting, he will. After departmental conferences, staff meetings, and compromises, he will pass on what is now labeled "his" budget to the city's legislative body or some committee of that body, for their perusal, hearings, meetings, and approval.

The March through April meetings of the city council, however, provide the department and agency heads with a second opportunity to promote their positions—a forum where they can underwrite or undermine the tacit agreements and understandings arrived at in those late winter meetings in the mayor's office. These legislative budgetary sessions also provide an opportunity for local legislative officials—particularly the ambitious—to demand changes or create alternatives which support or embarrass the administration.

- Proposition seven: That local chief executive leadership is constrained in its scope of leadership and popular responsiveness by the extent to which it lacks the political clout to reward or punish other elected officials, particular city legislators, their constituents, allies, or supporters.

- Proposition eight: That local chief executive leadership is constrained in its scope of leadership and popular responsiveness by the extent to which it lacks resources (fiscal, administrative, or political) with which to reward or punish department heads or agency administrators, their staffs, clients, or supporters.

One of the real dilemmas for urban leaders is how to get a power handle over those items which take the largest traditional chunk of their budget—public safety and education. Police have in most cities been professionalized to the point that they can act as an effective lobby, threat, and veto group, though the chief executive has a number of rewards and punishments he can use to try to check police department heads and staff. Another vital area, since it is usually the largest item in a city's budget, is public education. In this area, one readily see proposition eight in action. Public education has traditionally been removed from politics in America. Mayors and other urban chief executives, therefore, while they pay educational costs out of their budgets, rarely have control over how educational funds are expended or schools administered. School boards, whether elected or appointed, generally operate as semi-autonomous bodies. To many mayors, this is a good political situation since it keeps them directly out of the fire of this normally intense community concern. In this position, these mayors feel they can act as a court of last resort, should educational issues get out of hand. Other mayors, however, particularly those faced with strong value conflicts in the community, where the school administration acts as a bastion of anachronism, or where public safety in schools because of drug traffic, riot, strikes, and racial conflict is a prime issue, often wish they had greater control over educational resources and administration. In these latter instances, one notes that the city's chief executive wants control because the public and press are making him and his administration responsible for educational conditions which he has little control over. In the former (the normality situation?), the chief executive is happy not to have control, since his office has not been tagged by the public with responsibility. Increasingly, however, as educational issues and the society's problems mingle in the schoolyard, urban chief executives brood over their inability to affect this major area of budgetary and political concern.

After the city council has measured and adjusted the budgetary policy demands in terms of feasible tax levies, the expected level of state and federal aid, and made necessary compromises in light of local demands and pressures, the city's budget is approved by that body and passed on to the mayor for final approval. Thus, the next year's budget becomes law and the framework of future policy outputs is laid.

The mayor, if he possesses veto power through which he can make specific line-item cuts, can threaten the council to acquiesce to his policy desires, accept certain policy tradeoffs, or punish councilmen by eliminating pet projects or proposals. As the mayor cannot increase what the council cuts, this veto power is, in a sense, a negative weapon. It permits a mayor to punish in the short run in hopes of making long-run gains, but most frequently and critically, it facilitates the maintenance of routine and the "status quo" by limiting expenditures to essential areas. Thus, rather than being a weapon for system leadership, veto power is normally a weapon for system maintenance. Where the mayor possesses veto power over a total category of enactments and not the item scalpel described above, this power is often simply an empty threat. To use it means blanket cutbacks, and thus it is only useful to penalize major outgroups, specific nonthreatening constituencies, or to enforce the establishment austerity of management instead of change.

- Proposition nine: That local chief executive leadership is constrained in its scope of leadership and popular responsiveness by (a) its lack of veto power, and (b) the quality of that veto power. At the same time it must be recognized that the veto is essentially a negative weapon to block threat rather than initiate change. Lacking it, however, the chief executive has one less weapon with which to threaten other power blocs into accepting his leadership.

This, then, is the budgetary cycle as it typically takes place in a large number of American cities. There are, of course, many individual city differences, depending on the area's economic and political history, state and local statutes, formal structural constraints, tax base, revenue slack, homo- or heterogeneity of class makeup, community cleavages and conflicts, media access, group activity and mobilization potential, elite structure, and the like. But while these elements add spice and heat, which in a variety of ways can change the flavor and satisfaction of the budgetary meal, the basic meat and potatoes of the cycle remain the same.

The other key cycle affecting the opportunity structure for urban leadership is the electoral cycle. Indeed, many of the central aspects of the budgetary cycle only take on their full meaning in terms of leadership constraint when connected and seen in perspective with the electoral cycle.

The electoral cycle for mayors (in this section I will perforce concentrate on mayors) in American cities occurs from every one to six years, with two- to four-year terms being the most common. In small, nonpartisan cities, such as the "new towns" of suburban

California, one-year terms are not unusual. In these cities, the mayor is no more than another councilman, appointed to serve as titular head and chief ceremonial officer of the city. In such settings, the day-to-day administration and development of policy is carried out by an appointed city manager or chief administrative officer (CAO). In our larger metropolises, particularly our older and eastern ones, mayors are directly elected, often on partisan slates for two- to four-year terms. These mayors generally possess a full, or rather full, repertoire of the instruments of power. Tenure and power configuration generally depend, however, on the history of the particular setting or state. At present, most mayors in cities over 100,000 population serve four-year terms, while a majority of mayors in cities of under 100,000 serve two-year terms.

The importance of length of term—and of course the opportunity structure for reelection—is that it sets the parameters on the policy initiation and program risks that a mayor is likely to undertake.

- Proposition ten: That local chief executive leadership is constrained in its scope of leadership and popular responsiveness by its length of term. The shorter the term, the lower the likelihood of innovative leadership.

- Proposition eleven: That local chief executive leadership is constrained in its scope of leadership and popular responsiveness by the opportunity structure for reelection. The narrower the opportunity structure, the lower the likelihood of innovative leadership.

It is easily understandable that a mayor with a one-year term is not likely to try too hard to change his urban world. Rather, he is likely to defer risk-taking in favor of council advisement and CAO agenda-setting. A mayor with a four-year mandate, however, is likely to perceive greater opportunity and pressure (internal and external) for providing leadership policy initiative and directional change in his city. One would also expect that elite groups and citizens selecting a mayor for a four-year term, as well as the political organizations within that setting, would have higher expectations about the potential of the mayor for grasping the reins of power and realizing his potential for change. Even with high personal and public expectations the reality of interaction between the electoral and budgetary cycles—even with a four-year term—drastically limits the capability for change.

In order to understand this, it is useful to run through the typical scenario of a new mayor entering office for a four-year term. Before we take this step, however, we must make some recognitions. First, smooth transitions between urban office holders, particularly in partisan situations, or when party alternation is involved, are extremely rare. Even in same-party or nonpartisan turnover, cooperation is not common. The personalities and factions involved in creating turnover often breed, in the process, antagonisms rarely ended with election. Even when the outgoing incumbent can anoint his successor, transitions, if the new mayor appears to intend to assert himself and lead, are not easy. This, of course, should not surprise anyone familiar with the operation of American government, for even at the apex of the system—the presidency—attempts at smoothing incumbent transitions are a relatively recent phenomenon.

A second important recognition, and one which affects even the smoothest transition, is that most middle management in city government, protected by tenure, expertise, and often Civil Service regulation, believe that while mayors come and go, the bureaucracy, like the earth, abides. Civil Service personnel, secure in their jobs, can, particularly with supportive supervisors, covertly choose to sabotage those aspects of a mayor's program that they find disagreeable or threatening to their position. Mayor Lindsay of New York has on more than one occasion been forced to back down from enforcing policies he supported because entrenched elements in the bureaucracy, by use of selective under- and overenforcement of existing mandates, were able to "get the mayor off their back."

What we urban analysts at times forget is that not only is the city a highly interdependent system, it is a highly overloaded one. The typical American city is in many ways like the older home fitted to support the power use and demands of the 1920s but still operating with the same system in the electric and electronic world of the 1970s. Every time something is added or there is an increased pressure at one place in the system, the entire system is jeopardized, for a pressure at one place causes disruptions elsewhere or blows a fuse. Urban leadership is thus constrained by the recognition that any emphathetic action may not just rock the structure, it may cause the disruption or collapse of some unanticipated part of it.

- Proposition twelve: That over time local elected and chief executive leadership is constrained in its scope of leadership and popular responsiveness by its recognition of system overload and the fragile interdependence of the entire system.

Another critical area of system overload is the fiscal resource base of the American city. Most cities of the nation, even with the slack provided by revenue-sharing, are fiscal disaster areas. Many of the larger, older cities totter on bankruptcy with near confiscatory property tax rates, while many of the newer cities race with time—often buying growth packages and strategies they would rather not have—in order to keep their fiscal heads above water while inundated by a fast-surging and rapidly rising tide of demand. Enough has been written of fiscal imbalance, jurisdictional fragmentation, and the suburban noose around our cities that one need say little more (Advisory Commission on Intergovernmental Relations, 1967). For our purposes, the point of interest is that urban elected officials, readily aware of this fiscal reality, are constrained from innovative risk-taking. Even in times and places of system and resource slack—a variable that deserves much more attention by our profession—the thrust of my argument is that innovate leadership and change politics will be uncommon. In times of negligible resources and almost no slack, such leadership will be rather rare.

- Proposition thirteen: That elected and chief executive leadership is constrained in the scope of its leadership and popular responsiveness by the lack of systemic slack and flexibility, particularly reflected in revenue accumulation versus distribution ratios.

FURTHER PROPOSITIONS

THE SCENARIO

A new mayor is elected for a four-year term in a contested election. He has run on a set of policy commitments which he hopes to put into effect, and which the voters apparently want, as indicated by their turnout on election day. The new mayor takes office in January, so he has two months after that chilly day of election to recruit staff, settle his current affairs, and become acquainted with the powers and process of office. As the typical short ballot usually gives the mayor a number of administrative and department appointees, much of his preinaugural time is devoted to finding the right man for the right office.

Even with a fair amount of cooperation from the outgoing mayor and his staff, the new mayor and his people enter an ongoing system in which rules, procedures, and habits are set, and the vital cycle of budgetary planning is well in motion. Thus, after the photographers and well-wishers have gone, the new mayor settles into his office and finds he confronts department heads with their budgetary agendas in hand.

As is likely, some of these men will be his new appointees faced with the task of submitting their predecessors' agendas or—without time or information—devising their own programs. In such cases, some shifts of emphasis and compromises are normally made, but since the new man must rely on long-term, tenured, or Civil Service staff for his information and support, the agency agenda remains pretty much intact. With the long-term nonappointed department officials, the new mayor faces-off his prestige and limited staff knowledge and experience against the agency official's administrative expertise and long familiarity in this arena. It is not a very even match. If the new mayor wishes, he can seek to totally revise departmental budgets to meet his priorities. But, as the new mayor and his staff have not had time to find all the power levers in the system, it is generally seen as prudent to live the first year with the old administration's budget (trimming some fat, perhaps, or adding an election priority or two—if there is slack, and if it will not rock the system—but mostly accepting the initiatives of the ongoing officials, department heads, and agencies).

- Proposition fourteen: That local elected leadership (when newly elected) is constrained in the scope of its leadership and popular responsiveness by the variance between its prime needs of the moment—staffing, power-setting, and the like—and the recognition that the initiative in budgetary agenda-setting tends to be controlled by ongoing and long-term department heads and middle management staff.

As one sees from the preceding discussion of the budgetary process, it is not until the tenth or eleventh month in office, as the planning of his second year's budget begins, that the mayor can begin to exert his control and leadership and to implement the promises of his campaign. Even then, however, severe constraints limit the opportunities for exerting leadership. To more clearly illustrate this, I shall make two fairly reasonable assumptions about most local officials.

(1) That most local elected officials desire at least a second term.

(2) That, in terms of policy proposals, most local elected officials seek policy outputs that will reflect positively on them in office and that will—as a general tendency—have an impact or reach fruition before they leave office (Lifton, 1968).

This second assumption perhaps deserves some further elaboration. It rests in part on the electoral considerations and in part on the larger observation that most humans do not care what happens to the world after they are dead. While people often have positive desires to make this world a better place for their children, few are willing to sacrifice their good life and good times for the generations to be born long after they themselves are dead. To equate the end of a public life with death may, to some, seem extreme. But, I propose—moving with Robert Lifton (1968) that political leaders tend to seek historical immortality through deeds in office—that politicians, even those at the local level, are concerned with looking good to history. They therefore are under a psychological constraint to limit actions and policies to those that will reflect well on themselves and not to engage in long-range policies—however much needed—that will only serve to shine the historical spotlight on their successors rather than on themselves. Thus, policy proposals will tend to be focused on achieving a major impact within the mayor's tenure—branding it as his.

In the second year, then, I am suggesting that the mayor will begin to be constrained from engaging in active policy leadership and innovation because of his desire for renomination and reelection, and because he desires to have his policies impact during his remaining tenure.

- Proposition fifteen: That local elected leadership is constrained in the scope of its leadership and popular responsiveness by incumbent desires to be renominated and reelected for at least a second term.

- Proposition sixteen: That local elected leadership is constrained in its scope of leadership and popular responsiveness by incumbent desires and reelection need, to have policy outputs impact within their current term office.

The stress of this last proposition is to suggest that the policy horizons of local elected officials tend to end with their terms. The typical urban officeholder's policy vision is normally focused by the time of his second budgetary year. If he is ambitious and hopes for reelection, elements of this budget and larger portions of the third-year budget will be designed to gather in a winning electoral coalition. In short, the stress of these ideas is that the mayor's policy perspective becomes increasingly foreshortened as his term progresses. What this means is that risk avoidance increases, or stated somewhat differently, willingness to engage in innovative policy-making or change politics decreases, as the incumbent's term progresses.

- Proposition seventeen: That local elected leadership is constrained in its scope of leadership and popular responsiveness by a foreshortening of policy perspectives and a reduced willingness to engage in risk-taking policy innovation as an incumbent's term progresses.

From the urban politician's perspective, risk-taking policy innovations or change politics are a potential form of strong negative reinforcement. This is because such policies often possess the likelihood of: (1) disturbing existing arrangements and coalitions; (2) upsetting routine operations and procedures; (3) threatening (in a real or perceived way) certain groups or individuals; (4) involving immediate costs with no necessary guaranteed immediate payoff; (5) awakening and arousing presently dormant interests or groups, increasing their demand expectations. From this risk versus payoff perspective alone, it is not surprising that decisional avoidance rather than leadership is the urban norm.

The local political system provides very low positive reinforcement for risk-taking decisions and behavior and very strong negative reinforcement for risk-taking failures, while at the same time providing positive reinforcements for adaptive "get along-go along" behaviors, and for following system maintenance and routine strategies. It may therefore be of limited usefulness to decry non-decision-making, for until the constraints and reinforcement rations are changed, urban leadership must remain more a function of externalities, personality, and luck than the power and operation of elective office.

CONCLUSION

In this essay, I have tried to outline some of the ways in which the structural and operational constraints of office-holding limit the opportunities for creative policy leadership. Leadership was distinguished from managership and defined as the ability to innovate, anticipate, or change the issues facing the local polity. The process of initiative or agenda-setting was examined and the ways in which the two critical cycles of the local political process—the budgetary cycle and the electoral cycle—dissected. Seventeen propositions which could be subject to empirical examination were then developed. While creative, anticipatory policy leadership is one of the most needed resources in urban politics today, the message of this essay is that reactive politics and crisis policy will in all likelihood continue to be our main avenue to creative change.

REFERENCES

ADRIAN, C. (1972) State and Local Government. New York: McGraw-Hill.

Advisory Commission on Intergovernmental Relations (1967) Fiscal Imbalance in the American System. Washington, D.C.: Government Printing Office.

BACHRACH, P. and M. BARATZ (1970) Power and Poverty. New York: Oxford Univ. Press.

DAHL, R. (1961) Who Governs? New Haven, Conn.: Yale Univ. Press.

KAUFMAN, H. (1972) The Limits of Organizational Change. Tuscaloosa: Univ. of Alabama Press.

LIFTON, R. (1968) Revolutionary Immortality. New York: Vintage.

POWLEDGE, F. (1970) Model City. New York: Simon & Schuster.

PRESSMAN, J. (1972) "Preconditions of mayoral leadership." Amer. Pol. Sci. Rev. (June): 511-524.

TABLOT, A. (1967) The Mayor's Game. New York: Harper & Row.

WAHLKE, J., H. EULAU et al. (1962) The Legislative System. New York: John Wiley.

BIBLIOGRAPHY

PART I: ON KNOWING WHEN THINGS ARE GETTING BETTER

BLAIR, L. H. and A. I. SCHWARTZ (1972) How Clean Is Our City? A Guide for Measuring the Effectiveness of Solid Waste Collection Activities. Washington, D.C.: Urban Institute.

BOOTS, A., III, G. DAWSON, W. SILVERMAN, and H. P. HATRY (1972) Inequality in Local Government Services: A Case Study of Neighborhood Roads. Washington, D.C.: Urban Institute.

BURT, M. R. and L. H. BLAIR (1971) Options for Improving the Care of Neglected and Dependent Children: Nashville-Davidson, Tennessee. Washington, D.C.: Urban Institute.

Council of State Governments (1973) State Government Program Evaluation Activities. Lexington, Kentucky.

DEUTSCH, K. W. (1966) "On theories, taxonomies, and models as communication codes for organizing information." Behavioral Sci. 11 (January).

GOULD, P. R. (1970) "A prototype office of human statistics: context, strategy, and recommendations." RAND Corporation report P-4439.

HATRY, H. P. and D. R. DUNN (1971) Measuring the Effectiveness of Local Government Services: Recreation. Washington, D.C.: Urban Institute.

International City Management Association (1973) "Achieving quality local government." Public Management (September): 19-23.

――― (1972) "Local government budgeting, program planning, and evaluation." Urban Data Service Report, May.

――― (1970) "Measuring effectiveness of municipal services." Management Information Service, August.

KLAGES, H. (1973) "Assessment of an attempt at a system of social indicators." Policy Sciences 4 (September): 249-261.

MORRISON, P. A. (1971) "Demographic information for cities." RAND Corporation Paper R-618-HUD.

National Commission on Productivity (1972) The Challenge of Productivity Diversity: Improving Local Government Productivity Measurement and Evaluation. Washington, D.C.

――― (1971) Improving Productivity and Productivity Measurement in Local Governments. Washington, D.C.

ORCUTT, G. H. et al. (1968) "Data aggregation and information loss." Amer. Econ. Rev. 68 (September): 773-787.

OSTROM, E. (1973) "The need for multiple indicators in measuring the output of public agencies." Policy Studies J. (December).

――― (1971) "Institutional arrangements and the measurement of policy consequences in urban areas." Urban Affairs Q. 6: 447-475.

RIDLEY, C. E. and H. A. SIMON (1938) Measuring Municipal Activities: A Survey of Suggested Criteria for Appraising Administration. Chicago: International City Managers' Association.

SHELDON, E. B. and K. C. LAND (1972) "Social reporting for the 1970's: a review and programmatic assessment." Policy Sciences 3 (July): 137-151.
U.S. Department of the Interior, Bureau of Outdoor Recreation (1973) How Effective Are Your Community Recreation Services? Washington, D.C.: Government Printing Office.
U.S. General Accounting Office, Office of the Comptroller General (1972) Standards for Audit of Governmental Organizations, Programs, Activities and Functions. Washington, D.C.: Government Printing Office.
WEBB, K. and H. P. HATRY (1973) Obtaining Citizen Feedback: The Application of Citizen Surveys to Local Governments. Washington, D.C.: Urban Institute.
WILDAVSKY, A. (1972) "The self-evaluating organization." Public Administration Rev. (September/October): 509-520.
WINNIE, R. E. and H. P. HATRY (1972) Measuring the Effectiveness of Local Government services: recreation. Washington, D.C.: Urban Institute.
YOUNG, D. (1972) How Shall We Collect the Garbage? Washington, D.C.: Urban Institute.

PART II: TOWARD A MORE SOPHISTICATED URBAN MANAGEMENT

ABRAMOVITZ, M. et al. (1959) The Allocation of Economic Resources. Stanford: Stanford Univ. Press.
BERNSTEIN, S. et al. (1973) "The problems and pitfalls of quantitative methods in urban analysis." Policy Sciences 4 (March): 29-39.
BREWER, G. D. (1973) "On innovation, social change, and reality." Technological Forecasting and Social Change 5: 19-24.
——— (1973) "Documentation: an overview and design strategy." RAND Corporation Paper P-5052.
——— (1973) Politicians, Bureaucrats, and the Consultant: A Critique of Urban Problem-Solving. New York: Basic Books.
BRUNNER, R. D. and G. D. BREWER (1971) Organized Complexity: Empirical Theories of Political Development. New York: Free Press.
BURENSTAM LINDER, S. (1970) The Harried Leisure Class. New York: Columbia Univ. Press.
CAIDEN, N. and A. WILDAVSKY (forthcoming) A Constant Quality of Tears: Planning and Budgeting in Poor Countries. New York: John Wiley.
COOPER, W. [ed.] (1972) "Urban issues II." Management Sci. 19 (December).
CORNOG, G. et al. (1968) EDP Systems in Public Management. Chicago: Rand McNally.
CRECINE, J. P. [ed.] (1969) Governmental Problem Solving. Chicago: Rand McNally.
DRAKE, A. W. et al. [eds.] (1972) Analysis of Public Systems. Cambridge, Mass.: MIT Press.
DYCKMAN, J. W. (1963) "The scientific world of city planners." Amer. Behavioral Scientist 6 (February).
FITE, H. (1965) Computer Challenge to Urban Planners and State Administrators. Washington, D.C.: Spartan.
GERWIN, D. [ed.] (1969) Budgeting Public Funds. Madison: Univ. of Wisconsin Press.
GILMORE, J. et al. (1967) Defense Systems Research in the Civil Sector. Washington, D.C.: Arms Control and Disarmament Agency.
GREENBERG, S. B. [ed.] (1974) Politics and Poverty, Modernization and Response in Five Poor Neighborhoods. New York: John Wiley.
HAIMES, Y. Y. and D. MACKO (1970) "Hierarchical structures in water resources systems management." IEEE Transactions on Systems, Man, and Cybernetics 3.
HINRICHS, H. H. and G. M. TAYLOR (1969) Program Budgeting and Benefit-Cost Analysis. Pacific Palisades, Calif.: Goodyear.
JAMES, W. (1968 [1892]) The Writings of William James. New York: Modern Library.
KAPLAN, A. (1973) "Values, norms, and policies." Policy Sciences 4 (March): 103-111.
——— (1963) American Ethics and Public Policy. New York: Oxford Univ. Press.
KUENZLEN, M. (1972) Playing Urban Games. New York: George Braziller.
LA PORTE, T. R. [ed.] (forthcoming) Organized Social Complexity: Challenge to Politics and Policy. Princeton: Princeton Univ. Press.
LASSWELL, H. D. (1971) A Pre-View of Policy Sciences. New York: American Elsevier.

LEVIN, M. A. and H. D. DORNBUSCH (1973) "Pure and policy social science." Public Policy 21 (Summer): 383-423.

LEVINE, R. A. (1968) "Rethinking out social strategies." Public Interest (Winter): 88-92.

LINDBLOM, C. E. (1959) "The science of 'muddling through'." Public Administration Rev. 19 (Spring): 79-88.

LYDEN, F. J. and E. G. MILLER [eds.] (1967) Planning, Programming, Budgeting: A Systems Approach to Management. Chicago: Markham.

MARNEY, M. and N. M. SMITH (1972) "Interdisciplinary synthesis." Policy Sciences 3 (September): 299-323.

MICHELSON, W. (1965) "Most people don't want what architects want." Trans-action (July/August): 37-43.

MITCHELL, W. H. (1966) "SOGAMMIS—a systems approach to city administration." Public Automation (April): 1-4.

MUSHKIN, S. (1970) "PPB for the cities: problems and the next steps," chapter 10 in J. P. Crecine (ed.) Financing the Metropolis: Public Policy in Urban Economies. Beverly Hills: Sage Pubns.

NAY, J. N. et al. (1971) "Benefits and costs of manpower training programs: a synthesis of previous studies with reservations and recommendations." Urban Institute Paper 2400-1, June.

NOVICK, D. [ed.] (1965) Program Budgeting. Cambridge, Mass.: Harvard Univ. Press.

——— (1954) "Mathematics: logic, quantity, and method." Rev. of Economics and Statistics (November).

Public Automated Systems Service (1966) Automation in the Public Service: An Annotated Bibliography. Chicago.

QUADE, E. S. and W. I. BOUCHER [eds.] (1968) Systems Analysis and Policy Planning. New York: American Elsevier.

RAIFFA, H. (1968) Decision Analysis: Introductory Lectures on Choices Under Uncertainty. Reading, Mass.: Addison-Wesley.

RAWLS, J. (1971) A Theory of Justice. Cambridge, Mass.: Harvard Univ. Press.

RIVLIN, A. (1971) Systematic Thinking for Social Action. Washington, D.C.: Brookings Institution.

ROSENBLOOM, R. S. and J. R. RUSSELL (1971) New Tools for Urban Management. Cambridge, Mass.: Harvard Univ. Press.

SCHLAIFER, R. (1969) Analysis of Decisions Under Uncertainty. New York: McGraw-Hill.

SHUBIK, M. and G. D. BREWER (1972) "Models, simulations and games: a survey." RAND Corporation Paper R-1060-ARPA/RC.

SIMON, H. A. (1967) "The architecture of complexity," pp. 63-76 in L. von Bertalanffy and A. Rapoport (eds.) General Systems. New York: George Braziller.

——— (1957) Models of Man. New York: John Wiley.

STRAUCH, R. A. (forthcoming) "A critical assessment of qualitative methodology as a policy analysis tool." RAND Corporation Paper R-1423-PR/ARPA.

SZANTON, P. L. (1972) "Analysis and urban government: experiences of the New York City-Rand Institute." Policy Sciences 3 (July): 153-161.

Teachers College Record (1971) "Education vouchers: peril or panacea?" Volume 72 (February).

WAGNER, H. M. (1971) "The ABC's of OR." Operations Research 19: 1259-1281.

PART III: DECENTRALIZATION

ABERBACH, J. D. and J. L. WALKER (1970) "The attitudes of blacks and whites toward city services: implications for public policy," in J. P. Crecine (ed.) Financing the Metropolis: Public Policy in Urban Economies. Beverly Hills: Sage Pubns.

ALTSHULER, A. A. (1970) Community Control: The Black Demand for Participation in Large American Cities. New York: Pegasus.

ARNSTEIN, S. R. (1969) "A ladder of citizen participation." J. of Amer. Institute of Planners 35: 216-232.

Association of the Bar of the City of New York (1972) Decentralizing New York City Government. New York: Praeger.

BABCOCK, R. F. and F. BOSSELMAN (1967) "Citizen participation: a suburban suggestion for the central city." J. of Law and Contemporary Problems 32: 220-231.

BACHRACH, P. and M. BARATZ (1970) Power and Poverty. New York: Oxford Univ. Press.

BACON, E. (1968) "American houses and neighborhoods, city and county." Annals of the Amer. Academy of Pol. and Social Sci. (July): 117-129.

BELL, D. and V. HELD (1969) "The community revolution." Public Interest 16 (Summer).

BERGER, C. J. (1968) "Law, justice and the poor," in R. H. Connery (ed.) Urban Riots: Violence and Social Change. New York: Random House.

BREWER, G. D. (1973) "Professionalism: the need for standards." Interfaces 46 (November): 20-27.

CALLAHAN, J. and D. E. SHALALA (1969) "Some fiscal dimensions of three hypothetical decentralization plans." Education and Urban Society 2 (November): 40-53.

Center for Governmental Studies (1970) Public Administration and Neighborhood Control. Washington, D.C.

CLARK, T. M. (1970) "On decentralization." Polity 2: 508-514.

COSTIKYAN, E. N. and M. LEHMANN (1972) Re-Structuring the Government of New York City. New York: Praeger.

DAVIES, C. J. (1966) Neighborhood Groups and Urban Renewal. New York: Columbia Univ. Press.

ENNIS, P. H. (1967) Criminal Victimization in the United States: A Report of a National Survey. Washington, D.C.: Government Printing Office.

ERIE, S. P., J. J. KIRLIN, and F. F. RABINOVITZ (1972) "Can something be done? Propositions on the performance of metropolitan institutions," in L. Wingo (ed.) Reform of Metropolitan Governments. Washington, D.C.: Resources for the Future.

FITCH, L. C. and A. H. WALSH [eds.] (1970) Agenda for a City: Issues Confronting New York. Beverly Hills: Sage Pubns.

FREDERICKSON, H. G. [ed.] (1972) "Curriculum essays on citizens, politics and administration in urban neighborhoods." Public Administration Rev. 32 (October).

GITTELL, M. (1968) "Community control of education," pp. 63-75 in R. A. Connery (ed.) Urban Riots: Violence and Social Change. New York: Random House.

GREENSTONE, J. D. et al. (1973) Race and Authority in Urban Politics: Community Participation and the War on Poverty. New York: Russell Sage.

HAHN, H. (1971) "Local variations in urban law enforcement," in P. Orleans and W. R. Ellis (eds.) Race, Change and Urban Society. Beverly Hills: Sage Pubns.

HALLMAN, H. W. (1971) Administrative Decentralization and Citizen Control. Washington, D.C.: Center for Governmental Studies.

ITZKOFF, S. W. (1969) "Decentralization: dialectic and dilemma." Educ. Forum 34: 63-69.

JACOB, H. (1970) "Black and white perceptions of justice in the city." Law and Society Rev. 6: 69-90.

KAUFMAN, H. (1969) "Administrative decentralization and political power." Public Admin. Rev. 29 (January/February).

KRISTOL, I. (1968) "Decentralization for what?" Public Interest 11 (Spring): 17-25.

LARSON, R. C. [ed.] (1972) Urban Police Patrol Analysis. Cambridge, Mass.: MIT Press.

LEVY, B. (1968) "Cops in the ghetto: a problem of the police system," pp. 347-358 in L. Masotti and D. Bowen (eds.) Riots and Rebellion: Civil Violence in the Urban Community. Beverly Hills: Sage Pubns.

LIEBERSON, A. and A. R. SILVERMAN (1965) "The precipitants and underlying conditions of race riots." Amer. Soc. Rev. 31 (December): 887-898.

LIPSKY, M. (1970) Protest in City Politics. Chicago: Rand McNally.

--- and D. J. OLSON (1974) Riot Commission Politics: The Processing of Racial Crisis in America. New York: Dutton.

MARSHALL, D. R. (1971) The Politics of Participation in Poverty: A Case Study of the Board of the Economic and Youth Opportunities Agency of Greater Los Angeles. Berkeley: Univ. of California Press.

--- (1971) "Public participation and the politics of poverty," pp. 451-483 in P. Orleans and W. R. Ellis (eds.) Race, Change and Urban Society. Beverly Hills: Sage Pubns.

MAY, J. V. (1971) "Two model cities: negotiations in Oakland." Politics and Society (Fall): 57-88.

MAYER, A. (1971) "A new level of local government is struggling to be born." City (March/April): 60-64.

MELTZER, J. (1968) "A new look at the urban revolt." J. of the Amer. Institute of Planners 34 (December): 255-259.

MOGULOF, M. (1969) "Coalition to adversary: citizen participation in three federal programs." J. of the Amer. Institute of Planners 35: 225-232.

MURPHY, R. D. (1971) Political Entrepreneurs and Urban Poverty: The Formative Years of New Haven's Model Anti-Poverty Project. Lexington, Mass.: D. C. Heath.

ORBELL, J. M. and T. UNO (1972) "A theory of neighborhood problem solving: political actions vs. residential mobility." Amer. Pol. Sci. Rev. 66: 471-489.

OSTROM, E., R. PARKS, and G. WHITAKER (1973) "Do we really want to consolidate urban police forces? A reexamination of some old assertions." Public Administration Rev. (September/October).

OSTROM, E., W. BAUGH, R. GUARASCI, R. PARKS, and G. WHITAKER (1973) Community Organization and the Provision of Police Services. Beverly Hills: Sage Pubns.

PARENTI, M. (1970) "Power and pluralism: a view from the bottom." J. of Politics 32: 501-530.

PIVEN, F. F. and R. A. CLOWARD (1967) "Black control of the cities: heading it off by metropolitan government." New Republic (September 30): 19-21; (October 7): 15-19.

PRESS, C. (1963) "The cities within a great city: a decentralist approach to centralization." Centennial Rev. 7: 113-130.

RHODES, G. [ed.] (1973) The New Government of London: The First Five Years. London: Weidenfeld & Nicolson.

Royal Commission on Local Government in England (1968) The Lessons of the London Government Reforms. London: HMSO.

SHALALA, D. E. [ed.] (1971) Neighborhood Governance: Proposals and Issues. New York: Institute of Human Relations Press.

SMITH, G. A., Jr. (1958) Managing Geographically Decentralized Companies. Cambridge, Mass.: Harvard Business School.

SPIEGEL, H.V.C. (n.d.) Citizen Participation in Urban Development. Washington, D.C.: National Training Laboratory Institute for Applied Behavioral Science.

STRANGE, J. H. [ed.] (1972) "Citizens action in model cities and CAP programs: case studies and evaluation." Public Administration Rev. 32 (September).

TenHOUTEN, W., J. STERN, and D. TenHOUTEN (1971) "Political leadership in poor communities: applications of two sampling methodologies," pp. 215-254 in P. Orleans and W. R. Ellis (eds.) Race, Change and Urban Society. Beverly Hills: Sage Pubns.

TIEBOUT, C. M. (1956) "The pure theory of local expenditure." J. of Pol. Economy 64: 416-424.

WARREN, R. (1966) Government in Metropolitan Regions. Davis: University of California Institute of Governmental Affairs.

WASHNIS, G. (1972) Municipal Decentralization and Neighborhood Resources. New York: Praeger.

WILCOX, H. G. (1969) "Hierarchy, human nature, and participative panaceas." Public Administration Rev. (January/February).

WILSON, J. Q. (1968) Varieties of Police Behavior. Cambridge, Mass.: Harvard Univ. Press.

PART IV: REORGANIZATION

AIKEN, M. and J. HAGE (1966) "Organizational alienation: a comparative analysis." Amer. Soc. Rev. 31, 4: 497-507.

ANDERSON, J. G. (1968) Bureaucracy in Education. Baltimore: Johns Hopkins Press.

ARGYRIS, C. (1972) The Applicability of Organizational Sociology. Cambridge, Eng.: Cambridge Univ. Press.

——— (1970) Intervention Theory and Method. Reading, Mass.: Addison-Wesley.

——— (1968) "On the effectiveness of research and development organizations." Amer. Scientist 56, 4: 344-355.

——— (1968) "Organizations: effectiveness," in Volume 11 of International Encyclopedia of the Social Sciences. New York: Macmillan and Free Press.

——— (1967) "Today's problems with tomorrow's organizations." J. of Management Studies 4 (February): 31-55.

——— (1965) Organization and Innovation. Homewood, Ill.: Richard D. Irwin.

——— (1964) Integrating the Organization and the Individual. New York: John Wiley.

——— (1960) Understanding Organizational Behavior. Homewood, Ill.: Dorsey.

BACHMAN, J. G. and A. S. TANNENBAUM (1968) "The control-satisfaction relationship across varied areas of experience," pp. 241-249 in A. Tannenbaum (ed.) Control in Organizations. New York: McGraw-Hill.

BARNARD, C. (1938) The Functions of the Executive. Cambridge, Mass.: Harvard Univ. Press.

BASS, B. M. (1960) Leadership, Psychology and Organizational Behavior. New York: Harper.

BENDIX, R. (1968) "Bureaucracy," pp. 206-219 of Volume 2 of International Encyclopedia of the Social Sciences. New York: Macmillan and Free Press.

BENNIS, W. G. (1966) Changing Organizations. New York: McGraw-Hill.

——— and P. E. SLATER (1968) The Temporary Society. New York: Harper & Row.

BERLEW, D. E. and D. T. HALL (1964) "The management of tension in organizations: some preliminary findings." Industrial Management Rev. 6: 31-40.

BERLINER, J. (1967) "Russia's bureaucrats: why they're reactionary." Trans-action: 53-59.

BEVAN, W. et al. (1958) "Jury behavior as a function of the prestige of the foreman and the nature of his leadership." J. of Public Law 7: 419-449.

BLAU, P. (1955) The Dynamics of Bureaucracy. Chicago: Univ. of Chicago Press.

BRAEGER, G. (1969) "Commitment and conflict in a normative organization." Amer. Soc. Rev. 34 (August): 492-504.

BREWER, J. (1971) "Flow of communications, expert qualifications and organizational authority structures." Amer. Soc. Rev. 36: 475-484.

BROWN, W. (1960) Exploration in Management. New York: John Wiley.

CANGELOSI, V. E. and W. R. DILL (1965) "Organizational learning: observations toward a theory." Administrative Sci. Q. 10: 175-203.

CARPENTER, H. H. (1971) "Formal organizational structural factors and perceived job satisfaction of classroom teachers." Admin. Sci. Q. 16, 4: 460-466.

CARTWRIGHT, D. (1965) "Influence, leadership control," pp. 1-47 in J. G. March (ed.) The Handbook of Organizations. Chicago: Rand McNally.

CHILD, J. (1972) "Organization structure and strategies of control: a replication of the Aston study." Admin. Sci. Q. 17 (June): 163-177.

COCH, L. and J.R.P. FRENCH, Jr. (1948) "Overcoming resistance to change." Human Relations 1: 512-532.

CROZIER, M. (1964) The Bureaucratic Phenomenon. Chicago: Univ. of Chicago Press.

CYERT, R. M. and J. G. MARCH (1963) A Behavioral Theory of the Firm. Englewood Cliffs, N.J.: Prentice-Hall.

DAVIS, L. E. (1966) "The design of jobs." Industrial Relations 6 (October): 21-45.

——— and E. S. VALFER (1966) "Studies in supervisory job designs." Human Relations 19, 4: 339-352.

——— (1965) "Intervening responses to changes in supervisor job designs." Occupational Psychology 39, 3: 171-189.

DEUTSCH, M. (1968) "Groups: group behavior," pp. 265-276 in Volume 6 of International Encyclopedia of the Social Sciences. New York: Macmillan and Free Press.

DOWNS, A. (1966) Inside Bureaucracy. Boston: Little, Brown.

DUNCAN, R. B. (1972) "Characteristics of organizational environments." Admin. Sci. Q. 19: 313-327.

EMERY, F. E. and E. L. TRIST (1965) "The causal texture of organizational environments." Human Relations 18: 21-32.

EMERY, F. E., E. THORSRUD, and K. LANGE (1966) Field Experiments at Christiana Spigerwerk. London: Tavistock.

ETZIONI, A. (1965) "Dual leadership in complex organizations." Amer. Soc. Rev. 30 (October): 688-698.

——— (1964) Modern Organizations. Englewood Cliffs, N.J.: Prentice-Hall.

——— (1961) A Comparative Analysis of Complex Organizations. New York: Free Press.

EWELL, J. M. (1971) "The effect of change on organizations." Proctor & Gamble Co. Employee Relations.

EWING, D. W. (1971) "Who wants corporate democracy?" Havard Business Rev. 40, 5: 12-18 ff.

FRENCH, J.R.P., Jr. (1960) "An experiment on participation in a Norwegian factory." Human Relations 13: 3-19.

FRIEDLANDER, F. (1966) "Importance of work versus nonwork among socially and occupationally stratified groups." J. of Applied Psychology 50, 6: 437-441.

——— (1966) "Motivations to work and organizational performance." J. of Applied Psychology 50: 143-152.

GOLDTHORPE, J. H. (1969) The Affluent Worker in the Class Structure. Cambridge, Eng.: Cambridge Univ. Press.

——— D. L. LOCKWOOD, F. BECHOFER, and J. PLATT (1968) The Affluent Worker: Industrial Attitudes and Behaviour. Cambridge, Mass.: Cambridge Univ. Press.

GOLEMBIEWSKI, R. T. (1969) "Organization development in public agencies." Public Administration Rev. 29 (July): 367-377.

——— (1967) Organizing Men and Power: Patterns of Behavior and Line-Staff Models. Chicago: Rand McNally.

——— (1967) "The laboratory approach to organization development: the scheme of a method." Public Administration Rev. 27 (September): 211-220.

——— (1965) "Small groups and large organizations," pp. 87-141 in J. G. March (ed.) The Handbook of Organizations. Chicago: Rand McNally.

——— and S. B. CARRIGAN (1970) "Planned change in organization style based on laboratory approach." Administrative Sci. Q. 15: 79-93.

GROSS, N. et al. (1958) Explorations in Role Analysis. New York: John Wiley.

HACKMAN, R. and E. E. LAWLER III (1971) "Employee reactions to job characteristics." J. of Applied Psychology 55, 3: 259-286.

HAGE, J. and M. AIKEN (1969) "Routine technology, social structure, and organizational goals." Admin. Sci. Q. 14 (September): 366-378.

——— and C. B. MARRETT (1971) "Organization structure and communications." Amer. Soc. Rev. 36 (October): 860-871.

HALL, R. H. (1972) Organizations: Structure and Process. Englewood Cliffs, N.J.: Prentice-Hall.

HALPIN, A. (1966) Theory and Research in Administration. New York: Macmillan.

HARRIS, T. G. (1972) "Some idiot raised the ante." Psychology Today (February).

HAVELOCK, R. G. et al. (1969) Planning for Innovation through the Dissemination of Knowledge. Ann Arbor, Mich.: Institute for Social Research.

HERRICK, N. Q. (1972) Where Have All the Robots Gone? New York: Free Press.

HERTZ, D. B. (1971) "Has management science reached a dead end?" Innovation 25 (October): 12-17.

HERZBERG, F. (1966) Work and the Nature of Man. Cleveland: World.

HIRSCHMAN, A. O. and C. E. LINDBLOM (1962) "Economic development, research and development, policy making: some convering views." Behavioral Sci. 7: 211-222.

HOUGH, J. F. (1969) The Soviet Prefects: The Local Party Organs in Industrial Decision-Making. Cambridge, Mass.: Harvard Univ. Press.

HOUSE, R. J. et al. (1971) "Relation of leader consideration and initiating structure to R and D subordinates' satisfaction." Admin. Sci. Q. 16: 19-30.

JANIS, I. (1972) Victims of Groupthink. Boston: Houghton Mifflin.

JONES, S. C. and V. H. VROOM (1964) "Division of labor and performance under cooperative and competitive conditions." J. of Abnormal and Social Psychology 68: 313-320.

KAHN, R. L. (1972) "The meaning of work: interpretations and proposals for measurement," pp. 159-205 in A. Campbell and P. E. Converse (eds.) The Human Meaning of Social Change. New York: Russell Sage.

——— D. M. WOLFE, R. P. QUINN, J. D. SNECK, and R. ROSENTHAL (1964) Organizational Stress: Studies in Role Conflict and Ambiguity. New York: John Wiley.

KANTER, R. M. (1970) "Communes." Psychology Today (July): 53-57 ff.

KATZ, D. and B. S. GEORGEOPOULOS (1971) "Organizations in a changing world." J. of Applied Behavioral Sci. 7, 3: 342-370.

KATZ, D. and R. L. KAHN (1966) The Social Psychology of Organizations. New York: John Wiley.

KAUFMAN, H. (1972) The Limits of Organizational Change. Tuscaloosa: Univ. of Alabama Press.

KELMAN, H. C. (1972) "The problem solving workshop in conflict resolution," in R. L. Merritt (ed.) Communication in International Politics. Urbana: Univ. of Illinois.

KIRKPATRICK, F. et al. (1964) The Image of the Federal Service. Washington, D.C.: Brookings Institution.

KOHN, M. (1971) "Bureaucratic man: a portrait and an interpretation." Amer. Soc. Rev. 36: 461-474.

LAWRENCE, L. C. and P. C. SMITH (1955) "Group decision and employee participation." J. of Applied Psychology 39: 334-337.

LEAVITT, H. H. and T. L. WHISLER (1958) "Management in the 1980's." Harvard Business Rev. 36, 6: 41-48.

LIKERT, R. (1961) New Patterns of Management. New York: McGraw-Hill.

LINDBLOM, C. (1965) The Intelligence of Democracy. New York: Free Press.

LIPPITT, R. and R. K. WHITE (1947) "An experimental study of leadership and group life," pp. 315-330 in T. M. Newcomb and E. L. Hartley (eds.) Readings in Social Psychology. New York: Henry Holt.

LIPSKY, M. (1971) "Street-level bureaucracy and the analysis of urban reform." Urban Affairs Q. 6: 391-410.

LITWACK, E. (1968) "Technological innovation and theoretical functions of primary groups and bureaucratic structures." Amer. J. of Sociology 73 (January): 478-481.

––– and H. J. MEYER (1965) "Administrative styles and community linkages of public schools: some theoretical considerations," pp. 49-98 in A. J. Reiss, Jr. (ed.) Schools in a Changing Society. New York: Free Press.

McGREGOR, D. (1960) The Human Side of Enterprise. New York: McGraw-Hill.

MANN, F. C. and L. R. HOFFMAN (1960) Automation and the Worker. New York: Holt.

MARCH, J. and H. SIMON (1958) Organizations. New York: John Wiley.

MARGULIES, N. (1969) "Organizational culture and psychological growth." J. of Applied Behavioral Sci. 5, 4: 491-508.

MARINI, F. [ed.] (1971) Toward a New Public Administration. San Francisco: Chandler.

MARROW, A. J. et al. (1967) Management by Participation. New York: Harper & Row.

MASLOW, A. H. (1970) Motivation and Personality. New York: Harper & Row.

––– (1969) "Toward a humanistic biology." Amer. Psychologist 24, 8: 724-735.

––– (1965) Eupsychian Management. Homewood, Ill.: Richard D. Irwin.

MECHANIC, D. (1962) "Sources of power of lower participants in complex organizations." Admin. Sci. Q. 7: 349-364.

MENZIES, I.E.P. (1960) "A case study in the functioning of social systems as a defense against anxiety." Human Relations 13: 95-121.

MERTON, R. K. (1957) "Social theory and social structure." New York: Free Press.

MINER, J. B. (1971) "Changes in student attitudes toward bureaucratic role prescriptions during the 1960s." Admin. Sci. Q. 16, 3: 351-364.

MORSE, J. J. (1970) "Organizational characteristics and individual motivation," pp. 84-100 in J. Lorsch and P. Lawrence (eds.) Studies in Organizational Design. Homewood, Ill.: Irwin-Dorsey.

NEWCOMB, T. (1959) "The study of consensus," pp. 277-292 in R. K. Merton et al. (eds.) Sociology Today: Problems and Prospects. New York: Basic Books.

OSTROM, V. (1973) The Intellectual Crises in American Public Administration. University: Univ. of Alabama Press.

PAUL, W. T., K. B. ROBERTSON, and F. HERZBERG (1969) "Job enrichment pays off." Harvard Business Rev. 47, 2: 61-78.

PELZ, D. C. (1970) "The innovating organization: conditions for innovation," pp. 144-148 in W. Bennis (ed.) American Bureaucracy. Chicago: Aldine.

––– and F. M. ANDREWS (1966) "Autonomy, coordination, and stimulation in relation to scientific achievement." Behavioral Sci. 11 (March): 89-97.

PERROW, C. (1967) "A framework for the comparative analysis of organizations." Amer. Soc. Rev. 32: 194-208.

PORTER, L. W. and E. E. LAWLER III (1965) "Properties of organizational structure in relation to job attitudes and job behavior." Psych. Bull. 64: 23-31.

PUGH, D. S., C. R. HININGS, and C. TURNER (1968) "Dimensions of organizational structure." Admin. Sci. Q. 13 (March): 65-105.

PUNCH, K. F. (1969) "Bureaucratic structure in schools: towards redefinition and measurement." Educ. Administration Q. 5: 43-57.

RAVEN, B. H. (1968) "Groups: group performance," pp. 288-293 in Volume 6 of International Encyclopedia of the Social Sciences. New York: Macmillan and Free Press.

RICHARDS, C. B. and H. F. DOBRYNS (1957) "Topography and culture: the case of the changing cage." Human Organization 16 (Spring): 16-20.

RIDGWAY, V. F. (1956) "Dysfunctional consequences of performance measurements." Admin. Sci. Q. 1: 240-247.

SARASON, S. (1971) The Culture of the School and the Problem of Change. Boston: Allyn & Bacon.

SCHEIN, E. (1965) Organizational Psychology. Englewood Cliffs, N.J.: Prentice-Hall.

SHEPARD, J. M. (1971) Automation and Alienation. Cambridge, Mass.: MIT Press.

SHERIF, M. and C. W. SHERIF (1968) "Groups: group formation," pp. 276-283 of Volume 6 of International Encyclopedia of the Social Sciences. New York: Macmillan and Free Press.

SIMON, H. A. (1957) Administrative Behavior. New York: Free Press.

SMITH, C. and M. E. BROWN (1968) "Communication structure and control in a voluntary association," pp. 129-144 in A. Tannenbaum (ed.) Control in Organizations. New York: McGraw-Hill.

SMITH, C. and A. TANNENBAUM (1968) "Organizational control structure: a comparative analysis," pp. 73-90 in A. Tannenbaum (ed.) Control in Organizations. New York: McGraw-Hill.

SMITH, D. H. (1971) "Voluntary organization activity and poverty." Urban and Social Change Rev. 5 (Fall): 2-7.

SOEMERDJAN, S. (1957) "Bureaucratic organization in time of revolution." Admin. Sci. Q. 2: 182-199.

SVETLEK, B. et al. (1964) "Relationships between job difficulty, employee's attitude toward his job, and supervisory ratings of the employee effectiveness." J. of Applied Psychology 48: 320-324.

TANNENBAUM, A. S. (1968) Control in Organizations. New York: McGraw-Hill.

––– and J. G. BACHMAN (1968) "Attitude uniformity and role in a voluntary organization," pp. 229-238 in A. Tannenbaum (ed.) Control in Organizations. New York: McGraw-Hill.

TERREBERRY, S. (1968) "The evolution of organizational environments." Admin. Sci. Q. 12: 377-396.

THOMPSON, J. D. (1967) Organizations in Action. New York: McGraw-Hill.

THOMPSON, V. A. (1965) "Bureaucracy and innovation." Admin. Sci. Q. 10 (June): 1-20.

TRIST, E. L. and K. W. BAMWORTH (1951) "Some social and psychological consequences of the Longwall method of coal-getting." Human Relations 4: 3-38.

TRUMBO, D. A. (1961) "Individual and group correlates of attitudes toward work-related change." J. of Applied Psychology 45: 338-344.

UDY, S. H., Jr. (1965) "The comparative analysis of organizations," pp. 678-709 in J. G. March (ed.) Handbook of Organizations. Chicago: Rand McNally.

VERBA, S. (1961) Small Groups and Political Behavior: A Study of Leadership. Princeton: Princeton Univ. Press.

VROOM, V. H. (1964) Work and Motivation. New York: John Wiley.

––– (1962) "Ego-involvement, job-satisfaction, and job performance." Personnel Psychology 15: 159-177.

––– (1960) Some Personality Determinants of the Effects of Participation. Englewood Cliffs, N.J.: Prentice-Hall.

––– (1959) "Some personality determinants of the effects of participation." J. of Abnormal and Social Psychology 59: 322-327.

WAFFORD, J. C. (1971) "Managerial behavior, situational factors, productivity, and morale." Admin. Sci. Q. 16 (March).

WALDO, D. (1969) "Public administration and change." J. of Comp. Administration 1 (May): 94-113.

WALKER, C. R. (1954) "Work methods, working conditions and morale," in A. Kornhauser et al. (eds.) Industrial Conflict. New York: McGraw-Hill.

––– (1950) "The problem of the repetitive job." Harvard Business Rev. 28: 54-58.

––– and R. H. GUEST (1952 The Man on the Assembly Line. Cambridge, Mass.: Harvard Univ. Press.

WARREN, D. I. (1968) "Power, visibility and conformity in formal organizations." Amer. Soc. Rev. 33 (December): 951-970.

WARREN, R. L. (1967) "The interorganizational field as a focus for investigation." Admin. Sci. Q. 12.

WASHBURN, C. (1957) "Teacher in the authority system." J. of Educ. Sociology 30: 390-394.
WHYTE, M. K. (1973) "Bureaucracy and modernization in China: the Maoist critique." Amer. Soc. Rev. 38: 149-163.
WHYTE, W. F. (1969) Organizational Behavior. Homewood, Ill.: Irwin-Dorsey.
WILENSKY, H. (1964) "Varieties of work experience," pp. 125-154 in H. Borow (ed.) Man in a World of Work. Boston: Houghton Mifflin.
YUCHTMAN, E. (1972) "Reward distribution and work-role attractiveness in the kibbutzim: reflections on equity theory." Amer. Soc. Rev. 37 (October): 581-595.

PART V: LESSONS FROM THE PRIVATE SECTOR

ARGYRIS, C. (1970) Intervention Theory and Method. Reading, Mass.: Addison-Wesley.
CARPENTER, P. and G. R. HALL (1971) "Case studies in educational performance contracting: conclusions and implications." Vol. R-900/1, Department of Health, Education and Welfare, December.
Case Studies in Educational Performance Contracting (1971) Vols. R-900/1-R-900/6, Department of Health, Education and Welfare, RAND Corporation.
DOWNS, A. (1970) Urban Problems and Prospects. Chicago: Markham.
DRUCKER, P. F. (1968) The Age of Discontinuity. New York: Harper & Row.
Educational Vouchers (1970) Cambridge, Mass.: Center for the Study of Public Policy.
KAZAN, N. (1971) "Can free enterprise speed up our garbage collection?" New York Magazine (July 12): 41-43.
LaNOUE, G. R. (1972) Educational Vouchers: Concepts and Controversies. New York: Columbia University Teachers College Press.
PETERSON, G. E. (1972) "The distributional impact of performance contracting in schools." Urban Institute Working Paper 1200-22.

PART VI: BARRIERS TO CHANGE

ADRIAN, C. (1972) State and Local Government. New York: McGraw-Hill.
Advisory Commission on Intergovernmental Relations (1967) Fiscal Imbalance in the American System. Washington, D.C.: Government Printing Office.
ALFORD, R. R. [ed.] (1969) Bureaucracy and Participation: Political Cultures in Four Wisconsin Cities. Chicago: Rand McNally.
BARDACH, E. [ed.] (1972) The Skill Factor in Politics: Repealing the Mental Commitment Laws in California. Berkeley: Univ. of California Press.
BISH, R. L. (1971) The Public Economy of Metropolitan Areas. Chicago: Markham.
——— and R. WARREN (1972) "Scale and monopoly problems in urban government services." Urban Affairs Q. 8: 97-122.
CLARK, P. (1969) "Civic leadership: the symbols of legitimacy," pp. 350-366 in O. P. Williams and C. Press (eds.) Democracy in Urban America: Readings on Government and Politics. Chicago: Rand McNally.
CRECINE, J. P. [ed.] (1970) Financing the Metropolis. Beverly Hills: Sage Pubns.
CRENSON, M. (1971) The Unpolitics of Air Pollution. Baltimore: Johns Hopkins Press.
CRONIN, J. M. (1973) The Control of Urban Schools. New York: Harper & Row.
DAHL, R. (1961) Who Governs? New Haven, Conn.: Yale Univ. Press.
DERTHICK, M. (1972) New Towns in Town: Why a Federal Program Failed. Washington, D.C.: Urban Institute.
HAWKINS, B. W. (1971) Politics and Urban Policies. Indianapolis: Bobbs-Merrill.
HAWLEY, W. D. (1973) Nonpartisan Elections and the Case for Party Politics. New York: John Wiley.
JACOB, H. and M. LIPSKY (1968) "Outputs, structure, and power: an assessment of changes in the study of state and local politics." J. of Politics 30 (May): 510-538.
LEACH, R. H. (1970) American Federalism. New York: W. W. Norton.
LITTLE, A. D. (1966) Community Renewal Programming: A San Francisco Case Study. New York: Praeger.
LOWE, J. (1967) Cities in a Race with Time. New York: Harper & Row.
MARRIS, P. and M. REIN (1967) Dilemmas of Social Reform. Chicago: Aldine.

MELTSNER, A. J. (1971) The Politics of City Revenue. Berkeley: Univ. of California Press.

MURPHY, J. T. (1971) "Title I of ESEA: the politics of implementing federal educational reform." Harvard Educ. Rev. 41.

OSTROM, E. (1972) "Metropolitan reform: propositions derived from two traditions." Social Sci. Q. (December).

PIVEN, F. F. (1969) "Militant civil servants in NYC." Trans-action (November).

POWLEDGE, F. (1970) Model City. New York: Simon & Schuster.

PRESSMAN, J. L. (1972) "Preconditions of mayoral leadership." Amer. Pol. Sci. Rev. 66 (June): 511-524.

--- and A. B. WILDAVSKY (1963) Implementation. Berkeley: Univ. of California Press.

ROGERS, D. (1968) 110 Livingston Street: Politics and Bureaucracy in the New York City School System. New York: Random House.

ROSSI, P. H. (1963) "The middle-sized American city at mid-century." Library Q. 33: 3-13.

RUCHELMAN, L. I. [ed.] (1969) Big City Mayors: The Crisis in Urban Politics. Bloomington: Indiana Univ. Press.

RUTTENBERG, S. and J. GUTCHESS. (1970) The Federal-State Employment Service: A Critique. Baltimore: Johns Hopkins Press.

SUNDQUIST, J. L. (1969) Making Federalism Work. Washington, D.C.: Brookings Institution.

TALBOT, A. (1967) The Mayor's Game. New York: Harper & Row.

WILSON, J. Q. [ed.] (1968) The Metropolitan Enigma. Cambridge, Mass.: Harvard Univ. Press.

WOLFINGER, R. (1974) The Politics of Progress. Englewood Cliffs, N.J.: Prentice-Hall.

THE AUTHORS

CHRIS ARGYRIS is James Bryant Conant Professor of Education and Organizational Behavior at Harvard University. His more recent books include *Self-Scrutiny of Newspapers: An Organizational Analysis* (1974), *The Applicability of Organizational Sociology* (1972), *Intervention Theory and Method* (1970), and, with Donald Schon, *Education for Effectiveness* (1974).

ROBERT W. BLANNING is Assistant Professor of Management at the Wharton School, University of Pennsylvania. He received a B.S. in physics from the Pennsylvania State University, an M.S. in operations research from the Case Institute of Technology, and a Ph.D. from the University of Pennsylvania specializing in operations research and management information systems. He is co-author (with William F. Hamilton and Terrill A. Mast) of *Linear Programming for Management*.

GARRY D. BREWER is on the Senior Staff of the Social Science Department, the RAND Corporation. He did his doctoral work in political science at Yale University and has taught at the University of Southern California and UCLA. He is the editor of *Policy Sciences* and the author of numerous articles dealing with public policy and social science methodology. His books are: *Organized Complexity: Empirical Theories of Political Development; Politicians, Bureaucrats and the Consultant: A Critique of Urban Problem Solving;* and *A Policy Approach to the Study of Political Development and Change.*

LYLE C. FITCH is President of the Institute of Public Administration. He has been a member of the economic faculties of Columbia University, the City University of New York, and Wesleyan University and has had long experience in government, including service as First Deputy and then City Administrator of New York City. He has been a consultant for numerous federal, state, and local government agencies, as well as a number of governments overseas, the Committee for Economic Development, and other private organizations. He is author and editor of *Urban Transportation and Public Policy*, author or co-author of several other books, and author of numerous articles on urban planning, finance, and administration.

HARRY P. HATRY is Director of the State and Local Government Research Program at the Urban Institute. He is the author of numerous journal articles including contributions to *Public Administration Review, Public Management, Policy Sciences, National Tax Journal,* and *Governmental Finance.* He is an Associate Editor for *Operations Research* and a member of the International City Management Association's Committee on the Quality of Municipal Services.

WILLIS D. HAWLEY is Assistant Professor of Political Science and Policy Sciences, and Assistant Director of the Institute of Policy Sciences and Public Affairs at Duke University. He specializes in urban politics, organization theory and behavior, and the politics of education. His research interests include organizational and institutional change, the explanation of political learning and behavior, and education policy. He is the author of *Nonpartisan Elections and the Case for Party Politics* and co-author and co-editor of several books including *The Challenge of California, The Search for Community Power,* and *New Perspectives on Urban Politics* (forthcoming).

FREDERICK O'R. HAYES is a graduate of Hamilton College and received an M.P.A. and M.A. (political economy and government) from Harvard. He was Director of the Budget for the City of New York from 1966-1970; Deputy Director of the Community Action Program in the U.S. Office of Economic Opportunity, 1964-1966; and Assistant Commissioner of the U.S. Urban Renewal Administration from 1961-1964. He is currently a Visiting Professor in Urban and Policy Sciences at the State University of New York at Stony Brook. He is co-editor with John Rasmussen of *Centers for Innovation in The Cities and States.*

ARIE Y. LEWIN is Associate Professor of Management and Behavioral Science at the Graduate School of Business Administration, New York University. He received a B.S. (engineering) and M.S. (operations research) from UCLA and a Ph.D. (industrial administration) from Carnegie-Mellon University. He is co-author (with Michael Schiff) of *Behavioral Aspects of Accounting.*

PETER A. LUPSHA is Associate Professor of Political Science at the University of New Mexico and Director of the Division of Government Research. He has published articles in the areas of political violence and urban decision-making and leadership. He is currently engaged in research on water use and attitudes in rapidly urbanizing arid areas, and the politics of emergency medical systems and service delivery.

ELINOR OSTROM is an Associate Professor in the Department of Political Science and the Workshop in Political Theory and Policy Analysis at Indiana University. She is co-author of *Community Organization and the Provision of Police Services* and the author of articles published in the *Urban Affairs Quarterly, Social Science Quarterly, Public Choice, American Journal of Political Science, Public Administration Review* and *Journal of Criminal Justice.*

JEFFREY L. PRESSMAN is Assistant Professor of Political Science at Massachusetts Institute of Technology. His publications include the following books: *House vs. Senate: Conflict in the Appropriations Process; Implementation* (with Aaron Wildavsky); and *The Politics of Representation: The 1972 Democratic National Convention* (with Denis Sullivan, Benjamin Page, and John Lyons). In addition, he has written a number of articles on urban politics and federal-city relations.

DAVID ROGERS is Professor of Sociology and Management at New York University, Graduate School of Business Administration, and senior staff member of the Institute of Public Administration. Author of *110 Livingston Street, The Management of Big Cities, Big City Manpower Programs* (forthcoming), and many articles on big-city schools, politics, and management, he is currently working on a book evaluating the urban management programs of the Economic Development Council of New York. His chapter in this volume is a preliminary report from that study.

SUMNER M. ROSEN is Director of the Training Incentive Payments Program at the Institute of Public Administration. Formerly Research Director of the New Careers Development Center at New York University, he is author of many articles on labor economics, collective bargaining, civil service systems, upgrading, new careers, and other manpower programs for the disadvantaged.

E. S. SAVAS is Professor of Public Systems Management at the Graduate School of Business Administration, Columbia University, and author of *Computer Control of Industrial Processes* and of numerous articles on urban management. He received a Ph.D. in chemistry at Columbia and has served as First Deputy Administrator in the Office of the Mayor in New York and as manager of urban systems at IBM.

JONATHAN SUNSHINE was educated at Oxford University and received his doctorate from Columbia University. He has combined an academic career—at Columbia, Rutgers, and, most recently, Harvard—with public service. He was associated with New York City's Budget Bureau and is currently a fiscal economist with the U.S. Office of Management and Budget.

JAMES W. VAUPEL is on the faculty at Duke University, where he holds appointments in the Institute of Policy Sciences and Public Affairs and the Department of Management Sciences. He did his doctoral work in the Public Policy Program at Harvard University. His research and scholarship deals primarily with analytical methods, decision-making strategies and processes, and with the politics of international economic policy, especially policy toward multinational enterprise.

ANNMARIE h. WALSH is a senior staff member of the Institute of Public Administration and Associate Professor of Political Science at the City University of New York, Graduate Center. She is author of *Urban Government for the Paris Region, The Urban Challenge to Government,* co-editor of *Agenda for a City,* and author of many articles and reports. She is a member of the Subcommittee on Productivity in Government for the Committee for Economic Development. Her current research is on the management of public authorities.

GORDON P. WHITAKER is Visiting Assistant Professor of Political Science at the University of North Carolina (Chapel Hill). He has also taught at Brooklyn College/CUNY. He is co-author of *Community Organization and the Provision of Police Services* and has contributed to the *American Journal of Political Science, Public Administration Review,* and *Social Science Quarterly.*

DOUGLAS YATES is Chairman of the Center for the Study of the City and Its Environment at the Institution for Social and Policy Studies, Yale University, where he is also Assistant Professor of Political Science. He did graduate work at Oxford University and received his Doctorate from Yale. He is the author of *Neighborhood Democracy* and is at work on a book dealing with the politics of urban policy-making.

DENNIS R. YOUNG is currently a member of the faculty of the Program for Urban and Policy Sciences at the State University of New York at Stony Brook. He is on leave as a member of the senior research staff of the Urban Institute, where he developed and directed a project on the economic organization of public services. His principal research interest is in this area, including work in the areas of urban sanitation, day care, criminal justice, and education. Dr. Young received the B.A. and M.A. degrees in electrical engineering from the City College of New York and Stanford University, respectively, and the Ph.D. in the Engineering-Economic Systems Department at Stanford University.